*Analysis and Synthesis of
    Linear Active Networks*

ANALYSIS AND SYNTHESIS OF LINEAR ACTIVE NETWORKS

Sanjit Kumar Mitra

Department of Electrical Engineering
University of California, Davis

John Wiley & Sons, Inc.   New York/London/Sydney/Toronto

Copyright © 1969
by John Wiley & Sons, Inc.
All rights reserved. No part
of this book may be reproduced
by any means, nor transmitted, nor
translated into a machine language
without the written permission of the
publisher. *Library of Congress Catalog
Card Number:* 68-30919   SBN 471 61180 8
Printed in the United States of America

To my Mother and Father

/ *Preface*

For small-signal applications, the commonly used active elements like the transistor and the tunnel diode, when suitably biased, can be represented by linear equivalent circuits and thus can be considered as linear elements. Circuits built using these linearized active elements and the well-known passive elements like the resistor and the capacitor belong to the class of *linear*, *lumped*, and *finite* networks (usually abbreviated as LLF networks). The area of LLF networks is attractive to the circuit designer for numerous reasons, and this book is an introduction to some aspects of this very important and growing branch of network theory.

As the title suggests, this book is concerned with the analysis and synthesis of networks containing only linear passive and active elements. A majority of the contributions in this area have appeared in the past fifteen years and include some very significant theoretical and practical results. Space does not permit the inclusion of all of the contributions made in this area. Instead, I have tried to present, in a unified and comprehensive manner, some important results in order to provide the reader with a fairly complete background. Readers who want to specialize will benefit from the extensive list of references appended to each chapter.

This book may be used as a text in a first course on active networks at the senior or first-year graduate level. Alternatively, portions of it may serve as a companion text for a similar level course in network synthesis.

My purpose in writing this volume is to provide the student with a text which would also satisfy the needs of the practicing engineer. He may use it as a reference book or as a text for self-instruction. Mathematical rigor has been kept to a minimum. Examples have been used extensively, and the problems augment the text.

The reader is assumed to have a background of introductory network theory. More specifically, he should be familiar with an elementary level of theory of complex variables and matrix algebra. Knowledge of the properties of the network functions of *LC*, *RC*, and *RLC* one-port and two-port networks is a prerequisite. In addition, the reader should be familiar with synthesis methods of *LC*, *RC*, *RLC* one-ports and *RC*, *LC* two-ports. Some pertinent material from introductory network theory has been summarized in the appendices for a quick review.

It is appropriate to discuss briefly the contents of the book. The text is divided into twelve chapters.

Chapter 1 provides a short introduction to the major aspects of active network design. It discusses the advantages and disadvantages of linear active networks, along with the limitations of passive *RLC* networks. The chapter also includes a complete design example of inductorless active filter design.

The various linear active elements are introduced in *Chapter 2*. Here we discuss their properties, interrelationships, and use in forming equivalent circuits. No attempt is made in this chapter to describe their physical construction.

Some useful network theorems and their extensions are reviewed in *Chapter 3*. We discuss the various stability concepts and their implications, and several network transformations and their extensions that are particularly suitable for active networks. Also we discuss two important results on polynomial decompositions which are the keys to a large number of active network synthesis techniques.

A detailed explanation of the indefinite admittance matrix approach of analysis of linear networks is given in *Chapter 4*. This approach has been found to be more suitable than other methods available for analyzing networks containing active elements.

An important problem in the design of networks is the problem of variation of some network characteristic with respect to the variation of some parameters. This problem, which is more commonly known as a sensitivity problem, is more pronounced in the case of active filters. Various concepts of sensitivity and some related topics are discussed in *Chapter 5*.

Chapter 6 is concerned with the derivation of the realizability conditions of various classes of LLF networks. The contents and presentation of this chapter are somewhat academic in nature and can be omitted at the first reading.

The remaining chapters deal mainly with the synthesis aspects. *Chapter 7* considers the use of negative resistance as a network element in the synthesis of networks. The use of controlled sources in network synthesis is considered in *Chapter 8*. In *Chapter 9*, synthesis methods employing the negative impedance converter as the active element are discussed. The use of a gyrator as a network element is the main concern of *Chapter 10*. *Chapter 11* presents several synthesis methods based on the use of an operational amplifier as a network component. The final chapter summarizes three recent polynomial decomposition approaches to active network synthesis and is academic in nature.

The primary aim of this volume is to provide some basic tools in the design of inductorless filters. Detailed discussion on the actual design is beyond the scope of the book, although some practical considerations have been included for the purpose of motivation. Although each of the chapters is self-contained, an understanding of the last six chapters will be enhanced by reading, first, Chapters 1 through 3. Standard notations and conventions have been followed throughout the book. Unless otherwise stated, both in the text and in the figures, the units for element values are in ohms, henries, and farads. Also, the units for excitation and response variables are in volts and amperes.

This book evolved from a set of notes written for a first-year graduate-level course at Cornell University. It followed a course in network analysis and was given concurrently with a course in passive network synthesis. The book's present form is the result of two years' teaching at Cornell. Portions of these course notes were also used later in an out-of-hours course at the Bell Telephone Laboratories, Holmdel, New Jersey.

I am indebted to many of my friends and former colleagues for their help and advice without which the writing of this book would not have been possible. I thank Professor Glen Wade of the University of California, Santa Barbara, for his encouragement and for providing the necessary facilities at Cornell. A major revision of the preliminary notes was carried out at the Bell Telephone Laboratories, and I am grateful to a number of my former colleagues at the Laboratories who assisted in many ways. These include Drs. M. R. Aaron, I. Dostis, J. Mayo, R. A. McDonald, I. W. Sandberg, K. R. Swaminathan, and especially Dr. G. S. Moschytz who made numerous constructive criticisms regarding the organization of the materials and improvement of the presentation. Professors R. A. Rohrer and W. G. Howard, Jr., of the University of California, Berkeley; Mr. H. J. Orchard of Lenkurt Electric Co., San Carlos; Professor M. E. Van Valkenburg of Princeton University; and Dr. G. S. Moschytz reviewed critically the entire manuscript, and I acknowledge with gratitude their numerous helpful suggestions and comments. I also thank Professor Ronald F. Soohoo of the University of California, Davis, for his encouragement.

I was fortunate to have studied circuit theory under Professors E. S. Kuh and C. A. Desoer at the University of California, Berkeley; their influence is reflected in many portions of the book.

Finally, I owe a special debt of gratitude to my wife, Nandita, for her patience and understanding during the preparation of the book.

SANJIT K. MITRA

Davis, California
August 12, 1968

Contents

1 / INTRODUCTION 1
- 1-1 Limitations of the Behavior of Passive RLC Networks 3
- 1-2 Active Devices as Network Elements 7
- 1-3 An Active RC Filter Design Example 10
- 1-4 The Active Network Synthesis Problem 15
- 1-5 Summary 20

2 / ACTIVE NETWORK ELEMENTS 22
- 2-1 Controlled Sources 25
- 2-2 Negative Resistance 32
- 2-3 Negative Inductance and Capacitance 34
- 2-4 Impedance Converter 35
- 2-5 Impedance Inverter 45
- 2-6 Degenerate Circuit Elements 53
- 2-7 Equivalent Circuits 55
- 2-8 Summary 75

3 / USEFUL NETWORK THEOREMS 83
- 3-1 LC:RC Transformation 84
- 3-2 RC:CR Transformation 86
- 3-3 Generalized Inverse Networks 90
- 3-4 Conversion of Transfer Function Realization to Driving-Point Function Synthesis 93

3-5 Transformation of Driving-Point Function Synthesis to Transfer Function Realization 96
3-6 Theorems on Polynomial Decomposition 97
3-7 Stability Concepts 104
3-8 Summary 112

4 / ANALYSIS OF ACTIVE NETWORKS 117

4-1 The Indefinite Admittance Matrix 118
4-2 Elementary Operations 123
4-3 Classification of Multipoles 134
4-4 Network Functions of Multipole 135
4-5 Multipole Equivalent Circuits 139
4-6 Analysis of Networks Containing Ideal Active Elements 143
4-7 Analysis of Networks Containing Operational Amplifiers 148
4-8 Bilinear Form of Network Functions 155

5 / SENSITIVITY AND RELATED TOPICS 161

5-1 Estimation of Pole (Zero) Displacements due to Incremental Parameter Variations 163
5-2 Basic Concepts of Sensitivity Measure 165
5-3 Computation of Sensitivities 173
5-4 Sensitivity Function for Large Variation of a Parameter 182
5-5 Some General Remarks on Sensitivity Minimization 183
5-6 Incremental Variations of Several Parameters 187
5-7 Design Considerations 195

6 / REALIZABILITY CONDITIONS 208

6-1 Activity and Passivity of a Network 209
6-2 Linear Passive One-Port 213
6-3 Linear Passive Two-Port 220
6-4 Active Networks 226
6-5 Networks Containing Negative Resistances 227
6-6 Networks Containing Gyrators 237
6-7 Networks Containing Negative Impedance Converters 238
6-8 Other LLF Networks 238
6-9 Stability Conditions 240
6-10 Summary 248

7 / NEGATIVE RESISTANCE AS A CIRCUIT ELEMENT 254
- 7-1 Practical Considerations 254
- 7-2 Compensation of Losses in Reactive Elements 264
- 7-3 Synthesis of Driving-Point Functions 265
- 7-4 Synthesis of Transfer Functions 271
- 7-5 Summary 286

8 / SYNTHESIS USING CONTROLLED SOURCES 290
- 8-1 Practical Considerations 291
- 8-2 Synthesis of Driving-Point Functions 305
- 8-3 Synthesis of Transfer Functions 309
- 8-4 Network Design by Coefficient Matching 328
- 8-5 Sensitivity Considerations 339
- 8-6 Summary 340

9 / NEGATIVE-IMPEDANCE CONVERTER AS AN ACTIVE ELEMENT 348
- 9-1 Practical Considerations 348
- 9-2 Driving-Point Function Synthesis 364
- 9-3 Synthesis of Transfer Functions 381
- 9-4 Sensitivity Considerations: The Horowitz Decomposition 392
- 9-5 Summary 396

10 / GYRATOR AS A NETWORK ELEMENT 403
- 10-1 Practical Considerations 404
- 10-2 Synthesis of Driving-Point Functions 414
- 10-3 Transfer Function Synthesis 425
- 10-4 Sensitivity Considerations: The Calahan Decompositions 433
- 10-5 Inductance Simulation Using Gyrator 438
- 10-6 Summary 441

11 / THE OPERATIONAL AMPLIFIER AS A NETWORK ELEMENT 447
- 11-1 Practical Considerations 448
- 11-2 Stability Considerations 456
- 11-3 Realization of Ideal Active Devices 463
- 11-4 Synthesis of Driving-Point Functions 470
- 11-5 Synthesis of Transfer Functions 474
- 11-6 Network Design by Coefficient Matching Techniques 484

11-7 A Minimum Sensitive Realization of Transfer Function 489
11-8 Simulation of Inductance 492
11-9 Summary 495

12 / POLYNOMIAL DECOMPOSITION THEOREMS AND THEIR APPLICATIONS 502

12-1 A Unique Decomposition 503
12-2 The Odd-Part Partitioning 505
12-3 The Even-Part Partitioning 509
12-4 Synthesis of Driving-Point Functions 512
12-5 Synthesis of Transfer Functions 527

/ APPENDICES

A / USEFUL NETWORK DESCRIPTIONS 531

Table A-1 Two-Port Network Parameters 531
Table A-2 Two-Port Parameter Relations 532
Table A-3 Some Useful Network Specifications 533
Table A-4 Formulas for Interconnected Two-Ports 534
Table A-5 Summary of Normalization Relations 534

B / PROPERTIES OF PASSIVE RLC ONE-PORT NETWORKS 535

B-1 LC One-Port 536
B-2 RC One-Port 537
B-3 RL One-Port 539
B-4 RLC One-Port 540

C / PROPERTIES OF PASSIVE RLC TWO-PORT NETWORKS 542

C-1 General Two-Port 543
C-2 Two-Port without Mutual Coupling: Fialkow-Gerst Conditions 544
C-3 LC Two-Port 545
C-4 RC Two-Port 545

D / RC TWO-PORTS REALIZING SECOND-ORDER TRANSFER FUNCTIONS 547

Table D-1 Realizations of Short-Circuit Transfer Admittances 547
Table D-2 Realization of Voltage Transfer Ratio 551

/ INDEX 555

/ Symbols and Abbreviations

CVT — Current-to-voltage transducer. Also known as current-controlled voltage source.

CCT — Current-to-current transducer. Also known as current-controlled current source, current amplifier.

CGIC — Current inversion-type generalized impedance converter.

CNIC — Current inversion-type negative impedance converter.

Ev — Denotes "even part of," e.g., Ev $Z(s)$ represents the even part of $Z(s)$.

F_k — Return difference with respect to the parameter k.

F_k^0 — Null-return difference with respect to the parameter k.

F — Transmission ($ABCD$) matrix.

GIC — Generalized impedance converter.

G — Hybrid two-port parameter matrix—type I.

g_{ij} — The (i,j)th element of **G**.

H — Hybrid two-port parameter matrix—type II.

h_{ij} — The (i,j)th element of **H**.

IT — Ideal (real) transformer.

IVC — Ideal voltage converter.

IPC — Ideal power converter.

SYMBOLS AND ABBREVIATIONS

Im	Denotes "imaginary part of," e.g., Im $Z(j\omega)$ represents the imaginary part of $Z(j\omega)$.
IG	Ideal gyrator.
k	A vector, e.g., $\mathbf{k} = (k_1 \ k_2 \ k_3)$.
K	A matrix **K**.
$\mathbf{K}^t$	Transpose of matrix **K**.
k^*	Conjugate of k.
$|\mathbf{K}|$	Determinant of the matrix **K**.
NIC	Negative impedance converter.
NIV	(Reciprocal) negative impedance inverter.
NRNIV	Nonreciprocal negative impedance inverter.
Od	Denotes "odd part of," e.g., Od $Z(s)$ represents the odd part of $Z(s)$.
PIC	Positive impedance converter.
PIV	Positive impedance inverter.
p.r.	Positive real.
Re	Denotes the "real part of," e.g., Re $Z(j\omega)$ represents the real part of $Z(j\omega)$.
$\mathcal{R}(\omega, k)$	Represents Re $[F_k(j\omega)]$.
s	Complex frequency variable.
sgn(x)	Is equal to $+1$ if $x > 0$ and is equal to -1 if $x < 0$.
$S_k^{T(s)}$	Sensitivity function of $T(s)$ with respect to the variable parameter k.
$\hat{S}_k^{N_i}$	Coefficient sensitivity of the coefficient N_i of a polynomial $N(s) = \sum_i N_i s$ with respect to the variable parameter k.
$S_k^{p_i}$	Root sensitivity of the root at $s = p_i$ with respect to k.
$S_k^{\alpha(\omega)}$	Gain sensitivity of the gain function $\alpha(\omega)$ with respect to k.
$S_k^{\beta(\omega)}$	Phase sensitivity of the phase function $\beta(\omega)$ with respect to k.
$S_{e_i}^Q$	Q-sensitivity with respect to element e_i.
VCT	Voltage-to-current transducer. Also known as voltage-controlled current source.
VVT	Voltage-to-voltage transducer. Also known as voltage-controlled voltage source, voltage amplifier.
$V(s)°$	Degree of the polynomial $V(s)$.
$|x|$	Magnitude of x.

XIG	Ideal reactive gyrator.
XNIV	Ideal reactive negative impedance inverter.
$\mathfrak{X}(\omega, k)$	Represents Im $[F_k(j\omega)]$.
Y	Short-circuit admittance matrix.
y_{ij}	(i, j)th element of matrix **Y**.
$\hat{\mathbf{Y}}$	Indefinite admittance matrix.
$\hat{y}_{ij}$	(i, j)th element of $\hat{\mathbf{Y}}$.
$\mathbf{Y}_k^j$	Cofactor of (j, k) element of **Y**, i.e., $(-1)^{j+k}\mathbf{Y}_k^j$ is the minor obtained by deleting the jth row and kth column of **Y**.
$\mathbf{Y}_{ij}^{mn}$	Is equal to $(-1)^{m+n+i+j}$ $\begin{cases}\text{determinant of the submatrix ob-}\\ \text{tained by deleting the } m\text{th and } n\text{th} \\ \text{rows and } i\text{th and } j\text{th columns of } \mathbf{Y}.\end{cases}$
Z	Open-circuit impedance matrix.
z_{ij}	(i, j)th element of **Z**.
$Z^o(s)$	Generalized inverse of the impedance function $Z(s)$.
σ	Real part of the complex frequency variable s.
ω	Imaginary part of the complex frequency variable s.
η	Invariant stability factor.
$\sum_Q$	Sum of absolute values of Q-sensitivities of all the elements in a network.
VNIC	Voltage inversion-type negative impedance converter.
VGIC	Voltage inversion-type generalized impedance converter.

1 / *Introduction*

In 1915, Wagner and Campbell independently introduced the electric wave filter to meet the needs of the young communications industry. Later, Zobel conceived of the equalizer as a means of compensating for gain and phase distortion introduced by the transmission media. Soon, sophisticated techniques were developed to realize passive electrical networks capable of meeting the exacting performance demands of complex communication and control systems.

Active network design became a fruitful area of investigation after DeForest invented the Audion and Black introduced the feedback amplifier—the era of carrier communications had begun. The first active filters naturally arose out of the wedding of the feedback amplifier with passive RLC networks. Since then, extensive research on the theory and practice of active networks, coupled with the growing availability of cheap, reliable, and small passive and active components, has made the area of active networks one of the most attractive and promising branches of circuit theory.

What is an active network? Roughly speaking, an active network is a network composed of *passive* elements (like the inductor, the capacitor, and the resistor) and *active* elements (like the transistor and the tunnel diode). Even though most of the physical network elements are essentially nonlinear devices, over a band of frequencies, signal levels, etc., they can be replaced by their linearized models yielding satisfactory results for many applications. This leads to the class of linear active networks with which we are concerned in this book.[1]

There are numerous reasons and incentives behind the recent surge of interest in linear active networks. Probably the most significant stimulus has come from the rapidly growing area of integrated circuits where many network elements and their functional interconnections are fabricated in one chip. The main advantages of integrated microcircuits are as follows:[2] (1) increased system reliability and reduction of cost obtained by elimination of discrete joints and assembly operations, (2) reduction of size and weight, and increased equipment density obtained by reducing the packaging levels and electronically inactive structural materials, (3) increased functional performance obtained by regular and compact distribution of circuit elements, (4) increased operating speeds due to absence of parasitics and decreased propagation delay, and (5) reduction in power consumption. Unfortunately, associated with the above advantages, there are at present two major limitations of integrated circuits. First, building of inductors with reasonable value of inductance and quality factors by integrated circuit techniques have not been successful thus far.[3] On the other hand, conventional inductors, when miniaturized to be consistent in size with other integrated circuit components, are extremely poor in quality to be of any use for many applications.[4] Second, integrated thin film resistors and capacitors are limited to moderate values, and are available with wide tolerances only. In addition, circuit adjustments are difficult to achieve economically. These practical constraints have thus far limited the full integration of all linear circuitry.

One method, which offers a solution in eliminating the use of inductors in circuit design, is to design networks using linear active elements, resistors, and capacitors. Since active elements, resistors, and capacitors are easily integrable, this approach appears to be very promising and attractive.

There are other reasons why one may prefer to use an active *RC* network. These are discussed in the next two sections. In the first section we consider in detail the limitations of the functional and practical behavior of passive *RLC* networks. Advantages and disadvantages of active

---

[1] A formal definition of activity and passivity of a linear network is given in Section 6-1.

[2] A. E. Lessor, L. I. Maissel, and R. E. Thein, "Thin-film circuit technology: Part I—Thin film R-C networks," *IEEE Spectrum*, **1**, 72–80 (April 1964).

[3] R. M. Warner, Jr., Ed., *Integrated circuits—design principles and fabrication*, McGraw-Hill, New York, p. 267 (1965).

[4] W. E. Newell, "The frustrating problem of inductors in integrated circuits," *Electronics*, 50–52 (March 13, 1964).

networks are considered in the next section. A complete design example illustrating various aspects of active RC network design is presented in Section 1-3. A general discussion on synthesis of active networks is covered in Section 1-4. The main results of this chapter are summarized in the last section.

1-1 LIMITATIONS OF THE BEHAVIOR OF PASSIVE RLC NETWORKS

The synthesis of a lumped network invariably is based on the exact realization of a specified real rational function of the complex frequency variable s, $T(s)$, as a network function. The physically realizable network function $T(s)$ is usually obtained by approximating a given transmission characteristic.[5] Figure 1-1 shows three typical realizable transfer functions and their pole diagrams. Each of these transfer functions approximate an "ideal" brick-wall type of low-pass magnitude characteristic with different degrees of accuracy. Note from the pole-zero diagrams of Figure 1-1 that complex left half s-plane poles are present in each transfer function.

Complex left-half s-plane poles are almost a necessity to approximate any specified transmission characteristic, if the total number of poles of the approximating network function is required to be reasonably small. Now, from the properties of one-port and two-ports, it is known that the poles of network functions of networks composed of resistors and capacitors (usually abbreviated as RC networks) are always real and are situated on the negative real axis. Hence, RC networks alone are not able to generate complex poles. On the other hand, the network functions of RLC circuits (networks composed of resistors, capacitors, and inductors) are less restricted than the RC networks and can have complex poles in the left half s-plane. As a result, in passive network design, inductors are almost always used in addition to resistors and capacitors.

An inductor, which still is one of the major components in transmission network design, has some problems associated with it. Practical inductors, in contrast to resistors and capacitors, tend to deviate from the ideal inductance. An approximate lumped model of an inductor is shown in Figure 1-2a, where the capacitance C_L represents the total effect of the distributed capacitance between turns of the inductor coil, and the resistance r_L accounts for all the actual power losses observed in the inductor. For most practical purposes, the simpler model of the inductor indicated in Figure 1-2b is adequate and is used to define the quality of the inductor.

---

[5] For details on approximation procedures and tables of realizable transfer functions, see any book on network synthesis listed at the end of Appendices B and C.

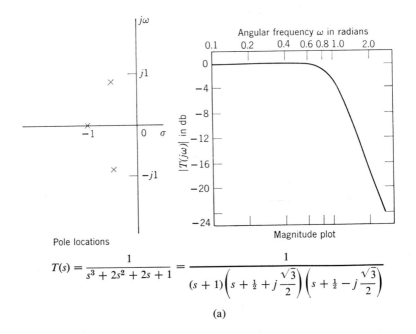

$$T(s) = \frac{1}{s^3 + 2s^2 + 2s + 1} = \frac{1}{(s+1)\left(s + \tfrac{1}{2} + j\tfrac{\sqrt{3}}{2}\right)\left(s + \tfrac{1}{2} - j\tfrac{\sqrt{3}}{2}\right)}$$

(a)

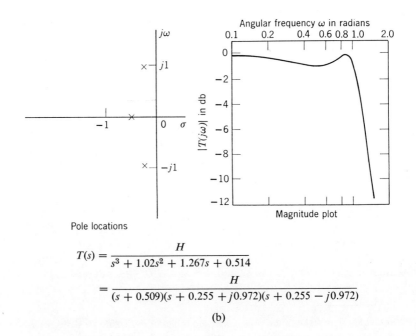

$$T(s) = \frac{H}{s^3 + 1.02s^2 + 1.267s + 0.514}$$
$$= \frac{H}{(s + 0.509)(s + 0.255 + j0.972)(s + 0.255 - j0.972)}$$

(b)

Figure 1-1 (a) Butterworth transfer function. (b) Chebyshev transfer function.

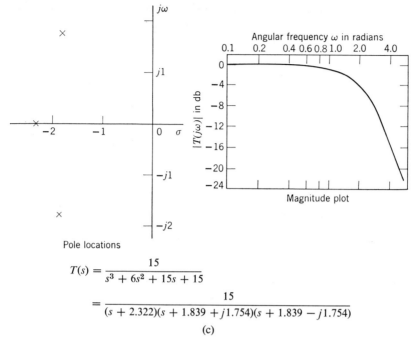

$$T(s) = \frac{15}{s^3 + 6s^2 + 15s + 15}$$

$$= \frac{15}{(s + 2.322)(s + 1.839 + j1.754)(s + 1.839 - j1.754)}$$

(c)

Figure 1-1 (continued) (c) Bessel transfer function.

The figure of merit of the inductor is given by its quality factor:

(1.1) $$Q = \frac{\omega L}{r_L}$$

where ω is the angular frequency. For an ideal inductor, r_L is equal to zero and hence Q is infinite. The distortion of the filter characteristic due to the presence of lossy inductors in some cases may be negligible and in some other cases may be severe. The latter is particularly true for filters having sharp frequency characteristics.

For most practical inductors, the total dissipative losses tend to increase with frequency, with the result that Q does not increase linearly with

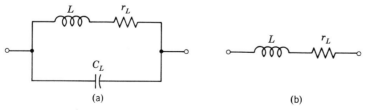

Figure 1-2 Equivalent circuits of an inductor.

frequency. The nonlinear frequency dependence of Q, coupled with the variation of Q factor from inductor to inductor, makes the formulation of an "exact" synthesis procedure of passive RLC networks fairly complicated.

The magnetic coupling between inductive elements in a circuit creates an additional problem. In particular, in satellites carrying instruments for measuring very weak magnetic fields, it is preferable not to use any inductive elements that might introduce errors in measurements, due to magnetic coupling.

Probably the major objection to the use of inductor as a network element arises in the case of low-frequency applications, like control systems and analog computers. At these frequencies (less than 1 kHz), practical inductors of reasonable Q tend to become bulky and expensive. The dissipative losses start increasing when an inductor is miniaturized. Roughly, the Q of a low-frequency inductor is proportional to the square of the scaling factor.[6] This means, an inductor having a Q of 100, when reduced to 12.5% of the original size will have approximately a Q of 25.

As indicated earlier, efforts to integrate inductors have not been successful. On the other hand, the demand for space-saving systems requires network miniaturization. This becomes a formidable problem when inductors are present in the circuit.

Use of quartz crystals helps the circuit designer to alleviate some of the above problems at higher frequencies. In addition, mechanical networks using piezoelectric ceramic plates have also been proposed to construct highly selective miniature filters.[7] However, for low-frequency work, these components are unsatisfactory.

In addition to the unavoidable use of inductors, passive networks have some fundamental limitations.[8] For example, it is well known that the driving-point function $Z(s)$ of an RLC network must satisfy the following conditions:

(1.2)
$$Z(s) \text{ is real for real values of } s,$$
$$\operatorname{Re} Z(s) \geq 0 \quad \text{for} \quad \operatorname{Re} s \geq 0$$

Thus, functions like

(1.3)
$$\frac{s-1}{s+3}, \quad -\frac{1}{s^2}, \quad \frac{s^2-s+1}{s^2+s+1}$$

---

[6] A. Rand, "Inductor size vs. Q: a dimensional analysis," *IEEE Trans. on Component Parts*, **CP-10**, 31–35 (March 1963).

[7] M. Kawakami, H. Tsuchiya, and H. Moreda, "H-shaped ceramic filter forms miniature I-F," *Electronics*, 55–57 (February 7, 1964).

[8] Properties of driving-point and transfer functions of LC, RC, RL and RLC networks are summarized in Appendices B and C.

cannot be realized as the driving-point impedance of an RLC one-port. In a similar fashion, the transfer functions of grounded transformerless RLC networks cannot have positive real transmission zeros (often encountered in equalizer design). For a transformerless RLC network, the transfer voltage (or current) ratio $T(s)$ is again restricted by the following condition:

(1.4) $\qquad\qquad T(s) < 1 \qquad$ for real values of s

Examples of transfer voltage ratios which cannot be realized by unbalanced transformerless RLC two-ports are

(1.5) $\qquad\qquad \dfrac{s-1}{s+1}, \quad \dfrac{6}{s^2+3s+3}$

1-2 ACTIVE DEVICES AS NETWORK ELEMENTS

One of the earliest active RC filters was introduced by Scott [9] in 1938, who showed that the use of an RC twin-T circuit in the feedback loop of a high-gain amplifier gave rise to a very high selective response. Feedback amplifiers surrounded by RC passive networks were known to be capable of realizing a restricted class of transfer functions by virtue of their early use in analog computers. Thus, techniques were available for the elimination of bulky inductors in low-frequency applications. Although frequency selective active filters had been in use in precision instruments, e.g., the audio-wave analyzer, speech octave-band analyzers, and variable electronic filters, little progress was made in the 1930's and the 1940's for designing high performance linear active circuits for several reasons. First, the vacuum tube with its relatively large size, high-power requirements, and cost did not always afford a favorable trade with the inductors. Second, existing analog computer-design techniques tended to proliferate both active components (amplifiers) and passive components (Rs and Cs).

The appearance of the transistor in 1948 lessened the disadvantages of active RC filters that had thus far been unavoidable with the vacuum tubes. The transistor was small, had low-power consumption, and promised to be inexpensive. Despite the initial parameter variability of the transistor, an inductorless active filter was soon to appear—the modern era of active networks was off and running. Effort was invested in the development of synthesis techniques for active RC networks to minimize the number of active devices required.[10] Applications to the relatively simple demands of

---

[9] H. H. Scott, "A new type of selective circuit and some applications," *Proc. IRE*, **26**, 226–235 (February 1938).

[10] In recent years the trend has changed somewhat, and the concern is often not so much with minimizing the number of active devices but rather on improving the functional performance of the overall circuit at the expense of more active devices.

8 / INTRODUCTION

some feedback control systems came first. The use of active networks to replace many of the passive filters in modern communication systems has been slower because of the more stringent cost and performance requirements imposed by this application. Thin film discrete and distributed *RC* networks, coupled with integrated amplifiers, have tended to make inroads into the communications field.

Use of active elements removes the two fundamental restrictions from *RLC* networks, passivity, and reciprocity. Thus, active networks can not only realize network functions which are realizable by passive *RLC* networks but also can be used to realize any driving-point or transfer characteristic not achievable with passive networks. For example, Figure 1-3 shows the active *RC* realizations of some of the driving-point impedances of (1.3), and Figures 1-4 and 1-7 show the realization of the transfer voltage ratios of (1.5) by means of grounded active *RC* two-port.

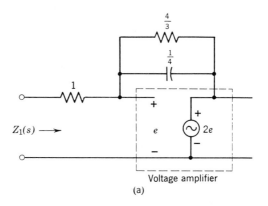

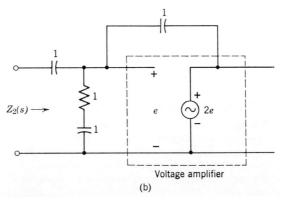

Figure 1-3 Active *RC* realization of driving-point functions: (a) $Z_1(s) = (s-1)/(s+3)$; (b) $Z_2(s) = -1/s^2$. Values in ohms and farads.

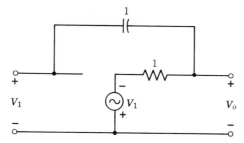

Figure 1-4 Active RC realization of the voltage transfer function $(V_o/V_1) = (s - 1)/(s + 1)$. Values in ohm and farad.

A major problem in the design of an active network is the *stability* of the complete circuit. Since poles of network functions of a passive RLC circuit are restricted to lie in the left-half s-plane, a passive filter can never become unstable with the change of element values or due to the presence of parasitics. On the other hand, an improperly designed active filter can start oscillating with a slight variation of passive components or active device parameters. Fortunately, the use of active elements often gives the designer freedom to avoid the stability problem. Stability concepts will be reviewed in Chapter 3.

Another problem encountered in both passive and active network design is the *sensitivity* of the network function due to the variation of network parameters. The characteristics of passive and active components may vary because of a change in the external (environmental) and the internal conditions. Such variations cause the poles and zeros to be displaced from their nominal positions. The effect of the displacements may cause the network not to exhibit the desired performance.[11] The sensitivity problem is not serious for most standard resistively terminated LC ladder filters.[12] However, active RC filters having extremely narrow frequency-selectivity are, in general, more sensitive to parameter variations than the conventional passive filters. The sensitivity problem will be discussed in detail in Chapter 5.

The problem of designing a *completely* insensitive network has not been solved.[13] However, for specific cases, methods do exist that minimize the effect of variation of one or more components.

---

[11] An excellent discussion on the effect of pole-zero displacements will be found in J. H. Mulligan, "The effect of pole and zero locations on the transient response of linear dynamic systems," *Proc. IRE*, **37**, 516–529 (May 1949).
[12] H. J. Orchard, "Inductorless filters," *Electronics Letters*, **2**, 224 (1966).
[13] Realistically, the design of a completely insensitive network will never be solved. Rather, sensitivity minimization relative to "cost" becomes a more meaningful problem.

10 / INTRODUCTION

It is interesting to indicate at this point an additional unique feature obtainable from active RC networks. This is the voltage dependent transmission characteristic that is not obtainable from passive RLC filters. An active RC filter can be designed to have an adjustable frequency response by varying an external voltage signal. Two examples of this type of active RC filters are shown in Figure 1-5.[14] Each filter shown employs a voltage

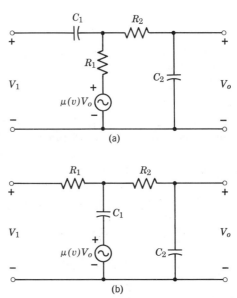

Figure 1-5 Voltage-controlled active RC filters. (a) Band-pass filter of constant bandwidth and variable Q and variable resonant frequency; (b) low-pass filter of variable bandwidth.

amplifier having a voltage variable gain. The circuit of Figure 1-5b is a low-pass filter of constant cutoff frequency and variable damping. The circuit of Figure 1-5a, on the other hand, can be used as a bandpass filter of constant bandwidth and variable center frequency. Design of these filters is beyond the scope of this book.

1-3 AN ACTIVE RC FILTER DESIGN EXAMPLE

Some general properties of active RC networks were discussed in the previous section. However, there are some very distinctive features of active filter design that are best illustrated by means of an example to be

---

[14] G. S. Moschytz, unpublished report.

considered next. This example also serves to illustrate several aspects of practical filter design with which one must be familiar.

Example 1-1 Design a low-pass inductorless filter having a second-order maximally flat delay characteristic. The input excitation is a voltage source and the output is terminated by a resistive load. Additional requirements are:

(1.6)
(1) Rise time[15] = 0.5 μ sec
(2) Load resistance = 1000 ohms

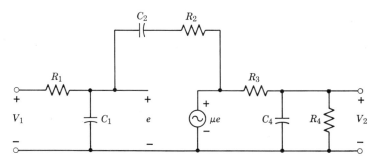

Figure 1-6 An active RC configuration suitable for the realization of a second-order low-pass voltage transfer ratio.

The normalized transfer function for a two-pole maximally flat delay characteristic is given as

(1.7) $$t_v(s) = \frac{V_2}{V_1} = \frac{H}{s^2 + 3s + 3}$$

An active RC network, which can be used to realize a second-order low-pass transfer function, is shown in Figure 1-6. Analysis yields, for this network,

(1.8) $$\frac{V_2}{V_1} = \frac{\mu(C_2 R_2 s + 1)}{[C_1 C_2 R_1 R_2 s^2 + (C_2 R_2 + C_1 R_1 + C_2 R_1 - \mu C_2 R_1)s + 1] \times \left[R_3 C_4 s + \frac{R_3}{R_4} + 1\right]}$$

The real zero and the real pole of V_2/V_1, as given above, can be made to cancel each other by selecting appropriate element values. One such choice

---
[15] The commonly used definition of *rise time* is the 10 to 90 percent rise time of the step response.

would be:

(1.9)
$$R_3 = R_4 = 1$$
$$C_2 R_2 = 1$$
$$C_4 = 2$$

The modified transfer voltage ratio then becomes

(1.10)
$$\frac{V_2}{V_1} = \frac{\mu/2}{C_1 R_1 s^2 + (1 + C_1 R_1 + C_2 R_1 - \mu C_2 R_1)s + 1}$$

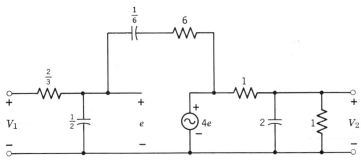

Figure 1-7 A second-order low-pass active RC Bessel filter (Example 1-1). Values in ohms and farads.

Comparing Equation 1.10 with Equation 1.7 and equating like coefficients we obtain

(1.11)
$$C_1 R_1 = \tfrac{1}{3}$$
$$1 + C_1 R_1 + C_2 R_1 - \mu C_2 R_1 = 1$$

Note that number of unknowns is larger than the number of equations, and hence the solution is not unique. One approach in solving for the element values is to select arbitrarily some of the element values and then to solve for the remaining ones by satisfying the constraint equations. This approach leads to one set of solutions as follows:

(1.12)
$$R_1 = \tfrac{2}{3} \quad C_1 = \tfrac{1}{2}$$
$$R_2 = 6 \quad C_2 = \tfrac{1}{6}$$
$$\mu = 4$$

The final normalized design is shown in Figure 1-7, which now has to be denormalized to satisfy the specifications given in (1.6). To this end, we first observe that the product of the rise time t_r and the 3-db bandwidth

ω_{3db} is given as[16]

(1.13) $$t_r \cdot \omega_{3db} = 2.2$$

Using (1.13), we note that the required rise time of $\tfrac{1}{2}$ μsec is equivalent to a 3-db bandwidth of 4.4×10^6 rad/sec. On the other hand, the normalized network of Figure 1-7 has a 3-db bandwidth of 1.362 rad/sec. In order to obtain the specified bandwidth (i.e., rise time of $\tfrac{1}{2}$ μsec), we shall have to denormalize the element values of the network of Figure 1-7 by the frequency denormalization factor Ω where[17]

(1.14) $$\Omega = \frac{\text{Normalized Bandwidth}}{\text{Desired Bandwidth}} = \frac{1.362}{4.4 \times 10^6} = 0.3095 \times 10^{-6}$$

In addition, to obtain the desired load impedance, the network will have to be scaled in impedance level by the impedance denormalization factor r_n where[17]

(1.15) $$r_n = \frac{\text{Normalized Load Impedance}}{\text{Desired Load Impedance}} = \frac{1}{10^3} = 10^{-3}$$

The element values of the final network can now be easily calculated using the Table A-5:

$$R_{1\text{actual}} = \frac{R_1}{r_n} = \frac{\tfrac{2}{3}}{10^{-3}} = 0.667 \text{ k}\Omega$$

$$R_{2\text{actual}} = \frac{R_2}{r_n} = 6 \times 10^3 = 6.0 \text{ k}\Omega$$

$$R_{3\text{actual}} = \frac{R_3}{r_n} = 1 \times 10^3 = 1.0 \text{ k}\Omega$$

$$R_{4\text{actual}} = R_{3\text{actual}} = 1.0 \text{ k}\Omega$$

$$C_{1\text{actual}} = r_n \Omega C_1 = 10^{-3} \times 0.3095 \times 10^{-6} \times \tfrac{1}{2} = 154.75 \text{ μμf}$$

$$C_{2\text{actual}} = r_n \Omega C_2 = 10^{-3} \times 0.3095 \times 10^{-6} \times \tfrac{1}{6} = 51.6 \text{ μμf}$$

$$C_{4\text{actual}} = r_n \Omega C_4 = 10^{-3} \times 0.3095 \times 10^{-6} \times 4 = 1238 \text{ μμf}$$

The corresponding final design is indicated in Figure 1-8. It should be noted that the scaling does not affect the gain of a voltage amplifier.

At this point, it would be interesting to compare the designed active RC filter with an equivalent passive filter. A passive RLC realization of (1.7)

---

[16] E. S. Kuh and D. O. Pederson, *Principles of circuit synthesis*, McGraw-Hill, New York, 27–30 (1959).

[17] The impedance and frequency normalization procedures are summarized in Table A-5 of Appendix A.

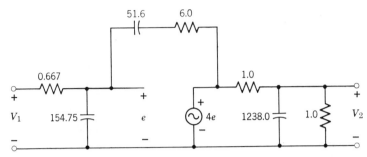

Figure 1-8 The active filter of Figure 1-7 after proper frequency and impedence scaling. Values in kΩ and μμf.

is shown in Figure 1-9. It is seen that the active *RC* realization uses more components (3 capacitors, 4 resistors, and 1 amplifier) than the corresponding passive realization (1 capacitor, 1 inductor, and 2 resistors). This is almost always the case. However, if the network is to be constructed by using thin film integrated circuit technology, this is not a serious problem.

Another feature of active filters is revealed by examining the d-c gains of both filters. The d-c gain of the active filter can be obtained from expression 1.10 by setting $s = 0$ and is thus given by $\mu/2$, which is equal to 2 (i.e., a *gain* of 6 db) for $\mu = 4$. In contrast, the passive *RLC* filter of Figure 1-9 has a d-c gain of $\frac{1}{3}$ (which is equivalent to a *loss* of 7.54 db).[18] Thus, in addition to providing the specified frequency response, the active filters can be designed to provide any amount of gain without employing additional amplifiers (as is usually done with passive filters).

Active *RC* networks are invariably designed to yield simpler network functions by cancellation of coincident poles and zeros. An example of such cancellation was presented in Example 1-1. When the element values

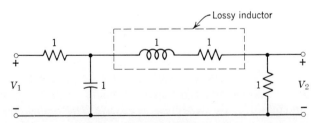

Figure 1-9 A second-order low-pass passive *RLC* Bessel filter. Values in ohms, henry, and farad.

[18] The maximum d-c gain obtainable by a grounded *RLC* network is 1 (i.e., 0 db).

change from their nominal values, the poles and zeros may not cancel with each other, thereby increasing the complexity of the transfer function. This may or may not be a problem, depending on the location of these critical frequencies.

Examination of expression 1.8 points out the "instability" feature associated with an active filter. We note that as μ, the gain of the amplifier, gets larger, the real part of the complex poles of the transfer function becomes smaller. For some large finite value of μ, the real part may become positive, leading to growing exponentials in the impulse response. Thus, with the active network realization, there is in general a chance of oscillation present. On the other hand, the passive RLC realization is always stable.

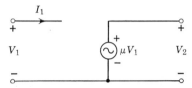

Figure 1-10 Model of a voltage-controlled voltage source.

1-4 THE ACTIVE NETWORK SYNTHESIS PROBLEM

An elementary definition of an active network given earlier is as follows. It is a circuit obtained by interconnecting passive elements (resistors, capacitors, etc.) and active elements (transistors, tunnel diodes, etc.). The transistor and other active components are complex devices characterized by a number of parameters. This complexity makes it difficult to formulate synthesis procedures using the available active components directly.

An alternative approach would be to define an "ideal" linear active two-port (or a one-port), describable by simple mathematical relations, and then to attempt to build a practical circuit whose performance approaches the behavior of its idealized counterpart using conventional active and passive components. Network synthesis procedures are next developed, assuming the existence of the "ideal" active devices. This latter approach has been followed by most authors of active filter synthesis techniques. The main purpose of this book is to introduce a representative sample of this type of active network design.

An example of the idealized linear active two-ports is the ideal "voltage amplifier" used earlier in this chapter. This active device (shown again in Figure 1-10 for convenience) is described by the simple input-output

16 / INTRODUCTION

relationships:

(1.16)
$$V_2 = \mu V_1$$
$$I_1 = 0$$

Other ideal active elements are introduced in Chapter 2.

With the above consideration in mind, let us try to formulate the basic steps involved in the synthesis of an active filter, from a specified network function. It is evident that an active network can be considered as a passive network into which some ideal active devices have been embedded. For example, Figure 1-11 shows the most general configuration of an active

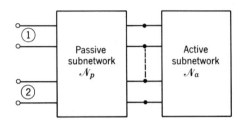

Figure 1-11 The general configuration of an active RC two-port.

two-port. For future reference, the passive subnetwork $\mathcal{N}_p$ will be designated as the *companion* network. The active subnetwork $\mathcal{N}_a$ shown in the figure may be one or more "ideal" active devices. An example of such a configuration is presented in Figure 1-12. This network was originally suggested by Hakim[19] and will be studied in a later chapter. The active subnetwork consists of an ideal voltage amplifier. The passive companion network consists of two RC one-port and one RC two-port networks.

A study of the transfer function of Figure 1-12 will lead us to our goal of this section. In order to determine the current transfer ratio, $t_I(s)$, of this network, we observe that at the input node:

(1.17) $$I_1 = Y_A(s)V_1 + Y_B(s)[V_1 - 2V_1] = [Y_A(s) - Y_B(s)]V_1$$

where $Y_A(s)$ and $Y_B(s)$ are the driving-point admittances of the RC one-ports. The output current I_2 is related to V_1 as follows

(1.18) $$-y_{21}(s) = \frac{I_2}{2V_1}$$

where $y_{21}(s)$ is the short-circuit transfer admittance of the RC two-port.

---

[19] S. S. Hakim, "*RC* active filters using an amplifier as the active element," *Proc. IEE (London)*, **112**, 901–912 (May 1965).

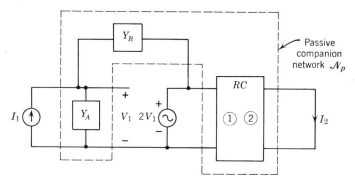

Figure 1-12 An RC: voltage-amplifier configuration for the realization of a current transfer ratio.

From Equations 1.17 and 1.18 we obtain

(1.19) $$t_I(s) = \frac{I_2}{I_1} = \frac{-2y_{21}(s)}{Y_A(s) - Y_B(s)}$$

The first problem is to determine the restrictions on $t_I(s)$ for realizability. This can be determined by examining expression 1.19. For example, the active RC configuration of Hakim cannot realize positive real transmission zeros. Let us denote

$$-y_{21}(s) = \frac{n_{21}(s)}{d(s)}$$

(1.20) $$Y_A(s) = \frac{p_A(s)}{q_A(s)}$$

$$Y_B(s) = \frac{p_B(s)}{q_B(s)}$$

Substituting these in Equation 1.19, we have

(1.21) $$t_I(s) = \frac{N(s)}{D(s)} = \frac{2n_{21}(s)q_A(s)q_B(s)}{d(s)[p_A(s)q_B(s) - p_B(s)q_A(s)]}$$

Expression 1.21 indicates the next problem in the realization of a specified real rational function $N(s)/D(s)$ as the current transfer ratio of the network of Figure 1-12 [assuming $N(s)/D(s)$ satisfies the necessary realizability conditions of $t_I(s)$]. The task is to identify the various polynomials $n_{21}(s)$, $d(s)$, $p_A(s)$, etc., by suitably decomposing and partitioning $N(s)$ and $D(s)$. A restriction associated with the identification of these polynomials is that the one-ports and the two-port must be RC realizable. After $n_{21}(s)$, $d(s)$,

$p_A(s)$, etc., are identified, the various parameters characterizing the companion RC network are obtained using (1.20). Finally, the RC one-ports and the RC two-port are realized, following standard passive RC synthesis procedures.

Now we are in a position to outline the five steps that are to be followed for a successful active network realization of a specified network function. These are:

1. Selection of a suitable active network configuration. The choice of the network will depend on the type of network function specified, location of its zeros and poles, and other design criteria,

2. Decomposition and partitioning of the specified network function into simpler functions of less complexity (to be called subfunctions for convenience),

3. Identification of the parameters characterizing the companion passive network from the subfunctions,

4. Realization of the companion network, using passive synthesis procedures,

5. Construction of the "ideal" active devices.

Step 5 is an additional burden on the designer of active filters. Fortunately, it is possible to build integrable practical active devices using transistors, resistors, and other passive and active components, whose functional performance can be reasonably close to the idealized requirements over a large range of frequencies, signal levels, etc. Some type of active devices are also available in the market in integrated forms. The nonideal properties of the practical, active devices, their physical limitations, and their actual departure from the idealized version determine the range of applicability of the synthesis methods based on these devices. These, in turn, determine the final performance of the realized active filters. Roughly speaking, the two major limitations of the practical, active device are the bandwidth and signal level requirements. These two limitations arise from the physical constraints imposed on the constituent active and passive elements—in particular, the transistor. A third limitation of the practical, active device is the necessity of using power supplies. These are needed to bias the active elements in the linear regions of their characteristics in order to enable the use of small signal linear models to represent the active elements for design and analysis purposes.

Another approach to the design of active filters, followed by a few authors, can be classified as the "coefficient-matching technique." The design procedure starts from the analysis of a particular circuit. Restrictions on realizable transfer functions are determined *a priori* and the element values are related to the parameters of the network function by

Table 1-1 GENERAL PROPERTIES OF ACTIVE AND PASSIVE NETWORKS

Feature	Passive *RLC* Network	Active *RC* Networks
1. Realizability condition	Can realize only a class of real rational functions which are analytic in the right half s-plane.	No restrictions. Can realize any real rational function.
2. Use of inductors	Inductors are required in almost all cases.	Inductors are *not* required.
3. Applicability	Used mostly for fixed frequency response type applications.	Can be used both for fixed and variable frequency response type applications.
4. Preferable frequency range	Above 200 Hz.	Below 100 kHz.
5. Type of signal source	Low-impedance source capable of supplying enough signal power. Source impedance must be resistive.	Can be designed with a high input impedance, to make the filter independent of source impedance. May not require much signal power.
6. Type of load impedance	Load impedance usually resistive. Filter operation usually critical of value of the load impedance.	Can be designed with low output impedance to make the filter independent of load impedance.
7. Insertion loss	Usually has significant insertion loss. Often require additional amplification for maintaining proper signal level.	Can be designed with insertion gain. Additional amplifiers to provide gain are not needed.
8. Buffer amplifier	Usually requires buffer amplifier for cascading more than one stage.	Can be cascaded without additional buffers, if designed with low output impedance.

20 / INTRODUCTION

Table 1-1 (*continued*)

9. Micro-miniaturization	Not possible, if inductors are used.	Possible, if resistors and capacitors are designed to have moderate values (range of values of resistors: 1 Ω to 5 MΩ; Maximum capacitance/substrate: 1 µf).
10. Stability	Absolutely stable.	Potentially unstable. Can be designed with a reasonable amount of stability margin.
11. Sensitivity	Most practical filter structures (ladder type) have low sensitivity.	Usually are more sensitive. Sensitivity can be minimized in specific cases.
12. Power supplies	Does not require any power supplies.	Power supplies required to bias the active elements for proper operation.

equating like coefficients. This short-cut approach (illustrated in Example 1-1) is useful in many cases.

1-5 SUMMARY

A general introduction to the area of active networks, in particular active *RC* networks, has been presented in this chapter. Various features of active and passive networks were discussed primarily with the aid of simple examples, which have been chosen to illustrate some typical and distinctive characteristics. Table 1-1 summarizes the main points established in this chapter along with some additional information.

Bibliography

Balabanian, N., "Active *RC* network synthesis." Tech. Rep. Contract No. AF 19 (604)-6142 AFCRC, Syracuse U. Res. Inst., December 1961.

Bennett, W. R., "Synthesis of active networks," Proc. Symp. on Modern Network Synthesis, MRI Symposia Series, Vol. V, Polytech Inst. of Brooklyn, pp. 45–61 (April 1955).

Bode, H. W., "Feedback—The history of an idea," Proc. Symp. on Active Networks and Feedback Systems, MRI Symposia Series, Vol. X, Polytech. Inst. of Brooklyn, pp. 1–17 (April 1960).

DeClaris, N., "Challenges and promises of active network theory," *1960 NEREM Record*, **II,** 80–81.

DePian L., *Linear active network theory*, Prentice-Hall, Englewood Cliffs, N.J., Ch. 1, 1962.

Field, R. K., "The tiny exploding world of linear microcircuits," *Electronic Design*, **15(15),** 49–66 (July 19, 1967).

Irons, F., "Active filters—properties and applications," *Frequency*, March–April 1964.

Kron-hite Catalog, Section I, "Variable electronic filters," Kron-hite Corporation, Cambridge, Mass.

Linvill, J. G., "Active networks—past, present and future," Proc. Symp. on Active Networks and Feedback Systems, MRI Symposia Series, Vol. X, Polytech. Institute of Brooklyn, pp. 19–26 (April 1960).

Mulligan, J. H., Jr., "The role of network theory in solid-state electronics—accomplishments and future challenges," *IEEE Trans. on Circuit Theory*, **CT-10,** 323–332 (September 1963).

Uzunoglu, V., "Six possible routes to non-inductive tuned circuitry," *Electronics*, **38(23),** 114–117 (November 15, 1965).

Zverev, A. I., "The golden anniversary of electric wave filters," *IEEE Spectrum*, **3(3),** 129–131 (March 1966).

2 / *Active Network Elements*

As mentioned earlier, the active components, commonly available to the circuit designer (like the transistor and the tunnel diode), are complex nonlinear devices. However, if the a-c signal is small enough in comparison to the d-c bias, an active component at a specified bias point can be represented by a linear equivalent circuit that is suitable for many applications. The parameters of the equivalent circuit are, in general, identified with the basic physical operation of the active component at the given operating point, and their values are usually dependent on the bias point, ambient temperature, etc. This linear-modelling approach is almost essential to the circuit designer before he can begin the design of complicated circuits for various functional requirements.

Depending on the type of application, operating range, and many other factors, a given active component can have more than one equivalent circuits. Unfortunately, a "good" linear equivalent representation is characterized by many parameters. For example, Figure 2-1*a* shows the small signal equivalent circuit of a biased junction transistor in the common base orientation, which is valid from low to medium frequencies. Similarly, Figure 2-1*b* shows a small-signal equivalent circuit of a biased tunnel diode. We observe that the transistor model is described by seven parameters and the tunnel diode model is characterized by four parameters including a negative resistance.

The complexity of the "exact" linear models of the available active components make it difficult to develop synthesis procedures, using these

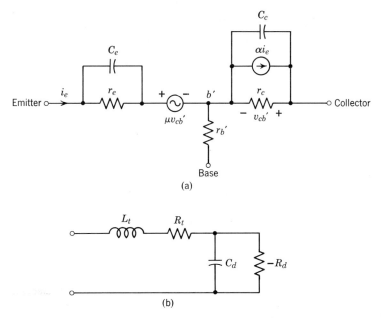

Figure 2-1 Small-signal equivalent circuits of biased semiconductor elements. (a) Transistor in the common base orientation. (b) Tunnel diode.

components directly and other passive elements. One approach to circumvent the above problem would be to reduce the original equivalent circuits to simpler models (using reasonable engineering judgments), which are easier to manipulate. Active network synthesis procedure can then be developed, based on the simpler models. Finally, the effect of the neglected parameters on the actual design can be corrected to a certain extent in the laboratory by a trial-and-error type of procedure or by using some optimization techniques on a digital computer. Based on the preceding approach, one can simplify the equivalent representation of the transistor and the tunnel diode described in Figure 2-1. Possible simplified models of these devices are indicated in Figure 2-2. The "idealized" two-port model of the common-base transistor (Figure 2-2a) has the following simple input-output relationship:

$$I_2 = -I_1$$
$$V_1 = 0$$

Likewise, the "idealized" model of the tunnel diode (Figure 2-2b) is a negative resistance having $v - i$ relationship as

$$V = (-R_d)I$$

24 / ACTIVE NETWORK ELEMENTS

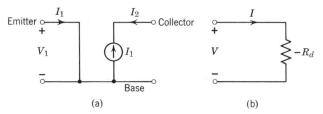

Figure 2-2 Simplified equivalent circuit of (a) common base transistor amplifier, and (b) tunnel diode.

Figure 2-2 suggests the existence of other types of ideal active one-ports and two-ports, which are the subject of discussion of this chapter. In the first six sections of this chapter, the properties of some commonly used ideal active elements are derived from fundamental considerations. In the next section, the relations between the ideal active elements and their use in forming equivalent circuits are studied.

Many of the "ideal" active elements discussed here have been found to be very useful in developing the inductorless active network configurations that are amenable for a systematic realization. These "idealized" devices, in general, can be constructed in practice, using transistors, resistors, and other passive and active components.[1] The transistor circuit can be designed to approach the ideal behavior over a large operating range. Like the transistor and the tunnel diode, these active elements in practice are "nonideal" devices, i.e., their respective equivalent representation has additional parasitic elements added to the original idealized model. The presence of the parasitics may lead to detrimental performance of the active filter designed, using the so-called idealized active devices. In some cases, the effect of the parasitics can be precorrected to a certain extent in the design of the filter.

The active one-port and two-port network elements introduced in this chapter are described by linear input-output relationships. Networks comprised of these linear active elements and the well-known linear passive elements (the resistance, the capacitance, the inductance, and the transformer) belong to the general class of *linear*, *lumped*, and *finite* networks, often abbreviated as *LLF* networks. It should be noted that the network functions of *LLF* networks are *real rational functions* of the complex frequency variable s, that is, are ratios of polynomials in s having real coefficients.[2]

---

[1] Practical realization of the active elements is considered in later chapters.

[2] This implies that for synthesis purposes, the specified transmission characteristic must first be approximated as a real rational function. The restrictions on the realizability of the real rational function are determined from the type of *LLF* structure used for realization. Physical realizability conditions of *LLF* networks are discussed in Chapter 6.

2-1 CONTROLLED SOURCES

Two versions of the ideal active two-port, called controlled sources, were introduced earlier. In Chapter 1, we made use of the voltage-controlled voltage source, more commonly known as the voltage amplifier. In particular, realization of driving-point and transfer functions that are not realizable by passive *RLC* structures were shown possible by means of active *RC* structures employing ideal voltage-controlled voltage source as

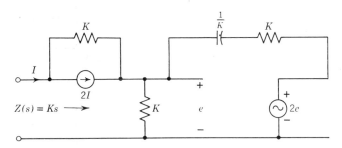

Figure 2-3 Realization of an inductance by an active *RC* network containing controlled sources.

the active device (Figures 1-3 and 1-4). In addition, realization of a second-order Bessel filter with the aid of *RC*-voltage-amplifier configuration was also illustrated (Example 1-1). The other type of controlled source was briefly introduced in this chapter, as a component in the modelling of a junction transistor (Figure 2-1a). To emphasize that inductorless realization is possible by employing controlled sources, we present in Figure 2-3, an active *RC* one-port which behaves like an ideal inductance.

In this section, we propose to study the properties of various types of controlled sources and their interrelationships. The use of controlled sources in the design of transmission network will be covered in Chapter 8, along with the development of transistor circuits for realizing these active devices.

Definition of a Controlled Source

The controlled source is a unidirectional, nonautonomous active two-port having two pairs of terminals, one controlled and one controlling. One of the terminal variables at the controlled port has a single valued dependence upon a terminal variable at the controlling port. Because of the unidirectional property, the controlling terminal-pair variables are insensitive or independent of the controlled terminal-pair variables.

There are four different types of controlled sources. For convenience, we shall designate these idealized two-ports as *transducers*. Alternate designations such as *transactors, unidirectional converters* will be found in the literature.

Current-to-Voltage Transducer. The current-to-voltage transducer is an ideal controlled source in which the output voltage is proportional to the input current. (For simplicity we shall designate such a device as CVT.)

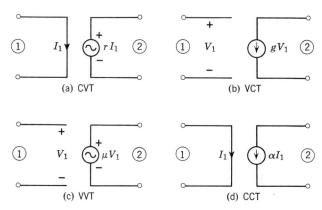

Figure 2-4 The four types of controlled sources.

The $v - i$ relationship of the CVT (Figure 2-4a) is given by

$$(2.1) \quad \begin{bmatrix} V_1 \\ V_2 \end{bmatrix} = \begin{bmatrix} 0 & 0 \\ r & 0 \end{bmatrix} \begin{bmatrix} I_1 \\ I_2 \end{bmatrix}$$

The CVT has infinite input and output admittances, a forward transfer impedance of r, and no reverse transmission. Although the input power is zero, the CVT can supply unlimited output power and as a result exhibits infinite power gain.

Voltage-to-Current Transducer. The voltage-to-current transducer (VCT) is an ideal voltage-controlled current source, i.e., the output current is dependent upon the input voltage (Figure 2-4b). The VCT is characterized by the following relation:

$$(2.2) \quad \begin{bmatrix} I_1 \\ I_2 \end{bmatrix} = \begin{bmatrix} 0 & 0 \\ g & 0 \end{bmatrix} \begin{bmatrix} V_1 \\ V_2 \end{bmatrix}$$

which indicates that it has infinite input and output impedances, a forward transfer admittance of g, and no reverse transmission. The VCT again can supply unlimited power with zero-input power, and thus exhibits infinite power gain.

Voltage-to-Voltage Transducer. The ideal unilateral two-port that transduces the input voltage to the output is the voltage-controlled voltage source or the voltage-to-voltage transducer (VVT) (Figure 2-4c). It is represented by the following voltage-current relation:

$$(2.3) \quad \begin{bmatrix} I_1 \\ V_2 \end{bmatrix} = \begin{bmatrix} 0 & 0 \\ \mu & 0 \end{bmatrix} \begin{bmatrix} V_1 \\ I_2 \end{bmatrix}$$

From Equation 2.3 we note that the VVT has a zero-input admittance and a zero-output impedance. The transmission is only in the forward direction with a voltage transfer ratio of μ. Like the CVT and VCT, the VVT has an infinite power gain. The VVT is also known as the *voltage amplifier*.

Current-to-Current Transducer. The output current is dependent upon the input current in the current-to-current transducer (CCT), a current-controlled current source. The input impedance and the output admittance are zero. The CCT is a unilateral two-port with a forward current transfer ratio of α. Like the previous elementary transducers, CCT consumes no input power and can supply any amount of output power, indicating an infinite power gain. The output-input relationship of the CCT (Figure 2-4d) is

$$(2.4) \quad \begin{bmatrix} V_1 \\ I_2 \end{bmatrix} = \begin{bmatrix} 0 & 0 \\ \alpha & 0 \end{bmatrix} \begin{bmatrix} I_1 \\ V_2 \end{bmatrix}$$

An alternate name of CCT is the *current amplifier*.

Some physical devices operate almost like ideal transducers. For example, an "ideal" pentode without interelectrode capacitances and with infinite plate resistance is a VCT. The cathode follower approximates a unity gain VVT. The grounded base transistor can be represented in some cases as an ideal CCT (see Figure 2-2a).

The constraint constants (also called the *gain*) r, g, μ, and α in the last four equations are usually assumed to be real. Transducers with complex constraint can be constructed, using the elementary controlled sources and *RLC* networks.

Interrelationships of the four elementary transducers are easy to derive. Figure 2-5a shows the construction of a CCT by cascading a CVT and a VCT. In a similar manner, if a VCT is followed by a CVT (Figure 2-5b), the overall two-port behaves like a VVT. It is possible to build a CVT or a VCT from the other two transducers if additional network elements are used as shown in Figures 2-5c and 2-5d. The two-port of Figure 2-5c is a

28 / ACTIVE NETWORK ELEMENTS

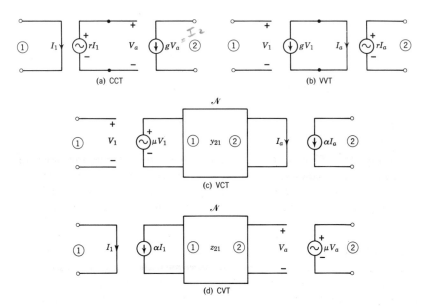

Figure 2-5 Illustration of the relations between various controlled sources.

VCT having a gain equal to $-\mu \alpha y_{21}$, where y_{21} is the forward short-circuit transfer admittance of the network $\mathcal{N}$. Figure 2-5d represents a CVT having a gain equal to $-\alpha \mu z_{21}$.

For convenience, the voltage-controlled voltage source (VVT) is occasionally represented by a triangle, the vertex of which indicates the direction of power transmission. For example, Figure 2-6a shows the

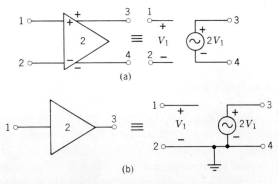

Figure 2-6 Commonly used symbolic representations of the voltage-controlled voltage source.

representation of a four-terminal VVT of gain 2. The polarities of each terminal are usually marked as indicated in the figure. A missing terminal implies that the corresponding terminal has been grounded. Thus, a grounded VVT of positive gain (also called a noninverting type VVT) of value 2 can be represented as shown in Figure 2-6b.

Isolation Amplifier

In many cases of active network design, electronic circuits are used as buffer stages between several two-port networks for isolation purposes. Essentially, these amplifiers are, in theory, good approximations of the ideal transducers. The cathode follower and the emitter follower are most often used as isolation amplifiers. An example of one used as a buffer stage is presented in Figure 2-7, where the overall transfer function is

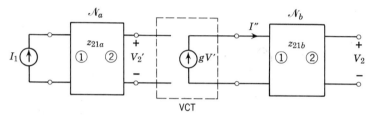

Figure 2-7 The use of a VCT as an isolation stage.

given by

(2.5) $$\frac{V_2}{I_1} = Z_{21} = \frac{V_2'}{I_1} \cdot \frac{I''}{V_2'} \cdot \frac{V_2}{I''} = (z_{21a}) \cdot (g) \cdot (z_{21b})$$

A higher-order transfer function can be factored into two simpler transfer functions, which then can be realized separately by the networks $\mathcal{N}_a$ and $\mathcal{N}_b$. Complete realization is obtained by isolating the two networks by a controlled source.

Operational Amplifier

Another electronic circuit that has found its application as an element in active network synthesis is the d-c amplifier usually known as an *operational amplifier* in analog computers. Theoretically, it is an infinite gain VVT, whose output voltage is of opposite polarity to that of one of the inputs and is of same polarity to that of the other input terminal. An ideal operational amplifier has an infinite input impedance, so the input excitation will not be affected by the power drawn by the amplifier. In addition, it has a zero-output impedance, implying that the amplifier can supply any amount

30 / ACTIVE NETWORK ELEMENTS

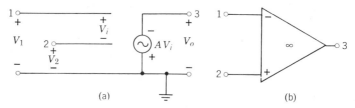

Figure 2-8 The operational amplifier (a) controlled-source representation, (b) symbolic representation.

of power as required by the load. The ideal operational amplifier is required to have "zero offset," i.e., the output voltage is zero when the input voltage is zero.

The usual representation of the operational amplifier is shown in Figure 2-8a. With reference to this figure, we can state the requirements of an ideal operational amplifier:

(2.6)
$$V_o = A(V_2 - V_1) = -AV_i$$
$$A \to \infty$$
$$V_o \to 0 \quad \text{when} \quad (V_2 - V_1) \to 0$$

The symbolic representation of such an active device is presented in Figure 2-8b. When both input terminals are used, it is called a differential input operational amplifier (because the output responds to the difference of voltages at the input terminals). In some applications, terminal 2 is grounded implying a "single-ended" operation.

For linear application, operational amplifiers are usually used with negative feedback. Negative feedback, in general, improves the functional performance of the operational amplifier. Two basic feedback arrangements are shown in Figure 2-9.

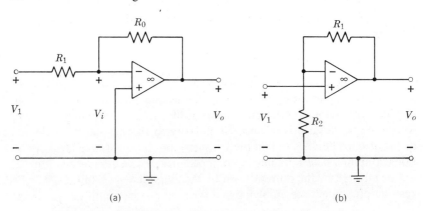

Figure 2-9 Operational amplifier circuits using negative feedback.

Consider the circuit of Figure 2-9a. The circuit equations are given as

$$\frac{V_1 - V_i}{R_1} = \frac{V_i - V_o}{R_0} = \frac{V_i + AV_i}{R_0}$$

or

$$V_i = \frac{V_1}{1 + \frac{R_1}{R_0}(1 + A)}$$

thus

$$V_i \to 0 \quad \text{as} \quad A \to \infty$$

Physically, a finite V_i will cause a large output voltage V_o; V_o, being of opposite polarity to V_i, will tend to increase until V_i becomes infinitesimal. A similar argument follows for the second arrangement (Figure 2-9b).

If terminal 2 is grounded, terminal 1 behaves as if it too were grounded, and is said to be at "virtual" ground. The use of operational amplifier as a network element is considered in Chapter 11.

Nonideal Controlled Sources

In reality, practical controlled sources are nonideal, i.e., the input, output, and feedback impedances are finite. Simplest approximations of the nonideal transducers are shown in Figure 2-10. In effect, nonideal transducers can be considered as ideal controlled sources to which parasitic impedances have been connected. The equivalent representations of Figure 2-10 are still unilateral devices. Presence of finite input and output

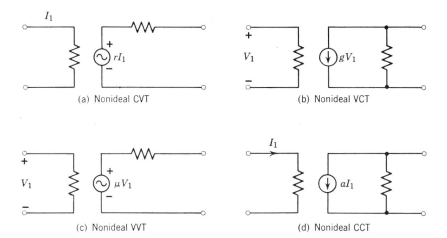

Figure 2-10 Nonideal controlled sources.

impedances make the power gain finite in these cases. If feedback impedance is present, then the device is no longer unilateral in power transmission.

The second type of nonidealness, which creates an additional problem, is that the gain (i.e., the transfer constant) of a practical controlled source is a frequency-dependent function instead of a real constant. This type of nonideal behavior appears mainly due to the charge storage effects in electronic circuit realizations of these active devices. Usually at low frequencies, the gain can be approximated by a real constant. However, at high frequencies, the complex nature of the gain is not negligible, and may well cause a realized active filter to oscillate or deviate markedly from the predicted performance. It should be noted that this type of phase-shift effects restricts the useful operation of most active filters to the lower end of the frequency spectrum, from direct current to several mHz.

2-2 NEGATIVE RESISTANCE

Conceptually, the negative resistance is the simplest of all ideal active elements. In the beginning of this chapter, the negative resistance was introduced in modelling the tunnel diode. The use of negative resistance as a circuit element allows one to realize many network functions that are not realizable by passive *RLC* networks alone. An example of this is the alternate realization of the driving-point impedance.

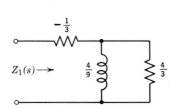

Figure 2-11 Use of negative resistance in realizing driving-point function not realizable with passive *R, L, C* elements.

$$Z_1(s) = \frac{s-1}{s+3}$$

as indicated in Figure 2-11 (compare with Figure 1-3a).

Some properties of the negative resistance are reviewed in this section. Construction of negative resistances, using transistors, is considered later in Chapter 7, which also includes a discussion of several synthesis techniques based on the use of negative resistance as a network element.

Definition of an Ideal Negative Resistance

Ideally, the negative resistance is a two-terminal reciprocal element characterized by a voltage-current relationship given by

(2.7) $$v(t) = -R \cdot i(t)$$

From the above definition, we note that an ideal negative resistance can deliver an infinite amount of power. The model and the $v - i$ characteristic of a negative resistance are shown in Figure 2-12.

Principles of conservation of energy dictate that it is physically impossible to obtain an ideal negative resistance. For small signal applications and at low frequencies, however, an ideal negative resistance characteristic is a good approximation of some physical devices and can be used as a network element.

Nonideal Negative Resistance

A large number of electronic devices and circuits show negative resistance characteristic and approach the idealized behavior (Equation

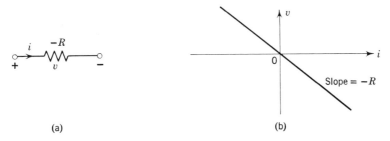

Figure 2-12 The ideal negative resistance.

2.7) over a limited range of voltage, current, and frequencies. The various negative resistance devices are usually divided into two groups, depending on the form of their $v - i$ characteristics (Figures 2-13a and b). The *S-type* negative resistance is a current-controlled element, i.e., the voltage across the element is a single-valued function of the current through it. The unijunction transistor exhibits such a characteristic. On the other hand, the *N-type* negative resistance is a voltage-controlled element. An example of such a device is the tunnel diode, mentioned earlier. The models of the *N*-type and the *S*-type negative resistance devices shown in Figures 2-13c and d represent the appropriate devices to a fairly good approximation over a small portion of their negative resistive region.[3]

Some practical realizations of the *N*-type and *S*-type negative resistance devices are included in Chapter 7.

---

[3] Based on the dynamic behavior of the negative resistance devices. See E. W. Herold, "Negative resistance and devices for obtaining it," *Proc. IRE*, **23**, 1201–1223 (October 1935).

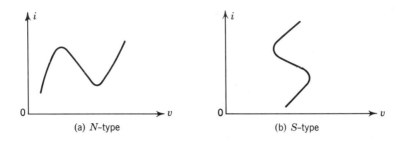

(a) N-type

(b) S-type

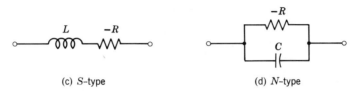

(c) S-type

(d) N-type

Figure 2-13 The i-v characteristic of practical negative resistances and their approximate small-signal equivalent circuits.

2-3 NEGATIVE INDUCTANCE AND CAPACITANCE

An idealized model of a negative inductance (Figure 2-14a) is characterized by the $v - i$ relationship:

$$v(t) = -L \cdot \frac{di}{dt} \tag{2.8}$$

Similarly, a negative capacitance (Figure 2-14b) is an active element that can be mathematically described as

$$i(t) = -C \cdot \frac{dv}{dt} \tag{2.9}$$

Both of these elements are 2-pole reciprocal elements. No distinct attempt has been made thus far to construct any electronic device that exhibits negative inductive or negative capacitive properties. The reason for this is that these models have not been widely used in the design of

(a) (b)

Figure 2-14 The negative inductance and the negative capacitance.

active circuits, although study of properties of network functions of linear, lumped circuits containing negative inductors and/or negative capacitors as circuit elements has been carried out.

2-4 IMPEDANCE CONVERTER

As the name implies, the impedance converter is a device to *convert* a terminating *impedance*. The most popular type of impedance converter is the negative-impedance converter, which can be used to convert an

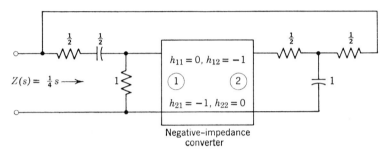

Figure 2-15 Realization of an inductance by an active RC one-port containing a negative-impedance converter.

impedance $Z(s)$ into a negative impedance $-Z(s)$. The impedance-converter-type active devices have been successfully used in inductorless filter design. Figure 2-15 illustrates the use of a negative-impedance converter in realizing an ideal inductance.[4]

Various types of impedance converters, which have been proposed for the design of filters without inductors, are discussed next. Practical realizations of the impedance converter will be found in Section 9-1 of Chapter 9. The use of the impedance converter as a network element is also covered in the same chapter.

Definition of an Ideal Converter

An ideal impedance converter is a two-port network which, when terminated at a one-port by a driving-point impedance $Z_L(s)$, presents at the other port an input impedance directly proportional to $Z_L(s)$ for all frequencies.

The necessary requirements of such a two-port can be derived from fundamental considerations. Consider an arbitrary two-port $\mathcal{N}$ terminated at its port 2 by an impedance $Z_L(s)$. (See Figure 2-16.) In terms

---

[4] J. M. Sipress, "Synthesis of active RC networks," *IRE Trans. on Circuit Theory*, CT-8, 260–269 (September 1961).

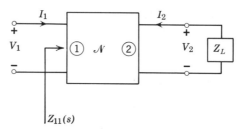

Figure 2-16 A general two-port $\mathcal{N}$ terminated at one-port by an impedance Z_L.

of its transmission matrix (**F**) parameters,[5] the input impedance at port 1 is

$$Z_{11}(s) = \frac{AZ_L(s) + B}{CZ_L(s) + D} \tag{2.10}$$

In order that the two-port may be an ideal impedance converter, the requirement is that for all possible termination impedance $Z_L(s)$

$$Z_{11}(s) = K(s) Z_L(s) \tag{2.11}$$

where $K(s)$ is a predetermined real rational function. Comparing Equations 2.10 and 2.11, we obtain the conditions[6]

$$\begin{aligned} B &= C = 0 \\ A &\neq 0 \\ D &\neq 0 \end{aligned} \tag{2.12}$$

where $(A/D) = K(s)$. Under the above condition, the driving-point impedance $Z_{22}(s)$ at port 2, obtained by terminating port 1 by $Z_L(s)$, is given by

$$Z_{22}(s) = \frac{DZ_L + B}{CZ_L + A} = \frac{D}{A} Z_L = \frac{Z_L}{K(s)} \tag{2.13}$$

Thus an ideal converter "converts" impedance in both directions.

Condition 2.12 can also be written in terms of hybrid parameters. For example, in terms of the h parameters, the necessary and sufficient conditions that a two-port be an ideal impedance converter are:

$$\begin{aligned} h_{11} &= h_{22} = 0 \\ h_{12} h_{21} &= -K(s) \end{aligned} \tag{2.14}$$

---

[5] The various two-port parameters are summarized in Table A-1 of Appendix A.
[6] W. Ralph Lundry, "Negative impedance circuits—some basic relations and limitations," *IRE Trans. on Circuit Theory*, **CT-4**, 132–139 (September 1957).

$K(s)$, the proportionality factor of the ideal converter, is known as the *conversion factor*. It is interesting to consider a few particular solutions of Equation 2.14.

Generalized Impedance Converter

When the conversion factor $K(s)$ is the negative of the ratio of two driving-point impedances, $-Z_\alpha/Z_\beta$, the ideal impedance converter is called a *generalized impedance converter* (GIC). From Equation 2.14, we note that many different types of GIC's are possible depending on the identification of h_{12} and h_{21}.

For example, one identification is:

(2.15)
$$h_{11} = h_{22} = 0$$
$$h_{12} = -Z_\alpha/Z_\beta$$
$$h_{21} = -1$$

In terms of the terminal variables, Equation 2.15 can be written as

(2.16)
$$V_1 = -\frac{Z_\alpha}{Z_\beta} V_2$$
$$I_2 = -I_1$$

From Equation 2.16, we conclude that the input current is transmitted through the converter without any change, whereas the output voltage is essentially of opposite polarity to the input voltage. This inversion of port voltages leads to the classification of this device as a *voltage inversion type* GIC.[7] For simplicity, we shall refer to the ideal two-port described by Equation 2.16 as a VGIC.

A second solution is given by

(2.17)
$$h_{11} = h_{22} = 0$$
$$h_{12} = 1$$
$$h_{21} = \frac{Z_\alpha}{Z_\beta}$$

The above equation implies the following relation between the port variables:

(2.18)
$$V_1 = V_2$$
$$I_2 = \frac{Z_\alpha}{Z_\beta} I_1$$

---

[7] S. K. Mitra and N. M. Herbst, "Synthesis of active *RC* one-ports using generalized impedance converters," *IEEE Trans. on Circuit Theory*, **CT-10**, 532 (December 1963).

Note that in the above case the output and input voltages are equal. On the other hand, there is an inversion of the port currents, which dictates the name *current inversion type* GIC.[8] An ideal active two-port described by Equation 2.18 is denoted as a CGIC for convenience.

Negative-Impedance Converter

If it is desired to have an active two-port that transforms an impedance terminated at one-port to its negative at the other port, the conversion factor $K(s)$ must be made a negative real constant. This type of ideal converter is known as a *negative-impedance converter* (NIC).[9]

Depending on the sign of the h parameters, there are two types of negative-impedance converter. If h_{12} and h_{21} are both negative *real* constants, the converter is called a *voltage inversion type negative-impedance converter* (VNIC). The VNIC is described by an h matrix:

$$(2.19) \qquad \begin{bmatrix} 0 & -k_1 \\ -k_2 & 0 \end{bmatrix}$$

where k_1 and k_2 are real positive constants. Alternately, for a *current inversion type negative-impedance converter* (CNIC), both h_{12} and h_{21} are positive real numbers. The h matrix of a CNIC is given by

$$(2.20) \qquad \begin{bmatrix} 0 & k_1 \\ k_2 & 0 \end{bmatrix}$$

where k_1 and k_2 are real positive numbers.

If k_1 and k_2 in Equations 2.19 and 2.20 are equal to unity, the notation UVNIC or UCNIC will be used as appropriate. Note that the generalized impedance converter and the negative-impedance converter are nonreciprocal active two-ports.

Nonideal Negative-Impedance Converter

In general, the NIC's built in practice are not ideal devices, i.e., h_{11} and h_{22} are not equal to zero. Nonzero values of these parameters may counteract the negative immittance conversion properties of the device. Usually h_{11} and h_{22}, termed parasitic impedances, have very small nonzero values. The representation of a nonideal NIC is shown in Figure 2-17a, showing the parasitic immittances as a series and shunt impedances.

---

[8] I. W. Sandberg, "Synthesis of driving-point impedance with active *RC* networks," *Bell System Tech. J.*, **39**, 947–962 (July 1960).

[9] J. L. Merill, Jr., "Theory of the negative impedance converter," *Bell System Tech. J.*, **30**, 88–109 (January 1951).

IMPEDANCE CONVERTER / 39

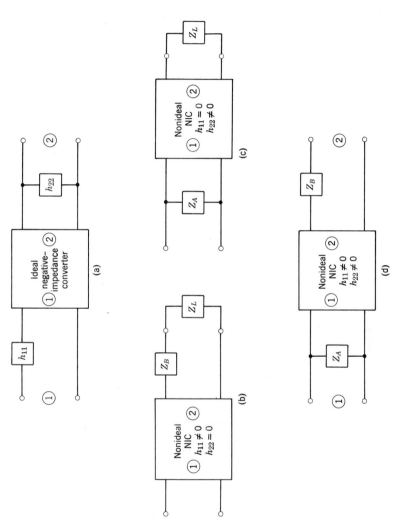

Figure 2-17 Compensation of a nonideal negative-impedance converter.

40 / ACTIVE NETWORK ELEMENTS

Let us now investigate the possibility of eliminating the effects of the parasitics.[10] The parameter h_{11} reduces the negative impedance produced at port 1 due to the terminating impedance Z_L at port 2.[11] If we assume, for the present, that $h_{22} = 0$, then the input impedance of the nonideal NIC due to Z_L at port 2 is

$$h_{11} - h_{12}h_{21}Z_L$$

If we insert an impedance Z_B in series with Z_L (Figure 2-17b), the input impedance then becomes

$$h_{11} - h_{12}h_{21}(Z_L + Z_B)$$

Choosing $Z_B = h_{11}/h_{12}h_{21}$, we can cancel the effect of nonzero h_{11}, thus obtaining ideal operation. Next consider the effect of h_{22} assuming $h_{11} = 0$. Connecting an impedance Z_A across the input port (Figure 2-17c), we obtain the input admittance as

$$\frac{1}{Z_A} - \left[\frac{h_{22}Z_L + 1}{h_{12}h_{21}Z_L}\right]$$

The above expression indicates that the selection of $Z_A = h_{12}h_{21}/h_{22}$ will cancel the effect of the parasitic admittance h_{22}. The foregoing consideration leads us to examine the possibility of cancelling the effects of h_{11} and h_{22} simultaneously by incorporating series and shunt impedances, Z_A and Z_B, with the nonideal two-port. The complete arrangement is shown in Figure 2-17d. It can be shown that the values of Z_A and Z_B as calculated above do not provide complete cancellation. In order to find their new values, we must first determine the h parameters of the composite network and apply the conditions given by Equation 2.14. Let us first define the compensating impedances in terms of the parasitic impedances of the nonideal negative-impedance converter. To this end, we define

(2.21) $$Z_A = \frac{A}{h_{22}}, \qquad Z_B = Bh_{11}$$

---

[10] A. I. Larky, "Negative-impedance converters," *IRE Trans. on Circuit Theory*, **CT-4**, 124–131 (September 1957).

[11] This can be seen from the following alternate expression of Equation 2.10

(2.10a) $$Z_{11}(s) = h_{11} - \frac{h_{12}h_{21}}{h_{22} + \dfrac{1}{Z_L}}$$

Simple calculation yields the following expressions for the hybrid parameters of the composite two-port:

(2.22) $$h'_{11} = \frac{Z_A\left[h_{11} - \dfrac{h_{12}h_{21}Z_B}{h_{22}Z_B + 1}\right]}{Z_A + h_{11} - \dfrac{h_{12}h_{21}Z_B}{h_{22}Z_B + 1}}$$

(2.23) $$h'_{22} = \frac{h_{22} - \dfrac{h_{12}h_{21}}{h_{11} + Z_A}}{Z_B\left[h_{22} - \dfrac{h_{12}h_{21}}{h_{11} + Z_A}\right] + 1}$$

Setting $h'_{11} = h'_{22} = 0$ and using Relation 2.21, we obtain the necessary conditions for compensation

(2.24) $$h_{12}h_{21} = h_{11}h_{22} + A$$
$$AB = 1$$

Further conditions can be derived by computing $h'_{12}h'_{21}$ and setting the product greater than zero to guarantee negative-impedance conversion action. This indicates, at least theoretically, that it is possible to design ideal NIC by using compensating impedances at appropriate places and by adjusting the values of the h parameters of the nonideal NIC.[12]

Positive-Impedance Converter: Ideal Transformer

The two-port for which the conversion factor $K(s)$ is a positive real constant, will be called a *"positive-impedance converter"* (PIC). Four types of positive-impedance converters can be defined.

The *"ideal real transformer"* (Figure 2-18a) is defined by the following

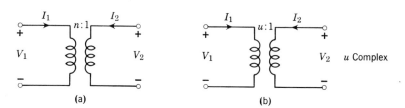

Figure 2-18 The ideal (a) real and (b) complex transformers.

---

[12] In general, the h parameters of a practical NIC circuit are complex functions of frequency. As a result, perfect compensation can be achieved only over a small band in the low-frequency range.

equations:

(2.25)
$$V_1 = nV_2$$
$$I_2 = -nI_1$$

where n is a real number. The ideal real transformer (more popularly known as the "ideal transformer") is a reciprocal PIC characterized by a single parameter "n" called the "turns ratio." Since the total input power is zero,

$$V_1 I_1 + V_2 I_2 = V_1 I_1 + \frac{V_1}{n} \cdot (-nI_1) = 0$$

it is also a lossless two-port. For convenience, we shall denote an ideal (real) transformer as IT.

The ideal transformer has been used freely in determining properties of network functions and in passive network synthesis.

A recent addition to network theory, the *"ideal complex transformer"* [13] is a useful element in deriving properties of linear active networks. Essentially, it is a generalization of an ideal transformer having the following constraint between its port voltages and currents:

(2.26)
$$V_1 = uV_2$$
$$I_2 = -u^* I_1$$

where the turns ratio u is a complex number and u^* is the complex conjugate of u. When u is real, the complex transformer becomes the conventional ideal transformer. Like its real counterpart, the complex transformer is also a lossless device. The generalized transformer with complex turns ratio can be regarded as a combination of an ideal real transformer and a nonbilateral phase shifter. Note that the ideal complex transformer (Figure 2-18b) is a nonreciprocal PIC.

The third type of positive-impedance converter is an alternate generalization of the ideal real transformer and is described by the following input-output relations:

(2.27)
$$V_1 = \gamma_1 V_2$$
$$I_2 = -\gamma_2 I_1$$

where γ_1 and γ_2 are real numbers and $\gamma_1 \neq \gamma_2$. Note that the total input power can be made negative by choosing suitable excitations. Hence this last type of positive-impedance converter is an active two-port, and will thus be designated as an *active transformer*.

[13] P. Bello, "Extension of Brune's energy function approach to the study of LLF networks," *IRE Trans. on Circuit Theory*, **CT-7**, 270–280 (September 1960).

An interesting generalization of the active transformer is obtained by making γ_1 and γ_2 functions of the complex frequency variable s. This type of device can be used to realize inductors by suitably choosing γ_1 and γ_2. Consider, for example, a *generalized active transformer* terminated at port 2 by a resistance R_L. The input impedance seen at port 1 is then given as

$$Z_{\text{in}} = \gamma_1 \gamma_2 R_L$$

If

$$\gamma_1 = sK_1, \qquad \gamma_2 = K_2$$

then the active one-port behaves like an inductance. More on generalized active transformer and their realizations will be discussed in Section 11.8.

Ideal Voltage, Current, and Power Converters

Three more general types of converters have proved useful in the analysis of active networks.

An *ideal voltage converter* (IVC) is a two-port network defined by the transmission matrix:[14]

(2.28)
$$\begin{bmatrix} k & 0 \\ 0 & 1 \end{bmatrix}$$

This circuit transduces the current without any change, but the input voltage is changed by the factor $1/k$, and as a result an IVC is an impedance converter. When $k < 0$, this is a VNIC. An ideal transformer is an IVC with $k = 1$.

An *ideal current converter* (ICC) is the dual of the IVC and is described by the transmission matrix:[15]

(2.29)
$$\begin{bmatrix} 1 & 0 \\ 0 & \beta \end{bmatrix}$$

As before, we note that an ICC is an impedance converter and if $\beta < 0$, it becomes a CNIC. For $\beta = 1$, an ICC is also an ideal transformer.

The two-port that transduces both the current and the voltage is called an *ideal power converter* (IPC).[16] It is defined by the following transmission matrix:

(2.30)
$$\begin{bmatrix} m & 0 \\ 0 & m \end{bmatrix}$$

[14] M. Kawakami, "Some fundamental considerations on active four-terminal linear networks," *IRE Trans. on Circuit Theory*, **CT-5**, 115–121 (June 1958).

[15] M. Kawakami, *loc. cit.*

[16] E. V. Zelyakh, "The ideal power converter—a new element of electric circuits," *Elektrozvyaz*, No. 1, 35–47 (1957); C. G. Aurell, "Representation of the general linear four-terminal network and some of its properties," *Ericsson Technics*, **11**, 155–179 (1955).

44 / ACTIVE NETWORK ELEMENTS

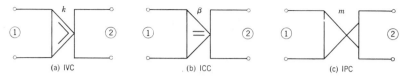

Figure 2-19 Symbolic representations of the ideal voltage, current, and power converters.

An ideal transformer of unity turns ratio is again a special case of an IPC if $m = 1$. From Equations 2.10 and 2.13 we note that if an IPC is terminated at either port by Z_L, the input impedance looking at the other port is also Z_L. This implies that an IPC is impedance-transparent. The symbolic representation of the three converters is shown in Figure 2-19.

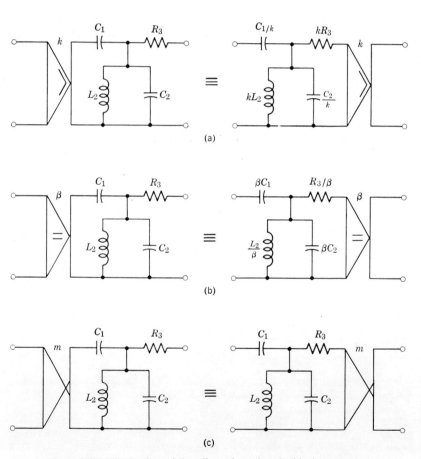

Figure 2-20 Illustration of the effect of moving the ideal converters.

The actions of these ideal converters on an arbitrary two-port will be apparent from the following transmission parameter identities:

(2.31) $$\begin{bmatrix} k & 0 \\ 0 & 1 \end{bmatrix}\begin{bmatrix} A & B \\ C & D \end{bmatrix} = \begin{bmatrix} A & kB \\ C/k & D \end{bmatrix}\begin{bmatrix} k & 0 \\ 0 & 1 \end{bmatrix}$$
$$\text{IVC} \qquad\qquad\qquad\qquad\qquad \text{IVC}$$

(2.32) $$\begin{bmatrix} 1 & 0 \\ 0 & \beta \end{bmatrix}\begin{bmatrix} A & B \\ C & D \end{bmatrix} = \begin{bmatrix} A & B/\beta \\ \beta C & D \end{bmatrix}\begin{bmatrix} 1 & 0 \\ 0 & \beta \end{bmatrix}$$
$$\text{ICC} \qquad\qquad\qquad\qquad\qquad \text{ICC}$$

(2.33) $$\begin{bmatrix} m & 0 \\ 0 & m \end{bmatrix}\begin{bmatrix} A & B \\ C & D \end{bmatrix} = \begin{bmatrix} A & B \\ C & D \end{bmatrix}\begin{bmatrix} m & 0 \\ 0 & m \end{bmatrix}$$
$$\text{IPC} \qquad\qquad\qquad\qquad\qquad \text{IPC}$$

Thus, an IVC to the left of a two-port when moved to the right increases all the impedances of the circuit elements by k times. Similarly, moving an ICC from left to right increases all the admittances of the circuit elements by β times. Finally, moving the IPC does not affect the elements of the network in any way. The effects of the converters are illustrated in Figure 2-20.

2-5 IMPEDANCE INVERTER

The impedance inverter is an ideal active two-port having an impedance inversion property. One well-known class of impedance inverter is the *gyrator*, which inverts an impedance $Z(s)$ into its reciprocal. Thus, an ideal gyrator, when terminated at a one-port by a capacitance C farads, has an input impedance at the other port proportional to Cs. As a result, the gyrator can be used in the design of inductorless networks.

Properties of several types of impedance inverters are derived in this section. Some synthesis applications of impedance inverters are considered in Chapter 10 along with several practical realizations of these devices.

Definition of an Ideal Impedance Inverter

A two-port, which when terminated at a one-port by an impedance $Z_L(s)$ presents at the other port an input impedance inversely proportional to $Z_L(s)$ at all frequencies, is called an *ideal impedance inverter*.

Like the ideal impedance converter, the necessary and sufficient conditions for a two-port to be an ideal inverter can be derived from fundamental considerations. With reference to Figure 2-16, we note that the input impedance of a terminated two-port is given by

(2.10) $$Z_{11}(s) = \frac{AZ_L(s) + B}{CZ_L(s) + D}$$

Ideal inverter action implies

(2.34) $$Z_{11}(s) = \frac{G(s)}{Z_L(s)}$$

In order that Equation 2.10 be identical to Equation 2.34, we must have[17]

(2.35) $$A = D = 0$$
$$B \neq 0$$
$$C \neq 0$$

where $B/C = G(s)$ is the *inversion factor*. The output impedance at port 2 when port 1 is terminated by $Z_L(s)$ is

(2.36) $$Z_{22}(s) = \frac{DZ_L + B}{CZ_L + A} = \frac{B}{CZ_L} = \frac{G(s)}{Z_L}$$

Thus, an ideal impedance inverter inverts impedance in both directions with the same proportionality factor.

Condition 2.35 can also be written in terms of the z or y parameters. For example, the necessary and sufficient conditions for an impedance inverter in terms of the open circuit parameters are

(2.37) $$z_{11} = z_{22} = 0$$
$$z_{12}z_{21} = -G(s)$$

We now consider a few solutions of Equation 2.37.

Positive-Impedance Inverter: Gyrator

An ideal impedance inverter, defined by

(2.38) $$z_{11} = z_{22} = 0$$
$$z_{12} = -r_1$$
$$z_{21} = r_2$$

where r_1 and r_2 are real positive numbers, will be called a *positive-impedance inverter*. Since $z_{12} \neq z_{21}$, it is a nonreciprocal two-port. In terms of the input and output variables, an ideal positive-impedance inverter (PIV) is

---

[17] W. R. Lundry, *loc. cit.*

described by the following equations:

$$V_1 = -r_1 I_2$$
$$V_2 = r_2 I_1$$

The total (real) power delivered to such a two-port is given as

$$V_1 I_1 + V_2 I_2 = (r_2 - r_1) I_1 I_2$$

If $r_2 = r_1$, the input power is always zero, and the device is a lossless (i.e., passive) nonreciprocal two-port. This type of positive-impedance inverter is more commonly known as the *ideal gyrator* (IG).[18] Thus the

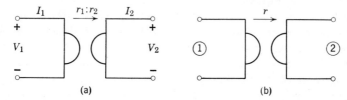

Figure 2-21 Symbolic representations of the (a) active gyrator and (b) the lossless gyrator.

ideal gyrator is characterized by the following open-circuit impedance matrix:

(2.39)
$$[z] = \begin{bmatrix} 0 & -r \\ r & 0 \end{bmatrix}$$

On the other hand, if $r_2 \neq r_1$, the total power delivered to the two-port can become negative for some suitably chosen excitations and, as a result, a PIV in general is an active nonreciprocal two-port. The latter type of PIV has also been designated as an *active gyrator* by several authors.[19]

The model of a PIV is indicated in Figure 2-21a. The significance of the arrow in the network model will be understood from a consideration of the operation of the two-port network of Figure 2-22, which shows a PIV with a resistor connected between its input and output ports. It can be

---

[18] First introduced as a network element by B. D. H. Tellegen in "The gyrator: a new network element," *Phillips Res. Rept.*, **3**, 81–101 (April 1948).

[19] T. Yanagisawa and Y. Kawashima, "Active gyrator," *Electronics Letters*, **3**, 105–107 (March 1967); W. H. Holmes, "A new method of gyrator-*RC* filter synthesis," *Proc. IEEE*, **54**, 1459 (1966).

shown that this network is described by the following short-circuit admittance parameters:

(2.40)
$$y_{11} = y_{22} = \frac{1}{r_2}$$
$$y_{12} = 0$$
$$y_{21} = -\left(\frac{1}{r_1} + \frac{1}{r_2}\right)$$

Therefore, the circuit of Figure 2-22 has a forward transmission path (coinciding with the direction of the arrow) and no backward transmission. Such a device is sometimes used as an isolator.

The network model of an ideal gyrator is the one shown in Figure 2-21b. The constant r in Equation 2.39 and in Figure 2-21b is called the *gyration impedance* of the gyrator. Because of its lossless nature, the ideal gyrator has become increasingly more popular to the network designer as an element in filter design. Although the ideal gyrator is a lossless and hence a passive device, it is generally built with vacuum tubes or transistors. Thus, for the purpose of this book, we shall treat it as an active device.

A simple application of an ideal PIV will now be discussed. Consider the one-port network of Figure 2-23. The input impedance is given by

(2.41)
$$Z_{in} = \frac{r_1 r_2}{Z(s)}$$

If $Z(s)$ represents a capacitance C, the input impedance becomes $Z_{in} = s(r_1 r_2 C)$, i.e., the one-port behaves like an ideal inductance of $r_1 r_2 C$ henries. This indicates a simple way to eliminate inductances in a passive RLC filter. For a large number of inductances, this approach may not be economically attractive.

Most of the practical gyrators and PIV circuits are grounded networks. This creates a problem for networks where one or more floating inductances are used. A situation like this is depicted in Figure 2-24a. The

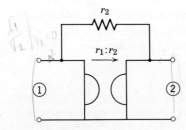

Figure 2-22   An isolator.

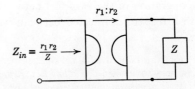

Figure 2-23   The active gyrator as a positive impedance inverter.

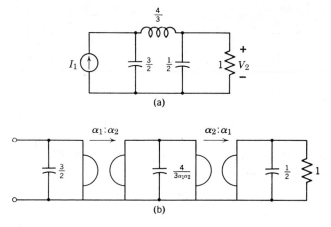

Figure 2-24 The use of two active gyrators to replace a floating inductor.

floating inductance can be replaced by means of two grounded PIV's and a capacitor as shown in Figure 2-24b.[20]

Another application of ideal gyrators and PIV's is the construction of ideal transformers. This is illustrated in Section 2-7.

Nonideal Gyrator

Several circuits are available for the practical realization of a gyrator. In general, a practical gyrator does not exactly satisfy Equation 2.39, i.e., it is not an ideal device. A gyrator can depart from its ideal behavior in many ways.

In one case, z_{11} and z_{22} may not be equal to zero. Nonzero z_{11} and z_{22} are essentially parasitic impedances, indicating a possible representation of such a nonideal gyrator as shown in Figure 2-25a. An alternate representation can be obtained on the y basis, as indicated in Figure 2-25b. Figure 2-25 suggests an obvious way of compensating the effect of parasitics using negative resistances. The complete compensation arrangement is shown in Figure 2-26.

A more general type of nonideal gyrator is the two-port for which

(2.42)
$$z_{11} \neq 0$$
$$z_{22} \neq 0$$
$$z_{12} \neq -z_{21}$$

[20] A. G. J. Holt and J. Taylor, "Method of replacing ungrounded inductors by grounded gyrators," *Electronics Letters*, **1**, 105 (June 1965). See also B. D. Anderson, W. New, and R. Newcomb, "Proposed adjustable tuned circuits for microelectronic structures," *Proc. IEEE*, **54**, 411 (March 1966).

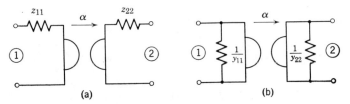

Figure 2-25 Nonideal gyrator.

Assuming that the parameters are real, such a nonideal gyrator can be described by an open-circuit impedance matrix,

$$(2.43) \qquad \begin{bmatrix} R_1 & -R_2 \\ R_3 & R_4 \end{bmatrix}$$

or a short-circuit admittance matrix

$$(2.44) \qquad \begin{bmatrix} R_1 & -R_2 \\ R_3 & R_4 \end{bmatrix}^{-1} = \begin{bmatrix} G_1 & G_2 \\ -G_3 & G_4 \end{bmatrix}$$

Expressing either the z matrix or the y matrix as a sum of one symmetric and one skew symmetric matrix, a representation of this type of nonideal gyrator can be obtained. We illustrate this by splitting the y matrix:

$$(2.45) \qquad \begin{bmatrix} G_1 & G_2 \\ -G_3 & G_4 \end{bmatrix} = \begin{bmatrix} G_1 & \dfrac{G_2 - G_3}{2} \\ \dfrac{G_2 - G_3}{2} & G_4 \end{bmatrix} + \begin{bmatrix} 0 & \dfrac{G_2 + G_3}{2} \\ -\dfrac{G_2 + G_3}{2} & 0 \end{bmatrix}$$

The symmetric matrix can be realized by a π network; the skew symmetric matrix represents an ideal gyrator of gyration constant $r = 2/(G_2 + G_3)$. The representation of Expression 2.44 is obtained by paralleling the π network and the ideal gyrator (Figure 2-27). Using negative and positive

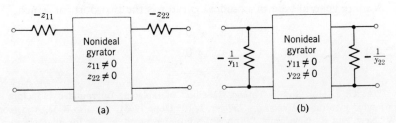

Figure 2-26 Compensation of nonideal gyrator.

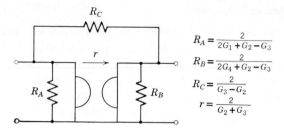

Figure 2-27 Representation of an arbitrary resistive two-port.

$$R_A = \frac{2}{2G_1 + G_2 - G_3}$$
$$R_B = \frac{2}{2G_4 + G_2 - G_3}$$
$$R_C = \frac{2}{G_3 - G_2}$$
$$r = \frac{2}{G_2 + G_3}$$

resistances, compensation can be achieved for the nonideal behavior. A compensation arrangement is illustrated in Figure 2-28. Note that in most cases it is not necessary to use three negative resistances. For example, if $G_2 > G_3$, $-R_C$ in fact becomes a positive resistance.

A second type of nonideal behavior is exhibited when the transfer parameters instead of being real become frequency-dependent. Such a gyrator will be called a *reactive gyrator* (XIG). More specifically, a reactive gyrator is defined by the following set of equations:

(2.46)
$$V_1 = -\alpha(s)I_2$$
$$V_2 = \alpha(s)I_1$$

where $\alpha(s)$ is the complex gyration impedance.[21] Consideration of the total input (real) power reveals that such a two-port is not a passive device. The symbol of Figure 2-29 will be used to represent a reactive gyrator.

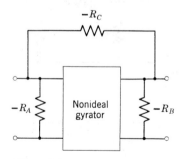

Figure 2-28 Construction of an ideal gyrator from a nonideal active gyrator.

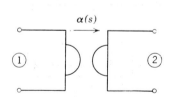

Figure 2-29 Symbolic representation of a reactive gyrator.

[21] T. J. Harrison, "A gyrator realization," *IEEE Trans. on Circuit Theory*, **CT-10**, 303 (June 1963).

Negative-Impedance Inverter

The ideal gyrator is a positive-impedance inverter, i.e., the inversion factor $G(s)$ in Equations 2.34 and 2.36 is positive and real. Another class of active two-port is obtained when $G(s)$ is negative and real. This type of inverter is more popularly known as a *negative-impedance inverter*.

An ideal *nonreciprocal negative-impedance inverter* (NRNIV)[22] is defined by the following open-circuit impedance parameters:

$$z_{11} = z_{22} = 0$$
(2.47) $$z_{12} = \pm R_1$$
$$z_{21} = \pm R_2$$

where R_1 and R_2 are real and positive numbers, and $R_1 \neq R_2$. The more popular type of negative-impedance inverter is defined by Equation 2.47 with $z_{12} = z_{21} = \pm R$ and will be denoted as NIV. Note that the NIV in contrast to NRNIV is a reciprocal two-port. Circuit realizations of the ideal NIV are indicated in Figure 2-30a.

An application of an NIV is illustrated next. Consider an NIV terminated at one port by an impedance Z_L. The input impedance is

(2.48) $$Z_{\text{in}} = z_{11} - \frac{z_{12}z_{21}}{z_{22} + Z_L}$$

(2.49) $$= -\frac{R^2}{Z_L}$$

If, for example, Z_L represents a capacitance, the terminated NIV behaves

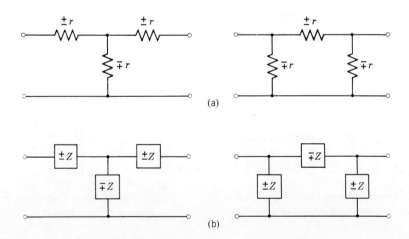

Figure 2-30 (a) Reciprocal NIV. (b) Reactive NIV.

[22] S. K. Mitra, "Non-reciprocal negative impedance inverter," *Electronics Letters*, **3**, 388 (August 1967).

like a negative inductance at its other port. Similarly, the NRNIV can be used to produce negative impedances. Another application of the NRNIV is in the realization of the active gyrator. This will be considered in Chapter 10.

Nonideal Negative-Impedance Inverter

Practical NIV and NRNIV circuits are nonideal devices, exhibiting nonzero values for z_{11} and z_{22}. In some cases, the nonideal NIV and NRNIV can be safely used in generating negative impedance. This is illustrated next for the case of the nonideal NIV.

Assuming $z_{11} = z_{22} = \delta$ and $z_{12} = \delta - R$, the input impedance of a terminated nonideal NIV is

(2.50)
$$Z_{in} = \delta - \frac{(\delta - R)^2}{\delta + Z_L}$$
$$\cong -\frac{1}{Z_L}\left[R^2 - \delta Z_L - 2\delta R - \frac{R^2 \delta}{Z_L}\right]$$

This implies that if

$$R^2 \gg \delta\left(Z_L + 2R + \frac{R^2}{Z_L}\right)$$

a nonideal NIV can be used to produce negative impedances. Synthesis techniques employing NIV circuits have not been proposed yet.

Reactive Negative-Impedance Inverter

An extension of the above idea leads to the concept of the *reactive negative-impedance inverter* (XNIV) circuits.[23] Ideally, an XNIV is an ideal impedance inverter described by a complex inversion ratio $G(s)$. More precisely, the voltage-current relationship of a XNIV is given by

(2.51)
$$V_1 = \pm Z(s)I_2$$
$$V_2 = \pm Z(s)I_1$$

The possible realizations are presented in Figure 2-30b. Since z_{12} is identical to z_{21}, an XNIV is an active reciprocal two-port. Similarly, the NRNIV can be generalized.

2-6 DEGENERATE CIRCUIT ELEMENTS[24]

Many authors have investigated the possible theoretical existence of networks consisting of physically realizable (in the idealized sense) elements, whose terminal behavior cannot be described by any of the existing

---

[23] K. L. Su, *Active network synthesis*, McGraw-Hill, New York, 49 (1965).
[24] H. J. Carlin, "Singular network elements," *IEEE Trans. on Circuit Theory*, **CT-11**, 67–72 (March 1964).

network descriptions, such as **Z, Y, H, G, F,** and **S**[25] matrices. Such networks, usually called "pathological" networks, are degenerate forms and are needed to complete the domain of network elements.

One such element is the *nullator*. It is a 2-terminal element defined by

(2.52) $$V = I = 0$$

The usual circuit model of a nullator is shown in Figure 2-31a. From the definition it is clear that this one-port is simultaneously an open and a short circuit. The nullator is a bilateral and lossless one-port.

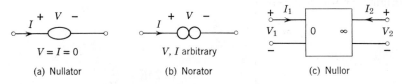

(a) Nullator (b) Norator (c) Nullor

Figure 2-31 Symbolic representations of the degenerate circuit elements.

Another pathological element is the *norator*. It is a 2-terminal element for which the terminal variables, V and I, are completely arbitrary. The circuit model used is shown in Figure 2-31b. By definition, the norator is a nonreciprocal 2-terminal element.

It has been shown[26] that the nullator and the norator cannot be derived from a limiting operation on physically realizable network components. There exists, however, a pathological two-port which is obtainable by a limiting process.[26] Such a two-port is called the *nullor*. One pair of terminals of this device acts as a short circuit and an open circuit at the same time, whereas the voltage and the current at the other port are arbitrary. The nullor is described by a transmission matrix which is a null-matrix, i.e.,

(2.53) $$\begin{bmatrix} V_1 \\ I_1 \end{bmatrix} = \begin{bmatrix} 0 & 0 \\ 0 & 0 \end{bmatrix} \begin{bmatrix} V_2 \\ -I_2 \end{bmatrix}$$

The nullator and the norator have been found useful in representing transistors and operational amplifiers. Formation of nullator-norator equivalent representations and their application in generating realizable circuits will be studied later in this book.

---

[25] Scattering matrices.
[26] H. J. Carlin, *loc. cit.*

2-7 EQUIVALENT CIRCUITS

For design and analysis considerations, it is useful to represent the various active elements and devices discussed in previous sections in an equivalent circuit form. The derivation of the equivalent circuit is based on the equivalent representation of an arbitrary two-port. Even though controlled sources are generally used for such purposes, we shall also discuss equivalent representations in terms of other active elements. This will enable us to indicate the relations between the active elements from a two-port point of view. In addition, alternate equivalent representations of the same device give the designer a choice of circuits that he may use to build the particular active device. It should be noted that although the main emphasis will be on the equivalent circuits of ideal active elements, discussion concerning arbitrary two-port will indicate the possible representation of the nonideal active element. A byproduct of the material of this section will be the formulation of the basic set of elements necessary and sufficient to represent and synthesize any active network.

Relations between Ideal Active Elements

Thus far, we have introduced many distinct ideal active devices. Our next problem is to find the connections that exist between these elements. To facilitate our objective, we have summarized in Table 2-1 the two-port parameters of some ideal active elements.

A careful study of this table reveals that these active elements can be classified into either one of two distinct groups: the impedance converter group or the impedance inverter group. The classification is based on the values of their transmission parameters. An *impedance converter type* active element is characterized by having $B = C = 0$; an *impedance inverter type*, by having $A = D = 0$.

The various active elements in the converter group can be defined by their respective positions in the A, D plane[27] (Figure 2-32). The two axes define the two controlled sources—the VVT and the CCT. The IVC is characterized by a line parallel to the A axis at a unit distance above it; the ICC, by a straight line parallel to the D axis at a unit distance to the right. The IPC can be located on the line which is at 45° to the axes. Points in the second quadrant describe a VNIC, whereas a point in the fourth quadrant will describe a CNIC. The two hyperbolas represent the ideal transformer.

---

[27] L. P. Huelsman, "A fundamental classification of negative-immittance converters," *1965 IEEE International Convention Record*, **13** (Part 7), 113–118 (March 1965).

Table 2-1 TWO-PORT PARAMETERS OF IDEAL ACTIVE ELEMENTS

Element	Z	Y	H	G	F
CVT	$\begin{matrix} 0 & 0 \\ r & 0 \end{matrix}$	—	—	—	$\begin{matrix} 0 & 0 \\ \frac{1}{r} & 0 \end{matrix}$
VCT	—	$\begin{matrix} 0 & 0 \\ g & 0 \end{matrix}$	—	—	$\begin{matrix} 0 & \frac{1}{g} \\ 0 & 0 \end{matrix}$
CCT	—	—	$\begin{matrix} 0 & 0 \\ \alpha & 0 \end{matrix}$	—	$\begin{matrix} 0 & 0 \\ 0 & \frac{1}{\alpha} \end{matrix}$
VVT	—	—	—	$\begin{matrix} 0 & 0 \\ \mu & 0 \end{matrix}$	$\begin{matrix} \frac{1}{\mu} & 0 \\ 0 & 0 \end{matrix}$
VNIC	—	—	$\begin{matrix} 0 & -k_1 \\ -k_2 & 0 \end{matrix}$	$\begin{matrix} 0 & -\frac{1}{k_2} \\ -\frac{1}{k_1} & 0 \end{matrix}$	$\begin{matrix} -k_1 & 0 \\ 0 & \frac{1}{k_2} \end{matrix}$
CNIC	—	—	$\begin{matrix} 0 & k_1 \\ k_2 & 0 \end{matrix}$	$\begin{matrix} 0 & \frac{1}{k_2} \\ \frac{1}{k_1} & 0 \end{matrix}$	$\begin{matrix} k_1 & 0 \\ 0 & -\frac{1}{k_2} \end{matrix}$
VGIC	—	—	$\begin{matrix} 0 & -\frac{Z_\alpha}{Z_\beta} \\ -1 & 0 \end{matrix}$	$\begin{matrix} 0 & -1 \\ -\frac{Z_\beta}{Z_\alpha} & 0 \end{matrix}$	$\begin{matrix} -\frac{Z_\alpha}{Z_\beta} & 0 \\ 0 & 1 \end{matrix}$
CGIC	—	—	$\begin{matrix} 0 & 1 \\ \frac{Z_\alpha}{Z_\beta} & 0 \end{matrix}$	$\begin{matrix} 0 & \frac{Z_\beta}{Z_\alpha} \\ 1 & 0 \end{matrix}$	$\begin{matrix} 1 & 0 \\ 0 & -\frac{Z_\beta}{Z_\alpha} \end{matrix}$
IVC	—	—	$\begin{matrix} 0 & k \\ -1 & 0 \end{matrix}$	$\begin{matrix} 0 & -1 \\ \frac{1}{k} & 0 \end{matrix}$	$\begin{matrix} k & 0 \\ 0 & 1 \end{matrix}$
ICC	—	—	$\begin{matrix} 0 & 1 \\ -\frac{1}{\beta} & 0 \end{matrix}$	$\begin{matrix} 0 & -\beta \\ 1 & 0 \end{matrix}$	$\begin{matrix} 1 & 0 \\ 0 & \beta \end{matrix}$

EQUIVALENT CIRCUITS / 57

Table 2-1 (*continued*)

IPC	—	—	$\begin{matrix} 0 & m \\ -\dfrac{1}{m} & 0 \end{matrix}$	$\begin{matrix} 0 & -m \\ \dfrac{1}{m} & 0 \end{matrix}$	$\begin{matrix} m & 0 \\ 0 & m \end{matrix}$
PIV[a]	$\begin{matrix} 0 & \mp\dfrac{1}{G_1} \\ \pm\dfrac{1}{G_2} & 0 \end{matrix}$	$\begin{matrix} 0 & \pm G_2 \\ \mp G_1 & 0 \end{matrix}$	—	—	$\begin{matrix} 0 & \pm\dfrac{1}{G_1} \\ \pm G_2 & 0 \end{matrix}$
NRNIV[b]	$\begin{matrix} 0 & \mp R_1 \\ \mp R_2 & 0 \end{matrix}$	$\begin{matrix} 0 & \mp\dfrac{1}{R_2} \\ \mp\dfrac{1}{R_1} & 0 \end{matrix}$	—	—	$\begin{matrix} 0 & \pm R_1 \\ \mp\dfrac{1}{R_2} & 0 \end{matrix}$
$IT_{(\pm)}$	—	—	$\begin{matrix} 0 & \pm n \\ \mp n & 0 \end{matrix}$	$\begin{matrix} 0 & \mp\dfrac{1}{n} \\ \pm\dfrac{1}{n} & 0 \end{matrix}$	$\begin{matrix} \pm n & 0 \\ 0 & \pm\dfrac{1}{n} \end{matrix}$

[a] The two-port description of the ideal gyrator (IG) is obtained from that of PIV by setting $G_1 = G_2$.
[b] The two-port descriptions of the ideal NIV are obtained from those of NRNIV by setting $R_1 = R_2$.

In a similar manner, the elements in the second group could be described by their positions in the B, C plane (Figure 2-33). The ideal gyrator and the negative-impedance inverter describe the four hyperbolas as shown in the figure. The VCT and the CVT can be considered as degenerate cases of an impedance inverter.

An alternate description of both types of active devices is obtained from their positions in the AD, BC plane[28] (Figure 2-34). The active elements in the impedance converter group lie on the AD axis, and the impedance inverter type elements must lie on the BC axis. Note that in this diagram, the positions of the controlled sources are at the origin. Thus the controlled sources can be considered as the basic ideal active elements.

The mathematical and physical connection between some ideal active four-terminal networks will now be derived. An excellent exposition to this topic will be found elsewhere.[28]

In Figure 2-5, the relations between the elementary controlled sources were shown. For example, we have the following relation between their

[28] M. Kawakami, *loc. cit.*

58 / ACTIVE NETWORK ELEMENTS

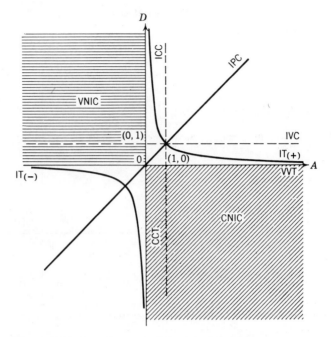

Figure 2-32 Locations of various impedance converter type elements in the A, D plane.

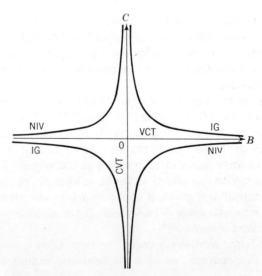

Figure 2-33 Locations of various impedance inverter type elements in the B, C plane.

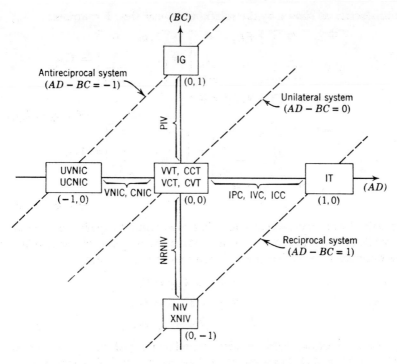

Figure 2-34 Locations of active elements in the (AD, BC) plane.

respective transmission matrices:

$$(2.54) \qquad \begin{bmatrix} 0 & 0 \\ \frac{1}{r} & 0 \end{bmatrix} \begin{bmatrix} 0 & \frac{1}{g} \\ 0 & 0 \end{bmatrix} = \begin{bmatrix} 0 & 0 \\ 0 & \frac{1}{rg} \end{bmatrix}$$
$$\text{CVT} \quad \text{VCT} \qquad \text{CCT}$$

$$(2.55) \qquad \begin{bmatrix} 0 & \frac{1}{g} \\ 0 & 0 \end{bmatrix} \begin{bmatrix} 0 & 0 \\ \frac{1}{r} & 0 \end{bmatrix} = \begin{bmatrix} \frac{1}{rg} & 0 \\ 0 & 0 \end{bmatrix}$$
$$\text{VCT} \quad \text{CVT} \qquad \text{VVT}$$

The negative-impedance converters can be cascaded to form the ideal transformer as shown by the relation between their **F** matrices:

(2.56) $$\begin{bmatrix} \mp k_1 & 0 \\ 0 & \dfrac{1}{\pm k_2} \end{bmatrix} \begin{bmatrix} \mp k_3 & 0 \\ 0 & \dfrac{1}{\pm k_4} \end{bmatrix} = \begin{bmatrix} k_1 k_3 & 0 \\ 0 & \dfrac{1}{k_2 k_4} \end{bmatrix} = \mathbf{F}_{\text{IT}(+)}$$

VNIC(CNIC)   VNIC(CNIC)

if $k_1 k_3 = k_2 k_4$

(2.57) $$\begin{bmatrix} -k_1 & 0 \\ 0 & \dfrac{1}{k_2} \end{bmatrix} \begin{bmatrix} k_3 & 0 \\ 0 & -\dfrac{1}{k_4} \end{bmatrix} = \begin{bmatrix} -k_1 k_3 & 0 \\ 0 & \dfrac{1}{k_2 k_4} \end{bmatrix} = \mathbf{F}_{\text{IT}(-)}$$

VNIC   CNIC

if $k_1 k_3 = k_2 k_4$

In a similar manner, we can show that two identical impedance inverters in cascade act as an ideal transformer. More precisely the following relations between their respective transmission matrices hold true:

(2.58)
$$\mathbf{F}_{\text{IG}} \times \mathbf{F}_{\text{IG}} = \mathbf{F}_{\text{IT}}$$
$$\mathbf{F}_{\text{NIV}} \times \mathbf{F}_{\text{NIV}} = \mathbf{F}_{\text{IT}}$$
$$\mathbf{F}_{\text{XNIV}} \times \mathbf{F}_{\text{XNIV}} = \mathbf{F}_{\text{IT}}$$

The connections between ideal transformer and the ideal converters (IVC, ICC, and IPC) are established below in terms of their transmission matrices:

(2.59) $$\begin{bmatrix} m & 0 \\ 0 & m \end{bmatrix} \begin{bmatrix} n & 0 \\ 0 & \dfrac{1}{n} \end{bmatrix} = \begin{bmatrix} mn & 0 \\ 0 & \dfrac{m}{n} \end{bmatrix}$$

IPC   IT

In Equation 2.59, if $n = m$, the resulting two-port behaves as an ideal voltage converter (IVC). On the other hand, if n is chosen as $1/m$, the cascade connection of an ideal power converter and an ideal transformer represents an ideal current converter (ICC).

The relation between the controlled sources and negative-impedance converters will be established later in this section; it will then be shown that a VNIC (CNIC) can be constructed by a proper orientation of a VVT (CCT). At that time, we shall also illustrate the construction of an ideal gyrator by the connection of two ideal voltage-current transducers suitably in parallel (and similarly, two ideal current-voltage transducers connected suitably in series produce an ideal gyrator).

Consider the series-parallel connection of an ideal transformer of unity turns ratio and an ideal current-to-current transducer (Figure 2-35a).

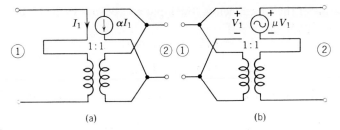

(a)   (b)

Figure 2-35 Representation of (a) ICC and (b) IVC, using controlled sources and ideal transformers.

We then have for the overall two-port the h matrix as

$$(2.60) \quad \begin{bmatrix} 0 & 0 \\ \alpha & 0 \end{bmatrix} + \begin{bmatrix} 0 & 1 \\ -1 & 0 \end{bmatrix} = \begin{bmatrix} 0 & 1 \\ \alpha - 1 & 0 \end{bmatrix}$$

The h matrix on the right-hand side of Equation 2.60 is seen to represent an ICC with a conversion factor $1/(1 - \alpha)$.[29] Taking the dual of the above circuit, i.e., connecting an ideal voltage-to-voltage transducer and an ideal transformer in a parallel-series fashion (Figure 2-35b), we obtain an IVC.

Controlled Source Representation

Any 2-terminal network is either a current-controlled device or a voltage-controlled device, and can be represented using either a VCT or a CVT. Figure 2-36 shows the transducer representation of negative resistances. Extension to complex impedances is obvious.

In general, an arbitrary two-port can be described mathematically in terms of any set of two-port parameters. As a result, an equivalent representation using any set of parameters is possible.[30]

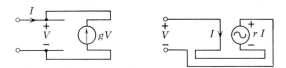

Figure 2-36 Controlled source representations of the negative resistance.

---

[29] D. A. Tsirel'son, "Ideal active elements for electric circuits," *Telecommunications*, No. 8, 49–61 (August 1961).

[30] L. C. Peterson, "Equivalent circuits of linear active four-terminal networks," *Bell System Tech. J.*, **27**, 593–622 (October 1948).

In terms of the z parameters, the input and output variables are related as

(2.61)
$$V_1 = z_{11}I_1 + z_{12}I_2$$
$$V_2 = z_{21}I_1 + z_{22}I_2$$

An equivalent circuit using two CVTs and one-port impedances (Figure

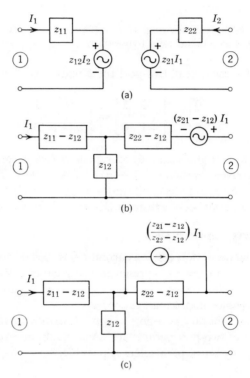

Figure 2-37 The equivalent representation of a two-port based on open-circuit impedance parameters.

2-37a) is obtained by expressing the z matrix as a sum of three matrices and making a series connection of the three two-ports representing the latter three matrices:

(2.62)
$$\begin{bmatrix} z_{11} & z_{12} \\ z_{21} & z_{22} \end{bmatrix} = \begin{bmatrix} z_{11} & 0 \\ 0 & z_{22} \end{bmatrix} + \begin{bmatrix} 0 & z_{12} \\ 0 & 0 \end{bmatrix} + \begin{bmatrix} 0 & 0 \\ z_{21} & 0 \end{bmatrix}$$

An alternate 2-mesh representation using one controlled source as shown

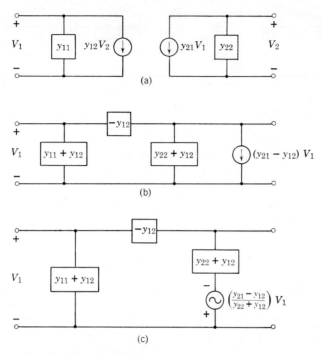

Figure 2-38 The equivalent representation of a two-port based on short-circuit admittance parameters.

in Figure 2-37b, is obtained if the second equation in Equation 2.61 is rewritten as

(2.63) $$V_2 = z_{12}I_1 + z_{22}I_2 + (z_{21} - z_{12})I_1$$

If the network is reciprocal, $z_{12} = z_{21}$ and the CVT in Figure 2-37b is then not required. The CVT in series with the impedance $(z_{22} - z_{12})$ can be replaced by its Norton's equivalent (Figure 2-37c). The last equivalent circuit is useful for representing transistors.

Figures 2-38 to 2-41 show the alternate representations of two-ports

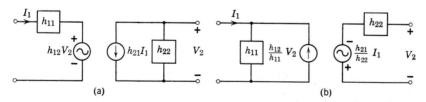

Figure 2-39 Equivalent representations based on h-parameters.

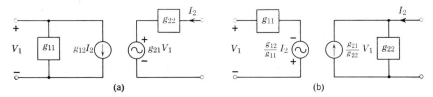

Figure 2-40 Equivalent representations based on g-parameters.

on **Y**, **H**, **G**,[31] and **F** basis.[32] The validity of each of these circuits can be checked by writing down the mesh or node equations.

Let us now investigate the transducer representation of several ideal active devices. These equivalent circuits follow directly from the basic two-port representation shown in Figures 2-37 through 2-41.

First consider the impedance converter defined by

$$h_{11} = h_{22} = 0$$

$$h_{12} \neq 0, \quad h_{21} \neq 0$$

These conditions when applied to Figure 2-39a yield the balanced equivalent circuit shown in Figure 2-42a. An unbalanced representation is given in Figure 2-42b, which can be easily adapted for the GIC (Figure 2-42c), the VNIC (Figure 2-42d), and the CNIC (Figure 2-42e). The circuit of Figure 2-42d, which uses only one VVT, and the circuit of Figure 2-42e, using a single CCT, clearly point out the relation between the negative-impedance converters and the transducers.

The transducer equivalent circuit (Figure 2-43) of an ideal transformer (IT) is derived from Figure 2-41a by setting $B = C = 0$, $A = n$, and $D = 1/n$.[32]

Equivalent circuits of an impedance inverter are easily obtained from the two-port representations on the z and y basis, by setting $z_{11} = z_{22} = 0$

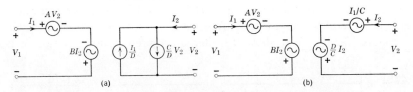

Figure 2-41 Equivalent representations based on F-parameters.

---

[31] L. DePian, *Linear active network theory*, Prentice-Hall, Englewood Cliffs, N.J., pp. 82–88 (1962).

[32] L. P. Huelsman, *Circuits, matrices and linear vector spaces*, McGraw-Hill, New York, p. 115 (1963).

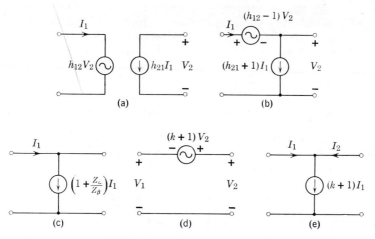

Figure 2-42 Controlled source representations of various impedance converters.

in Figure 2-37a and by setting $y_{11} = y_{22} = 0$ in Figure 2-38a. For example, the circuits of Figure 2-44 represent the ideal active gyrators. Figures 2-44a and b also bring out in better perspective the relation between ideal gyrators and the basic elementary transducers.

The equivalent circuit of the NIV, if derived from Figure 2-37b or 2-38b, will result in the circuits of Figure 2-30. The absence of controlled sources is evident from the reciprocal character of the device.

Some of the circuits presented above can be used to construct active devices, e.g., the negative-impedance converters, gyrators. More practical circuits are presented in later chapters.

Small signal low-frequency equivalent circuits of some well-known electronic devices are shown in Figure 2-45.

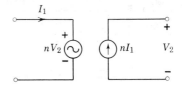

Figure 2-43 Controlled source representation of an ideal transformer.

The controlled source equivalent representation of active components are useful in analyzing electronic circuits. A better representation of the common base transistor was shown earlier in Figure 2-1.

Gyrator:Negative-Resistance Representation

It has been shown that any arbitrary resistive two-port has an equivalent form that employs positive and negative resistances, and ideal gyrators. This leads us to investigate the possibility of representing other ideal active

66 / ACTIVE NETWORK ELEMENTS

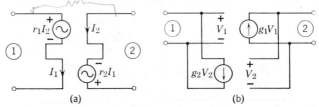

Figure 2-44 Controlled source representations of an ideal gyrator.

elements in this way. As CVT and the VCT are the basic controlled sources, let us first derive their equivalent circuits. Writing the z matrix of a CVT as a sum of two matrices:

$$(2.64) \quad \begin{bmatrix} 0 & 0 \\ R & 0 \end{bmatrix} = \begin{bmatrix} 0 & \frac{R}{2} \\ \frac{R}{2} & 0 \end{bmatrix} + \begin{bmatrix} 0 & -\frac{R}{2} \\ \frac{R}{2} & 0 \end{bmatrix}$$

$$\text{CVT} \qquad \text{NIV} \qquad \text{IG}$$

we obtain the equivalent representation shown in Figure 2-46a by connecting an NIV in series with an IG. In a dual manner, the equivalent circuit of a VCT shown in Figure 2-46b is obtained.

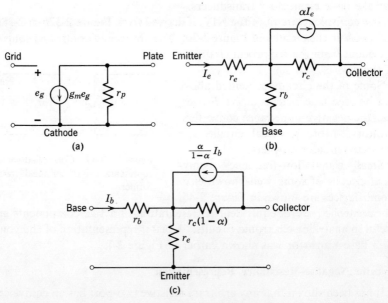

Figure 2-45 Controlled source representations of (a) common cathode triode, (b) common base transistor, and (c) common emitter transistor.

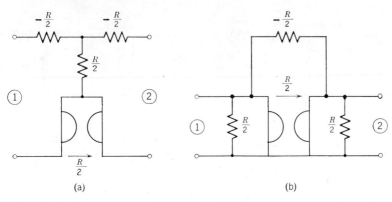

Figure 2-46 Representation of (a) CVT and (b) VCT using gyrator and negative resistances.

The gyrator:negative-resistance representations of several active and passive elements are given in Figure 2-47. The validity of these representations can be justified by writing down the terminal voltage-current relationships.

Ideal Converter Representation

Three types of ideal converters were introduced earlier. Equivalent circuits of an arbitrary two-port can be formulated in terms of these ideal converters. These equivalent representations are particularly useful in the analysis of cascaded two-ports.

The development of an IPC equivalent circuit is as follows.[33] The specified two-port must be identical to a reciprocal two-port in tandem with an ideal power converter (in any order). In terms of their respective transmission matrices, this implies

$$(2.65) \qquad \underset{\text{Original}}{\begin{bmatrix} A & B \\ C & D \end{bmatrix}} = \underset{\text{Reciprocal}}{\begin{bmatrix} A' & B' \\ C' & D' \end{bmatrix}} \underset{\text{IPC}}{\begin{bmatrix} m & 0 \\ 0 & m \end{bmatrix}}$$

where the prime denotes the reciprocal two-port and hence $A'D' - B'C' = 1$. Equation 2.65 can be solved for m yielding

$$(2.66) \qquad m = \sqrt{AD - BC}$$

---

[33] E. V. Zelyakh, *loc. cit.*

68 / ACTIVE NETWORK ELEMENTS

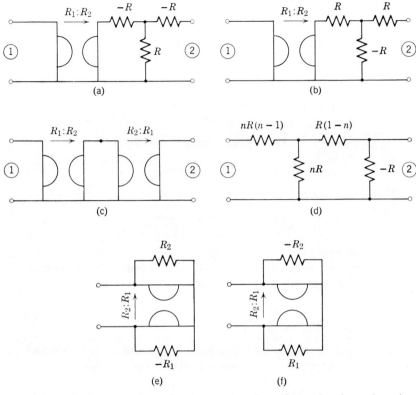

Figure 2-47 Gyrator: negative-resistance representations of several active and passive elements: (a) VNIC, (b) CNIC, (c) ideal transformer, (d) ideal transformer, (e) nullator, and (f) norator.

The transmission parameters of the reciprocal two-port is thus given by

$$(2.67) \quad \begin{bmatrix} A' & B' \\ C' & D' \end{bmatrix} = \begin{bmatrix} \dfrac{A}{\sqrt{AD-BC}} & \dfrac{B}{\sqrt{AD-BC}} \\ \dfrac{C}{\sqrt{AD-BC}} & \dfrac{D}{\sqrt{AD-BC}} \end{bmatrix}$$

from which we easily obtain its open-circuit impedance matrix and short-circuit admittance matrix:

$$(2.68) \quad \begin{bmatrix} z'_{11} & z'_{12} \\ z'_{12} & z'_{22} \end{bmatrix} = \begin{bmatrix} z_{11} & \sqrt{z_{12}z_{21}} \\ \sqrt{z_{12}z_{21}} & z_{22} \end{bmatrix}$$

$$(2.69) \quad \begin{bmatrix} y'_{11} & y'_{12} \\ y'_{12} & y'_{22} \end{bmatrix} = \begin{bmatrix} y_{11} & \sqrt{y_{12}y_{21}} \\ \sqrt{y_{12}y_{21}} & y_{22} \end{bmatrix}$$

EQUIVALENT CIRCUITS / 69

The conversion factor of the ideal power converter can also be related to the z or y parameters of the original two-port:

$$(2.70) \qquad m = \sqrt{\frac{z_{12}}{z_{21}}} = \sqrt{\frac{y_{12}}{y_{21}}}$$

The final equivalent circuits shown in Figure 2-48 result on representing the reciprocal two-port by its π- or T-equivalent representation.

An application of the IPC equivalent circuit is discussed next. Consider, for simplicity, a cascade connection of two arbitrary two-ports as indicated in Figure 2-49a. Each of the individual two-ports $\mathcal{N}_j$ can be replaced by a tandem connection of a reciprocal two-port $\mathcal{N}_j{}^R$ and an IPC m_j (Figure 2-49b). The new voltage and current variables are given as

$$(2.71) \qquad I_2' = m_1 I_2; \qquad V_2' = m_1 V_2; \qquad I_3' = m_2 I_3; \qquad V_3' = m_2 V_3$$

Since the IPC's are impedance transparent, they all can be considered as one, as indicated in Figure 2-49c, where

$$(2.72) \qquad I_3'' = m_1 m_2 I_3, \qquad V_3'' = m_1 m_2 V_3$$

For analysis purposes, the complete network can be considered as a reciprocal network, as shown in Figure 2-49d, and can be analyzed by conventional ladder network analysis methods after replacing each reciprocal two-port by its T- or π-equivalent. The actual current and voltage variables are obtained from the dummy variables using Equations 2.71

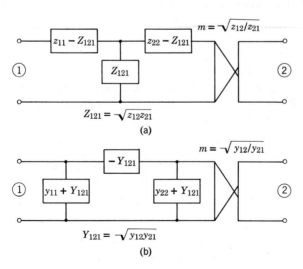

Figure 2-48 Equivalent representations of an arbitrary two-port using an IPC.

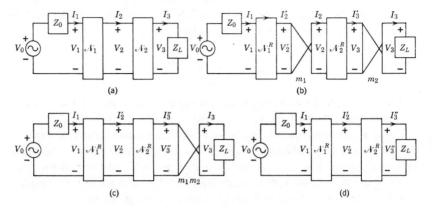

Figure 2-49 Use of IPC equivalent representations in the analysis of cascaded two-ports.

and 2.72. Extension of the above approach to more than two networks in cascade should be obvious.

Equivalent representation using an IVC or an ICC as a basic nonreciprocal element can be easily developed. Figure 2-50 shows four such equivalent circuits,[34] the development of which follows the same lines as the IPC equivalent circuit and is left as an exercise to the reader. Since the ideal voltage or current converter does convert impedances, analysis of cascaded two-ports by using the IVC or ICC equivalent circuits for the individual two-ports is not as simple as in the case of IPC equivalent representation.

Nullator-Norator Representation[35]

The nullator and the norator also form a basic set of active elements. Consequently, equivalent circuits using such devices can be formed. However, in a physically realizable circuit all voltages and currents are always uniquely and definitely determined. This in turn implies that in the equivalent representation of a physically realizable circuit the nullator and the norator must occur in a pair. Note that the nullator-norator combination can be considered as a nullor, which is theoretically physically realizable.

The sufficiency of the nullator and the norator in representing an arbitrary network is best illustrated by considering the representation of ideal

---

[34] D. A. Tsirel'son, *loc. cit.*

[35] S. K. Mitra, "Nullator-norator equivalent circuits of linear active elements and their applications," *Proc. Asilomar Conference on Circuits and Systems*, Monterey, Calif. **1**, 267–276 (November 1967).

controlled sources. It should be noted that the dual of a nullator is a nullator and the dual of a norator is a norator. Hence dual representations of a planar network are easy to construct.

In Figure 2-51, basic representations of each of the four types of controlled sources are shown.[36] Each basic circuit employs a single nullator

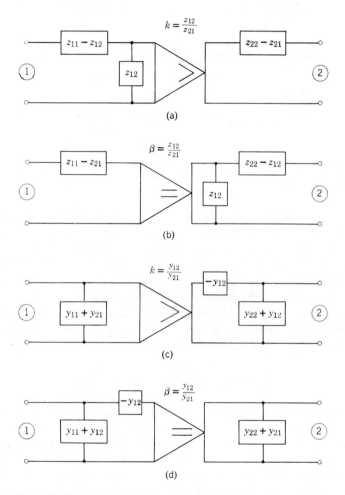

Figure 2-50 The equivalent representations of an arbitrary two-port using an IVC and an ICC.

---

[36] These equivalent circuits and some of the circuits of Figure 2-53 were recently independently proposed by A. C. Davies, "Nullator-norator equivalent networks for controlled sources," *Proc. IEEE*, **55**, 722–723 (May 1967).

72 / ACTIVE NETWORK ELEMENTS

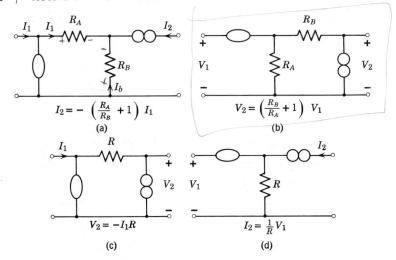

Figure 2-51 The four basic nullator-norator models of controlled sources: (a) CCT, (b) VVT, (c) CVT, and (d) VCT.

and a single norator. It is a simple exercise to show that the VVT circuit of Figure 2-51b is the dual of the CCT circuit of Figure 2-51a, and the VCT circuit of Figure 2-51d is the dual of the CVT circuit of Figure 2-51c.

We now proceed to analyze these circuits. Consider the circuit of Figure 2-51a first. Since the current through a nullator is zero, the input current I_1 also flows through the resistance R_A. The voltage drop across R_A, $I_1 R_A$ volts, also appears across the resistance R_B (because the voltage across the nullator is zero) causing a current $I_b = I_1 R_A / R_B$ to flow in the direction indicated. The output current is thus given by

$$I_2 = -I_1 - I_1 \frac{R_A}{R_B} = -\left(1 + \frac{R_A}{R_B}\right) I_1$$

This indicates that the pertinent network behaves as a CCT. A similar analysis can be carried out to justify the operation of the remaining circuits.

Observe that the gain of the CCT and the VVT circuits of Figure 2-51 are always greater than one. The model of the unity gain CCT shown in Figure 2-52a is derived from the circuit of Figure 2-51a by letting $R_A = 0$ and R_B infinite. Likewise, the model of a unity gain VVT of Figure 2-52b can be obtained from the parent VVT circuit of Figure 2-51b if R_A is set equal to infinite and R_B equal to zero.

Using the equivalence relations of the controlled sources (Figure 2-5), other nullator-norator equivalent circuits can be constructed from the basic networks of Figures 2-51 and 2-52. A collection of these circuits, each

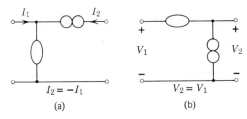

Figure 2-52 Special cases: (a) CCT and (b) VVT.

employing two nullators and two norators, is presented in Figure 2-53. The verification of these circuits is left as an exercise.

Additional equivalent circuits can also be generated with the aid of the identities illustrated in Figure 2-54. We note from Figure 2-54a that the series combination of a nullator and a norator is equivalent to an open circuit. Similarly, the parallel combination of a nullator and a norator is equivalent to a short circuit (Figure 2-54b).

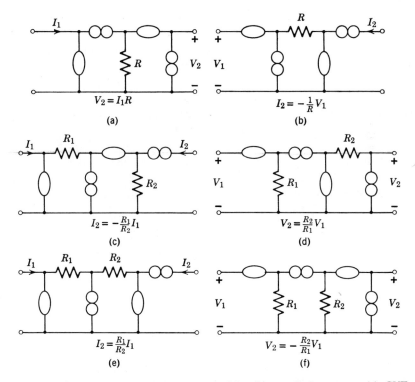

Figure 2-53 Additional nullator-norator models of controlled sources: (a) CVT, (b) VCT, (c) CCT, (d) VVT, (e) CCT, and (f) VVT.

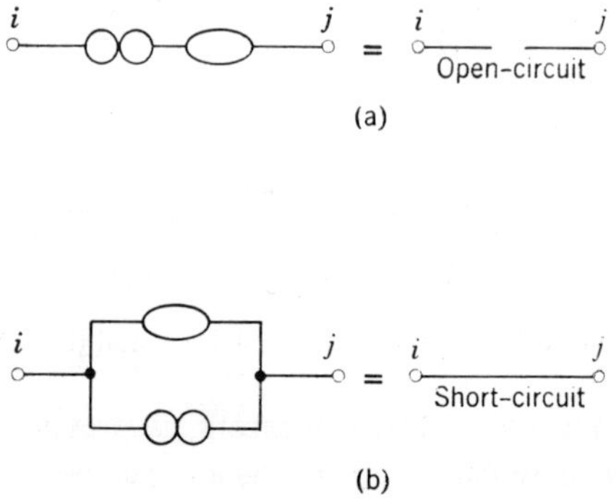

74 / ACTIVE NETWORK ELEMENTS

(a)

(b)

Figure 2-54 Two nullator-norator identities.

To illustrate the use of these identities, we indicate in Figure 2-55 a modified version of the CCT circuit of Figure 2-51a. Similar modifications of the basic circuits of Figure 2-51 and some of the circuits of Figure 2-53 will be found in Chapter 8.

The equivalent representation of the controlled sources can be used to construct the equivalent circuits of other ideal active devices. In Sections 9-1 and 10-1, we develop the nullator-norator representations of the negative-impedance converter and the gyrator, respectively.

One useful application of the nullator-norator equivalent representation is the construction of a transistor circuit realization from the original equivalent circuit. Note from Figure 2-2a that the controlled source model of an ideal grounded base transistor is a CCT, whose output current is equal and opposite to the input current. Thus, an alternate first-order representation of a grounded base transistor will be the one shown in Figure 2-52a.[37] This equivalence suggests a method of converting a nullator-norator circuit into a transistor circuit. The method is first to group

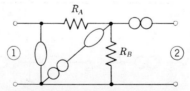

Figure 2-55 One method of modification of the CCT model of Figure 2-51a to allow transistorized realization.

---

[37] B. Myers, "Nullor model of the transistor," *Proc. IEEE*, **53**, 758–759, July 1965. See also, J. Braun, "Equivalent NIC networks with nullators and norators," *IEEE Trans. on Circuit Theory*, **CT-12**, 441–442, September 1965.

together all the nullators and the norators into pairs of a nullator and a norator having a common junction. Each pair is replaced by a transistor as follows. The junction of the pair corresponds to the emitter, the terminal of the nullator which is not connected at the junction corresponds to the base, and finally the terminal of the norator not connected to the junction corresponds to the collector. Based on this approach, the CCT circuit

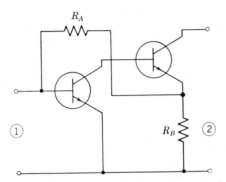

Figure 2-56 Transistorized realization of the nullator-norator model of Figure 2-55.

of Figure 2-55 is transformed into a transistor circuit as indicated in Figure 2-56. Details on transistor realizations of the controlled sources, negative-impedance converters, and the gyrators will be found in later chapters.

2-8 SUMMARY

There are basically two types of ideal active two-ports: (a) impedance converter type, and (b) impedance inverter type. Some commonly used ideal active devices in the impedance converter group are: the current-controlled current source and the voltage-controlled voltage source (Section 2-1, p. 27), the negative-impedance converter (Section 2-4, p. 38), and the generalized-impedance converter (Section 2-4, p. 37). Examples of ideal active two-ports belonging to the impedance inverter type group are positive-impedance inverter and ideal gyrator (Section 2-5, p. 46), negative impedance inverter (Section 2-5, p. 52), and current-controlled voltage source and voltage-controlled current source (Section 2-1, p. 26). Two-port parameters of various ideal active two-ports are cataloged in Table 2-1.

Controlled sources can be used to form equivalent circuits of ideal active devices (Section 2-7, p. 61). Another useful equivalent representation of active network elements are obtained using the degenerate network elements, the nullator and the norator (Section 2-7, p. 70).

Electronic circuits can be designed to have its small signal equivalent circuit approach one of the ideal active elements over a band of frequency. However, the actual model of the practical realization of an active network element has additional parasitics and nonideal effects. The nonideal behavior may be negligible for some applications. On the other hand, amount of the nonidealness of a practical active element may be crucial for some other applications, and their effect must be minimized either by careful design and construction or by precorrecting the companion network.

Additional References

Arouca, M., "Two new theorems on active network manipulations," Summaries of Papers, ICMCI, Tokyo, Japan, pp. 79–80 (September 1964).

Brodie, J. H. and R. S. Crocker, "The active transformer," *Proc. IEEE*, **54**, 1125–1127 (August 1966).

Carlin, H. J. and D. C. Youla, "Network synthesis with negative resistors," *Proc. IRE*, **49**, 907–920 (May 1961).

Debart, H., "Physical realizability of an active impedance," Proc. Symp. On Active Networks and Feedback Systems, MRI Symposia Series, Vol. X, Polytech. Inst. of Brooklyn, pp. 379–386 (April 1960).

DeClaris, N., "Transformations of active networks," *Proc. NEC*, **15**, 707–712 (1959).

DePian, L., *Linear Active Network Theory*, Prentice–Hall, Englewood Cliffs, N.J., 1962.

Fjällbrant, T., "Linear active network theory," *Ericsson Technics*, **20**, 231–265 (1964).

Freeman, J. R., "The general two-port T-π transformations," *Proc. IEEE*, **53**, 1799 (November 1965).

Handbook of operational amplifier applications, Burr–Brown Research Corp., Tucson, Arizona, 1963.

Herold, E. W., "Negative resistance and devices for obtaining it," *Proc. IRE*, **23**, 1201–1223 (October 1935).

Huelsman, L. P., *Circuits, matrices and linear vector spaces*, McGraw–Hill, New York, 1963.

Keen, A. W., "Transactive network elements," *J. Inst. Elec. Engrs.* (*London*), **3** (New Series), 213–214 (April 1957).

Keen, A. W., "The transactor, an idealised network element," *Electron. and Radio Engr.*, **34** (New Series), 459–461 (December 1957).

————, "A topological nonreciprocal element," *Proc. IRE*, **47**, 1148–1150 (June 1959).

————, "Ideal three-terminal active networks," Proc. Symp. Active Networks and Feedback Systems, MRI Symposia Series, Vol. X, Polytech. Inst. Brooklyn, pp. 201–240 (April 1960).

Moad, M. F., "Two-port networks, with independent sources," *Proc. IEEE*, **54**, 1008–1009 (July 1966).

RCA "Linear integrated circuit fundamentals," Technical Series IC-40, RCA, Harrison, N.J., 1966.

Sankiewicz, M., "Resistance-stable negative resistance," *Proc. IEEE*, **54**, 1598–1599 (November 1966).

Sharpe, G. E., "Ideal active elements," *J. Inst. Elec. Engrs. (London)*, **3** (New Series), 33–34 (January 1957); 430–431 (July 1957).

―――――, "Transactors," *Proc. IRE*, **45**, 692–693 (May 1957).

―――――, "Axioms on transactors," *IRE Trans. on Circuit Theory*, **CT-5**, 189–196 (September 1958).

Su, K. L., "Realization of the ideal transformer with active elements," *Proc. IEEE*, **54**, 1083–1084 (August 1966).

Su, K. L., *Active networks synthesis*, McGraw–Hill, New York, Ch. 2 (1965).

Tellegen, B. D. H., "On nullators and norators," *IEEE Trans. on Circuit Theory*, **CT-13**, 466 (December 1966).

Weiss, C. D., "Ideal transformer realizations with negative resistors," *Proc. IEEE*, **54**, 302–303 (February 1966).

Problems

2.1 Verify the circuits of Figure 2-5.

2.2[38] Figure 2-57 depicts a configuration suitable for designing an active

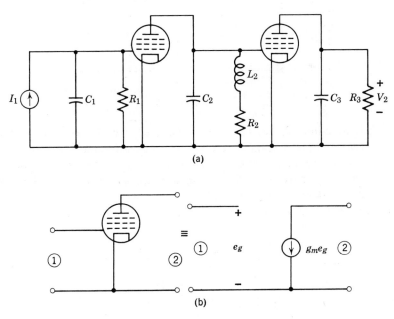

Figure 2-57

[38] M. E. Van Valkenburg, *Network Analysis*, Prentice–Hall, Englewood Cliffs, N.J., 1st ed., Ch. 14 (1955).

low-pass filter. Compute the overall transfer function after replacing the pentode by its equivalent circuit. For what values of the passive components will the network have a maximally flat magnitude characteristic?

2.3 Determine the transfer voltage ratio of the circuit of Figure 2-58. Explain the operation of the circuit when the input impedance Z_i is a resistance and the feedback impedance Z_f is (a) a resistance, (b) a capacitance.

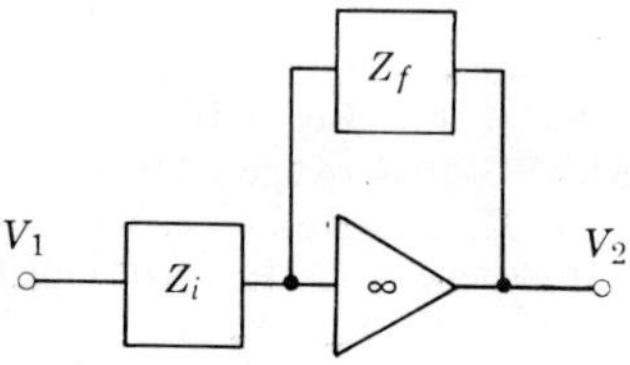

2.4 Compute the input impedance of the active one-ports of Figures 2-3 and 2-15.

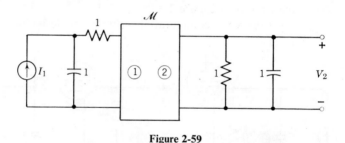

Figure 2-58

2.5 Compute the transfer impedance Z_{21a} of the two-port of Figure 2-59 where $\mathcal{M}$ is an ideal negative-impedance converter ($h_{11} = h_{22} = 0$, $h_{12}h_{21} = 2$). Replace the ideal impedance converter by a nonideal

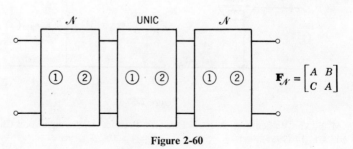

Figure 2-59

negative-impedance converter described by $h_{11} = h_{22} = 1$ and $h_{12}h_{21} = 2$ and recompute the new transfer impedance Z_{21b}. If Z_{21b} is different from Z_{21a}, how would you change the element values of the companion network in the second situation so that the overall transfer impedance is equal to Z_{21a}?

2.6 Compute the transmission matrix (**F**) of the network of Figure 2-60

Figure 2-60

where $\mathcal{N}$ is an arbitrary two-port, described by an **F** matrix as indicated. Next, compute the **F** matrix of two such networks in cascade. Any comments?

2.7 Show that a reactive gyrator is in general an active network.

2.8 The three-port of Figure 2-61 is known as a *circulator* and is described

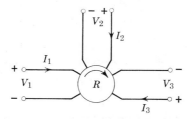

Figure 2-61

by the following open-circuit impedance matrix:

$$\begin{bmatrix} 0 & R & -R \\ -R & 0 & R \\ R & -R & 0 \end{bmatrix}$$

(a) Show that the three-port is a lossless device.[39]
(b) Show that the network is matched at all ports and power can flow only in one direction from port 1 to port 2, port 2 to port 3, etc. (Note that the direction of power transmission coincides with the direction of arrow in the figure.)

2.9 Show that the structure of Figure 2-62 is a four-port circulator.

2.10 Compute the voltage transfer ratio of the network of Figure 2-63. If it is desired to use a nonideal gyrator characterized by $y_{11} = y_{22} = 1$ and $y_{12} = -y_{21} = 1$, how would you modify the companion RC

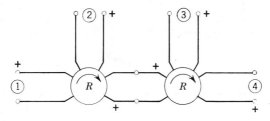

Figure 2-62

---

[39] For a lossless device, the total real power input is equal to zero for all frequencies.

network of the two-port of Figure 2-63 to keep the transfer voltage ratio invariant?

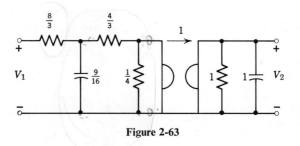

Figure 2-63

2.11 Show that the two circuits of Figure 2-64 act as three-port circulators.

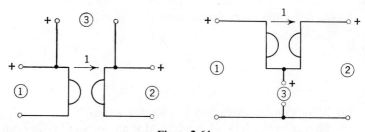

Figure 2-64

2.12 Verify the equivalent circuits of Figures 2-38 through 2-41.
2.13 Compute the **F** matrix of each of the two-ports of Figure 2-65 and determine their operation for large values of K.[40]

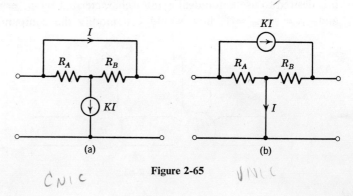

Figure 2-65

[40] See Problem 9.1.

2.14 Using an ideal gyrator, and positive and negative resistors, find a realization of a resistive two-port, described by the following y matrix:

$$[y] = \begin{bmatrix} 1 & 3 \\ 2 & 7 \end{bmatrix}$$

2.15 Verify the equivalent circuits of Figure 2-47.

2.16 Figure 2-66 shows two other basic ways to convert or invert an

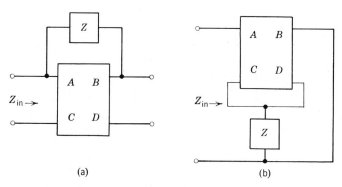

Figure 2-66

impedance Z. Compute the input impedance of each network. Determine the necessary and sufficient conditions in terms of the **F** parameters of the two-port, under which the input impedance Z_{in} of each circuit is directly proportional to Z and inversely proportional to Z.

2.17 Verify the IVC and ICC equivalent circuits of a transistor shown in Figure 2-67.[41]

2.18 Show that for any two-port, the following relations between its various parameters always hold true:

$$\frac{z_{11}}{z_{22}} = \frac{y_{22}}{y_{11}}, \qquad \frac{z_{12}}{z_{21}} = \frac{y_{12}}{y_{21}}$$

$$\frac{h_{11}}{h_{22}} = \frac{g_{22}}{g_{11}}, \qquad \frac{h_{12}}{h_{21}} = \frac{g_{12}}{g_{21}}$$

$$\frac{z_{11}}{z_{12}} = -\frac{y_{22}}{y_{12}}, \qquad \frac{h_{11}}{h_{12}} = -\frac{g_{22}}{g_{12}}$$

2.19 Verify the nullator-norator representations of the controlled sources shown in Figures 2-51b, c, d, and Figure 2-53.

[41] D. A. Tsirel'son, *loc. cit.*

82 / ACTIVE NETWORK ELEMENTS

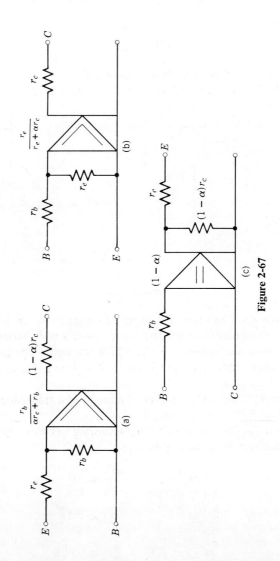

Figure 2-67

3 / *Useful Network Theorems*

There are many results of linear network and systems theory which are rather useful in active network synthesis. For example, some well-known network transformation techniques and their generalization have been found to lead to alternate approaches to active RC network design. Likewise, the stability concepts of linear system theory are equally applicable in the case of active networks.

The purpose of this chapter is to introduce these results and their extensions. In addition, we develop two very useful polynomial decomposition techniques, which are almost basic to the development of active RC synthesis procedures. Briefly, three types of network transformations are considered in the first three sections. The $RC:LC$ transformation (Section 3-1) relates the network functions of an RC network and its "equivalent" LC network. The $RC:CR$ transformation (Section 3-2), on the other hand, relates an RC network and a "transformed" RC network obtained by replacing the resistors and the capacitors in the first network by capacitors and resistors, respectively. The usual duality concepts are generalized in Section 3-3. In the next two sections, methods are introduced to convert a driving-point function realization problem into a problem of transfer function synthesis and vice versa. Section 3-6 is concerned with the development of $RC:-RC$ and $RC:RL$ type decompositions of polynomials. A review of various stability concepts and some related topics are included in Section 3-7. Main results are summarized in the concluding section.

3-1 *LC:RC* TRANSFORMATION

The *LC:RC* transformation relates the network function of an *RC* network $\mathcal{N}_0$ to the corresponding network function of an "equivalent" *LC* network $\hat{\mathcal{N}}$. The equivalent network $\hat{\mathcal{N}}$ is obtained from the original *RC* network $\mathcal{N}_0$ according to the following rule: Each resistance R_i in $\mathcal{N}_0$ is replaced by an inductance L_i of value R_i henries. Thus the topology of $\hat{\mathcal{N}}$ and $\mathcal{N}_0$ are the same, i.e., both have the same graph.

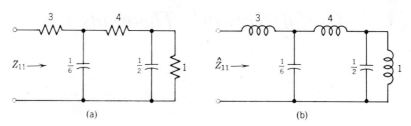

Figure 3-1 An *RC* one-port and its equivalent *LC* one-port.

To develop the pertinent relationship, let us first consider for simplicity an *RC* one-port network as shown in Figure 3-1a. The input impedance $Z_{11}(s)$ of this network can be written in the form of a continued fraction expansion:

$$(3.1) \quad Z_{11}(s) = 3 + \cfrac{1}{\frac{1}{6}s + \cfrac{1}{4 + \cfrac{1}{\frac{1}{2}s + \frac{1}{1}}}} = 3\left(\frac{s^2 + 6s + 8}{s^2 + 4s + 3}\right)$$

The "equivalent" *LC* one-port of this *RC* structure obtained by applying the *RC:LC* transformation outlined earlier is shown in Figure 3-1b. For this new one-port, the input impedance can be expressed as

$$(3.2) \quad \hat{Z}_{11}(s) = 3s + \cfrac{1}{\frac{1}{6}s + \cfrac{1}{4s + \cfrac{1}{\frac{1}{2}s + \frac{1}{s}}}} = 3s\left(\frac{s^4 + 6s^2 + 8}{s^4 + 4s^2 + 3}\right)$$

Comparing Equations 3.1 and 3.2 we obtain the following relation:

$$(3.3) \quad \hat{Z}_{11}(s) = sZ_{11}(s^2)$$

Consider next the RC two-port of Figure 3-2a. For this network, the transfer impedance $z_{21}(s)$ can be computed as follows:

(3.4)
$$I_0 = (\tfrac{2}{3}s + 1)V_2$$
$$V_1 = 3I_0 + V_2 = 3(\tfrac{2}{3}s + 1)V_2 + V_2 = (2s + 4)V_2$$
$$I_1 = \tfrac{1}{6}sV_1 + I_0 = \frac{s}{6}(2s + 4)V_2 + (\tfrac{2}{3}s + 1)V_2$$

Hence,

(3.5)
$$z_{21}(s) = \frac{V_2}{I_1} = \frac{3}{s^2 + 4s + 3}$$

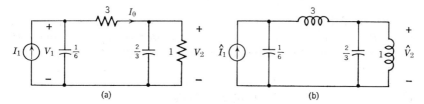

(a)   (b)

Figure 3-2 An RC two-port and its equivalent LC two-port.

Making an $RC:LC$ transformation on this network, i.e., replacing each resistance by an inductance of equal value, we obtain the LC two-port of Figure 3-2b for which one can show:

(3.6)
$$\hat{z}_{21}(s) = \frac{3s}{s^4 + 4s^2 + 3}$$

Again, comparison of Equations 3.5 and 3.6 yields the relation:

(3.7)
$$\hat{z}_{21}(s) = sz_{21}(s^2)$$

The above results can be generalized as:

Theorem 3-1 *Let a given RC network be characterized by a driving-point (transfer) impedance function $Z_{RC}(s)$. The corresponding network function $\hat{Z}_{LC}(s)$ of the equivalent LC network obtained by an RC:LC transformation is given by:*

(3.8)
$$\hat{Z}_{LC}(s) = sZ_{RC}(s^2)$$

A rigorous proof of the above theorem will be found elsewhere.[1] One

---

[1] E. S. Kuh and D. O. Pederson, *Principles of circuit synthesis*, McGraw-Hill, New York, pp. 148–149 (1959).

simple application of Relation 3.8 is to derive the physical realizability conditions of RC networks from the properties of LC networks. It should be noted that Theorem 3-1 is equally valid if active two-terminal elements like negative resistances, negative capacitances, and negative inductances are considered along with their passive counterpart. The inverse transformation relationship follows from Equation 3.8:

$$(3.9) \qquad Z_{RC}(s) = \frac{1}{\sqrt{s}} \hat{Z}_{LC}(\sqrt{s})$$

Since transfer voltage (current) ratio can be expressed as a ratio of impedance functions, a consequence of Theorem 3-1 is the theorem below.

Theorem 3-2 *Let $T_{RC}(s)$ be the transfer voltage (current) ratio of a given RC network. The corresponding transfer voltage (current) ratio $\hat{T}_{LC}(s)$ of the equivalent LC network obtained by an RC:LC transformation is given as:*

$$(3.10) \qquad \hat{T}_{LC}(s) = T_{RC}(s^2)$$

The above relationships will be used frequently in the chapter on the polynomial decomposition approach to active RC network synthesis (Chapter 12).

3-2 RC:CR TRANSFORMATION

In the RC:LC transformation, the network functions of an RC network and the equivalent LC network are related. However, it is convenient to define network transformations that transform an RC network into another RC network, and then to derive the relations between network functions of the original network and the transformed network. This type of transformation is particularly useful in the area of active RC networks where only allowable passive components are the resistances and the capacitances.

One such transformation is the RC:CR transformation, which is achieved by replacing each resistance R_i in the original network by a capacitance of value $1/R_i$ and replacing each capacitance C_j in the original network by a resistance of value $1/C_j$. Note that the topological graph of the transformed and the original network are identical.

Consider first the RC one-port of Figure 3-1a. Application of RC:CR transformation on this network yields the RC one-port of Figure 3-3a,

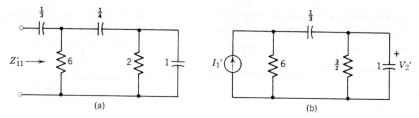

Figure 3-3 RC networks obtained using RC:CR transformation from the networks of Figures 3-1a and 3-2a.

which has the following input impedance:

(3.11) $$Z'_{11}(s) = \frac{3}{s} + \cfrac{1}{\cfrac{1}{6} + \cfrac{1}{\cfrac{4}{s} + \cfrac{1}{\frac{1}{2}+s}}} = \frac{3}{s}\left(\frac{8s^2 + 6s + 1}{3s^2 + 4s + 1}\right)$$

Comparing Equations 3.1 and 3.11 we obtain the relation:

(3.12) $$Z'_{11}(s) = \frac{1}{s} Z_{11}\left(\frac{1}{s}\right)$$

Similar results can be derived for the case of transfer impedances. To illustrate this, we now consider the RC two-port of Figure 3-2a. The transformed RC two-port obtained from this network by applying an RC:CR transformation is shown in Figure 3-3b, for which one readily obtains

(3.13) $$z'_{21}(s) = \frac{V'_2}{I'_1} = \frac{3s}{3s^2 + 4s + 1}$$

We observe that $z'_{21}(s)$ of Equation 3.13 is related to $z_{21}(s)$ of Equation 3.5 as

(3.14) $$z'_{21}(s) = \frac{1}{s} z_{21}\left(\frac{1}{s}\right)$$

Generalizing the preceding results, we can state:

Theorem 3-3 Let a given RC network be characterized by a driving-point (transfer) impedance function $Z_{RC}(s)$. The corresponding network function $Z'_{RC}(s)$ of the transformed RC network obtained by an RC:CR transformation is given by:

(3.15) $$Z'_{RC}(s) = \frac{1}{s} Z_{RC}\left(\frac{1}{s}\right)$$

Analogous results can be stated on the admittance basis. Now, the transfer voltage or current ratio of a network can be expressed as the ratio of impedance functions. Consequently, we have the following theorem:

Theorem 3-4 *The transfer voltage (current) ratio $T_{RC}(s)$ of an RC network and the corresponding transfer function $T'_{RC}(s)$ of the transformed RC network obtained by an RC:CR transformation on the original RC network are related as:*

(3.16) $$T'_{RC}(s) = T_{RC}\left(\frac{1}{s}\right)$$

The transformation given by Equation 3.16 is also known as *low-pass to high-pass transformation*. A topological proof of Theorems 3-3 and 3-4 has been advanced by Hakimi and Cruz.[2] It should be pointed out here that Theorems 3-3 and 3-4 are also valid if active components like negative resistances, negative capacitances, voltage-controlled voltage sources and current-controlled current sources are included in the *RC* network. Note that in the process of transformation, the controlled sources, VVT, and CCT, are left intact.[3]

The synthesis procedures to be described later in this book are, in most cases, applicable to network functions of any complexity. However, occasionally we may be faced with the following situation. The active *RC* realization of a transfer (driving-point) impedance function $Z(s)$ is known. It is desired to realize a transfer (driving-point) impedance function $Z'(s) = (1/s)Z(1/s)$. Realization of $Z'(s)$ is simply obtained by applying an *RC:CR* transformation on the known network realization of $Z(s)$. Actual steps of synthesizing $Z'(s)$ need not be followed. A similar situation may arise for the realization of a transfer voltage (current) ratio as illustrated in the following example.

Example 3-1 The active *RC* two-port of Figure 3-4a realizes a second-order low-pass Butterworth voltage transfer ratio:[4]

(3.17) $$t_v(s) = \frac{V_2}{V_1} = \frac{1}{s^2 + 1.4142s + 1}$$

In order to design a second-order high-pass Butterworth filter, we apply an *RC:CR* transformation to the network of Figure 3-4a. The resultant structure is shown in Figure 3-4b, and has the desired transfer voltage

---

[2] S. L. Hakimi and J. B. Cruz, Jr., "On minimal realization of *RC* two-ports," *Proc. NEC*, **16**, 258–267 (1960).

[3] S. K. Mitra, "A network transformation for active *RC* networks," *Proc. IEEE*, **55**, 2021–2022 (November 1967).

[4] The method of realization is outlined in Section 8-4.

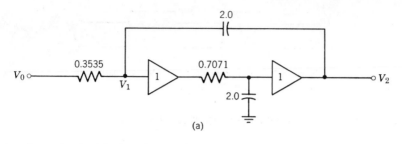

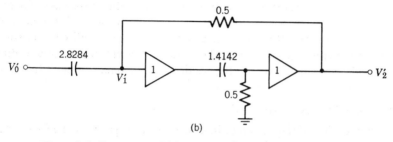

Figure 3-4 Low-pass to high-pass transformation (Example 3-1).

ratio:

$$(3.18) \qquad t'_v(s) = t_v\left(\frac{1}{s}\right) = \frac{s^2}{s^2 + 1.4142s + 1}$$

One useful application of the $RC:CR$ transformation is that it can be used to transform a low-pass active RC filter to a high-pass active RC filter, provided that the pertinent network function is a voltage (or current) transfer ratio. In addition, the components of the active RC filter can be only passive resistances, capacitances, and voltage (current) amplifiers.

It should be pointed out here that no such network transformation exists to modify an active RC low-pass filter to an active RC band-pass filter vice versa. The usual low-pass to band-pass transformation

$$s \to s + \frac{1}{s}$$

can be carried out *only* on the network function. However, a quick way of realizing a band-pass filter is to cascade (with proper isolation) a low-pass network with a high-pass network as depicted in Figure 3-5.

Another network transformation that preserves the RC character is the capacitive inverse relation. This is discussed in the next section.

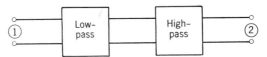

Figure 3-5 Band-pass filter obtained by cascading a low-pass and a high-pass section.

3-3 GENERALIZED INVERSE NETWORKS[5]

Duality and inverse networks play an important role in network theory. Usual definitions of these concepts are not that useful in the case of active *RC* networks. This is because dual of an active *RC* network becomes an active *RL* network. Recently, these concepts were generalized, and some of the problems associated with the classical concept of inverse and dual networks were eliminated. In this section, we introduce some of the basic ideas concerning the generalization.

Concept of Inverse Networks

An extension of the notion of inverse network is given by the following:

DEFINITION 3-1   *Two impedances $Z(s)$ and $Z^o(s)$ are defined as* generalized inverses *of each other with respect to a real rational function $f(s)$ if*

$$(3.19) \qquad Z(s) \cdot Z^o(s) = f(s)$$

When $f(s) = K/s$, the networks will be defined as capacitive inverse *of each other. If $f(s) = Ks$, then they are* inductive inverse *networks*.

The capacitive and inductive inverses of some passive elements are shown in Figure 3-6a, and 3-6b, from which it is seen that the capacitive inverse of an *RC* impedance is an *RC* impedance and that the inductive inverse of an *RL* impedance is an *RL* impedance.

To distinguish the above two types of generalized inverse from the usual definition of inverse network [$f(s) = K$], the latter will be called the *resistive inverse* relation. It can be shown that the resistive, capacitive, and inductive inverse relations are the only three inverses possible if we also require that Equation 3.19 be satisfied on an element-by-element basis.

The existence of other solutions of $f(s)$ can be shown if the element-by-element inverse requirement is relaxed.[6] We shall presently show that

---

[5] S. K. Mitra, N. Herbst, and N. DeClaris, "Generalized dual and inverse networks—an extended definition," *IEEE International Convention Record*, **11** (Part 2), 76–82 (1963).

[6] S. K. Mitra and W. G. Howard, Jr., "On generalized equivalent networks," *IEEE Trans. on Circuit Theory*, **CT-12**, 613–615 (December 1965).

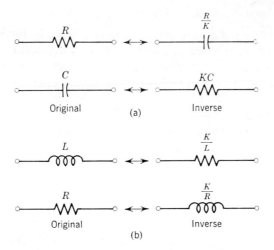

Figure 3-6 (a) Capacitive inverse transformation. (b) Inductive inverse transformation.

under certain conditions two *RC* driving-point impedance $Z(s)$ and $Z^o(s)$ can be generalized inverse of each other with respect to $f(s) = K/(s + \alpha)$,[6] i.e.,

(3.20) $$Z(s) \cdot Z^o(s) = \frac{K}{s + \alpha}, \quad \alpha > 0$$

Let us apply a frequency transformation

(3.21) $$s \to s - \alpha$$

to Equation 3.20. Writing $Z_1(s) = Z(s - \alpha)$ and $Z_1^o(s) = Z^o(s - \alpha)$, we have

(3.22) $$Z_1(s) \cdot Z_1^o(s) = \frac{K}{s}$$

Equation 3.22 implies that in order for $Z(s)$ and $Z^o(s)$ to be generalized inverses of each other with respect to $K/(s + \alpha)$, their transformed impedances $Z_1(s)$ and $Z_1^o(s)$ must be capacitive inverses of each other with respect to K. The restriction on the impedances $Z(s)$ and $Z^o(s)$ satisfying (3.20) is that the rightmost pole σ_1 of them must satisfy the following relation:

(3.23) $$\sigma_1 \geq \alpha$$

The above idea can be extended to more complex types of $f(s)$.[7]

---

[7] S. K. Mitra and W. G. Howard, Jr., "The reactive gyrator—a new concept and its application to active network synthesis," *IEEE International Convention Record*, **14** (Part 7), 319–326 (March 1966).

Generalized Dual Networks

The construction of inverse network can be effected by forming the dual network, which is obtained as follows.[8] From the graph of the original network $\mathcal{N}$, the dual graph is formed. Each branch of the dual graph is then replaced by an element that is the generalized inverse of the network in the corresponding branch of the graph of $\mathcal{N}$. The new network, thus obtained, is the generalized dual network of $\mathcal{N}$.

The duality relationship of CVT and VCT is indicated in Figure 3-7. Construction of generalized dual of other active devices follows similar lines.

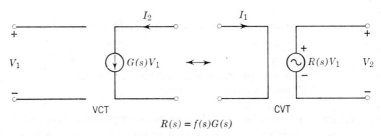

Figure 3-7 Generalized inverse relations between active elements.

We illustrate the application of this new concept by considering two simple examples.

Example 3-2 Realize an impedance $Z^{o}(s) = 1/s^2$ by forming the capacitive dual of the network of Figure 3-8a, which has an input impedance equal to s.

Figure 3-8b illustrates the formation of dual graph from which the final realization shown in Figure 3-8c is obtained by replacing each branch of the dual graph by an element that is capacitive dual of the corresponding branch in the graph of the original network of Figure 3-8a.

Example 3-3 Construct the generalized inverse of $Z_1(s) = (s + 3)/(s + 2)(s + 4)$ with respect to $f(s) = 1/(s + 1)$.

Form the transformed impedance $Z_1^{o}(s)$,

$$Z_1^{o}(s) = Z_1(s - 1) = \frac{(s + 2)}{(s + 1)(s + 3)}$$

---

[8] Construction of dual network is possible if the original network has a planar graph. For methods of constructing dual graphs, see M. E. Van Valkenburg, *Network Analysis*, 2nd ed., Prentice-Hall, Englewood Cliffs, N.J., p. 79, (1964) or other texts on network analysis.

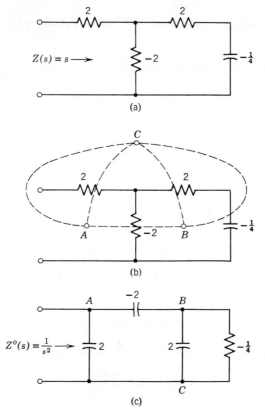

Figure 3-8 Construction of a capacitive dual active RC network (Example 3-2).

The realization of $Z_1^o(s)$ is given in Figure 3-9a and its capacitive dual in Figure 3-9b. Final realization (Figure 3-9c) of the desired inverse is obtained by replacing each capacitor C in Figure 3-9b with a parallel combination of a capacitor C and a resistor of value $1/C$.

3-4 CONVERSION OF TRANSFER FUNCTION REALIZATION TO DRIVING-POINT FUNCTION SYNTHESIS

Realization of one-port networks is often simpler than the synthesis of a two-port network. In addition, there exist many methods of active network realization of driving-point functions. In this section, a simple technique for converting the problem of two-port realization to synthesis of a one-port will be outlined.

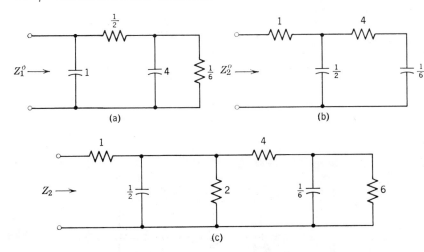

Figure 3-9 Realization of a generalized dual network (Example 3-3).

Consider the network of Figure 3-10a for which the voltage transfer ratio is given by

(3.24) $$\frac{V_o}{V_i} = \frac{N(s)}{D(s)} = \frac{Z_1(s)}{Z(s) + Z_1(s)}$$

If $Z(s)$ is a known impedance, then we can write $Z_1(s)$ as

(3.25) $$Z_1(s) = \frac{N(s)Z(s)}{D(s) - N(s)}$$

Usually $Z(s)$ can be preselected to represent a simple 2-terminal network, like a resistor or a capacitor. Then, using expression 3.25, $Z_1(s)$ can be calculated.

The dual network of Figure 3-10a, shown in Figure 3-10b, can be used in the case of transfer current ratio $I_o/I_i = N(s)/D(s)$. It can be shown that in this case $Z_1(s)$ is again given by expression 3.25, where $Z(s)$, as before, is a preselected impedance.

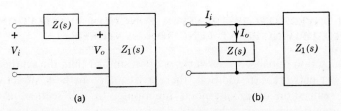

Figure 3-10 Conversion of two-port synthesis problem to one-port realization.

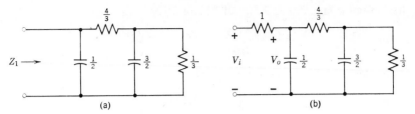

Figure 3-11 Example 3-4.

In each of the above two cases, realization of the transfer function is achieved by realizing first the driving-point function $Z_1(s)$ and then by connecting the known impedance $Z(s)$ as indicated.

Example 3-4 Using the above method, realize the following transfer voltage ratio:

$$\text{(3.26)} \qquad \frac{V_o}{V_i} = \frac{N(s)}{D(s)} = \frac{2s + 5}{s^2 + 6s + 8}$$

Let $Z(s)$ be chosen as a resistance of 1 ohm. Then from Equation 3.25 we obtain

$$\text{(3.27)} \qquad Z_1(s) = \frac{2s + 5}{s^2 + 4s + 3}$$

The realization of $Z_1(s)$ is shown in Figure 3-11a, from which the final network (Figure 3-11b) is obtained by connecting a series resistance of 1 ohm.

An alternate approach to this problem was suggested by Kinariwala.[9] Let $Z_{11}(s)$ be a driving-point impedance realized in the form of Figure 3-12. If we transform the current source into a voltage source V_i, the network of Figure 3-10a is obtained from Figure 3-12, where

$$\text{(3.28)} \qquad V_i = Z(s)I_o$$

The voltage transfer ratio V_o/V_i of Figure 3-10a can thus be related to the

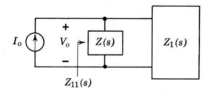

Figure 3-12 Realization of a voltage transfer ratio by synthesizing $Z_{11}(s)$.

[9] B. K. Kinariwala, "Synthesis of active *RC* networks," *Bell System Tech. J.*, **38**, 1269–1316 (September 1959).

driving-point impedance $Z_{11}(s)$ of Figure 3-12.

$$(3.29) \qquad \frac{V_o}{V_i} = \frac{N(s)}{D(s)} = \frac{V_o}{Z(s)I_o} = \frac{Z_{11}(s)}{Z(s)}$$

from which results the desired expression

$$(3.30) \qquad Z_{11}(s) = \frac{N(s)}{D(s)} Z(s)$$

There are two cases of interest. If $N(s)/D(s)$ is realized as a driving-point impedance in the form of Figure 3-12 so that $Z(s)$ is a resistance of R ohms, then by modifying the network as indicated in Figure 3-10a, we obtain the realization of $N(s)/D(s)$ as the transfer voltage ratio within a multiplier constant, i.e.,

$$(3.31) \qquad \frac{V_o}{V_i} = \frac{1}{R} \cdot \frac{N(s)}{D(s)}$$

Alternately, $N(s)/sD(s)$ may be realized as a driving-point impedance as indicated in Figure 3-12 where $Z(s)$ represents a capacitance of C farads. With reference to Figure 3-10a, we now observe that the voltage transfer ratio is given as

$$(3.32) \qquad \frac{V_o}{V_i} = C \frac{N(s)}{D(s)}$$

3-5 TRANSFORMATION OF DRIVING-POINT FUNCTION SYNTHESIS TO TRANSFER FUNCTION REALIZATION

Two simple methods were suggested by Sandberg for converting a driving-point function synthesis problem to the realization of transfer impedance (admittance). The methods use ideal controlled sources.

Consider, for example, the one-port active network of Figure 3-13.[10] Simple analysis yields the following voltage-current relation at the input port:

$$(3.33) \qquad (1 - \mu)E_0 = \left(\frac{z_{11}z_{22} - z_{12}^2}{z_{22}}\right)I_0 - \frac{z_{12}}{z_{22}} \mu E_0$$

where μ is the gain of the voltage amplifier, and $\{z_{ij}\}$ represents the open-circuit parameters of the passive 3-terminal network. If μ is set equal to unity, we get

$$(3.34) \qquad \frac{E_0}{I_0} = Z_{in} \doteq \frac{z_{11}z_{22} - z_{12}^2}{z_{12}} \doteq -\frac{1}{y_{12}}$$

---

[10] I. W. Sandberg, "Active RC networks," Res. Rept. R-662-58, PIB-590, Polytechnic Inst. of Brooklyn, Microwave Res. Inst., p. 19 (May 28, 1958).

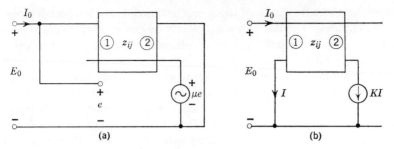

Figure 3-13 Conversion of driving-point function synthesis to transfer-function realization.

Thus realization of driving-point impedance Z_{in} is obtained by synthesizing the reciprocal function as the short-circuit transfer impedance of a 3-terminal network. If the 3-terminal network is restricted to being an RC network, then it is seen that a single network of the form of Figure 3-13a can realize arbitrary complex poles (excluding the positive real axis poles) and negative real zeros only. By connecting two such networks in series, complex zeros can be realized.

An alternate structure, which uses a current amplifier, is shown in Figure 3-13b.[11] Analysis yields for this configuration:

$$(3.35) \qquad \frac{E_0}{I_0} = Z_{in} = \frac{z_{11}}{1+K} + \frac{Kz_{12}}{1+K}$$

If we let $K \to \infty$, i.e., use an infinite gain current amplifier, then Equation 3.35 reduces to

$$(3.36) \qquad Z_{in} \doteq z_{12}$$

In this case, we observe that realization of Z_{in} can be accomplished by realizing it as an open-circuit transfer impedance of a 3-terminal network and then embedding an infinite gain current amplifier in the manner shown in Figure 3-13b.

3-6 THEOREMS ON POLYNOMIAL DECOMPOSITION

In Section 1-4, we observed that synthesis of an active RC network essentially boils down to an identification of the parameters of the companion RC network. There are basically three approaches which have been followed to yield the companion network parameters. The first approach is the coefficient matching technique illustrated in Example 1-1. The

---

[11] I. W. Sandberg, unpublished report.

second approach makes use of root-locus techniques, and has been followed by few authors. The last approach, which we can classify as polynomial decomposition approach, has been followed extensively in developing active *RC* synthesis procedures.

The polynomial decomposition approach can usually be divided into two parts. In the first part, the polynomial of interest (in general, the denominator polynomial of the specified network function) is decomposed suitably into simpler polynomials. Then these polynomials are manipulated to yield the parameters of the companion network. Thus, an aim of the decomposition technique would be to obtain realizable companion network parameters.

For active *RC* realization, there are two different methods of polynomial decomposition. These are discussed next.

RC:–RC Decomposition

In the *RC:–RC* decomposition, the pertinent polynomial $N(s)$ is decomposed into a difference of two polynomials, $E(s)$ and $F(s)$:

(3.37) $$N(s) = E(s) - F(s)$$

The polynomials $E(s)$ and $F(s)$ are required to have negative real roots. Additional properties of these polynomials and the method of obtaining the decomposition are discussed next. We now state the main theorem.

Theorem 3-5 *Let $N(s)$ be the specified polynomial, and let $Q(s)$ be an arbitrary polynomial having only distinct negative real roots. Then $N(s)/Q(s)$ can be expressed in either of the following forms:*

(i) $$\frac{N(s)}{Q(s)} = Z_{RC}{}^A - Z_{RC}{}^B \quad \text{if} \quad N(s)^0 \leq Q(s)^0$$

(ii) $$\frac{N(s)}{Q(s)} = Y_{RC}{}^a - Y_{RC}{}^b \quad \text{if} \quad N(s)^0 \leq Q(s)^0 + 1$$

where $Z_{RC}{}^A (Z_{RC}{}^B)$ is an RC driving-point impedance, and $Y_{RC}{}^a (Y_{RC}{}^b)$ is an RC driving-point admittance.[12]

Let us prove part i of the theorem first. Since $N(s)$ and $Q(s)$ are polynomials with real coefficients, a partial fraction expansion of $N(s)/Q(s)$ will be of the form:

(3.38) $$\frac{N(s)}{Q(s)} = k_\infty{}^+ + \frac{k_0{}^+}{s} + \sum_i \frac{k_i{}^+}{s + \sigma_i{}^+} - k_\infty{}^- - \frac{k_0{}^-}{s} - \sum_j \frac{k_j{}^-}{s + \sigma_j{}^-}$$

where the residues $k_i{}^+$, $k_j{}^-$ are real and nonnegative. Grouping together

---
[12] The degree of a polynomial $Q(s)$ will be denoted by $Q(s)^0$.

all the positive terms, we have

(3.39) $$Z_{RC}{}^A = k_\infty{}^+ + \frac{k_0{}^+}{s} + \sum_i \frac{k_i{}^+}{s + \sigma_i{}^+}$$

A similar grouping of the negative terms yields:

(3.40) $$Z_{RC}{}^B = k_\infty{}^- + \frac{k_0{}^-}{s} + \sum_j \frac{k_j{}^-}{s + \sigma_j{}^-}$$

Note from expressions 3.39 and 3.40, $Z_{RC}{}^A$ and $Z_{RC}{}^B$ are of the form of RC driving-point impedance functions.[13]

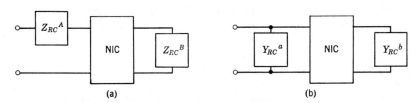

Figure 3-14 Physical interpretation of $RC:-RC$ decomposition.

If the degree condition of part ii is satisfied, then we can write

$$\frac{N(s)}{sQ(s)} = Z_{RC}{}^A - Z_{RC}{}^B$$

or

(3.41) $$\frac{N(s)}{Q(s)} = sZ_{RC}{}^A - sZ_{RC}{}^B = Y_{RC}{}^a - Y_{RC}{}^b$$

where $Y_{RC}{}^a \; (= sZ_{RC}{}^A)$ and $Y_{RC}{}^b \; (= sZ_{RC}{}^B)$ are RC driving-point admittance functions.

A network interpretation of the results of Theorem 3-5 can be given. If $N(s)/Q(s)$ is considered as a driving-point impedance, results of part i imply an active network representation of $N(s)/Q(s)$ as shown in Figure 3-14a. Part ii indicates a dual network representation of $N(s)/Q(s)$ when considered as a driving-point admittance function (Figure 3-14b).

Consider the decomposition indicated in part i. Let

(3.42) $$Z_{RC}{}^A = \frac{n_A(s)}{d_A(s)}, \quad Z_{RC}{}^B = \frac{n_B(s)}{d_B(s)}$$

---

[13] See Appendix B for properties of RC driving-point functions.

100 / USEFUL NETWORK THEOREMS

Thus we have

(3.43) $$\frac{N(s)}{D(s)} = \frac{n_A(s)}{d_A(s)} - \frac{n_B(s)}{d_B(s)}$$

This implies $N(s)$ can be expressed as indicated in (3.37) where

(3.44) $$E(s) = n_A(s)d_B(s)$$
$$F(s) = d_A(s)n_B(s)$$

Several interesting properties of $E(s)$ and $F(s)$ follow immediately from expression 3.44.[14] Since poles and zeros of an RC impedance are simple and lie on the negative real axis, it is evident the roots of $E(s)$ and $F(s)$ are negative real and at most of multiplicity 2. If we denote by R_E the number of roots of $E(s)$ to the right of a point $s = \alpha$ on the real axis and by R_F the number of roots of $F(s)$ to the right of α, then

(3.45) $$|R_E - R_F| \leq 1$$

To prove (3.45), let us denote by z_A, the number of zeros of polynomial $n_A(s)$ to the right of α, and let p_A, z_B, and p_B denote the same for the other three polynomials, $d_A(s)$, $n_B(s)$, and $d_B(s)$, respectively. Hence

(3.46) $$R_E = z_A + p_B, \qquad R_F = p_A + z_B.$$

We have

(3.47) $$R_E - R_F = (p_B - z_B) - (p_A - z_A)$$

Now it follows from expression 3.42 that each quantity inside the brackets on the right-hand side of Equation 3.47 is either one or zero. Thus maximum value of $|R_E - R_F|$ can be at most one.

Since $N(s)/Q(s)$ is expressed as the difference of two RC impedances (admittances), the type of decomposition of $N(s)$ illustrated above will be called an $RC{:}{-}RC$ type decomposition.

Example 3-5 Obtain an $RC{:}{-}RC$ type decomposition of $N(s) = s^4 + 6s^3 + 12s^2 + 9s + 3$. Choose $Q(s) = s(s+1)(s+2)(s+3)$. Then, the partial fraction expansion of $N(s)/Q(s)$ will be of the form:

(3.48) $$\frac{N(s)}{Q(s)} = k_\infty + \frac{k_0}{s} + \frac{k_1}{s+1} + \frac{k_2}{s+2} + \frac{k_3}{s+3}$$

---

[14] I. M. Horowitz, "Optimization of negative-impedance conversion methods of active RC synthesis," *IRE Trans. on Circuit Theory*, **CT-6**, 296–303 (September 1959).

From Equation 3.48, we note

$$k_\infty = \left.\frac{N(s)}{Q(s)}\right|_{s\to\infty} = \left.\frac{s^4 + 6s^3 + 12s^2 + 9s + 3}{s(s+1)(s+2)(s+3)}\right|_{s\to\infty} = 1$$

$$k_0 = \left.\frac{sN(s)}{Q(s)}\right|_{s=0} = \left.\frac{s^4 + 6s^3 + 12s^2 + 9s + 3}{(s+1)(s+2)(s+3)}\right|_{s=0} = \frac{1}{2}$$

(3.49)
$$k_1 = \left.\frac{(s+1)N(s)}{Q(s)}\right|_{s=-1} = -\frac{1}{2}$$

$$k_2 = \left.\frac{(s+2)N(s)}{Q(s)}\right|_{s=-2} = \frac{1}{2}$$

$$k_3 = \left.\frac{(s+3)N(s)}{Q(s)}\right|_{s=-3} = -\frac{1}{2}$$

By substituting Equation 3.49 in Equation 3.48 and rearranging, we obtain

(3.50)
$$\begin{aligned}\frac{N(s)}{Q(s)} &= \frac{s^4 + 6s^3 + 12s^2 + 9s + 3}{s(s+1)(s+2)(s+3)} \\ &= 1 + \frac{\frac{1}{2}}{s} + \frac{\frac{1}{2}}{s+2} - \frac{\frac{1}{2}}{s+1} - \frac{\frac{1}{2}}{s+3} \\ &= \frac{s^2 + 3s + 1}{s(s+2)} - \frac{s+2}{(s+1)(s+3)}\end{aligned}$$

Thus one possible decomposition of the specified $N(s)$ will be

(3.51) $E(s) - F(s) = (s^2 + 3s + 1)(s+1)(s+3) - s(s+2)^2$

A plot of the roots of $E(s)$ and $F(s)$ is indicated in Figure 3-15. We observe that condition 3.45 is satisfied. Furthermore, from Equation 3.50, we

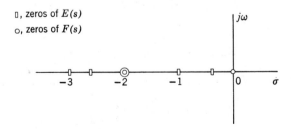

Figure 3-15

identify

$$Z_{RC}{}^A = \frac{s^2 + 3s + 1}{s(s+2)}, \quad Z_{RC}{}^B = \frac{s+2}{(s+1)(s+3)}$$

$Z_{RC}{}^A$ and $Z_{RC}{}^B$, as given above, are seen to be RC driving-point impedances.

$RC:RL$ Decomposition

In the second type of decomposition, a given polynomial $N(s)$ is expressed as a sum of two polynomials:

(3.52) $$N(s) = E(s) + F(s)$$

The polynomials $E(s)$ and $F(s)$, which are required to have negative real roots, also have some additional restrictions. The development of the $RC:RL$ decomposition closely parallels the first one. Let $N(s)$ be restricted to have positive leading coefficient and nonnegative constant term. We choose a polynomial $Q(s)$ having all distinct negative real roots and positive leading coefficient. In addition, $Q(s)$ is restricted to the following degree requirement:

(3.53a)
(3.53b) $$\text{deg } Q(s)^0 = \begin{cases} N(s)^0 \\ N(s)^0 - 1 \end{cases}$$

From the results of the previous section, it immediately follows that $N(s)/Q(s)$ can be expressed as

(3.54) $$\frac{N(s)}{D(s)} = h_\infty + \frac{k_0^+}{s} + \sum_i \frac{k_i^+}{s + \sigma_i^+} - \sum_j \frac{k_j^-}{s + \sigma_j^-} + k_\infty s$$

where all the residues and h_∞ are real and nonnegative. Note that k_∞ is equal to zero if condition 3.53a is satisfied. Rearranging (3.54) we have

(3.55) $$\frac{N(s)}{Q(s)} = \left(h'_\infty + \frac{k_0^+}{s} + \sum_i \frac{k_i^+}{s + \sigma_i^+} \right) + \left(k_\infty s + h''_\infty - \sum_j \frac{k_j^-}{s + \sigma_j^-} \right)$$

where $h_\infty = h'_\infty + h''_\infty$. If

(3.56) $$h_\infty \geq h''_\infty \geq \sum_j \frac{k_j^-}{\sigma_j^-}$$

then

(3.57) $$Z_{RL} = k_\infty s + h''_\infty - \sum_j \frac{k_j^-}{s + \sigma_j^-}$$

is realizable as an RL driving-point admittance. On the other hand, impedance

(3.58) $$Z_{RC} = h'_\infty + \frac{k_0^+}{s} + \sum_i \frac{k_i^+}{s + \sigma_i^+}$$

is guaranteed to be an RC driving-point impedance.

Thus, under the conditions stated above, $N(s)/Q(s)$ can be expressed as the sum of an RC driving-point impedance and an RL driving-point impedance. Hence the name, the $RC:RL$ decomposition, is given to this type of decomposition. A possible network interpretation of the second type of decomposition is given in Figure 3-16. If we let

$$(3.59) \qquad Z_{RC} = \frac{n_{RC}(s)}{d_{RC}(s)}, \qquad Z_{RL} = \frac{n_{RL}(s)}{d_{RL}(s)}$$

it follows that a restricted class of polynomials can be expressed as (3.52),

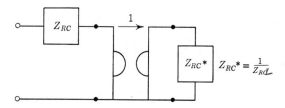

Figure 3-16 Physical interpretation of $RC:RL$ decomposition.

where

$$(3.60) \qquad \begin{aligned} E(s) &= n_{RC}(s)d_{RL}(s) \\ F(s) &= n_{RL}(s)d_{RC}(s) \end{aligned}$$

We shall now state two interesting properties of $E(s)$ and $F(s)$.[15] It follows from (3.59) that the roots of $E(s)$ and $F(s)$ are negative real and at most of multiplicity two. If R_E and R_F denote the number of roots of $E(s)$ and $F(s)$, respectively, to the right of any point $s = \alpha$ on the negative real axis, then

$$(3.61) \qquad 0 \le R_F - R_E \le 2$$

Proof of (3.61) is similar to the proof of condition 3.45 for the $RC:-RC$ decomposition and is left as an exercise. Sufficient condition for the $RC:RL$ decomposition of a polynomial is given in Theorem 10-2.3

Example 3-6 We consider the $RC:RL$ decomposition of the polynomial of Example 3-5:

$$N(s) = s^4 + 6s^3 + 12s^2 + 9s + 3$$

If $Q(s) = s(s+1)(s+2)(s+3)$, the partial fraction expansion of

---

[15] D. A. Calahan, "Sensitivity minimization in active RC synthesis," *IEEE Trans. on Circuit Theory*, **CT-9**, 38–42 (March 1962).

104 / USEFUL NETWORK THEOREMS

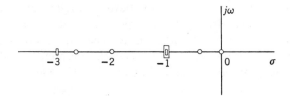

▯, zeros of $E(s)$

○, zeros of $F(s)$

Figure 3-17

$N(s)/Q(s)$ is given in Equation 3.50. Note that $h''_\infty = h_\infty = 1$ and

hence,
$$\sum_j \frac{k_j^-}{\sigma_j^-} = \left[\frac{1}{2} + \frac{\frac{1}{2}}{3}\right]$$

$$1 \geq \frac{\frac{1}{2}}{1} + \frac{\frac{1}{2}}{3} = \frac{2}{3}$$

i.e., condition 3.56 is satisfied. We can now write

(3.62)
$$\frac{N(s)}{Q(s)} = \left(\frac{\frac{1}{2}}{s} + \frac{\frac{1}{2}}{s+2}\right) + \left(1 - \frac{\frac{1}{2}}{s+1} - \frac{\frac{1}{2}}{s+3}\right)$$

$$= \frac{s+1}{s(s+2)} + \frac{s^2+3s+1}{(s+1)(s+3)}$$

It is seen that $Z_{RC} = (s+1)/s(s+2)$ is an RC impedance and $Z_{RL} = (s^2 + 3s + 1)/(s+1)(s+3)$ is an RL impedance.

From (3.62) we also obtain

$$E(s) = (s+1)^2(s+3)$$

$$F(s) = s(s^2 + 3s + 1)(s+2) = s(s + 0.385)(s + 2.625)(s+2)$$

The plot of the roots of $E(s)$ and $F(s)$ are shown in Figure 3-17, from which it is seen that condition 3.61 is also satisfied.

3-7 STABILITY CONCEPTS

It was mentioned in Chapter 1 that an active network can become unstable due to a change in the element values of the passive and active components. This change in component values usually occurs because of a change of internal and external conditions. The instability feature of the

STABILITY CONCEPTS / 105

active network is an inherent property arising from the use of feedback in the design of the circuit.

This section reviews some of the fundamental concepts of stability. The stability of a network is intimately related with the mode of operation of the network. For example, it is possible for a network to be unstable when excited by an ideal current source and stable when the current source has some finite internal impedance. It may also be possible for the same network to be stable if excited by a voltage source instead.

With the above considerations in mind, let us now discuss some of the well-known definitions of stability and their implications.

Strict Stability and Marginal Stability

The stability of a linear network $\mathcal{N}$ can be conveniently defined in terms of its *impulse response*. We recall that impulse response is the response of $\mathcal{N}$ (under the condition of no initial stored energy) due to an impulse function $\delta(t)$ as the excitation applied at $t = 0$.

Let us denote the excitation by $\hat{x}(t)$ and the response of initially relaxed network $\mathcal{N}$ by $y(t)$. Note that the nature of $\hat{x}(t)$ and $y(t)$ determine the mode of operation of $\mathcal{N}$. The response $y(t)$ is related to the excitation $\hat{x}(t)$ and the impulse response $h_{y\hat{x}}(t)$ by the convolution integral:

$$(3.63) \qquad y(t) = \int_0^t h_{y\hat{x}}(t - \tau)\hat{x}(\tau)\,d\tau = \int_{-t}^{\infty} h_{y\hat{x}}(\tau)\hat{x}(t - \tau)\,d\tau$$

This implies that knowing the impulse response, response to any arbitrary excitation can be determined using Equation 3.63. Hence the impulse response is a unique characteristic of the network which depends only on the network elements, their values and the specified manner of interconnection.

There are two distinct concepts of stability which are defined as follows:

DEFINITION 3-2 *A linear network is* strictly stable, *if its impulse response* $h_{y\hat{x}}(t)$ *is absolutely integrable, i.e.,*

$$(3.64) \qquad \int_0^{\infty} |h_{y\hat{x}}(\tau)|\,d\tau = M < +\infty$$

DEFINITION 3-3 *A linear network is* marginally stable *or simply* stable, *if its impulse response remains bounded except possibly for a finite number of impulses.*

From the above definitions, we conclude that the impulse response of a strictly stable *LLF* network will contain only such terms as

$$(3.65) \qquad At^n e^{-\alpha t} \sin \omega t, \qquad At^n e^{-\alpha t} \cos \omega t$$

where α is a positive real number, n is a nonnegative integer, and ω is a nonnegative real number. All of these terms go to zero as $t \to \infty$.

In addition to the terms given in expression 3.65, the impulse response of a (marginally) stable *LLF* network can have only such terms as

(3.66) $$A \cos \Omega t, \quad A \sin \Omega t$$

where Ω is a nonnegative real number.

If terms such as

(3.67) $$t^n, \quad e^{\beta t}, \quad e^{\beta t} \cos \omega t \quad \beta > 0$$

are present in the impulse response, then the network will be unstable.

The above definitions imply that if the excitation $\hat{x}(t)$ is bounded,

(3.68) $$|\hat{x}(t)| \leq M_1 < +\infty$$

then the response $y(t)$ of a strictly stable network is also bounded. In the case of a stable network, if the excitation is in the form of a decaying exponential, the response is still bounded.

Definitions 3-2 and 3-3 express the stability of a network in the time domain. Equivalent conditions in the frequency domain can be easily developed. To this end we note that the network function $H_{y\hat{x}}(s)$ characterizing the specified network is given by the Laplace Transform of its impulse response $h_{y\hat{x}}(t)$, i.e.

(3.69) $$H_{y\hat{x}}(s) = \int_0^\infty h_{y\hat{x}}(\tau) e^{-s\tau} \, d\tau$$

The following two theorems give the necessary and sufficient conditions of stability in terms of network function:

Theorem 3-6 *An LLF network under given mode of operation will be strictly stable if, and only if, its pertinent network function has no poles in the right half s-plane including the jω-axis.*

Theorem 3-7 *An LLF network under a given mode of operation is stable if, and only if, the corresponding network function has poles in the left half s-plane and simple jω-axis poles (if any).*

The proofs of the above theorems are simple and are left as an exercise (Problem 3.14).

In general, the network specifications, in the frequency domain or in the time domain, is approximated by a network function of the complex frequency variable, which is then used to synthesize the network. For most networks, the network functions are easily obtained by analysis. Thus the results of the above theorems are extremely useful in determining stability margin of a designed network.

In the following example, we test the stability of the active Bessel filter of Figure 1-8 using the above concepts.

Example 3-7 The transfer voltage ratio of the active RC two-port is given by Equation 1.8:

(1.8)
$$\frac{V_2}{V_1} = \frac{\mu(C_2 R_2 s + 1)}{[C_1 C_2 R_1 R_2 s^2 + (C_2 R_2 + C_1 R_1 + C_2 R_1 - \mu C_2 R_1)s + 1]} \times \left[R_3 C_4 s + \frac{R_3}{R_4} + 1 \right]$$

and the *nominal* values of the components are as follows:

$$R_1 = \tfrac{2}{3}, \quad R_2 = 6, \quad R_3 = 1, \quad R_4 = 1$$
$$C_1 = \tfrac{1}{2}, \quad C_2 = \tfrac{1}{6}, \quad C_4 = 2, \quad \mu = 4$$

Let us investigate the stability when the gain of the voltage amplifier μ varies from its nominal value. Then Equation 1.8 reduces to

(3.70)
$$\frac{V_2}{V_1} = \frac{6}{s^2 + \tfrac{1}{3}(13 - \mu)s + 3}$$

From Equation 3.70 we observe that the poles of the pertinent network function are given by the roots of

(3.71)
$$s^2 + \frac{13 - \mu}{3} s + 3 = 0$$

and are at

(3.72)
$$s_1, s_2 = \frac{\mu - 13 \pm \sqrt{(\mu - 13)^2 - 108}}{6}$$

Note that if $\mu > 13$, the roots are in the right-half s-plane and, hence, the network is unstable. For $\mu = 13$, both the roots are on the $j\omega$-axis and are simple, indicating marginal stability. The network is strictly stable for all values of $\mu < 13$.

Since the network has been designed for a nominal value of the gain, μ equal to 4, the complete circuit has a gain margin of stability equal to

$$20 \log_{10} (\tfrac{13}{4}) = 10.237 \text{ db}$$

A plot of the roots of all nonnegative values of μ is shown in Figure 3-18. Such *root-locus* plots[16] are useful in the study of network stability. Some active network synthesis methods are based on the root locus approach.

---

[16] For details in root-locus plots, see R. G. Brown and J. W. Nilsson, *Introduction to linear systems analysis*, John Wiley, New York, pp. 161–178 (1962), or other books on systems analysis.

108 / USEFUL NETWORK THEOREMS

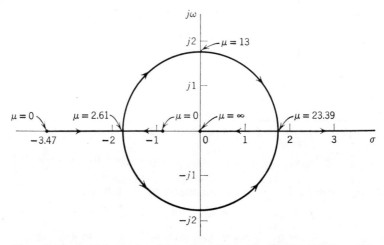

Figure 3-18 Root-locus plot of the poles of the active *RC* filter of Figure 1-8 as a function of the gain of active element (Example 3-7).

Open-Circuit and Short-Circuit Stability

Two theorems were presented earlier, which give the necessary and sufficient conditions for the two types of stability of a network. In both of the theorems, the terms "mode of operation" and "pertinent network function" were used. Let us now examine these terms more closely.

Consider a one-port first. For a one-port, either the voltage or the current can be the excitation. Thus, there are two types of network functions describing the one-port:

(1) The driving-point impedance function,

$$Z(s) = \frac{V(s)}{I(s)}$$

(2) The driving-point admittance function,

$$Y(s) = \frac{I(s)}{V(s)}$$

If the one-port is being driven by a current source (this is the mode of operation), then the "pertinent" network function is $Z(s)$. On the other hand, if the mode of operation is described by a voltage source as the excitation, then the relevant network function would be $Y(s)$. These are illustrated in Figure 3-19.

DEFINITION 3-4 *If $Z(s)$ has all its poles in the left-half s-plane excluding the $j\omega$-axis, then the one-port would be called* open-circuit stable (OCS).

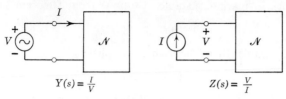

Figure 3-19

DEFINITION 3-5 *A one-port is short-circuit stable* (SCS) *if* $Y(s)$ *has all its poles in the left-half s-plane excluding the $j\omega$-axis.*

Of course, it may be possible for a network to be stable under one mode of operation but unstable under the other. This is illustrated in Example 3-8.

Example 3-8 A one-port is characterized by the following driving-point impedances:

$$Z(s) = \frac{s^2 - 2s + 5}{s^2 + 4s + 5}$$

From the pole-zero plot of $Z(s)$, as shown in Figure 3-20, we conclude that the one-port is open-circuit stable and short-circuit unstable.

The stability of a two-port can be discussed in a similar manner. For example, if $z_{21}(s)$ is the pertinent network function, then the mode of operation is described by open-circuit input and output ports. It is seen that for a two-port there can be four different kinds of modes of operation.

Hurwitz Polynomials

Whether a network is stable or not is determined from the locations of the poles of the pertinent network function

$$T(s) = \frac{N(s)}{D(s)}$$

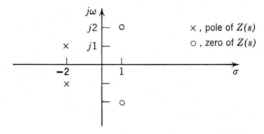

Figure 3-20 Pole-zero plot of the driving-point function of Example 3-8.

If the network is (marginal) stable, it follows from Theorem 3-7 that

$$D(s)^0 \geq N(s)^0$$

If it is strictly stable, then Theorem 3-6 implies

$$D(s)^0 > N(s)^0$$

Once either of these restrictions is satisfied, stability is determined by the roots of $D(s)$, usually referred to as the *characteristic polynomial* of the network.

DEFINITION 3-6 *A polynomial with real coefficients is called a* Hurwitz *polynomial if each of its roots is located in the left-half s-plane excluding the $j\omega$-axis. If the polynomial has all roots in the left-half s-plane and $j\omega$-axis zeros (if any) are simple, then it is called a* pseudo-Hurwitz *polynomial.*

A polynomial $D(s)$ is Hurwitz if the continued fraction expansion of Ev $D(s)$/Od $D(s)$ has all the coefficients present and they are positive. The total number of coefficients in the expansion is equal to $D(s)^0$. If the expansion ends prematurely [the number of coefficients less than $D(s)^0$] with positive coefficients and $Q(s)$ as common factor, then $D(s)$ is pseudo-Hurwitz provided that $Q(s) + (d/ds)Q(s)$ is Hurwitz.

Therefore, for a strictly stable network, the characteristic polynomial must be Hurwitz and for a (marginal) stable network, the characteristic polynomial can be pseudo-Hurwitz. It is clear from Definition 3-6, a necessary (not sufficient) condition for a Hurwitz polynomial is that all coefficients must be present and positive.

Stability of a network can be tested by means of any one of the several algebraic and graphical methods. The reader unfamiliar with the methods is referred to the textbooks listed at the end of this chapter.

Potential Instability

In general, voltage or current sources have some finite internal impedances. A two-port often works into a load impedance. Thus, any stability investigation must also take into account the effect of generator impedance and possibly load impedances. These terminating impedances are usually passive in nature and, as a result, this type of conditional stability always refers to passive terminations.

For simplicity, consider first a one-port described by a driving-point impedance $Z_A(s)$. If the one-port is excited by a source (voltage or current) of internal impedance $Z_p(s)$, as shown in Figure 3-21, the overall system stability would be determined by the roots of

(3.73) $$Z_A(s) + Z_p(s) = 0$$

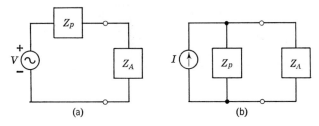

Figure 3-21 Illustration of the potential instability concept.

The validity of the above statement is easy to verify. We now make the following definitions:

DEFINITION 3-7 *A one-port $Z_A(s)$ is* potentially unstable *if there exists a single passive terminating impedance $Z_p(s)$ such that*

$$Z_A(s) + Z_p(s) = 0$$

will have at least one root in the right-half s-plane including the $j\omega$-axis.

DEFINITION 3-8 *A one-port is* absolutely stable *if it is strictly stable for all possible passive terminations.*

It is clear that if the one-port is either open-circuit unstable or short-circuit unstable, it can never be absolutely stable.

Example 3-9 The input impedance of the one-port of Figure 3-22 is

$$Z_A(s) = -\frac{s+2}{2s+1}$$

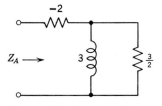

Figure 3-22 A potentially unstable one-port.

Observe that $Z_A(s)$ is both open-circuit stable and short-circuit stable. If the generator impedance $Z_p(s)$ is 1 ohm, then we find that

$$Z_p(s) + Z_A(s) = 1 - \frac{s+2}{2s+1} = \frac{s-1}{2s+1}$$

which has a right-half plane zero. Therefore, the one-port of Figure 3-22 is potentially unstable.

Next, we investigate the stability of a terminated two-port (Figure 3-23). The input impedance of the two-port terminated at port 2 by $Z_2(s)$ is given as

(3.74) $$Z_{11}(s) = z_{11}(s) - \frac{z_{12}(s)z_{21}(s)}{z_{22}(s) + Z_2(s)}$$

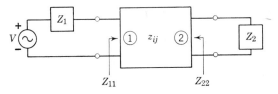

Figure 3-23

where $\{z_{ij}\}$ are the open-circuit parameters of the two-port. If we consider the terminated two-port as a one-port, then the system stability is determined by the roots of

(3.75) $$Z_1(s) + Z_{11}(s) = 0$$

We now state the following:

DEFINITION 3-9 *A two-port network is* potentially unstable *if there exists at least a pair of passive terminating impedances, $Z_1(s)$ and $Z_2(s)$, which will make the equation*

$$Z_1(s) + Z_{11}(s) = 0$$

to have one or more roots in the right-half s-plane including the $j\omega$ axis.

DEFINITION 3-10 *A two-port is* absolutely stable *if it is strictly stable for all possible input and output passive terminating impedances.*

It is a simple exercise to show that the roots of

$$Z_1(s) + Z_{11}(s) = 0$$

are identical to the roots of[17]

(3.76) $$Z_2(s) + Z_{22}(s) = 0$$

where $Z_{22}(s)$ is the input impedance at port 2 when port 1 is terminated by the impedance $Z_1(s)$:

(3.77) $$Z_{22}(s) = z_{22}(s) - \frac{z_{12}(s)z_{21}(s)}{z_{11}(s) + Z_1(s)}$$

Potential instability conditions are derived in Chapter 6.

3-8 SUMMARY

The network functions of an *RC* network and its "equivalent" *LC* network obtained by an *RC:LC* transformation (i.e., by replacing each

---

[17] Assuming that $z_{12}z_{21}$ is not equal to zero.

resistor R_i in the RC network by an inductance of value R_i henries) are related according to Equations 3.8 and 3.10 as appropriate.

The network functions of an RC network and a "transformed" RC network obtained by an $RC:CR$ transformation (i.e., by replacing each resistor R_i and each capacitor C_j in the original RC network by a capacitor of value $1/R_i$ farads and a resistor of value $1/C_j$ ohms, respectively) are related according to Equations 3.15 and 3.16. The $RC:CR$ transformation can be used to transform an active RC low-pass network into an active RC high-pass network, provided that the pertinent network function is a transfer voltage or current ratio.

Two networks realizing the driving-point (transfer) impedances $Z(s)$ and $Z^o(s)$ are generalized inverses of each other with respect to a real rational function $f(s)$ if Relation 3.19 holds true. If $f(s) = K/s$, then the relation is called the capacitive inverse relation. To construct the generalized inverse of a planar network, first the dual graph of the original network is formed. Then each branch of the dual graph is replaced by an element that is a generalized inverse of the dual branch in the original network.

A transfer voltage (current) ratio $N(s)/D(s)$ can be realized in two steps. First, a driving-point impedance $Z_1(s)$, as given by Equation 3.25, is synthesized with $Z(s)$ as a preselected one-port network. Then, $Z_1(s)$ and $Z(s)$ are connected in the way shown in Figures 3-10a and 3-10b to yield the realization of the transfer function.

A specified driving-point admittance can be realized by synthesizing it as a short-circuit transfer admittance in the form of a grounded two-port network and then by connecting a unity gain voltage amplifier as indicated in Figure 3-13a. Similarly, a driving-point impedance can be realized by synthesizing it as an open-circuit impedance in the form of a grounded two-port network and then by connecting an infinite gain current amplifier as shown in Figure 3-13b.

In the $RC:-RC$ decomposition, a function $N(s)/Q(s)$ is decomposed as the differences of two RC driving-point impedances (admittances), where $N(s)$ is an arbitrary polynomial and $Q(s)$ is a polynomial with distinct negative real roots such that $N(s)^0 \leq Q(s)^0$ [or $N(s)^0 \leq Q(s)^0 + 1$] (Theorem 3-5).

In the $RC:RL$ decomposition, a function $N(s)/Q(s)$ is expressed as a sum of an RC driving-point impedance and an RL driving-point impedance, where $N(s)$ has certain restrictions and $Q(s)$ is a polynomial with distinct negative real roots satisfying the degree condition 3.53.

A network function is strictly stable if all of its poles are in the left-half s-plane excluding $j\omega$-axis (Theorem 3-6). Potential instability conditions (i.e., stability of a network with finite terminations) are given according to Definitions 3-7 through 3-10.

Additional References

Brown, R. G., and J. W. Nilsson, *Introduction to linear systems analysis*, John Wiley, New York, 1962.
DePian, L., *Linear active network theory*, Prentice-Hall, Englewood Cliffs, N.J., 1962.
Ghausi, M. S., *Principles and design of linear active circuits*, McGraw-Hill, New York, 1965.
Lathi, B. P., *Signals, systems and communication*, John Wiley, New York, 1965.
Su, K. L., *Active network synthesis*, McGraw-Hill, New York, 1965.
Van Valkenburg, M. E., *Network analysis*, 2nd ed., Prentice-Hall, Englewood Cliffs, N.J., 1964.

Problems

3.1[18] Let $\mathcal{N}$ be an active RC network containing a single CVT described by a transimpedance r ohms. Let $\mathcal{N}$ be characterized by a driving-point (transfer) impedance function $Z(s)$. Show that if each resistor R_i in $\mathcal{N}$ is replaced by a capacitor of value $1/R_i$ farads, each capacitor C_j is replaced by a resistor of value $1/C_j$ ohms, and the CVT is replaced by a CVT of transimpedance of value r/s, then this transformed network $\mathcal{N}'$ is characterized by a driving-point (transfer) impedance function $Z'(s)$ where

$$Z'(s) = \frac{1}{s} Z\left(\frac{1}{s}\right)$$

Hint: Use a signal flow-graph representation of $\mathcal{N}$.

3.2 Generalize the above results for other types of active elements.

3.3 Design an active RC filter using a voltage amplifier having the following transfer function:

$$Z_{21} = \frac{6s^2}{3s^2 + 3s + 1}$$

3.4 If $\mathcal{N}_o$ represents the generalized dual of a two-port $\mathcal{N}$ characterized by a voltage transfer ratio $T(s)$, show that the current transfer ratio of $\mathcal{N}_o$ is also $T(s)$.

3.5 A natural extension of the concept of generalized inverse networks is the concept of generalized equivalent networks,[19] as defined next:

---

[18] S. K. Mitra, *loc. cit.*
[19] S. K. Mitra and W. G. Howard, *loc. cit.*

DEFINITION 3-11 *The generalized equivalent impedance $Z(s)$ of an impedance $Z'(s)$ with respect to a real rational function $g(s)$ is defined as*

$$Z(s) = g(s)Z'(s)$$

If $Z'(s)$ is an RC driving-point impedance and it is required that $Z(s)$ be an RC impedance, develop the restrictions on $g(s) = (s + \sigma_1)/(s + \sigma_2)$.

3.6 An alternate method[20] of converting a driving-point function realization problem into a transfer-function synthesis problem is by means of the structure of Figure 2-24 (which is dual of Figure 3-13a).

(a) Show that for this new structure $Z_{in} = z_{12}$.
(b) Derive the configuration of Figure 3-13b from the structure of Figure 3-24.

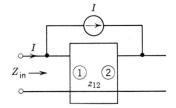

Figure 3-24

3.7 Obtain an $RC:-RC$ type decomposition of the following polynomials:
(a) $1 - s^6$
(b) $s^3 + 2s^2 + 2s + 1$
(c) $s^3 + 6s^2 + 15s + 15$
(d) $s^2 + 3s + 3$

3.8 Obtain an $RC:RL$ type decomposition of the following polynomials:
(a) $s^2 + 3s + 3$
(b) $2s^3 + 15s^2 + 35s + 27$
(c) $s^3 + 6s^2 + 11s + 7$
(d) $s^3 + 5s^2 + 8s + 5$

3.9 Is it possible to obtain an $RC:RL$ decomposition of a polynomial with one or more zeros in the right-half s-plane? Justify your answer.

3.10 Show that if a network has an impulse response of the form $At^n e^{-\alpha t} \sin \omega t$ ($n > 0$, $\alpha > 0$), then it is strictly stable.

3.11 Prove that if an LLF network is strictly stable, then a bounded excitation results in a bounded response.

3.12 Determine which of the following network functions are strictly stable, stable, and unstable.

(a) $\dfrac{(s + 1)(s + 3)}{(s + 2)(s + 4)}$
(b) $\dfrac{(s^2 + 1)(s^2 + 5)}{s(s^2 + 3)(s^2 + 10)}$
(c) $\dfrac{(s + 1)(s + 3)}{(s^2 + 2)^2}$

(d) $\dfrac{(s + 1)(s + 7.9)(s + 10)}{(s + 2)}$
(e) $\dfrac{s^2 + 4s + 3}{s^4 + 6s^3 + 8s^2}$

(f) $\dfrac{(s + 1)(s - 3)}{(s + 2)(s + 4)}$
(g) $\dfrac{(s + 1)(s + 3)}{(s - 2)(s + 4)}$

[20] I. W. Sandberg, unpublished report.

3.13 A fundamental property of a strictly stable network is described by the following theorem:

Theorem 3-8 *If an LLF network characterized by a network function $T(s)$ is strictly stable, then for an input*

$$\hat{x}(t) = \cos \omega_0 t = \text{Re}\,[e^{j\omega_0 t}]$$

corresponds an output

$$y(t) \simeq \text{Re}\, T(j\omega_0)e^{j\omega_0 t}$$
$$= |T(j\omega_0)| \cos(\omega_0 t + \phi) \quad as \quad t \to \infty$$

where $\quad \phi = \arg T(j\omega_0)$

Prove the above theorem.

3.14 Prove Theorems 3-6 and 3-7.

3.15 A two-port network was designed using a voltage amplifier of gain α. The pertinent network function is

$$t_v = \frac{H}{s^2 + 1.2s + 0.1(1 - \alpha)}$$

Study the network stability for positive values of the gain α.

3.16 Determine the conditions for open-circuit stability of a biased tunnel diode for small signal applications (use Figure 2-1b).

3.17 A resistive two-port described by

$$y_{11} = s + 2 \qquad y_{12} = -1$$
$$y_{21} = 4 \qquad y_{22} = -3$$

is terminated at the output port by a load resistance of 0.25 ohms. Study the stability of the network when excited by a generator of internal impedance (1) 0.1 ohms, (2) 1 ohm.

3.18 Test the following polynomials and determine which are Hurwitz and which are pseudo-Hurwitz polynomials.

(a) $(s^2 + 2)(s^2 + 4) + s(s^2 + 3)$
(b) $s^4 + s^3 + 4s^2 + 2s + 3$
(c) $s^4 + 3s^2 + 2s + 1$
(d) $s^5 + s^4 - 2s^3 + s^2 + s + 1$
(e) $s^6 + 6s^4 + 11s^2 + 6$
(f) $s^6 + s^5 + 9s^4 + 4s^3 + 26s^2 + 3s + 8$

4 / *Analysis of Active Networks*

A major part of network theory is circuit analysis. The knowledge of analysis techniques is a necessity for the circuit designer, particularly the designer of active circuits. As mentioned earlier, there are basically two aspects of active filter design: (i) design of the active elements and (ii) design of the passive "companion" sub-network. With respect to both of these aspects, the designer may either conceive his own circuit or use a well-known circuit, but before the circuit is built, he must evaluate the network. The designed circuit can be evaluated in two ways—by actually constructing it or by analyzing it. In many cases, it is profitable to evaluate a circuit by analysis.

Consider, for example, the design of active devices. The design is usually based on somewhat idealized behavior of the available solid-state element. The operation of the circuit in practice, however, depends on many factors like biasing, parasitics, actual behavior of the solid-state components, etc., and can be predicted theoretically to some extent by analysis of an equivalent representation of the designed circuit which takes into account all the associated external and internal factors. Analysis is particularly helpful in comparing several alternate designs from the point of view of practical operation. Similar observations can be made with respect to the design of the passive portion of the filter and the operation of the complete active filter.

This chapter discusses the multipole-analysis approach, which is more suitable than other methods available for analyzing networks containing active elements. Sections 4-1 through 4-6 develop the indefinite admittance matrix analysis of multiterminal networks and discuss various aspects of such an approach. In Section 4-7, the equivalent circuits of multiterminal networks are developed on the admittance basis. Networks containing elements with "pure constraints" (e.g., the ideal VVT, CCT, NIC) can be handled by means of a simple modification of the indefinite admittance matrix approach, which is discussed in Section 4-8. Networks containing operational amplifiers are considered in Section 4-9. Derivation of the bilinear form of network function concludes Chapter 4.

4-1 THE INDEFINITE ADMITTANCE MATRIX[1]

Any linear network element can be considered as a multiterminal element or simply a *multipole*, where each terminal is associated with a current I_k and a potential V_k (defined with respect to a reference terminal not belonging to the network element). The circuit representation of a multipole is shown in Figure 4-1, which also indicates the assumed positive direction of the currents. By definition, the currents $\{I_k\}$ and the potentials $\{V_k\}$ satisfy the following postulates:

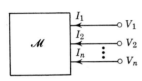

Figure 4-1 A multipole $\mathcal{M}$.

(i) $\sum_k I_k = 0$

(ii) Each current I_k depends linearly only on the potential difference between each set of terminals.

The two postulates will be explained in terms of a three pole (an n pole is an n terminal multipole) as shown in Figure 4-2. From Figure 4-2, it is clear that by Kirchhoff's current law,

(4.1) $$I_1 + I_2 + I_3 = 0$$

Equation 4.1 indicates that postulate i is a restatement of Kirchhoff's current law. In both circuits of Figure 4-2, the potential difference between corresponding pairs of terminals is identical; therefore, the currents $\{I_k\}$ remain in-variant when all of the potentials V_k are increased by the same amount V_0.

The resistor is an example of a two pole (Figure 4-3a); the transistor, a three pole (Figure 4-3b); the negative-impedance converter, a four pole (Figure 4-3c).

---

[1] There are many references on this topic. This section and the two following sections are based on L. A. Zadeh, "Multipole analysis of active networks," *IRE Trans. on Circuit Theory*, **CT-4**, 97–105 (September 1957).

THE INDEFINITE ADMITTANCE MATRIX / 119

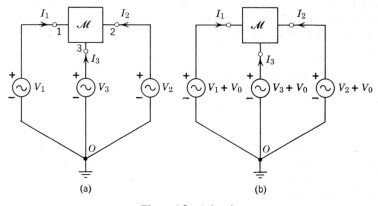

Figure 4-2 A 3-pole.

Based on the linearity of a multipole and the two postulates governing its terminal behavior, relations between the terminal currents and the potentials can be written. For simplicity, let us consider a three pole (Figure 4-2a). In general, it will be described by a set of equations of the form:

(4.2)
$$I_1 = \hat{y}_{11}V_1 + \hat{y}_{12}V_2 + \hat{y}_{13}V_3 + I_1^0$$
$$I_2 = \hat{y}_{21}V_1 + \hat{y}_{22}V_2 + \hat{y}_{23}V_3 + I_2^0$$
$$I_3 = \hat{y}_{31}V_1 + \hat{y}_{32}V_2 + \hat{y}_{33}V_3 + I_3^0$$

where the coefficients $\{\hat{y}_{ij}\}$ and $\{I_i^0\}$ are, in general, real rational functions.[2] It is, of course, assumed here that $\mathcal{M}$ is not degenerate, i.e., no short circuit exists between any two terminals of $\mathcal{M}$. From the equations of (4.2), we note

(4.3) $$I_k\big|_{V_1=V_2=V_3=0} = I_k^0, \qquad k = 1, 2, 3$$

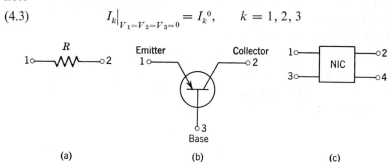

Figure 4-3 Examples of multipoles: (a) resistor, (b) transistor, and (c) NIC.

[2] Assuming that $\{I_j\}$ and $\{V_j\}$ are the Laplace transforms of the current and voltage variables.

indicating that I_k^0 is the current through the kth terminal when all the terminals of the multipole are connected directly to the reference node 0. Thus, because of postulate i,

(4.4) $$I_1^0 + I_2^0 + I_3^0 = 0$$

Similarly, the coefficients $\hat{y}_{ij}$ can be expressed as[3]

(4.5) $$\hat{y}_{ij} = \left. \frac{I_i - I_i^0}{V_j} \right|_{\substack{V\mu = 0 \\ \mu \neq j}}$$

The coefficient $\hat{y}_{ij}$ can be interpreted as the difference of the currents I_i and I_i^0, where I_i is the current flowing through terminal i when all the terminals of $\mathcal{M}$ except the jth terminal are grounded and a potential of 1 volt is applied to j with respect to the ground. I_i^0 has already been defined in Equation 4.3. It is evident that $\hat{y}_{ij}$ has the dimension of an admittance. It is also evident from Equation 4.2 that the coefficients $\hat{y}_{ij}$ and I_i^0 completely specify the external behavior of $\mathcal{M}$. Equation 4.2 can be written as

(4.6) $$\begin{bmatrix} I_1 \\ I_2 \\ I_3 \end{bmatrix} = \begin{bmatrix} \hat{y}_{11} & \hat{y}_{12} & \hat{y}_{13} \\ \hat{y}_{21} & \hat{y}_{22} & \hat{y}_{23} \\ \hat{y}_{31} & \hat{y}_{32} & \hat{y}_{33} \end{bmatrix} \begin{bmatrix} V_1 \\ V_2 \\ V_3 \end{bmatrix} + \begin{bmatrix} I_1^0 \\ I_2^0 \\ I_3^0 \end{bmatrix}$$

or in matrix notation

(4.7) $$\mathbf{I} = \hat{\mathbf{Y}} \cdot \mathbf{V} + \mathbf{I}^0$$

The matrix $\hat{\mathbf{Y}}$ is called the *indefinite admittance matrix* of $\mathcal{M}$. It is termed indefinite because the reference (ground) terminal for the potential is an arbitrary node outside the multipole $\mathcal{M}$.

Example 4-1 Let us determine the indefinite admittance matrix description of the resistor of Figure 4-3a.

It is apparent from the figure that

$$I_1^0 = I_2^0 = 0$$

because of absence of independent internal sources. Determination of the coefficients $\hat{y}_{ij}$ is illustrated in Figure 4-4. We note that $\hat{y}_{11}$ and $\hat{y}_{21}$ is determined by setting $V_2 = 0$. From Figure 4-4a we thus obtain

$$\hat{y}_{11} = \left. \frac{I_1}{V_1} \right|_{V_2 = 0} = \frac{1}{R}$$

$$\hat{y}_{21} = \left. \frac{I_2}{V_1} \right|_{V_2 = 0} = -\frac{1}{R}$$

---

[3] The overscript "^" on the admittance coefficient is to emphasize that these parameters are different from the more well-known short-circuit admittance coefficients.

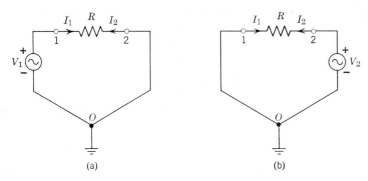

Figure 4-4 On the calculation of the indefinite admittance matrix coefficients of a resistive 2-pole.

Similarly, $\hat{y}_{12}$ and $\hat{y}_{22}$ are determined by shorting terminal 1 to ground (Figure 4-4b), and thus

$$\hat{y}_{12} = \frac{I_1}{V_2}\bigg|_{V_1=0} = -\frac{1}{R}$$

$$\hat{y}_{22} = \frac{I_2}{V_2}\bigg|_{V_1=0} = \frac{1}{R}$$

The complete indefinite description is thus given as

$$\begin{bmatrix} I_1 \\ I_2 \end{bmatrix} = \begin{bmatrix} \frac{1}{R} & -\frac{1}{R} \\ -\frac{1}{R} & \frac{1}{R} \end{bmatrix} \begin{bmatrix} V_1 \\ V_2 \end{bmatrix}$$

We note from above that for this two-pole,

$$\hat{y}_{11} + \hat{y}_{12} = 0$$
$$\hat{y}_{21} + \hat{y}_{22} = 0$$
$$\hat{y}_{11} + \hat{y}_{21} = 0$$
$$\hat{y}_{12} + \hat{y}_{22} = 0$$

A rigorous proof of the above property of the elements of the indefinite admittance matrix is given below.

Two very important properties of the indefinite admittance matrix are next derived. Adding the three equations of (4.2) and employing Equations 4.1 and 4.4, we arrive at

(4.8)

$$(\hat{y}_{11} + \hat{y}_{21} + \hat{y}_{31})V_1 + (\hat{y}_{12} + \hat{y}_{22} + \hat{y}_{32})V_2 + (\hat{y}_{13} + \hat{y}_{23} + \hat{y}_{33})V_3 = 0$$

Equation 4.8 holds for all values of V_1, V_2, and V_3, which implies that

(4.9)
$$\hat{y}_{11} + \hat{y}_{21} + \hat{y}_{31} = 0$$
$$\hat{y}_{12} + \hat{y}_{22} + \hat{y}_{32} = 0$$
$$\hat{y}_{13} + \hat{y}_{23} + \hat{y}_{33} = 0$$

Let us now consider the second equation of (4.2). Postulate ii in effect means

(4.10) $\quad I_2 = \hat{y}_{21}(V_1 + V_0) + \hat{y}_{22}(V_2 + V_0) + \hat{y}_{23}(V_3 + V_0) + I_2^0$

Subtracting the second equation of (4.2) from Equation 4.10, we obtain

(4.11) $\quad (\hat{y}_{21} + \hat{y}_{22} + \hat{y}_{23})V_0 = 0$

which must hold for all values of V_0. This indicates that

(4.12a) $\quad \hat{y}_{21} + \hat{y}_{22} + \hat{y}_{23} = 0$

Similar reasoning will show that

(4.12b)
$$\hat{y}_{11} + \hat{y}_{12} + \hat{y}_{13} = 0$$
$$\hat{y}_{31} + \hat{y}_{32} + \hat{y}_{33} = 0$$

Thus we have established the two major properties of the indefinite admittance matrix: the sum of the elements in any row and sum of the elements in any column is zero. As a consequence of the zero-sum property, the admittance matrix is singular.

Let us denote by $\hat{\mathbf{Y}}_k^j$ the cofactor of the (j, k)th element of $\hat{\mathbf{Y}}$ or, in other words, $(-1)^{j+k}\hat{\mathbf{Y}}_k^j$ is the minor obtained by deleting the jth row and the kth column of the admittance matrix. We shall now show that for a three pole the first cofactors are identical, i.e.,

$$\hat{\mathbf{Y}}_k^j = \hat{\mathbf{Y}}_3^3, \quad \begin{array}{l} j = 1, 2, 3 \\ k = 1, 2, 3 \end{array}$$

Now,

$$\hat{\mathbf{Y}}_1^1 = \begin{vmatrix} \hat{y}_{22} & \hat{y}_{23} \\ \hat{y}_{32} & \hat{y}_{33} \end{vmatrix} = -\begin{vmatrix} \hat{y}_{22} & \hat{y}_{21} \\ \hat{y}_{32} & \hat{y}_{31} \end{vmatrix}$$ [Adding first column to the second and using Equation 4.12]

$$= \begin{vmatrix} \hat{y}_{22} & \hat{y}_{21} \\ \hat{y}_{12} & \hat{y}_{11} \end{vmatrix}$$ [Adding first row to second and using Equation 4.9]

$$= -\begin{vmatrix} \hat{y}_{12} & \hat{y}_{11} \\ \hat{y}_{22} & \hat{y}_{21} \end{vmatrix} = \begin{vmatrix} \hat{y}_{11} & \hat{y}_{12} \\ \hat{y}_{21} & \hat{y}_{22} \end{vmatrix} = \hat{\mathbf{Y}}_3^3$$

In the same way we obtain

$$\hat{Y}_2^1 = -\begin{vmatrix} \hat{y}_{21} & \hat{y}_{23} \\ \hat{y}_{31} & \hat{y}_{33} \end{vmatrix} = \begin{vmatrix} \hat{y}_{21} & \hat{y}_{22} \\ \hat{y}_{31} & \hat{y}_{32} \end{vmatrix} = -\begin{vmatrix} \hat{y}_{21} & \hat{y}_{22} \\ \hat{y}_{11} & \hat{y}_{12} \end{vmatrix} = \begin{vmatrix} \hat{y}_{11} & \hat{y}_{12} \\ \hat{y}_{21} & \hat{y}_{22} \end{vmatrix} = \hat{Y}_3^3$$

Similarly, the other first cofactors can be shown to be equal to $\hat{Y}_3^3$.

The previous results for a three pole can easily be generalized to the case of an n pole. Before concluding this section, let us summarize the properties of the indefinite admittance matrix:

1. It is a zero-sum matrix, i.e., the sum of the elements in each row (column) is zero.
2. Determinant of the matrix is zero.
3. The first cofactors are equal to each other.

For a general proof of the third property, the reader is referred to Weinberg.[4]

4-2 ELEMENTARY OPERATIONS

Several operations that enable the designer to manipulate the analysis of circuits to his advantage can be performed on a multipole.

Specification of a Voltage Reference Terminal

The first operation is concerned with the relation between the admittance matrix of an n pole and the corresponding description when one of its terminals is grounded. Note that a grounded n pole can be considered as an $(n-1)$-port. We are interested in the relation between an n pole in its indefinite form and its definite form.

As before, let us establish the relation for the case of a three pole (Figure 4-2). Suppose terminal 3 is grounded. Since the three equations of (4.2) are linearly dependent, we can remove the last equation. Setting V_3 equal to zero in the remaining equations of (4.2), we arrive at the desired description of the grounded three pole (Figure 4-5):

(4.13)
$$I_1 = \hat{y}_{11}V_1 + \hat{y}_{12}V_2 + I_1^0$$
$$I_2 = \hat{y}_{21}V_1 + \hat{y}_{22}V_2 + I_2^0$$

If $I_1^0 = I_2^0 = 0$, the admittance description of expression 4.13 is identical to the short-circuit admittance matrix of the three pole considered as a two-port, where terminals 1, 3 constitute one port, and terminals 2, 3 constitute the other port. This process can be reversed. From the grounded

---
[4] L. Weinberg, *Network analysis and synthesis*, McGraw-Hill, New York, p. 57 (1962).

124 / ANALYSIS OF ACTIVE NETWORKS

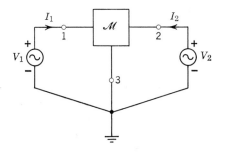

Figure 4-5 A 3-pole with its terminal 3 considered as the reference node.

port description of a three pole on the admittance basis, the indefinite admittance matrix can be obtained by adding one extra row and one extra column to satisfy Equations 4.9 and 4.12. Similarly, I_3^0 is chosen to satisfy Equation 4.4. The indefinite admittance matrix description of the three-pole is then given as:

$$(4.14) \quad \begin{bmatrix} I_1 \\ I_2 \\ I_3 \end{bmatrix} = \begin{bmatrix} \hat{y}_{11} & \hat{y}_{12} & -\hat{y}_{11} - \hat{y}_{12} \\ \hat{y}_{21} & \hat{y}_{22} & -\hat{y}_{21} - \hat{y}_{22} \\ -\hat{y}_{11} - \hat{y}_{21} & -\hat{y}_{12} - \hat{y}_{22} & (\hat{y}_{11} + \hat{y}_{12} + \hat{y}_{21} + \hat{y}_{22}) \end{bmatrix} \begin{bmatrix} V_1 \\ V_2 \\ V_3 \end{bmatrix}$$

The following two examples will illustrate the above ideas.

Example 4-2 We are interested in determining the indefinite admittance matrix of the transistor. The equivalent representation of a biased transistor, considered as a two-port with the base terminal as the common terminal, is shown in Figure 2-45b. This equivalent circuit is seen to be identical in form to that of Figure 2-37c. Hence, we make the following identifications:

$$z_{11} - z_{12} = r_e$$

$$z_{22} - z_{12} = r_c$$

$$z_{12} = r_b$$

$$\frac{z_{21} - z_{12}}{z_{22} - z_{12}} = \alpha$$

Thus the z-matrix of the common-base transistor is given as

$$\begin{bmatrix} z_{11} & z_{12} \\ z_{21} & z_{22} \end{bmatrix} = \begin{bmatrix} r_b + r_e & r_b \\ \alpha r_c + r_b & r_b + r_c \end{bmatrix}$$

ELEMENTARY OPERATIONS / 125

The next step is to calculate the short-circuit admittance matrix from the z-matrix given above, using the conversion table of Appendix A. This yields

(4.15)
$$\begin{bmatrix} y_{11} & y_{12} \\ y_{21} & y_{22} \end{bmatrix} = \begin{bmatrix} z_{11} & z_{12} \\ z_{21} & z_{22} \end{bmatrix}^{-1}$$

$$= \frac{1}{r_b r_e + r_e r_c + (1 - \alpha) r_b r_c} \begin{bmatrix} r_b + r_c & -r_b \\ -\alpha r_c - r_b & r_b + r_e \end{bmatrix}$$

The desired indefinite admittance matrix (expression 4.16) of the transistor

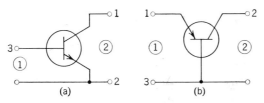

Figure 4-6 Transistor in its common-emitter and common-base orientation.

is obtained from expression 4.15 by adding a third column and a third row satisfying the zero-sum property:

(4.16)
$$\frac{1}{r_b r_e + r_e r_c + (1 - \alpha) r_b r_c} \begin{bmatrix} r_b + r_c & -r_b & -r_c \\ -\alpha r_c - r_b & r_b + r_e & \alpha r_c - r_e \\ (\alpha - 1) r_c & -r_e & r_e - (\alpha - 1) r_c \end{bmatrix} \begin{matrix} E \\ C \\ B \end{matrix}$$
$$\quad\quad\quad E \quad\quad\quad C \quad\quad\quad B$$

For clarity, the rows and columns corresponding to the various terminals of the transistor are marked in expression 4.16.

Example 4-3 A biased common emitter transistor (Figure 4-6a) can be described by an h-matrix for small-signal analysis:

(4.17)
$$[h_e] = \begin{bmatrix} h_{11e} & h_{12e} \\ h_{21e} & h_{22e} \end{bmatrix}$$

Determine the h-parameters of the same transistor in the common-base orientation (Figure 4-6b) in terms of $[h_e]$.

Using the conversion table, we first obtain the short-circuit admittance matrix from expression 4.17:

$$(4.18) \quad [y_e] = \begin{bmatrix} \dfrac{1}{h_{11e}} & -\dfrac{h_{12e}}{h_{11e}} \\ \dfrac{h_{21e}}{h_{11e}} & -\dfrac{\Delta h_e}{h_{11e}} \end{bmatrix} \begin{matrix} B \\ \\ C \end{matrix}$$

$$\begin{matrix} B & \quad C \end{matrix}$$

where $\Delta h_e = h_{11e}h_{22e} - h_{12e}h_{21e}$.

If we consider the transistor as a three pole, its indefinite admittance matrix is readily obtained from Equation 4.18 by adding a third row and a third column so that the overall matrix has the zero-sum property:

$$(4.19) \quad \hat{Y} = \begin{bmatrix} \dfrac{1}{h_{11e}} & -\dfrac{h_{12e}}{h_{11e}} & \dfrac{h_{12e}-1}{h_{11e}} \\ \dfrac{h_{21e}}{h_{11e}} & -\dfrac{\Delta h_e}{h_{11e}} & -\dfrac{h_{21e}+\Delta h_e}{h_{11e}} \\ \dfrac{-1-h_{21e}}{h_{11e}} & \dfrac{h_{12e}-\Delta h_e}{h_{11e}} & \dfrac{1+h_{21e}+\Delta h_e-h_{12e}}{h_{11e}} \end{bmatrix} \begin{matrix} B \\ \\ C \\ \\ E \end{matrix}$$

$$\begin{matrix} B & \qquad C & \qquad\quad E \end{matrix}$$

In Equations 4.18 and 4.19 the rows and columns associated with each terminal have been marked for clarity. From Equation 4.19, the y matrix of grounded-base transistor is obtained by deleting the row and the column that correspond to the base terminal. The result, after rearrangement, is as follows:

$$(4.20) \quad [y_b] = \begin{bmatrix} \dfrac{1+h_{21e}+\Delta h_e-h_{12e}}{h_{11e}} & \dfrac{h_{12e}-\Delta h_e}{h_{11e}} \\ \dfrac{-h_{21e}-\Delta h_e}{h_{11e}} & \dfrac{\Delta h_e}{h_{11e}} \end{bmatrix} \begin{matrix} E \\ \\ C \end{matrix}$$

$$\begin{matrix} E & \qquad\quad C \end{matrix}$$

from which (again using the conversion table) the desired answer is obtained:

$$(4.21) \quad [h_b] = \begin{bmatrix} \dfrac{h_{11e}}{1+h_{21e}+\Delta h_e-h_{12e}} & \dfrac{\Delta h_e-h_{12e}}{1+h_{21e}+\Delta h_e-h_{12e}} \\ \dfrac{-h_{21e}-\Delta h_e}{1+h_{21e}+\Delta h_e-h_{12e}} & \dfrac{h_{22e}}{1+h_{21e}+\Delta h_e-h_{12e}} \end{bmatrix}$$

In a typical transistor,

$$h_{21e} \gg \Delta h_e$$

$$h_{21e} \gg h_{12e}$$

Thus Equation 4.21 reduces to:

(4.22) $$[h_b] \cong \begin{bmatrix} \dfrac{h_{11e}}{1 + h_{21e}} & \dfrac{h_{11e}h_{22e}}{1 + h_{21e}} - h_{12e} \\ -\dfrac{h_{21e}}{1 + h_{21e}} & \dfrac{h_{22e}}{1 + h_{21e}} \end{bmatrix}$$

which is fairly accurate for many applications.

Consider, for example, the following values for the h parameters in the grounded-emitter orientation of GE2N525 at some operating point:

$$h_{11e} = 1400 \; \Omega$$
$$h_{12e} = 3.37 \times 10^{-4}$$
$$h_{21e} = 44$$
$$h_{22e} = 27 \times 10^{-6} \text{ mhos}$$

Use of Equation 4.22 gives the h parameters for the grounded-base operation as

$$h_{11b} = 31.11 \; \Omega$$
$$h_{12b} = 5.03 \times 10^{-4}$$
$$h_{21b} = -0.978$$
$$h_{22b} = 0.6 \times 10^{-6} \text{ mhos}$$

Reduction of a Multipole

In many applications, it is often convenient to reduce the order of a multipole. The indefinite admittance matrix of an $(n - m)$ pole can be derived from the admittance description of the original n pole. There are two operations that lead to a reduced multipole: (1) *contraction* and (2) *suppression*.

Contraction of a multipole is the joining of two or more of its terminals to form a single terminal (Figure 4-7). This union forces the potential of the connected terminals to be equal. Consider the four pole whose terminals 3 and 4 have been connected together to form a three pole. The

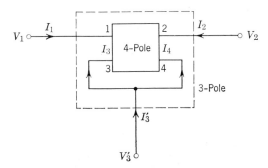

Figure 4-7 Illustration of contraction of terminals of a multipole.

indefinite description of the four pole is

(4.23)
$$I_1 = \hat{y}_{11}V_1 + \hat{y}_{12}V_2 + \hat{y}_{13}V_3 + \hat{y}_{14}V_4 + I_1^0$$
$$I_2 = \hat{y}_{21}V_1 + \hat{y}_{22}V_2 + \hat{y}_{23}V_3 + \hat{y}_{24}V_4 + I_2^0$$
$$I_3 = \hat{y}_{31}V_1 + \hat{y}_{32}V_2 + \hat{y}_{33}V_3 + \hat{y}_{34}V_4 + I_3^0$$
$$I_4 = \hat{y}_{41}V_1 + \hat{y}_{42}V_2 + \hat{y}_{43}V_3 + \hat{y}_{44}V_4 + I_4^0$$

Contraction of terminals 3 and 4 to constitute a new terminal 3′ implies

(4.24)
$$I_3' = I_3 + I_4$$
$$I_3^{0'} = I_3^0 + I_4^0$$
$$V_3' = V_3 = V_4$$

Adding the last two equations of (4.23) and employing Equation 4.24, we obtain the desired description of the new three pole as:

(4.25)
$$I_1 = \hat{y}_{11}V_1 + \hat{y}_{12}V_2 + (\hat{y}_{13} + \hat{y}_{14})V_3' + I_1^0$$
$$I_2 = \hat{y}_{21}V_1 + \hat{y}_{22}V_2 + (\hat{y}_{23} + \hat{y}_{24})V_3' + I_2^0$$
$$I_3' = (\hat{y}_{31} + \hat{y}_{41})V_1 + (\hat{y}_{32} + \hat{y}_{42})V_2$$
$$+ (\hat{y}_{33} + \hat{y}_{34} + \hat{y}_{43} + \hat{y}_{44})V_3' + I_3^0 + I_4^0$$

Thus the indefinite admittance matrix of the new three pole is:

$$\begin{bmatrix} \hat{y}_{11} & \hat{y}_{12} & \hat{y}_{13} + \hat{y}_{14} \\ \hat{y}_{21} & \hat{y}_{22} & \hat{y}_{23} + \hat{y}_{24} \\ \hat{y}_{31} + \hat{y}_{41} & \hat{y}_{32} + \hat{y}_{42} & \hat{y}_{33} + \hat{y}_{34} + \hat{y}_{43} + \hat{y}_{44} \end{bmatrix}$$

The result can now be generalized: the indefinite admittance matrix of a contracted multipole is obtained by adding the respective elements of the rows and the columns corresponding to the terminals being joined together.

Suppression of a multipole is the operation by which some terminals are made inaccessible. Thus, the currents associated with the suppressed terminals are zero. For illustration purposes, consider the two pole of Figure 4-8 obtained by suppressing terminal 3 of a three pole. In the

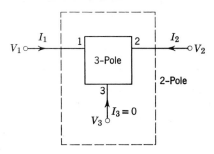

Figure 4-8 Illustration of suppression of terminals of a multipole.

indefinite admittance description of the three pole given by Equation 4.2, we set I_3 equal to zero. Solving for V_3 we obtain

(4.26) $$V_3 = -\frac{\hat{y}_{31}}{\hat{y}_{33}} V_1 - \frac{\hat{y}_{32}}{\hat{y}_{33}} V_2 - \frac{I_3^0}{\hat{y}_{33}}$$

assuming $\hat{y}_{33}$ is not equal to zero. By substituting the above expression for V_3 in the first two equations of Equation 4.2 and rearranging the results, we finally obtain the desired admittance description for the suppressed three pole:

(4.27)
$$I_1 = \left(\hat{y}_{11} - \frac{\hat{y}_{31}\hat{y}_{13}}{\hat{y}_{33}}\right) V_1 + \left(\hat{y}_{12} - \frac{\hat{y}_{32}\hat{y}_{13}}{\hat{y}_{33}}\right) V_2 + \left(I_1^0 - \frac{\hat{y}_{13}}{\hat{y}_{33}} I_3^0\right)$$
$$I_2 = \left(\hat{y}_{21} - \frac{\hat{y}_{23}\hat{y}_{31}}{\hat{y}_{33}}\right) V_1 + \left(\hat{y}_{22} - \frac{\hat{y}_{23}\hat{y}_{32}}{\hat{y}_{33}}\right) V_2 + \left(I_2^0 - \frac{\hat{y}_{23}}{\hat{y}_{33}} I_3^0\right)$$

Suppression of more than one terminal can be considered as a series of suppressions each involving a single terminal and, proceeding as above, we can derive the resultant admittance description. A compact form for the general case can be derived using matrix notation. By a proper rearrangement, the admittance equations for an n pole can be written as

(4.28)
$$\mathbf{I}_1 = \hat{\mathbf{Y}}_{11}\mathbf{V}_1 + \hat{\mathbf{Y}}_{12}\mathbf{V}_2 + \mathbf{I}_1^0$$
$$\mathbf{I}_2 = \hat{\mathbf{Y}}_{21}\mathbf{V}_1 + \hat{\mathbf{Y}}_{22}\mathbf{V}_2 + \mathbf{I}_2^0$$

where $\mathbf{I}_1$ is a column vector of order k representing the currents associated with the k accessible terminals and $\mathbf{I}_2$ is a column vector of order $(n - k)$ representing the currents associated with the terminals being suppressed.

The interpretation of $\hat{\mathbf{Y}}_{ij}$, $\mathbf{V}_i$, and $\mathbf{I}_i^0$ should now be evident to the reader. Setting $\mathbf{I}_2 = 0$, we have

(4.29) $$\mathbf{V}_2 = -\hat{\mathbf{Y}}_{22}^{-1}\hat{\mathbf{Y}}_{21}\mathbf{V}_1 - \hat{\mathbf{Y}}_{22}^{-1}\mathbf{I}_2^0$$

Substitution of the above expression for $\mathbf{V}_2$ in the first equation of Equation 4.28 yields the admittance description of the reduced multipole as

(4.30) $$\mathbf{I}_1 = (\hat{\mathbf{Y}}_{11} - \hat{\mathbf{Y}}_{12}\hat{\mathbf{Y}}_{22}^{-1}\hat{\mathbf{Y}}_{21})\mathbf{V}_1 + (\mathbf{I}_1^0 - \hat{\mathbf{Y}}_{12}\hat{\mathbf{Y}}_{22}^{-1}\mathbf{I}_2^0)$$

where $(\hat{\mathbf{Y}}_{11} - \hat{\mathbf{Y}}_{12}\hat{\mathbf{Y}}_{22}^{-1}\hat{\mathbf{Y}}_{21})$ is the new indefinite admittance matrix. The reader should compare Equations 4.30 and 4.27. In some cases, $\hat{\mathbf{Y}}_{22}$ may be singular. Then this one step reduction procedure may not be followed. However, suppression of one terminal at a time may work.

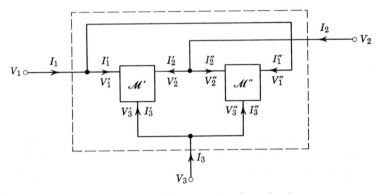

Figure 4-9 Parallel connection of two 3-poles.

Parallel Connection of Multipoles

In some cases, it is easier to write down the indefinite admittance matrix of a multipole by considering it as a parallel connection of simpler multipoles whose admittance characterization can be found by inspection. Consider the three pole obtained by connecting two three poles $\mathcal{M}'$ and $\mathcal{M}''$ in parallel as shown in Figure 4-9. Multipoles $\mathcal{M}'$ and $\mathcal{M}''$ are described by the following equations:

(4.31) $$\begin{aligned} I_1' &= \hat{y}_{11}'V_1' + \hat{y}_{12}'V_2' + \hat{y}_{13}'V_3' + I_1^{0'} \\ I_2' &= \hat{y}_{21}'V_1' + \hat{y}_{22}'V_2' + \hat{y}_{23}'V_3' + I_2^{0'} \\ I_3' &= \hat{y}_{31}'V_1' + \hat{y}_{32}'V_2' + \hat{y}_{33}'V_3' + I_3^{0'} \end{aligned}$$

(4.32) $$\begin{aligned} I_1'' &= \hat{y}_{11}''V_1'' + \hat{y}_{12}''V_2'' + \hat{y}_{13}''V_3'' + I_1^{0''} \\ I_2'' &= \hat{y}_{21}''V_1'' + \hat{y}_{22}''V_2'' + \hat{y}_{23}''V_3'' + I_2^{0''} \\ I_3'' &= \hat{y}_{31}''V_1'' + \hat{y}_{32}''V_2'' + \hat{y}_{33}''V_3'' + I_3^{0''} \end{aligned}$$

ELEMENTARY OPERATIONS / 131

The terminal variables of the composite three pole are related to the same of the original three poles,

$$I_1 = I'_1 + I''_1; \qquad I_2 = I'_2 + I''_2; \qquad I_3 = I'_3 + I''_3$$
(4.33) $\quad I_1{}^0 = I_1{}^{0'} + I_1{}^{0''}; \qquad I_2{}^0 = I_2{}^{0'} + I_2{}^{0''}; \qquad I_3{}^0 = I_3{}^{0'} + I_3{}^{0''}$
$$V_1 = V'_1 = V''_1; \qquad V_2 = V'_2 = V''_2; \qquad V_3 = V'_3 = V''_3$$

Adding the proper equations of expressions 4.31 and 4.32 and employing Equation 4.33, the admittance description of the new three pole can be stated

$$I_1 = (\hat{y}'_{11} + \hat{y}''_{11})V_1 + (\hat{y}'_{12} + \hat{y}''_{12})V_2 + (\hat{y}'_{13} + \hat{y}''_{13})V_3 + (I_1{}^{0'} + I_1{}^{0''})$$
(4.34) $\quad I_2 = (\hat{y}'_{21} + \hat{y}''_{21})V_1 + (\hat{y}'_{22} + \hat{y}''_{22})V_2 + (\hat{y}'_{23} + \hat{y}''_{23})V_3 + (I_2{}^{0'} + I_2{}^{0''})$
$$I_3 = (\hat{y}'_{31} + \hat{y}''_{31})V_1 + (\hat{y}'_{32} + \hat{y}''_{32})V_2 + (\hat{y}'_{33} + \hat{y}''_{33})V_3 + (I_3{}^{0'} + I_3{}^{0''})$$

In general, if m multipoles are connected in parallel, the admittance description of the composite multiple is given in a matrix form as

(4.35) $$\mathbf{I} = \left[\sum_{k=1}^{m} \hat{\mathbf{Y}}^{(k)}\right] \mathbf{V} + \left[\sum_{k=1}^{m} \mathbf{I}^{0(k)}\right]$$

A p pole, where $p < n$, can be considered as an n pole by adding $(n - p)$ terminals to the p pole; such additional terminals are floating, i.e., not connected to any of the original terminals of the p pole. In the admittance description, this amounts to adding rows and columns of zeros at appropriate places. For example, a three pole considered as a four pole (Figure 4-10) will have the following description on the admittance basis:

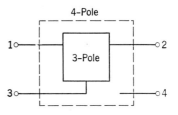

Figure 4-10 A 3-pole considered as a 4-pole.

(4.36) $$\begin{bmatrix} I_1 \\ I_2 \\ I_3 \\ I_4 \end{bmatrix} = \begin{bmatrix} \hat{y}_{11} & \hat{y}_{12} & \hat{y}_{13} & 0 \\ \hat{y}_{21} & \hat{y}_{22} & \hat{y}_{23} & 0 \\ \hat{y}_{31} & \hat{y}_{32} & \hat{y}_{33} & 0 \\ 0 & 0 & 0 & 0 \end{bmatrix} \begin{bmatrix} V_1 \\ V_2 \\ V_3 \\ V_4 \end{bmatrix} + \begin{bmatrix} I_1{}^0 \\ I_2{}^0 \\ I_3{}^0 \\ 0 \end{bmatrix}$$

From the above discussion, it is clear that multipoles having an unequal number of terminals can be connected in parallel.

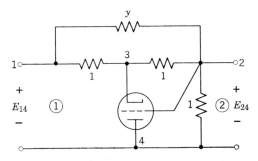

Figure 4-11 An active two-port (Example 4-4).

The following example will illustrate the various elementary operations discussed in this section.

Example 4-4[5] Suppose we are interested in the short-circuit admittance matrix of the two-port of Figure 4-11.

The first step in solving this problem will be to obtain the indefinite admittance matrix of the two-port considered as a four pole, where the terminals are indicated by the numbered nodes in the figure. Since the composite four pole can be obtained by connecting in parallel the five elementary four poles shown in Figure 4-12, the problem reduces to deriving the indefinite admittance description of the elementary four poles. The

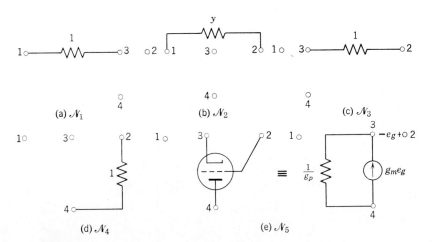

Figure 4-12 Constituent multipoles of the active two-port of Figure 4-11.

[5] Adapted from L. A. Zadeh, *loc. cit.*

respective admittance matrices, which can be written by inspection, are

(4.37)
$$\hat{\mathbf{Y}}^{(1)} = \begin{bmatrix} 1 & 0 & -1 & 0 \\ 0 & 0 & 0 & 0 \\ -1 & 0 & 1 & 0 \\ 0 & 0 & 0 & 0 \end{bmatrix}; \quad \hat{\mathbf{Y}}^{(2)} = \begin{bmatrix} y & -y & 0 & 0 \\ -y & y & 0 & 0 \\ 0 & 0 & 0 & 0 \\ 0 & 0 & 0 & 0 \end{bmatrix}$$

$$\hat{\mathbf{Y}}^{(3)} = \begin{bmatrix} 0 & 0 & 0 & 0 \\ 0 & 1 & -1 & 0 \\ 0 & -1 & 1 & 0 \\ 0 & 0 & 0 & 0 \end{bmatrix}; \quad \hat{\mathbf{Y}}^{(4)} = \begin{bmatrix} 0 & 0 & 0 & 0 \\ 0 & 1 & 0 & -1 \\ 0 & 0 & 0 & 0 \\ 0 & -1 & 0 & 1 \end{bmatrix}$$

$$\hat{\mathbf{Y}}^{(5)} = \begin{bmatrix} 0 & 0 & 0 & 0 \\ 0 & 0 & 0 & 0 \\ 0 & -g_m & g_p + g_m & -g_m \\ 0 & g_m & -g_p - g_m & g_p \end{bmatrix}$$

where $\hat{\mathbf{Y}}^{(k)}$ is the indefinite admittance matrix of the four pole $\mathcal{N}_k$. $\hat{\mathbf{Y}}^{(5)}$ has been obtained after replacing the triode by its low-frequency equivalent circuit (Figure 4-12e). Adding the matrices of (4.37), we obtain the indefinite description of the composite four pole of Figure 4-11.

(4.38)
$$\hat{\mathbf{Y}} = \sum_{k=1}^{5} \hat{\mathbf{Y}}^{(k)} = \begin{bmatrix} 1+y & -y & -1 & 0 \\ -y & 2+y & -1 & -1 \\ -1 & -1-g_m & 2+g_p+g_m & -g_p \\ 0 & -1+g_m & -g_p-g_m & 1-g_p \end{bmatrix}$$

Grounding terminal 4 implies deleting the fourth row and fourth column of Equation 4.38. Such action yields the short-circuit admittance matrix of the three-port whose ports are the terminal pairs 1-4, 2-4, and 3-4:

(4.39)
$$\begin{bmatrix} 1+y & -y & -1 \\ -y & 2+y & -1 \\ \hdashline -1 & -1-g_m & 2+g_p+g_m \end{bmatrix}$$

Next, partition the matrix of Equation 4.39 to identify:

(4.40)
$$\mathbf{Y}_{11} = \begin{bmatrix} 1+y & -y \\ -y & 2+y \end{bmatrix}; \quad \mathbf{Y}_{12} = \begin{bmatrix} -1 \\ -1 \end{bmatrix}$$
$$\mathbf{Y}_{21} = [-1 \quad -1-g_m]; \quad \mathbf{Y}_{22} = [2+g_p+g_m]$$

Thus the short-circuit admittance matrix obtained by suppressing port 3 is given by

(4.41)
$$\mathbf{Y}_{11} - \mathbf{Y}_{12}\mathbf{Y}_{22}^{-1}\mathbf{Y}_{21}$$
$$= \begin{bmatrix} 1+y & -y \\ -y & 2+y \end{bmatrix} - \frac{1}{(2+g_p+g_m)} \begin{bmatrix} -1 \\ -1 \end{bmatrix} \begin{bmatrix} -1 & -1-g_m \end{bmatrix}$$
$$= \frac{1}{(2+g_p+g_m)} \begin{bmatrix} \{(1+y)(2+g_p+g_m)-1\} & \{-y(2+g_p+g_m)-(1+g_m)\} \\ \{-y(2+g_p+g_m)-1\} & \{(2+y)(2+g_p+g_m)-(1+g_m)\} \end{bmatrix}$$

It should be noted that, for suppression on the admittance matrix for the three-port, we have used the same formula as was derived for the multipole suppression. The reader can easily verify that the two expressions are identical in form.

An alternate approach to this problem would have been to first suppress terminal 3 and then ground terminal 4. It can be shown that an identical result is obtained in this second approach.

4-3 CLASSIFICATION OF MULTIPOLES

Network elements are usually classified as active or passive elements. As mentioned in Chapter 6, the activity or passivity of network elements is defined in terms of the ability or inability to supply power to a load connected across it. A more general and often used useful classification of multipoles whereby they are categorized as self-generative or nongenerative types is given below.

DEFINITION 4-1   *A multiterminal element is defined as a* self-generative *multipole if all the terminal currents are not identically zero when the terminals are shorted together.*

In the admittance description (4.7), at least one of the I_k^0 is different from zero for a self-generative multipole indicating the presence of internal current (voltage) sources within the multipole that are independent of terminal variables.

DEFINITION 4-2   *A* nongenerative *multipole is defined by the condition of vanishing terminal currents with vanishing potential difference between each pair of terminals.*

Definition 4-2 implies that $I^0 = 0$ for such a multipole, i.e., the indefinite admittance description is given as

(4.42) $$I = \hat{Y} \cdot V$$

In most cases we shall be concerned with the small signal a-c operation of an electronic circuit, i.e., the network will be studied after neglecting the bias and power supplies. Under these conditions, the circuit becomes a connection of nongenerative multipoles. If the supplies are considered, then some multipoles will become a self-generative type. For example, an ideal negative resistance for multipole analysis is a nongenerative two pole;

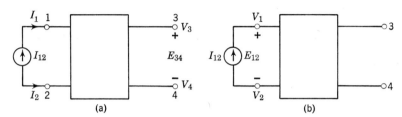

Figure 4-13

however, a positive resistance in series with a battery can be considered as a self-generative two pole.

4-4 NETWORK FUNCTIONS OF MULTIPOLE

The ultimate aim of network analysis is to derive an expression for some network function in terms of the element values of the circuit. The zero-sum property of the indefinite admittance matrix makes it possible to express various network functions in simple compact forms. It is tacitly assumed that the multipoles are of the nongenerative type.

The development of the necessary formulas is best illustrated by considering a four pole (Figure 4-13). Initially, we are interested in computing the transfer impedance E_{34}/I_{12}, where E_{34} is defined as the voltage across terminals 3 and 4 when a current source I_{12} is connected across terminals 1 and 2 (Figure 4-13a). In terms of the terminal variables,

(4.43)
$$I_{12} = I_1 = -I_2$$
$$E_{34} = V_3$$
$$V_4 = 0$$
$$I_3 = I_4 = 0$$

Substituting the above in the indefinite admittance matrix description of the four pole and deleting the fourth row and fourth column of matrix (since terminal 4 is the reference), we arrive at

$$(4.44) \quad \begin{bmatrix} I_{12} \\ -I_{12} \\ 0 \end{bmatrix} = \begin{bmatrix} \hat{y}_{11} & \hat{y}_{12} & \hat{y}_{13} \\ \hat{y}_{21} & \hat{y}_{22} & \hat{y}_{23} \\ \hat{y}_{31} & \hat{y}_{32} & \hat{y}_{33} \end{bmatrix} \begin{bmatrix} V_1 \\ V_2 \\ E_{34} \end{bmatrix}$$

Solving the above for E_{34}:

$$E_{34} = \frac{\begin{vmatrix} \hat{y}_{11} & \hat{y}_{12} & I_{12} \\ \hat{y}_{21} & \hat{y}_{22} & -I_{12} \\ \hat{y}_{31} & \hat{y}_{32} & 0 \end{vmatrix}}{\hat{Y}_4^4} = \frac{\{(\hat{y}_{21}\hat{y}_{32} - \hat{y}_{31}\hat{y}_{22}) + (\hat{y}_{11}\hat{y}_{32} - \hat{y}_{31}\hat{y}_{12})\}I_{12}}{\hat{Y}_4^4}$$

Therefore,

$$(4.45) \quad \frac{E_{34}}{I_{12}} = \frac{\hat{y}_{32}(\hat{y}_{11} + \hat{y}_{21}) - \hat{y}_{31}(\hat{y}_{12} + \hat{y}_{22})}{\hat{Y}_4^4}$$

Zero-sum property implies

$$(4.46) \quad \begin{aligned} \hat{y}_{11} + \hat{y}_{21} &= -\hat{y}_{31} - \hat{y}_{41} \\ \hat{y}_{12} + \hat{y}_{22} &= -\hat{y}_{32} - \hat{y}_{42} \end{aligned}$$

Use of Equation 4.46 in Equation 4.45 results in

$$(4.47) \quad \frac{E_{34}}{I_{12}} = \frac{\hat{y}_{31}\hat{y}_{42} - \hat{y}_{32}\hat{y}_{41}}{\hat{Y}_4^4}$$

For convenience, let us denote by $\hat{Y}_{ij}^{mn}$ the second cofactor of $\hat{Y}$ obtained by deleting the mth row, nth row, ith column, and jth column of $\hat{Y}$, i.e.,

(4.48)

$$\hat{Y}_{ij}^{mn} = (-1)^{m+n+i+j} \times \begin{Bmatrix} \text{Determinant of the submatrix obtained by} \\ \text{omitting the } m\text{th row, } n\text{th row, } i\text{th column} \\ \text{and } j\text{th column of } \hat{Y}. \end{Bmatrix}$$

Using the above notation, Equation 4.47 can be rewritten as

$$(4.49)^6 \quad \frac{E_{34}}{I_{12}} = \frac{\begin{vmatrix} \hat{y}_{31} & \hat{y}_{32} \\ \hat{y}_{41} & \hat{y}_{42} \end{vmatrix}}{\hat{Y}_4^4} = \text{sgn}(1-2)\,\text{sgn}(3-4)\,\frac{\hat{Y}_{34}^{12}}{\hat{Y}_4^4}$$

[6] $\text{sgn}(x) = 1$ if $x > 0$ and $\text{sgn}(x) = -1$ if $x < 0$.

Next, we obtain an expression for the driving-point impedance E_{12}/I_{12}, where E_{12} is the potential difference between terminals 1, 2 (Figure 4-13b). We can now consider terminal 2 to be the reference node, which leads to the following:

(4.50)
$$I_{12} = I_1 = -I_2$$
$$E_{12} = V_1$$
$$V_2 = 0$$
$$I_3 = I_4 = 0$$

By substituting Equation 4.50 in the indefinite admittance matrix description of the four pole and deleting the second column and second row, we arrive at

(4.51)
$$\begin{bmatrix} I_{12} \\ 0 \\ 0 \end{bmatrix} = \begin{bmatrix} \hat{y}_{11} & \hat{y}_{13} & \hat{y}_{14} \\ \hat{y}_{31} & \hat{y}_{33} & \hat{y}_{34} \\ \hat{y}_{41} & \hat{y}_{43} & \hat{y}_{44} \end{bmatrix} \begin{bmatrix} E_{12} \\ V_3 \\ V_4 \end{bmatrix}$$

By Cramer's rule we obtain

(4.52)
$$E_{12} = \frac{\begin{vmatrix} I_{12} & \hat{y}_{13} & \hat{y}_{14} \\ 0 & \hat{y}_{33} & \hat{y}_{34} \\ 0 & \hat{y}_{43} & \hat{y}_{44} \end{vmatrix}}{\hat{Y}_2^2}$$

Thus,

(4.53)
$$Z_{12} = \frac{E_{12}}{I_{12}} = \frac{\begin{vmatrix} \hat{y}_{33} & \hat{y}_{34} \\ \hat{y}_{43} & \hat{y}_{44} \end{vmatrix}}{\hat{Y}_4^4} = \frac{\hat{Y}_{12}^{12}}{\hat{Y}_4^4}$$

because the first cofactors of $\hat{Y}$ are all equal.

An expression for the transfer voltage ratio E_{34}/E_{12} follows from Equations 4.49 and 4.53:

(4.54)
$$\frac{E_{34}}{E_{12}} = t_v \begin{vmatrix} 34 \\ 12 \end{vmatrix} = \operatorname{sgn}(3-4)\operatorname{sgn}(1-2)\frac{\hat{Y}_{34}^{12}}{\hat{Y}_{12}^{12}}$$

Equations 4.49, 4.53, and 4.54 can be generalized for an n pole. Suppose we are interested in the transfer impedance determined by the voltage E_{ij} across terminals i, j and the current source I_{mn} across terminals m, n as shown in Figure 4-14. Then

(4.55)
$$Z_{mn}^{ij} = \frac{E_{ij}}{I_{mn}} = \operatorname{sgn}(m-n)\operatorname{sgn}(i-j)\frac{\hat{Y}_{ij}^{mn}}{\hat{Y}_n^n}$$

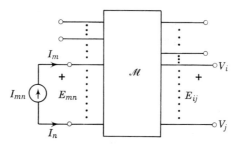

Figure 4-14

where

$$E_{ij} = V_i - V_j \quad \text{and} \quad I_{mn} = I_m = -I_n$$

If $i = m$, and $j = n$, then from Equation 4.55 we obtain the expression for the driving-point impedance looking into the multipole from terminal pair m, n:

(4.56) $$Z_{mn} = \frac{E_{mn}}{I_{mn}} = \frac{\hat{Y}_{mn}^{mn}}{\hat{Y}_n^n}$$

From Equations 4.55 and 4.56 we easily obtain the expression for the transfer voltage ratio between terminal pair i, j and terminal pair m, n:

(4.57) $$t_v \bigg|_{mn}^{ij} = \frac{E_{ij}}{E_{mn}} = \text{sgn}\,(m - n)\,\text{sgn}\,(i - j)\,\frac{\hat{Y}_{ij}^{mn}}{\hat{Y}_{mn}^{mn}}$$

It should be noted that in Equations 4.55 and 4.56, $\hat{Y}_n^n$ represents the first cofactor of $\hat{Y}$ and hence can be replaced by any other first cofactor of $\hat{Y}$. This is because the first cofactors of a zero-sum matrix are equal to each other.

A more rigorous proof of the last three formulas will be found elsewhere.[7] Let us now illustrate the usefulness of the above formulas by means of an example.

Example 4-5 Consider the active two-port of Figure 4-11. For this network, the short-circuit matrix was derived earlier and is given by Equation 4.41. From the short-circuit admittance matrix, the voltage-transfer ratio E_{24}/E_{14} is given as

(4.58) $$t_v = \frac{E_{24}}{E_{14}} = -\frac{y_{21}}{y_{22}} = \frac{y(2 + g_p + g_m) + 1}{(2 + y)(2 + g_p + g_m) - (1 + g_m)}$$

[7] G. E. Sharpe and B. Spain, "On the solution of networks by means of equi-cofactor matrix," *IRE Trans. on Circuit Theory*, **CT-7**, 230–239 (September 1960).

Alternatively, the voltage transfer ratio can be computed using Equation 4.57:

$$(4.59) \quad \frac{E_{24}}{E_{12}} = \text{sgn}(2-4)\,\text{sgn}(1-4)\,\frac{\hat{Y}_{24}^{14}}{\hat{Y}_{14}^{14}} = \frac{\hat{Y}_{24}^{14}}{\hat{Y}_{14}^{14}}$$

$$(4.60) \quad \hat{Y}_{24}^{14} = -\begin{vmatrix} -y & -1 \\ -1 & 2+g_p+g_m \end{vmatrix} = y(2+g_p+g_m)+1,$$

$$(4.61) \quad \hat{Y}_{14}^{14} = \begin{vmatrix} 2+y & -1 \\ -1-g_m & 2+g_p+g_m \end{vmatrix}$$

$$= (2+y)(2+g_p+g_m) - (1+g_m)$$

Therefore, using Equations 4.61 and 4.60 in Equation 4.59, we obtain

$$(4.62) \quad t_v = \frac{E_{24}}{E_{14}} = \frac{y(2+g_p+g_m)+1}{(2+y)(2+g_p+g_m)-(1+g_m)}$$

which as expected is the same result as Equation 4.58. The driving-point impedance at port 1 is given by

$$(4.63) \quad Z_{14} = \frac{E_{14}}{I_{14}} = \frac{\hat{Y}_{14}^{14}}{\hat{Y}_3^3} = \frac{(2+y)(2+g_p+g_m)-(1+g_m)}{(1+y)(1+2g_p+g_m)+y(1+g_p)}$$

If we let $g_m \to \infty$, i.e., replace the triode by an infinite gain VCT, then from Equation 4.63 the input impedance is

$$(4.64) \quad Z_{14} = \frac{(2+y)g_m - g_m}{(1+y)g_m} = 1$$

and the voltage transfer ratio given by Equation 4.58 reduces to

$$(4.65) \quad t_v = \frac{y}{1+y}$$

Equations 4.65 and 4.64 indicate that the two-port of Figure 4-11, under the condition $g_m \to \infty$, behaves like a constant-resistance network.

4-5 MULTIPOLE EQUIVALENT CIRCUITS

The controlled source equivalent representation of a multipole can be developed in a simple manner. First we note that any multipole of the

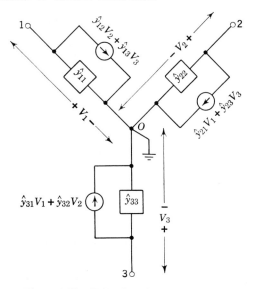

Figure 4-15 General equivalent circuit of a 3-pole.

self-generative type, described by

(4.66) $$I_k = \sum_{j=1}^{n} \hat{y}_{kj} V_j + I_k^0 \qquad k = 1, 2, \ldots, n$$

can be represented by a parallel connection of a nongenerative n pole characterized by the admittance coefficients $\hat{y}_{kj}$ and an n pole current source characterized by $\{-I_k^0\}$. Therefore, for the purpose of discussion in this section, we shall restrict our attention to multipoles of the nongenerative type.

Consider, for convenience, a nongenerative three pole described by

(4.67) $$\begin{aligned} I_1 &= \hat{y}_{11} V_1 + \hat{y}_{12} V_2 + \hat{y}_{13} V_3 \\ I_2 &= \hat{y}_{21} V_1 + \hat{y}_{22} V_2 + \hat{y}_{23} V_3 \\ I_3 &= \hat{y}_{31} V_1 + \hat{y}_{32} V_2 + \hat{y}_{33} V_3 \end{aligned}$$

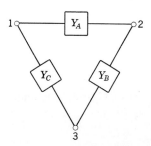

Figure 4-16 A delta network.

The equivalent circuit of such a three pole is shown in Figure 4-15, which uses six voltage-current transducers (VCT). Reduction of the number of controlled sources is necessary for a more useful representation. Let us now consider the parallel connection of the original three pole and the delta network of Figure 4-16, as shown in Figure 4-17. The reason for doing

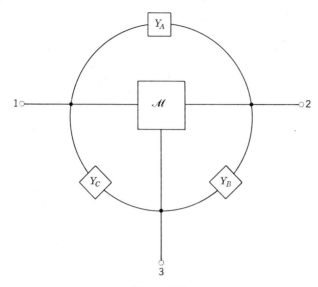

Figure 4-17

this will be clear a little later. The indefinite admittance matrix of the composite multipole of Figure 4-17 is

(4.68) $\begin{bmatrix} \hat{y}_{11} + Y_A + Y_C & \hat{y}_{12} - Y_A & \hat{y}_{13} - Y_C \\ \hat{y}_{21} - Y_A & \hat{y}_{22} + Y_B + Y_A & \hat{y}_{23} - Y_B \\ \hat{y}_{31} - Y_C & \hat{y}_{32} - Y_B & \hat{y}_{33} + Y_B + Y_C \end{bmatrix} = [\hat{y}'_{ij}]$

The representation of the above matrix will also be of the form of Figure 4-15, except that the admittance coefficients $\hat{y}_{ij}$ should be replaced by $\hat{y}'_{ij}$. In order to reduce the number of transducers, the new admittances can be chosen arbitrarily. For example, if we choose

(4.69)
$$Y_A = \hat{y}_{12}$$
$$Y_B = \hat{y}_{23}$$
$$Y_C = \hat{y}_{13}$$

then expression 4.68 becomes

(4.70) $\begin{bmatrix} 0 & 0 & 0 \\ \hat{y}_{21} - \hat{y}_{12} & \hat{y}_{22} + \hat{y}_{23} + \hat{y}_{12} & 0 \\ \hat{y}_{31} - \hat{y}_{13} & \hat{y}_{32} - \hat{y}_{23} & 0 \end{bmatrix}$

The equivalent representation of expression 4.70, derived from Figure 4-15, is shown in Figure 4-18, from which the equivalent circuit of the original

142 / ANALYSIS OF ACTIVE NETWORKS

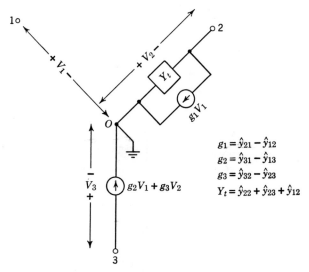

Figure 4-18

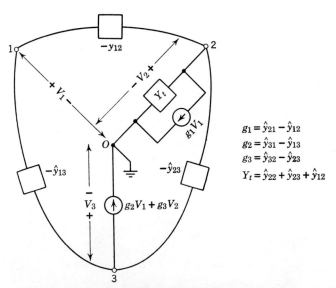

Figure 4-19 A simplified equivalent circuit of a 3-pole.

ANALYSIS OF NETWORKS CONTAINING IDEAL ACTIVE ELEMENTS / 143

three pole is obtained by connecting $-Y_A$, $-Y_B$, and $-Y_C$ between appropriate terminal pairs. The final circuit is shown in Figure 4-19,[8] which uses three transducers of the VCT type. It should be noted that the choice of the admittances Y_A, Y_B, and Y_C is by no means unique; other equivalent representations are possible.

Several observations can be made at this point. If the three pole is reciprocal, then $\hat{y}_{ij} = \hat{y}_{ji}$ and the equivalent circuit of Figure 4-19 reduces to that of Figure 4-20, which was originally proposed by Campbell.[9] Next, observe that if we consider the three pole as a grounded two-port by grounding terminal 3 (this amounts to short-circuiting terminals 3 and 0), the familiar π-equivalent of the two-port results from Figure 4-19 as shown in Figure 4-21.

Extension of the above treatment to higher order multipoles is straightforward. It can be shown, for example, that the equivalent circuit of a reciprocal nongenerative four pole is of the form of Figure 4-22; verification of this circuit is left as an exercise.

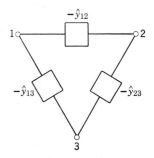

Figure 4-20 Equivalent representation of a reciprocal 3-pole.

4-6 ANALYSIS OF NETWORKS CONTAINING IDEAL ACTIVE ELEMENTS

The indefinite matrix approach can be applied to a network containing various types of multipoles, provided that the short-circuit admittance description exists for each multipole. Although the transistor and the

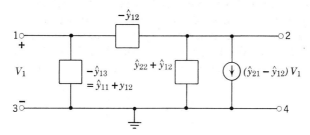

Figure 4-21 Development of the equivalent circuit of a two-port on the y-basis.

---

[8] J. L. Stewart, "An equivalent circuit for linear N-port active networks," *IRE Trans. on Circuit Theory*, **CT-6**, 234–235 (June 1959).

[9] G. A. Campbell, "Cisoidal oscillations," *Trans. AIEE*, **30**, 873–909 (April 1911).

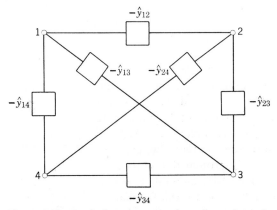

Figure 4-22 Equivalent representation of a reciprocal 4-pole.

vacuum tubes can be characterized by the admittance matrix, some of the ideal active elements discussed in Chapter 2 fail in this respect. This inadequacy of the multipole-analysis approach can be circumvented by a simple modification to be described next.

The basic idea behind the modification is that the addition of parasitics results in a nonideal active element for which the short-circuit admittance matrix exists. The original ideal device is then obtained by connecting suitably negative elements. Consider, for example, an ideal active two-port (Figure 4-23a) described by an **F** matrix of the form

(4.71) $$\begin{bmatrix} A & 0 \\ 0 & D \end{bmatrix}$$

Note that the CCT, VVT, NIC, GIC, IVC, ICC, IPC, and IT are in this class. By connecting a series combination of two resistances of values R and $-R$ ohms at the input as shown in Figure 4-23b, an equivalent network is obtained. The two-port inside the dotted box $\mathcal{N}$ in Figure 4-23b can be considered a nonideal active two-port characterized by the following y matrix

(4.72) $$\begin{bmatrix} -\dfrac{1}{R} & \dfrac{A}{R} \\ \dfrac{1}{RD} & -\dfrac{A}{RD} \end{bmatrix}$$

An alternate equivalent network is obtained by connecting parasitics at the output (Figure 4-23c) where the network $\mathcal{M}$ inside the dotted box is

ANALYSIS OF NETWORKS CONTAINING IDEAL ACTIVE ELEMENTS / 145

described by the following short-circuit admittance matrix:

(4.73)
$$\begin{bmatrix} -\dfrac{D}{AR} & \dfrac{D}{R} \\ \dfrac{1}{AR} & -\dfrac{1}{R} \end{bmatrix}$$

Any of the two modifications suggested in Figure 4-23b and c can be used in cases of all impedance converter type active elements, except for

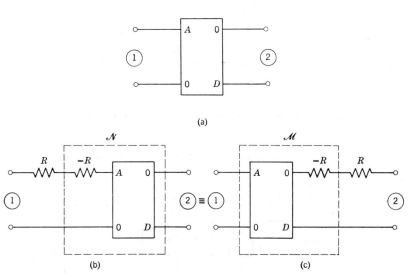

Figure 4-23 Alternate representations of an impedance converter type active device to enable admittance analysis.

the cases of the VVT and the CCT. If an ideal voltage-controlled voltage source is used, then the approach suggested in Figure 4-23c must be used (because $D = 0$). On the other hand, in case of a CCT, Figure 4-23b is the only approach (because $A = 0$).

If the network contains a CVT, then parasitics must be added to both input and output terminals (Figure 4-24). The dotted box of Figure 4-24 is a nonideal CCT having a y matrix:

(4.74)
$$\begin{bmatrix} -\dfrac{1}{R_1} & 0 \\ \dfrac{r}{R_1 R_2} & -\dfrac{1}{R_2} \end{bmatrix}$$

146 / ANALYSIS OF ACTIVE NETWORKS

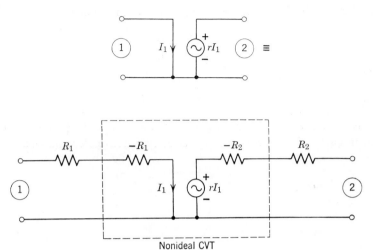

Figure 4-24 Alternate representation of the CVT to aid admittance analysis.

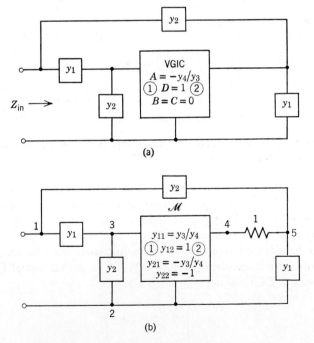

Figure 4-25 An active one-port containing a VGIC (Example 4-6).

ANALYSIS OF NETWORKS CONTAINING IDEAL ACTIVE ELEMENTS / 147

In all of the above modification schemes, any suitable numerical value of the parasitics can be chosen as dictated by the actual network.

Example 4-6 Compute the driving-point impedance of the one-port shown in Figure 4-25a.[10] Following the approach suggested by Figure 4-23c, the VGIC is replaced by its equivalent network as indicated in Figure 4-25b, where the short-circuit admittance matrix of two port $\mathcal{M}$ has been derived from Equation 4.73 by substituting $A = -y_4/y_3$, $D = R = 1$.

The indefinite admittance matrix of the network shown in Figure 4-25b can now be written by inspection:

(4.75)

$$\begin{array}{c} \\ 1 \\ 2 \\ 3 \\ 4 \\ 5 \end{array} \begin{array}{cccccc} 1 & 2 & 3 & 4 & 5 \\ \end{array}$$

$$\begin{bmatrix} y_1 + y_2 & 0 & -y_1 & 0 & -y_2 \\ 0 & y_1 + y_2 & -y_2 & 0 & -y_1 \\ -y_1 & -\dfrac{y_3}{y_4} - y_2 - 1 & \dfrac{y_3}{y_4} + y_1 + y_2 & 1 & 0 \\ 0 & 1 + \dfrac{y_3}{y_4} & -\dfrac{y_3}{y_4} & 1-1 & -1 \\ -y_2 & -y_1 & 0 & -1 & y_1 + y_2 + 1 \end{bmatrix}$$

Using the results of Section 4-4, we have

(4.76) $$Z_{in} = \frac{E_{12}}{I_{12}} = \frac{\hat{\mathbf{Y}}^{12}_{12}}{\hat{\mathbf{Y}}^{3}_{3}}$$

From expression 4.75,

$$\hat{\mathbf{Y}}^{12}_{12} = \begin{vmatrix} \dfrac{y_3}{y_4} + y_1 + y_2 & 1 & 0 \\ -\dfrac{y_3}{y_4} & 0 & -1 \\ 0 & -1 & y_1 + y_2 + 1 \end{vmatrix}$$

$$= (y_1 + y_2)\left(\dfrac{y_3}{y_4} - 1\right)$$

---

[10] S. K. Mitra and N. M. Herbst, "Synthesis of active *RC* one-ports using generalized impedance converters," *IEEE Trans. on Circuit Theory*, **CT-10**, 532 (December 1963).

and

$$\hat{Y}_3^3 = \begin{vmatrix} y_1 + y_2 & 0 & 0 & -y_2 \\ 0 & y_1 + y_2 & 0 & -y_1 \\ 0 & \frac{y_3}{y_4} + 1 & 0 & -1 \\ -y_2 & -y_1 & -1 & y_1 + y_2 + 1 \end{vmatrix}$$

$$= (y_1 + y_2)\left(\frac{y_3}{y_4} y_1 - y_2\right)$$

Thus,

(4.77) $$Z_{in} = \frac{(y_1 + y_2)\left(\frac{y_3}{y_4} - 1\right)}{(y_1 + y_2)\left(\frac{y_3}{y_4} y_1 - y_2\right)} = \frac{y_3 - y_4}{y_3 y_1 - y_4 y_2}$$

4-7 ANALYSIS OF NETWORKS CONTAINING OPERATIONAL AMPLIFIERS

The method of the previous section lends itself easily to the analysis of networks containing operational amplifiers. This will be clear from the following example.

Example 4-7 Let us compute the voltage transfer ratio V_o/V_{in} of the circuit of Figure 4-26. This circuit was originally proposed by Bridgman and Brennan[11] for synthesis of transfer voltage ratio.

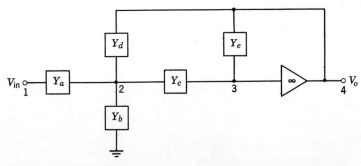

Figure 4-26 An active two-port containing an operational amplifier.

---

[11] A. Bridgman and R. Brennan, "Simulation of transfer functions using only one operational amplifier," *1957 IRE WESCON Convention Record*, **1** (Part 4), 273–278.

ANALYSIS OF NETWORKS CONTAINING OPERATIONAL AMPLIFIERS / 149

First we can replace the operational amplifier by a voltage amplifier of gain μ and then let $\mu \to -\infty$ after V_o/V_{in} has been computed.

The indefinite admittance matrix of the five-pole network without the voltage amplifier is written by inspection:

(4.78)
$$\begin{array}{c} \\ 1 \\ 2 \\ 3 \\ 4 \\ 5 \end{array} \begin{bmatrix} 1 & 2 & 3 & 4 & 5 \\ Y_a & -Y_a & 0 & 0 & 0 \\ -Y_a & (Y_a + Y_b + Y_c + Y_d) & -Y_c & -Y_d & -Y_b \\ 0 & -Y_c & Y_c + Y_e & -Y_e & 0 \\ 0 & -Y_d & -Y_e & Y_d + Y_e & 0 \\ 0 & -Y_b & 0 & 0 & Y_b \end{bmatrix}$$

Following the technique outlined in the previous section, a voltage amplifier of gain μ is identical to the network of Figure 4-27, which has an indefinite admittance matrix as given below:

(4.79)
$$\begin{array}{c} \\ 3 \\ 4 \\ 5 \\ 6 \end{array} \begin{bmatrix} 3 & 4 & 5 & 6 \\ 0 & 0 & 0 & 0 \\ 0 & 1 & 0 & -1 \\ -\mu & 0 & \mu - 1 & 1 \\ \mu & -1 & 1 - \mu & -1 + 1 \end{bmatrix}$$

Adding the matrices of expressions 4.78 and 4.79, we obtain the indefinite admittance matrix of the composite network:

(4.80)
$$\hat{Y} = \begin{array}{c} \\ 1 \\ 2 \\ 3 \\ 4 \\ 5 \\ 6 \end{array} \begin{bmatrix} 1 & 2 & 3 & 4 & 5 & 6 \\ Y_a & -Y_a & 0 & 0 & 0 & 0 \\ -Y_a & Y & -Y_c & -Y_d & -Y_b & 0 \\ 0 & -Y_c & Y_c + Y_e & -Y_e & 0 & 0 \\ 0 & -Y_d & -Y_e & Y_d + Y_e + 1 & 0 & -1 \\ 0 & -Y_b & -\mu & 0 & Y_b + \mu - 1 & 1 \\ 0 & 0 & \mu & -1 & 1 - \mu & 0 \end{bmatrix}$$

where for convenience we have let $Y = Y_a + Y_b + Y_c + Y_d$.

The transfer voltage ratio is now given as

(4.81) $$\frac{V_o}{V_{in}} = \frac{E_{45}}{E_{15}} = \text{sgn}(4 - 5)\,\text{sgn}(1 - 5)\,\frac{\hat{Y}^{15}_{45}}{\hat{Y}^{15}_{15}}$$

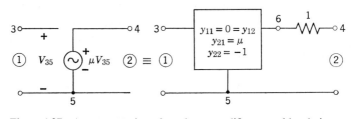

Figure 4-27 A representation of a voltage amplifier to enable admittance analysis.

From Equation 4.80 we compute $\hat{\mathbf{Y}}_{45}^{15}$ and $\hat{\mathbf{Y}}_{15}^{15}$:

$$\hat{\mathbf{Y}}_{15}^{15} = \begin{vmatrix} Y & -Y_c & -Y_d & 0 \\ -Y_c & Y_c + Y_e & -Y_e & 0 \\ -Y_d & -Y_e & Y_d + Y_e + 1 & -1 \\ 0 & \mu & -1 & 0 \end{vmatrix}$$

$$= \mu(YY_e + Y_c Y_d) - [Y(Y_c + Y_e) - Y_c^2]$$

$$\hat{\mathbf{Y}}_{45}^{15} = - \begin{vmatrix} -Y_a & Y & -Y_c & 0 \\ 0 & -Y_c & Y_c + Y_e & 0 \\ 0 & -Y_d & -Y_e & -1 \\ 0 & 0 & \mu & 0 \end{vmatrix}$$

$$= -\mu Y_a Y_c$$

Thus, as $\mu \to -\infty$, we have

$$\frac{V_o}{V_{in}} = -\frac{Y_a Y_c}{(Y_a + Y_b + Y_c + Y_d)Y_e + Y_c Y_d}$$

which is the desired expression.

An alternate elegant approach to the problem of analysis of networks constrained by infinite gain operational amplifiers was recently advanced by Nathan.[12] Consider, for convenience, a 5-terminal nongenerative network with its terminal 5 as the reference node (Figure 4-28a). The short-circuit admittance matrix description of this four-port network is given as:

$$(4.83) \quad \begin{bmatrix} I_1 \\ I_2 \\ I_3 \\ I_4 \end{bmatrix} = \begin{bmatrix} y_{11} & y_{12} & y_{13} & y_{14} \\ y_{21} & y_{22} & y_{23} & y_{24} \\ y_{31} & y_{32} & y_{33} & y_{34} \\ y_{41} & y_{42} & y_{43} & y_{44} \end{bmatrix} \begin{bmatrix} V_1 \\ V_2 \\ V_3 \\ V_4 \end{bmatrix}$$

[12] A. Nathan, "Matrix analysis of networks having infinite gain operational amplifiers," *Proc. IEEE*, **49**, 1577–1578 (October 1961).

ANALYSIS OF NETWORKS CONTAINING OPERATIONAL AMPLIFIERS / 151

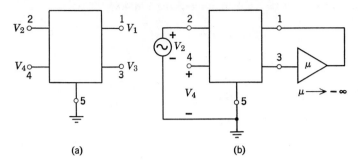

Figure 4-28

Observe that in Equation 4.83 and in Figure 4-28a, V_j refer to the voltage at terminal j with respect to terminal 5.

Let us connect now a single-ended operational amplifier to this network between terminal 3 and terminal 1. We wish to derive an expression for the voltage transfer ratio V_4/V_2 with terminal 4 open-circuited. The constrained network is shown in Figure 4-28b. Note that due to the presence of the operational amplifier, a constraint

(4.84) $$V_1 = \mu V_3, \qquad \mu \to -\infty$$

is introduced. Moreover, the operational amplifier has infinite input impedance, hence terminal 3 is not loaded by its presence (i.e., $I_3 = 0$).

In order to satisfy condition 4.84, when $\mu \to -\infty$,

(4.84) $$V_3 \to 0$$

implying that terminal 3 is at "virtual" ground, which indicates that the current I_1 going into terminal 1 must be such that condition 4.84 holds. Or in other words, I_1 is the only dependent current variable, hence the first equation from Equation 4.83 must be eliminated. Because of condition 4.84, we can also eliminate the third column of the admittance matrix. The final admittance description of the network of Figure 4-28b is then given as:

(4.85) $$\begin{bmatrix} I_2 \\ 0 \\ 0 \end{bmatrix} = \begin{bmatrix} y_{21} & y_{22} & y_{24} \\ y_{31} & y_{32} & y_{34} \\ y_{41} & y_{42} & y_{44} \end{bmatrix} \begin{bmatrix} V_1 \\ V_2 \\ V_4 \end{bmatrix}$$

Equation 4.85 can be written in the matrix notation as:

(4.86) $$\tilde{\mathbf{I}} = \tilde{\mathbf{Y}} \cdot \tilde{\mathbf{V}}$$

152 / ANALYSIS OF ACTIVE NETWORKS

Cramer's rule can be used to solve Equation 4.85. Thus,

(4.87) $$V_2 = \frac{\begin{vmatrix} y_{21} & I_2 & y_{24} \\ y_{31} & 0 & y_{34} \\ y_{41} & 0 & y_{44} \end{vmatrix}}{|\tilde{Y}|} = \frac{(y_{41}y_{34} - y_{31}y_{44})I_2}{|\tilde{Y}|}$$

(4.88) $$V_4 = \frac{\begin{vmatrix} y_{21} & y_{22} & I_2 \\ y_{31} & y_{32} & 0 \\ y_{41} & y_{42} & 0 \end{vmatrix}}{|\tilde{Y}|} = \frac{(y_{31}y_{42} - y_{41}y_{32})}{|\tilde{Y}|} I_2$$

Therefore, the desired transfer voltage ratio is given as

(4.89) $$\frac{V_4}{V_2} = \frac{(y_{31}y_{42} - y_{41}y_{32})}{(y_{41}y_{34} - y_{31}y_{44})}$$

Expression 4.89 can be written in a compact form. To this end, we rewrite Equation 4.87 and 4.88 as:

(4.90) $$V_2 = \frac{\tilde{Y}_2^1}{|\tilde{Y}|} I_2$$

$$V_4 = \frac{\tilde{Y}_3^1}{|\tilde{Y}|} I_2$$

Thus,

(4.91) $$\frac{V_4}{V_2} = \frac{\tilde{Y}_3^1}{\tilde{Y}_2^1}$$

We illustrate Nathan's approach by means of an example.

Example 4-8 Consider the same network (Figure 4-26) of the previous example.

From the indefinite admittance matrix of the five pole as given by Equation 4.78, we arrive at the short-circuit admittance matrix of the four-port network by deleting the fifth row and the fifth column:

(4.92)

$$Y_a = \begin{array}{c} \\ 1 \\ 2 \\ 3 \\ 4 \end{array} \begin{bmatrix} \overset{1}{Y_a} & \overset{2}{-Y_a} & \overset{3}{0} & \overset{4}{0} \\ -Y_a & (Y_a + Y_b + Y_c + Y_d) & -Y_c & -Y_d \\ 0 & -Y_c & Y_c + Y_e & -Y_e \\ 0 & -Y_d & -Y_e & Y_d + Y_e \end{bmatrix}$$

ANALYSIS OF NETWORKS CONTAINING OPERATIONAL AMPLIFIERS / 153

Note that the admittance matrix of Equation 4.92 could have been written without forming the indefinite admittance matrix given by Equation 4.78. In Equation 4.92, Y_b being connected between terminal 2 and the reference terminal 5 appears only in $(Y_d)_{22}$ position. All other admittances appear symmetrically at four different positions in Equation 4.92. Next, observe that terminal 3 is the driving terminal and terminal 4 is the driven terminal. This implies that the admittance description of the constrained network is obtained by deleting the third column and fourth row of Equation 4.92.

(4.93) $$\tilde{Y} = \begin{bmatrix} Y_a & -Y_a & 0 \\ -Y_a & Y_a + Y_b + Y_c + Y_d & -Y_d \\ 0 & -Y_c & -Y_e \end{bmatrix}$$

The desired voltage transfer ratio is obtained from the above equation, using Cramer's rule as:

$$t_v = \frac{V_4}{V_1} = \frac{Y_3^1}{Y_1^1} = \frac{Y_a Y_c}{-Y_e(Y_a + Y_b + Y_c + Y_d) - Y_c Y_d}$$

If the constraint is imposed by means of a differential input operational amplifier (Figure 4-29), the technique outlined above must be modified. Note that for a differential input amplifier $V_i = V_k$. This is satisfied by adding the kth column to the ith column and deleting the original kth column of the admittance matrix Y_d. Similarly, since I_j takes on a value to make the differential input voltage zero, we can delete the jth row of the resultant admittance matrix. These two steps enable us to derive the admittance description of the constrained network from the (definite) admittance matrix of the unconstrained network.[12]

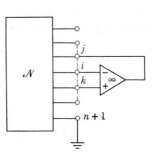

Figure 4-29

Example 4-9 Let us determine the voltage transfer ratio V_4/V_1 of the differential-input operational amplifier network of Figure 4-30.

The short-circuit admittance matrix of the four-port "companion" network of Figure 4-30 without the operational amplifier can be written by

---

[12] A. S. Morse, "The use of operational amplifiers in active network theory," M.S. Thesis, University of Arizona, Tucson, Arizona, January 1964.

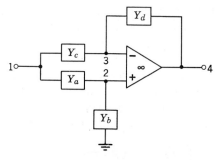

Figure 4-30

inspection as:

$$
(4.94) \quad \mathbf{Y}_d = \begin{bmatrix} Y_a + Y_c & -Y_a & -Y_c & 0 \\ -Y_a & Y_a + Y_b & 0 & 0 \\ -Y_c & 0 & Y_c + Y_d & -Y_d \\ 0 & 0 & -Y_d & Y_d \end{bmatrix} \begin{matrix} 1 \\ 2 \\ 3 \\ 4 \end{matrix}
$$

$$\begin{matrix} 1 & 2 & 3 & 4 \end{matrix}$$

The operational amplifier imposes the constraint $V_2 = V_3$. Furthermore, the current going into terminal 4 is the dependent current. Thus, adding the third column of (4.94) to its second column and eliminating the third column and the fourth row, we arrive at the admittance description of the constrained network of Figure 4-30:

$$
(4.95) \quad \tilde{\mathbf{Y}} = \begin{bmatrix} Y_a + Y_c & -Y_a - Y_c & 0 \\ -Y_a & Y_a + Y_b & 0 \\ -Y_c & Y_c + Y_d & -Y_d \end{bmatrix}
$$

The desired voltage transfer ratio is given as:

$$
(4.96) \quad \frac{V_4}{V_1} = \frac{\tilde{\mathbf{Y}}_3^1}{\tilde{\mathbf{Y}}_1^1} = \frac{\begin{vmatrix} -Y_a & Y_a + Y_b \\ -Y_c & Y_c + Y_d \end{vmatrix}}{\begin{vmatrix} Y_a + Y_b & 0 \\ Y_c + Y_d & -Y_d \end{vmatrix}}
$$

$$
= \frac{Y_a Y_d - Y_b Y_c}{Y_d(Y_a + Y_b)}
$$

Extension of the above techniques to the case of two or more operational amplifiers is straightforward.

4-8 BILINEAR FORM OF NETWORK FUNCTIONS

In Chapter 5, we shall be interested in the effect of variation of a single passive or active parameter on the overall transfer function of the network. To this end, it is necessary to express the pertinent network function as a bilinear function of the parameter of interest.

As an example, let us reconsider the two-port of Figure 4-26, where the operational amplifier has been replaced by a voltage amplifier of gain μ. The voltage transfer ratio is then given as

(4.97)
$$\frac{V_o}{V_{in}} = \frac{-\mu Y_a Y_c}{\mu[(Y_a + Y_b + Y_c + Y_d)Y_e + Y_c Y_d] - [(Y_a + Y_b + Y_d)Y_c + (Y_a + Y_b + Y_c + Y_d)Y_e]}$$

Expression 4.97 alternately can be written as

(4.98) $$\frac{V_o}{V_{in}} = \frac{N(s)}{D(s)} = \frac{\mu N_2(s)}{D_1(s) + \mu D_2(s)}$$

where

$$N_2(s) = -Y_a Y_c$$
$$D_1(s) = -(Y_a + Y_b + Y_d)Y_c - (Y_a + Y_b + Y_c + Y_d)Y_e$$
$$D_2(s) = (Y_a + Y_b + Y_c + Y_d)Y_e + Y_c Y_d$$

If Y_a is of interest, we can rewrite expression 4.97 as

(4.99) $$\frac{V_o}{V_{in}} = \frac{N(s)}{D(s)} = \frac{Y_a N_2(s)}{D_1(s) + Y_a D_2(s)}$$

where now

$$N_2(s) = -\mu Y_c$$
$$D_1(s) = \mu(Y_b + Y_c + Y_d)Y_e + \mu Y_c Y_d - (Y_b + Y_d)Y_c - (Y_b + Y_c + Y_d)Y_e$$
$$D_2(s) = \mu Y_e - Y_c - Y_e$$

It is now clear that, in general, we can write any network function as

(4.100) $$\frac{N(s)}{D(s)} = \frac{N_1(s) + k N_2(s)}{D_1(s) + k D_2(s)}$$

where k is the passive or active parameter of interest. Expression 4.100 is known as the *bilinear form* of the network function $N(s)/D(s)$,[13] a more general proof of which follows.

The general expressions for various network functions were given in Section 4-4. Consider the expression for transfer impedance $Z_{mn}{}^{ij}$ as given by Equation 4.55:

$$(4.55) \qquad Z_{mn}^{ij} = \text{sgn}\,(m-n)\,\text{sgn}\,(i-j)\,\frac{\hat{Y}_{ij}^{mn}}{\hat{Y}_n^n}$$

Without loss of any generality, let us assume that $m > n$, $i > j$, and $(m + n + i + j)$ is an even number.

Suppose that we are interested in expressing Z_{mn}^{ij} in terms of an admittance y connected between nodes p and q. We can then write $\hat{Y}_{ij}^{mm}$ as

$$(4.101) \qquad \hat{Y}_{ij}^{mn} = \begin{array}{c} \\ p \\ \\ q \\ \end{array} \begin{vmatrix} \cdot & \cdot & \overset{p}{\cdot} & \cdot & \overset{q}{\cdot} & \cdot \\ \cdot & \cdot & \cdot & \cdot & \cdot & \cdot \\ \cdot & \cdot & y_{pp}{}^0 + y & \cdot & y_{pq}{}^0 - y & \cdot \\ \cdot & \cdot & \cdot & \cdot & \cdot & \cdot \\ \cdot & \cdot & y_{qp}{}^0 - y & \cdot & y_{qq}{}^0 + y & \cdot \\ \cdot & \cdot & \cdot & \cdot & \cdot & \cdot \end{vmatrix}$$

which can be reexpressed by determinant manipulation as

$$(4.102) \qquad \begin{array}{c} \\ p \\ \\ q \\ \end{array} \begin{vmatrix} \cdot & \cdot & \overset{p}{\cdot} & \cdot & \overset{q}{\cdot} & \cdot \\ \cdot & \cdot & \cdot & \cdot & \cdot & \cdot \\ \cdot & \cdot & y_{pp}{}^0 + y & \cdot & y_{pp}{}^0 + y_{pq}{}^0 & \cdot \\ \cdot & \cdot & \cdot & \cdot & \cdot & \cdot \\ \cdot & \cdot & y_{pp}{}^0 + y_{qp}{}^0 & \cdot & \sum y & \cdot \\ \cdot & \cdot & \cdot & \cdot & \cdot & \cdot \end{vmatrix}$$

where $\sum y = y_{pp}{}^0 + y_{pq}{}^0 + y_{qp}{}^0 + y_{qq}{}^0$. Let us denote the above determinant as $\triangle$; then from expression 4.102 we note that

$$(4.103) \qquad \triangle = \triangle^0 + (y_{pp}{}^0 + y)\triangle_p{}^p$$

where $\triangle^0$ is the value of $\triangle$ when $y_{pp}{}^0 + y = 0$, and $\triangle_p{}^p$ is the determinant obtained by deleting the pth row and column of expression 4.102. Therefore, we can write $\hat{Y}_{ij}^{mn}$ as

$$(4.104) \qquad \hat{Y}_{ij}^{mn} = \triangle = N_1(s) + yN_2(s)$$

where $N_1(s) = \triangle^0 + y_{pp}{}^0 \triangle_p{}^p$, and $N_2(s) = \triangle_p{}^p$. In a similar manner

---

[13] H. W. Bode, *Network analysis and feedback amplifier design*, Van Nostrand, New York, p. 10 (1945).

we can express $\hat{\mathbf{Y}}_n^n$ as

(4.105) $$\hat{\mathbf{Y}}_n^n = D_1(s) + yD_2(s)$$

Therefore,

(4.106) $$Z_{mn}^{ij} = \frac{N_1(s) + yN_2(s)}{D_1(s) + yD_2(s)}$$

If the element of interest is an impedance converter type active device, we note from expressions 4.72, 4.73, and 4.74 that the pertinent parameter appears linearly in the indefinite admittance matrix. For an impedance inverter type active device, similar observations follow, and we thus conclude that any network function can be expressed in a bilinear form in terms of a single parameter of the given network.

Additional References

Barabaschi, S. and E. Gatti, "Modern methods of analysis for active electrical networks with particular regard to feedback systems, Part I," *Energia Nucleare*, **2**, 105–119 (December 1954).

Baranov, V., "Méthode de calcul des réseaux d'impédances," *Rev. gén. élec.*, **37**, 339–351 (March 16, 1935).

Davies, A. C., "Matrix analysis of networks containing nullators and norators," *Electronics Letters*, **2**, 48–49 (February 1966).

Deards, S. R., "Matrix theory applied to thermionic valve circuits," *Electronic Engg.*, **24**, 264–277 (June 1952).

DePian, L., *Linear active network theory*, Prentice-Hall, Englewood Cliffs, N. J., 1964.

Huelsman, L. P., *Circuits, matrices and linear vector spaces*, McGraw-Hill, New York, 1963.

Nathan, A., "Matrix analysis of constrained networks," *Proc. IEE (London)*, **108** (Part C), 98–106 (March 1961).

Puckett, T. H., "A note on the admittance and impedance matrices of an n-terminal network," *IRE Trans. on Circuit Theory*, **CT-3**, 70–75 (March 1956).

Shekel, J., "Matrix representation of transistor circuits," *Proc. IRE*, **40**, 1493–1497 (November 1952).

Shekel, J., "Voltage reference node," *Wireless Engr.*, **31**, 6–10 (January 1954).

Weinberg, L., *Network analysis and synthesis*, McGraw-Hill, New York, pp. 44–59 (1962).

Youla, D., "Some remarkable formulae occurring in the theory of *n*-terminal networks," Memo. No. 31, MRI, Polytechnic Institute of Brooklyn, October 22, 1959 (unpublished).

Zadeh, L. A., "On passive and active networks, and generalized Norton's and Thevenin's theorems," *Proc. IRE*, **44**, 378 (March 1956).

Problems

4.1 (a) Determine the indefinite admittance matrix of the 3-terminal gyrator of Figure 4-31.

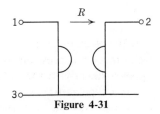

Figure 4-31

(b) Discuss the operation of the device when terminal-pair (2, 1) and terminal-pair (3, 1) are considered as ports.

(c) Repeat part (b) with terminal-pairs (1, 2) and (3, 2) considered as the ports.

4.2 Compute the h parameters of a transistor in the common collector orientation ($[h_c]$) in terms of $[h_e]$, the h parameters in the common emitter orientation.

4.3 Relate $[h_c]$ and $[h_e]$ of a transistor to the h parameters in the common-base orientation ($[h_b]$).

4.4 From Expression 4.16, determine the short-circuit admittance matrix of a common-emitter transistor.

4.5 (a) Compute the y matrix of the two-port of Figure 4-32.

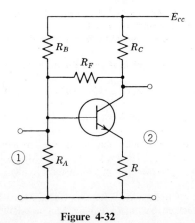

Figure 4-32

(b) If the transistor approaches idealized behavior ($r_c \to \infty$, $\alpha \to 1$), show that it is possible to make $y_{12} = -y_{21}$, i.e., that the two-port behaves like a nonideal gyrator.

4.6 (a) A 4-terminal ideal transformer can be represented as shown in Figure 4-33. Determine the indefinite admittance matrix of the network inside the dotted box; from that, formulate the equivalent representation of a 4-terminal ideal transformer using positive and negative resistances.[14]

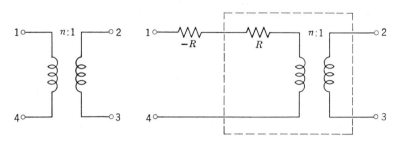

Figure 4-33

(b) Derive the equivalent circuit of a 3-terminal ideal transformer from the results of part (a).

4.7 (a) Compute the indefinite admittance matrix of the 4-terminal network of Figure 4-34 by assuming ideal pentodes and show that it represents a VCT.[15]

(b) Construct an ideal gyrator using two such VCT's.

4.8 Compute the input impedances of the networks of Figures 2-3 and 2-15 using the modified indefinite admittance matrix approach.

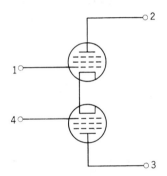

Figure 4-34

4.9 Using the indefinite admittance matrix analysis, determine the voltage transfer ratio of the network of Figures 8-37a, 8-38, and 8-40b.

4.10 Using Nathan's approach outlined in Section 4-7, show that for the active network of Figure 4-35:

$$\frac{V_2}{V_1} = \frac{Y_1 Y_3 Y_5}{Y_6 Y_A (Y_3 + Y_B) + Y_3 Y_6 Y_B + Y_5 Y_7 (Y_3 + Y_A) + Y_3 Y_5 Y_8}$$

[14] C. D. Weiss, "Ideal transformer realizations with negative resistors," *Proc. IEEE*, **54**, 302–303 (February 1966).

[15] G. E. Sharpe, "The pentode gyrator," *IRE Trans. on Circuit Theory*, **CT-4**, 321–323 (December 1957).

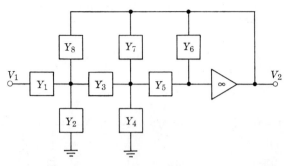

Figure 4-35

where

$$Y_A = Y_1 + Y_2 + Y_B \quad \text{and} \quad Y_B = Y_4 + Y_5 + Y_7$$

4.11 The network shown in Figure 11-31 is suitable for transfer function realization.[16] Show that its voltage transfer ratio is given by Equation 11.59.

[16] Section 11–6.

5 / *Sensitivity and Related Topics*

Characteristics of physical devices are subject to change for many reasons. For example, the transistor parameters may vary due to a change in temperature, bias levels, or humidity; also, parameters may drift because of aging. A typical variation of the β, the d-c current gain in the common emitter orientation, of a transistor with respect to temperature is shown in Figure 5-1. Likewise, the parameters of the passive components may change from their nominal values because of variations of external and internal conditions. Even though components with very low drift are available, economy and size limit their use.

There are basically two types of element variations. The first and most frequently considered case is the so-called incremental variations case where the percent change of element values are assumed to be very small. The second situation, where element values change rather drastically in comparison to their nominal values, is of practical significance in relatively fewer applications and has not been treated as extensively as the former case. In this chapter, we are primarily concerned with the case of incremental variations of element values.

A detrimental effect of parameter variation is the displacement of the poles and zeros of the network function from their nominal positions. If the natural frequencies move to the right-half s-plane, the network will become unstable. (An example of such a network was presented in Section

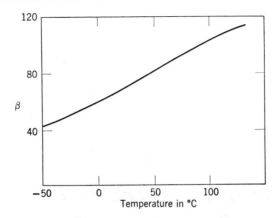

Figure 5-1 A typical variation of the parameter β of a transistor with temperature.

3-7.) Even if the critical frequency displacements do not cause instability, the overall system response may differ considerably from the desired response.

It is thus desirable to determine a priori the effect on the network performance due to an incremental variation of one or more of the network component values. One method of estimating the effect would be to perform a *deviation study* by means of a digital computer. In this method, each element is independently perturbed, and the network is analyzed at a set of prescribed frequencies repeatedly for each element change. The difference between the nominal and modified values of the transfer function is the deviation. If the element changes are small, e.g., of the order of 1 percent, a worst case study can be made by adding the magnitudes of the deviations. This process is particularly useful in cases of complex networks in order to determine critical elements and their allowable tolerances.

A more commonly used quantitative measure of the change in network characteristic due to a change in network parameters is called the *sensitivity*. Knowledge of sensitivity helps the circuit designer in selecting the proper network configuration and its element values for a specified application. Sensitivity calculations also enable him to determine the critical elements and their allowable tolerances. It should be noted that the problem of designing a completely insensitive network has not only been solved but, realistically, it never may be solved. For specific cases, it is, however, possible to minimize the effect of parameter variations to an extent that is satisfactory for many applications.

Various concepts of sensitivity and their interrelationships are introduced in this chapter. Methods of computing the different types of sensitivity measures are also included here, along with some helpful remarks concerning the actual design of a network from a sensitivity point of view.

5-1 ESTIMATION OF POLE (ZERO) DISPLACEMENTS DUE TO INCREMENTAL PARAMETER VARIATIONS

A simple analytical method was proposed by Papoulis[1] for approximate determination of displacements of poles and zeros due to the incremental variation(s) of one or more network parameters. The method is based on the assumption that for very small change in element values, the poles (and zeros) move by a very small amount and are in the neighborhood of the original locations.

Let $F(s)$ denote the polynomial (or the rational function) of interest. The coefficients of $F(s)$ are multilinear functions of network parameters. If $F(s)$ is a polynomial, then an incremental change in network elements will change it to a new polynomial $F'(s)$ where

(5.1) $$F'(s) = F(s) + \delta f(s)$$

$\delta f(s)$ in Equation 5.1 can be considered as the effect of the small change in element values. If $F(s)$ is a rational function,

(5.2) $$F(s) = \frac{N(s)}{D(s)}$$

then the new rational function $F'(s)$ due to the change in element values can be expressed as

(5.3) $$F'(s) = \frac{N(s) + \delta n(s)}{D(s) + \delta d(s)}$$

which can again be reexpressed as:

(5.4) $$F'(s) \cong \frac{N(s)}{D(s)} + \frac{D(s)\delta n(s) - N(s)\delta d(s)}{D^2(s)}$$

$$= F(s) + \delta f(s)$$

Let $s = p_j$ be a zero of multiplicity n of $F(s)$ which moves to a new position at $s = p'_j$ when $F(s)$ is changed incrementally by $\delta f(s)$, i.e.,

(5.5) $$F(p_j) = 0$$
and
(5.6) $$F(p'_j) + \delta f(p'_j) = 0$$

---

[1] A. Papoulis, "Displacement of the zeros of the impedance $Z(p)$ due to an incremental variation in the network elements," *Proc. IRE*, **43**, 79–82 (January 1955).

164 / SENSITIVITY AND RELATED TOPICS

In the neighborhood of $s = p_j$, we can make a Laurent expansion of $1/F(s)$:

$$\text{(5.7)} \qquad \frac{1}{F(s)} = \frac{L_n}{(s - p_j)^n} + \frac{L_{n-1}}{(s - p_j)^{n-1}} + \cdots$$

The coefficients $L_n, L_{n-1}, \ldots$ of the expansion can be expressed in terms of $F(s)$. For example,

$$\text{(5.8)} \qquad L_n = \frac{(s - p_j)^n}{F(s)} \bigg|_{s = p_j}$$

Now if $s = p'_j$ is close to $s = p_j$, we obtain from Equation 5.7,

$$\text{(5.9)} \qquad \frac{1}{F(p'_j)} \cong \frac{L_n}{(p'_j - p_j)^n}$$

Therefore

$$\text{(5.10)} \qquad (p'_j - p_j) = \delta p_j \cong \sqrt[n]{L_n F(p'_j)} = \sqrt[n]{-L_n[\delta f(p'_j)]}$$
$$\cong \sqrt[n]{-L_n[\delta f(p_j)]}$$

assuming $\delta f(p'_j)$ is approximately equal to $\delta f(p_j)$. Expression 5.10 is the desired result. Physically it can be interpreted as follows. Each zero of multiplicity n becomes n simple zeros situated equidistant from each other on a circle of radius $|p'_j - p_j|$ centered at $s = p_j$.

For very small displacement of the zero, Equation 5.10 provides a fairly satisfactory answer. A more accurate measure can be obtained by using p'_j as obtained as the first approximation to evaluate $\delta f(s)$ in the expression 5.10.

We now consider an example to illustrate the use of expression 5.10.

Example 5-1 Suppose we are interested in computing the displacement of the poles of the active *RC* Bessel filter of Figure 1-7 because of a 10 percent change in the value of the capacitor C_1 across the input port. From Equation 1.10 we observe that the pertinent polynomial $F(s)$ is given as

$$\text{(5.11)} \qquad F(s) = C_1 R_1 s^2 + (1 + C_1 R_1 + C_2 R_1 - \mu C_2 R_1) s + 1$$

The nominal element values are given by Equation 1.12 and hence

$$\text{(5.12)} \qquad F(s) = \tfrac{1}{3} s^2 + s + 1$$

It is seen from Equation 5.11 that the change $\delta f(s)$, due to a change in C_1, is given as

$$\text{(5.13)} \qquad \delta f(s) = 0.1 C_1 R_1 s^2 + 0.1 C_1 R_1 s = \frac{0.1}{3} s(s + 1)$$

BASIC CONCEPTS OF SENSITIVITY MEASURE / 165

The nominal pole positions are given by the roots of (5.12) and are at

(5.14) $$p_1 = -\frac{3}{2} - j\frac{\sqrt{3}}{2}, \quad p_1^* = -\frac{3}{2} + j\frac{\sqrt{3}}{2}$$

Consider the movement of p_1. Using Equation 5.8, we obtain

(5.15)
$$L_1 = \frac{\left(s + \frac{3}{2} + j\frac{\sqrt{3}}{2}\right)}{\left(\frac{1}{3}s^2 + s + 1\right)}\bigg|_{s=-\frac{3}{2}-j(\frac{\sqrt{3}}{2})} = \frac{3}{\left(s + \frac{3}{2} - j\frac{\sqrt{3}}{2}\right)}\bigg|_{s=-\frac{3}{2}-j(\frac{\sqrt{3}}{2})}$$

$$= -\frac{3}{j\sqrt{3}} = j\sqrt{3}$$

Hence, from Equation 5.10

(5.16)
$$\delta p_1 = p_1' - p_1 = -(j\sqrt{3})\left(\frac{0.1}{3}\right)\left(-\frac{3}{2} - j\frac{\sqrt{3}}{2}\right)\left(-\frac{1}{2} - j\frac{\sqrt{3}}{2}\right) = 0.1$$

Approximate new pole position is then

(5.17) $$p_1' = p_1 + \delta p_1 = -\frac{3}{2} - j\frac{\sqrt{3}}{2} + 0.1 = -1.4 + j0.866$$

Actual locations of the new pole positions are given by the roots of $F(s) + \delta f(s) = 0$, i.e., of

$$\frac{1}{3}s^2 + s + 1 + \frac{0.1}{3}s(s + 1) = \frac{1}{3}[1.1s^2 + 3.1s + 3] = 0$$

These are at

(5.18) $$p_1' = 1.409 - j0.861; \quad p_1'^* = -1.409 + j0.861$$

Comparison of Equations 5.17 and 5.18 shows that there is a good agreement between the approximate solution and the exact location.

5-2 BASIC CONCEPTS OF SENSITIVITY MEASURE

The purpose of this section is to review some of the useful concepts associated with the notion of sensitivity. Some of the ideas were originally introduced by Bode and later successfully extended and used in the design of feedback control systems. In this section, we shall be concerned only with the variation of a single parameter.

Signal Flow Graph Representation

Since linear active networks are essentially feedback systems, it is convenient to represent the network in terms of a signal flow graph[2] in which the variable parameter is represented by a single branch only. Such a representation follows directly from the bilinear form of a network function introduced in the previous chapter. If $T(s, k)$ is the network function where k is the network parameter of interest, then we can write

$$(5.19) \quad T(s, k) = \frac{Y(s, k)}{\hat{X}(s)} = \frac{N(s, k)}{D(s, k)} = \frac{N_1(s) + kN_2(s)}{D_1(s) + kD_2(s)}$$

Equation 5.19 can be expressed in the following form:

$$(5.20) \quad T(s, k) = T_0(s) + \frac{kA(s)B(s)}{1 - kC(s)}$$

where

$$(5.21a) \quad T_0(s) = \frac{N_1(s)}{D_1(s)} = T(s, 0)$$

$$(5.21b) \quad C(s) = \frac{-D_2(s)}{D_1(s)}$$

$$(5.21c) \quad A(s)B(s) = \frac{N_2(s)D_1(s) - N_1(s)D_2(s)}{D_1^2(s)}$$

The signal flow graph representation of (5.20) is shown in Figure 5-2. $T_0(s)$ is the value of the transfer function when $k = 0$ and is called the *leakage transmission*. $kC(s)$ is called the *loop gain*.

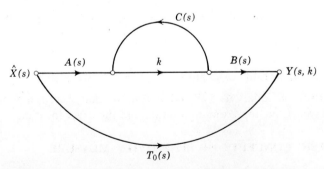

Figure 5-2 Signal flow-graph representation of a single loop feedback system.

---

[2] S. J. Mason, "Feedback theory—some properties of signal flow graphs," *Proc. IRE*, **41**, 1144–1156 (September 1953).

BASIC CONCEPTS OF SENSITIVITY MEASURE / 167

Example 5-2 The active RC one-port of Figure 5-3a was proposed by Kinariwala[3] for the realization of driving-point impedance function. The terminal variables are related as:

(5.22)
$$I_1 = y_{11}V_1 + y_{12}V_2$$
$$I_2 = y_{12}V_1 + y_{22}V_2$$
$$Y_L V_2 = kI_2$$

where $\{y_{ij}\}$ are the short-circuit parameters of the RC two-port. Using a new variable $V_2' = Y_L V_2$, a signal flow graph representation of (5.22) is easily obtained as shown in Figure 5-3b.

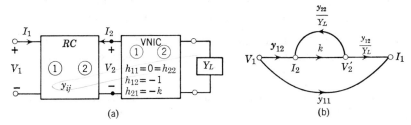

(a)   (b)

Figure 5-3 The cascade RC:NIC configuration and the signal flow-graph representation of its driving-point function.

Return Difference and Null-Return Difference[4]

Two useful functions can be defined from the signal flow graph representation of Figure 5-2.

DEFINITION 5-1   *The return difference $F_k(s)$ of the network of Figure 5-2 is defined as*

(5.23)
$$F_k(s) \triangleq 1 - kC(s)$$

A physical interpretation of the return difference is as follows. The branch of interest is broken at the input (Figure 5-4). A signal of unit strength is applied at node "a." The difference between the transmitted signal at node "a" and the returned signal at node "b" under the condition of zero input ($\hat{X}(s) = 0$) is the return difference. Note that to compute the return difference, it is not at all necessary to draw the signal flow graph. In terms of the polynomials of the network function, it follows that

(5.24)
$$F_k(s) = 1 + k\frac{D_2(s)}{D_1(s)} = \frac{D_1(s) + kD_2(s)}{D_1(s)} = \frac{D(s,k)}{D_1(s)}$$

[3] B. K. Kinariwala, "Synthesis of active RC networks," *Bell System Tech. J.*, **38**, 1269–1316, September 1959.

[4] J. G. Truxal, *Automatic feedback control system synthesis*, McGraw-Hill, New York, pp. 114–115, 123 (1955).

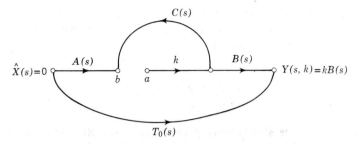

Figure 5-4 Physical interpretation of the concept of return difference.

where use of Equation 5.21b has been made. We conclude from Equation 5.24 that the zeros of $F_k(s)$ are identical to the poles of $T(s, k)$.

DEFINITION 5-2 *The return difference calculated under the condition of zero output $[Y(s, k) = 0]$ by choosing suitable input signal $\hat{X}(s)$ is called the* null-return difference.

It can be easily shown that the null-return difference $F_k^0(s)$ is given as [Problem 5.4]:

$$(5.25) \qquad F_k^0(s) = 1 - kC(s) + \frac{kA(s)B(s)}{T_0(s)}$$

Use of Equation 5.21 in Equation 5.25 yields

$$(5.26) \qquad F_k^0(s) = 1 + k\frac{N_2(s)}{N_1(s)} = \frac{N_1(s) + kN_2(s)}{N_1(s)} = \frac{N(s, k)}{N_1(s)}$$

The above expression indicates that the zeros of the null-return difference are identical to the zeros of $T(s, k)$.

The Sensitivity Function

The usual definition of the sensitivity of a network function due to an incremental variation of a single parameter is generally taken as the reciprocal of Bode's sensitivity function.[5]

DEFINITION 5-3 *The sensitivity function $S_k^{T(s)}$ of the network function $T(s, k)$ due to the variation of the parameter k is defined as*

$$(5.27) \qquad S_k^{T(s)} \triangleq \frac{d[\ln T(s)]}{d[\ln k]} = \frac{dT(s, k)/T(s, k)}{dk/k}$$

Expression 5.27 indicates that $S_k^{T(s)}$ is equal to the percentage change in $T(s, k)$ divided by the percentage change in k, assuming that all changes

---

[5] S. J. Mason, *loc. cit.*

are incremental. For example, if S_k^T is equal to 0.5, then a 1 percent change in the parameter will cause a 0.5 percent change in the network function. Therefore, from the design point of view, it is necessary to keep S_k^T as small as possible. Ideally, an insensitive network will have a sensitivity function S_k^T equal to zero.

From the definition of the sensitivity function, the following two identities immediately follow:

(5.28) $$S_{1/k}^T = -S_k^T$$

(5.29) $$S_{k_1}^T = S_{k_2}^T \cdot S_{k_1}^{k_2}$$

The sensitivity function can be easily related to the return difference. From Equation 5.20 we note

$$\frac{dT}{dk} = \frac{A(s)B(s)}{[1 - kC(s)]^2}$$

and hence

(5.30) $$S_k^{T(s)} = \frac{kA(s)B(s)}{T(s,k)[1 - kC(s)]^2} = \frac{1}{F_k(s)}\left[1 - \frac{T_0(s)}{T(s,k)}\right]$$

Observe from Equation 5.30 that $S_k^{T(s)}$ is a rational function of the complex frequency variable s. If $T_0(s)$ is negligible, Equation 5.30 simplifies to

(5.31) $$S_k^{T(s)} = \frac{1}{F_k(s)}$$

The sensitivity function can also be expressed in terms of $F_k(s)$ and $F_k^0(s)$. From Equation 5.20 and 5.23 we have

(5.32) $$kA(s)B(s) = F_k(s)[T(s,k) - T_0(s)]$$

Use of Equation 5.32 in Equation 5.25 yields

(5.33) $$F_k^0(s) = F_k(s) + \frac{F_k(s)[T(s,k) - T_0(s)]}{T_0(s)} = F_k(s) \cdot \frac{T(s,k)}{T_0(s)}$$

which when used in Equation 5.30 results in the desired expression:

(5.34) $$S_k^{T(s)} = \frac{1}{F_k(s)} - \frac{1}{F_k^0(s)}$$

An important property of $S_k^{T(s)}$ can be deduced from expression 5.34: the poles of the sensitivity function are given by the zeros and poles of the original network function $T(s, k)$. Hence, if $T(s, k)$ has multipole poles (or zeros), then $S_k^{T(s)}$ will also have multiple poles.

Gain Sensitivity and Phase Sensitivity

The sensitivity function $S_k^{T(s)}$ is concerned with the change of the overall network function $T(s)$. For sinusoidal inputs, we are more concerned with the variation of $T(j\omega)$ with respect to the variation of the parameter k. An alternate more useful definition will then be

$$(5.35) \qquad S_k^{T(j\omega)} = \frac{d[\ln T(j\omega)]}{d[\ln k]} = \frac{k}{T(j\omega)} \cdot \frac{dT(j\omega)}{dk}$$

Let us write

$$(5.36) \qquad T(j\omega) = |T(j\omega)| \, e^{\arg T(j\omega)}$$

Then

$$(5.37) \qquad \ln T(j\omega) = \ln |T(j\omega)| + j \arg T(j\omega)$$
$$= \alpha(\omega) + j\beta(\omega)$$

In Equation 5.37, $\alpha(\omega)$ and $\beta(\omega)$ are known as the *gain function* and the *phase function*, respectively.

The *gain sensitivity* and the *phase sensitivity* are defined as follows:

$$(5.38) \qquad S_k^{\alpha(\omega)} \triangleq \frac{d\alpha(\omega)}{dk/k}$$

and

$$(5.39) \qquad S_k^{\beta(\omega)} \triangleq \frac{d\beta(\omega)}{dk/k}$$

The gain and phase sensitivities are related to the real and imaginary parts of the sensitivity function $S_k^{T(j\omega)}$. Using Equation 5.37 in Equation 5.35, we obtain

$$(5.40) \qquad S_k^{T(j\omega)} = \frac{d[\ln T(j\omega)]}{d[\ln k]} = \frac{d[\alpha(\omega) + j\beta(\omega)]}{dk/k}$$
$$= \frac{d\alpha(\omega)}{dk/k} + j \frac{d\beta(\omega)}{dk/k}$$

If k is considered real (which in general is true), then we can write

$$(5.41) \qquad S_k^{\alpha(\omega)} = \mathrm{Re} \, [S_k^{T(j\omega)}]$$

$$(5.42) \qquad S_k^{\beta(\omega)} = \mathrm{Im} \, [S_k^{T(j\omega)}]$$

where "Re" denotes "the real part of" and "Im" the "imaginary part of." Denoting the even part of $S_k^{T(s)}$ by Ev $[S_k^{T(s)}]$ and the odd part by Od $[S_k^{T(s)}]$, we observe

$$(5.43) \qquad \mathrm{Re} \, [S_k^{T(j\omega)}] = \mathrm{Ev} \, [S_k^{T(s)}]\Big|_{s=j\omega}$$

$$(5.44) \qquad \mathrm{Im} \, [S_k^{T(j\omega)}] = \frac{1}{j} \mathrm{Od} \, [S_k^{T(s)}]\Big|_{s=j\omega}$$

The following relations should be clear:

(5.45) $$S_k^{\alpha(\omega)} = S_k^{|T(j\omega)|}$$

(5.46) $$S_k^{\beta(\omega)} = S_k^{\exp \beta(\omega)}$$

Note that the gain and phase sensitivities are functions of ω.

A special case of interest is when the leakage transmission is negligible. The overall sensitivity function $S_k^{T(j\omega)}$ is then given by the reciprocal of the return difference $F_k(j\omega)$ (see Equation 5.31). Let us write $F_k(j\omega)$ as

(5.47) $$F_k(j\omega) = \mathcal{R}(\omega, k) + j\mathcal{X}(\omega, k)$$

where
$$\mathcal{R}(\omega, k) = \text{Re}\,[F_k(j\omega)]$$

and
$$\mathcal{X}(\omega, k) = \text{Im}\,[F_k(j\omega)]$$

Then

(5.48) $$S_k^{\alpha(\omega)} = \text{Re}\left[\frac{1}{F_k(j\omega)}\right] = \frac{\mathcal{R}(\omega, k)}{\mathcal{R}^2(\omega, k) + \mathcal{X}^2(\omega, k)}$$

The significance of Equation 5.48 will be clear in Section 5-4.

Polynomial Sensitivity

For some applications, one may be interested in the performance of a network over a specified band of frequencies. For example, in filters, the denominator polynomial of the transfer function usually determines the passband response. On the other hand, in a notch filter, the numerator polynomial is more important for the network's behavior around the notch frequency. For this and similar applications, the concept of *polynomial sensitivity* as introduced by Herbst[6] may be useful.

DEFINITION 5-4 *The* polynominal sensitivity *of a polynomial $D(s, k)$ with respect to the variable parameter k is defined as*

(5.49) $$S_k^{D(s,k)} \triangleq \frac{d[\ln D(s, k)]}{d[\ln k]}$$

For sinusoidal applications, it is convenient to set $s = j\omega$ in the above expression:

(5.50) $$S_k^{D(j\omega,k)} = \frac{d[\ln D(j\omega, k)]}{d[\ln k]} = \frac{dD(j\omega, k)/D(j\omega, k)}{dk/k}$$

[6] N. M. Herbst, "Optimization of pole sensitivity in active *RC* networks," Res. Rept. EE 569, Cornell University, Ithaca, N.Y., 1963.

Zero Sensitivity and Pole Sensitivity

In some situations, one or more poles (or zeros as the case may be) could be of interest and a measure of their displacements due to an incremental change in the network parameter would thus be useful. For example, in a band-pass filter having a high selectivity, the movement of the resonant frequency could be critical. An estimate of the change in the zero (pole) location is given by the *zero (pole) sensitivity*,[7] which is defined as follows:

DEFINITION 5-5 *Let $s = s_i$ be a zero (pole) of $T(s, k)$ when k takes its nominal value. The* zero (pole) sensitivity *of $T(s, k)$ is then defined as*

$$(5.51) \qquad \mathcal{S}_k^{s_i} \triangleq \left. \frac{ds_i}{dk/k} \right|_{s=s_i}$$

Expression 5.51 can be considered as defining the root sensitivity of the polynomial $N(s, k)$ if $s = s_i$ is a zero of $T(s, k)$. Similarly if $s = s_i$ is a pole, then expression 5.51 defines the root sensitivity of the polynomial $D(s, k)$.

Observe that both the pole and zero sensitivities are numbers. If the root is real, the corresponding root sensitivity is real; otherwise they are complex numbers.

Coefficient Sensitivity

The coefficients of the numerator and the denominator polynomials of a network function are implicit functions of the variable parameter k. We can thus introduce the concept of coefficient sensitivity that would give us an estimate of the change in coefficients due to an incremental change in the parameter k.

DEFINITION 5-6 *Let*

$$(5.52) \qquad Q(s, k) = \sum_{j=0}^{n} Q_j s^j$$

be the polynomial of interest. The coefficient sensitivity *is then defined as*[8]

$$(5.53) \qquad \mathcal{S}_k^{Q_j} \triangleq \frac{d(\ln Q_j)}{d(\ln k)} = \frac{dQ_j/Q_j}{dk/k}$$

---

[7] J. G. Truxal and I. M. Horowitz, "Sensitivity considerations in active network synthesis," *Proc. 2nd Midwest Symp. on Circuit Theory*, Michigan State Univ., East Lansing, Mich., December 1956.

[8] I. M. Horowitz, "Optimization of negative impedance conversion methods of active *RC* synthesis," *IRE Trans. on Circuit Theory*, **CT-6**, 296–303 (September 1959).

COMPUTATION OF SENSITIVITIES / 173

Thus, the coefficient sensitivity is a measure of the fractional change of each coefficient of a polynomial as a result of the fractional change of the parameter k. Note that the coefficient sensitivity is again a number in contrast to $S_k^{T(s)}$, which is a function of s. The coefficient sensitivity, unlike the sensitivity function and the pole (zero) sensitivity is not a physically measurable quantity. It is introduced as a convenient aid in sensitivity analysis.

5-3 COMPUTATION OF SENSITIVITIES

Four distinct concepts of sensitivity were introduced in the previous section. Our next problem is to formulate some methods that will facilitate the computations of the sensitivities.

Calculation of Sensitivity Function

The bilinear form of network function can be used to obtain a simple expression for $S_k^{T(s)}$. From Equations 5.24, 5.26, and 5.34 we arrive at

(5.54) $$S_k^{T(s)} = \frac{D_1(s)}{D(s, k)} - \frac{N_1(s)}{N(s, k)}$$

Equation 5.54 can be reexpressed as

(5.55) $$S_k^{T(s)} = k\left[\frac{N_2(s)}{N(s, k)} - \frac{D_2(s)}{D(s, k)}\right]$$

which follows from Equation 5.54 if the following identities are used:

$$D_1(s) = D(s, k) - kD_2(s)$$
$$N_1(s) = N(s, k) - kN_2(s)$$

If we are interested in $S_k^{T(j\omega)}$, then we can use

(5.56) $$S_k^{T(j\omega)} = \frac{D_1(j\omega)}{D(j\omega, k)} - \frac{N_1(j\omega)}{N(j\omega, k)}$$

from which one arrives at expressions for calculating the gain and phase sensitivities

(5.57a) $$S_k^{\alpha(\omega)} = \text{Re}\left[\frac{D_1(j\omega)}{D(j\omega, k)} - \frac{N_1(j\omega)}{N(j\omega, k)}\right]$$

and

(5.57b) $$S_k^{\beta(\omega)} = \text{Im}\left[\frac{D_1(j\omega)}{D(j\omega, k)} - \frac{N_1(j\omega)}{N(j\omega, k)}\right]$$

174 / SENSITIVITY AND RELATED TOPICS

Example 5-3 The inverting amplifier of Figure 5-5a is characterized by the following transfer voltage ratio:

(5.58) $$T(s) = \frac{-\mu R_2}{R_1 + R_2 + \mu R_1}$$

which is derived using the equivalent representation shown in Figure 5-5b. If we are interested in computing S_μ^T, it is convenient to recast Equation 5.58 in the form of Equation 5.19. By inspection, we obtain for this case:

(5.59) $\quad N_1 = 0, \quad N_2 = -R_2, \quad D_1 = R_1 + R_2, \quad D_2 = R_1$

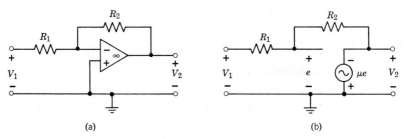

(a)   (b)

Figure 5-5 An inverting type voltage amplifier.

Use of Equation 5.59 in Equation 5.54 yields

(5.60) $$S_\mu^T = \frac{R_1 + R_2}{R_1 + R_2 + \mu R_1}$$

In a like manner, we readily obtain the other two sensitivities:

(5.61a) $$S_{R_1}^T = -\frac{R_1(\mu + 1)}{R_1 + R_2 + \mu R_1}$$

and

(5.61b) $$S_{R_2}^T = \frac{R_1(\mu + 1)}{R_1 + R_2 + \mu R_1}$$

The plots of the magnitudes of these sensitivities with respect to μ are shown in Figure 5-6 for typical values of R_1 and R_2. We note that $|S_{R_i}^T|$ is a monotonically increasing function of μ and $|S_\mu^T|$ is a monotonically decreasing function.

It is a common practice to use an amplifier with a very large gain because then the transfer voltage ratio becomes insensitive to variations in μ. This can be seen from Equation 5.60. Thus, as $\mu \to \infty$, $S_\mu^T \to 0$, and the transfer voltage ratio is given by $-R_2/R_1$ or, in other words, the overall gain of the inverting amplifier depends only on the resistor values.

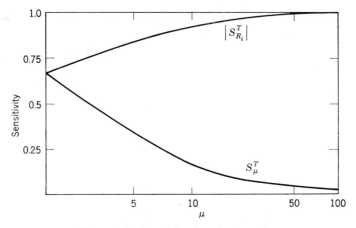

Figure 5-6 Sensitivity plots for $R_1 = R_2$.

However, we note from Figure 5-6 that $|S_{R_i}^T|$ approaches the maximum value of unity as $\mu \to \infty$. Note from Equation 5.61, the minimum value of $|S_{R_i}^T|$ is 0.5 which is obtained when $\mu = 0$. $|S_{R_i}^T|$ equal to one implies that a 10 percent change in the resistor values will cause a 20 percent change in the transfer voltage ratio in the worst case, if the gain of the operational amplifier is infinitely large.

This example points out the penalty that often one must pay in desensitizing the network performance with respect to one parameter only. This is the increase in sensitivity with respect to other parameters. For some applications, if one parameter is more critical in comparison to others, it may be profitable to minimize network sensitivity with respect to the most critical parameter only. To minimize the effect of variation of the remaining parameters, good quality components can be used in the design to keep the variations in the remaining parameters negligible.

Calculation of Pole and Zero Sensitivities

Formulas for calculating pole and zero sensitivities are easily obtained from Equation 5.10. Let us first consider the movement of the poles. Let $s = p_j$ be a multiple pole of order m of $T(s, k)$, i.e., $s = p_j$ is a zero of multiplicity m of the return difference $F_k(s)$. It is clear from Equation 5.24 that at $s = p_j$

$$(5.62) \qquad D_1(p_j) = -kD_2(p_j)$$

It is also seen from Equation 5.24, an incremental change δk in the parameter k will cause a change

$$(5.63) \qquad \delta f_k(s) = \frac{\delta k D_2(s)}{D_1(s)}$$

in the return difference. Hence from Equation 5.10, the change in the pole location is given by

$$\delta p_j = \sqrt[m]{-L_m[\delta f_k(p_j)]} = \sqrt[m]{-L_m\left[\frac{\delta k D_2(p_j)}{D_1(p_j)}\right]}$$

which by using Equation 5.62 reduces to

(5.64) $$\delta p_j = \sqrt[m]{L_m\left[\frac{\delta k}{k}\right]}$$

In Equation 5.64 L_m is given as

(5.65) $$L_m = \frac{(s-p_j)^m}{F_k(s)}\bigg|_{s=p_j}$$

which follows from Equation 5.8. Expression 5.64 indicates that a multiple pole of order m becomes m simple poles situated equidistant on a circle centered at $s = p_j$.

If we use Equation 5.64 in the definition equation (5.51) to calculate the pole sensitivity, we observe that the pole sensitivity of a multiple pole is infinite even though the actual displacement δp_j due to an incremental change in k is finite and small. As a result, the definition equation (5.51) should not be used to calculate root sensitivity of multiple roots. For the case of simple poles, Equations 5.64 and 5.65 lead to an interesting result when used in Equation 5.51. For $m = 1$, we note from (5.65)

$$L_1 = \frac{(s-p_j)}{F_k(s)}\bigg|_{s=p_j} = \left\{\begin{array}{l}\text{Residue of } 1/F_k(s) \\ \text{at a pole at } s = p_j\end{array}\right\}$$

Therefore, from Equation 5.64,

(5.66a) $$S_k^{p_j} = \frac{\delta p_j}{\delta k/k} = L_1 = \left\{\begin{array}{l}\text{Residue of } 1/F_k(s) \\ \text{at a pole at } s = p_j\end{array}\right\}$$

Use of Equation 5.24 in Equation 5.66a leads to

(5.66b) $$S_k^{p_j} = \frac{(s-p_j)D_1(s)}{D(s,k)}\bigg|_{s=p_j}$$

which is equivalent to

(5.66c) $$S_k^{p_j} = \frac{-k(s-p_j)D_2(s)}{D(s,k)}\bigg|_{s=p_j}$$

The last expression follows from Equation 5.62. Equations 5.66 can be used to calculate the pole sensitivity of a simple pole of $T(s, k)$.

It should be clear from the above, if $s = p_j$ and $s = p_j^*$ are a pair of complex conjugate poles, then

(5.67) $$S_k^{p_j^*} = (S_k^{p_j})^*$$

The preceding discussion can be duplicated for the calculation of the zero sensitivity of $T(s, k)$. For example, it can be shown easily that if $s = z_i$ is a simple zero of $T(s, k)$, then the corresponding zero sensitivity is given as

(5.68) $$S_k^{z_i} = \frac{(s - z_i)}{F_k^0(s)}\bigg|_{s=z_i} = \left\{\begin{array}{l}\text{Residue of } 1/F_k^0(s) \\ \text{at a pole at } s = z_i\end{array}\right\}$$

The pole and zero sensitivities are simply related to the sensitivity function as demonstrated next. For convenience we write

(5.69) $$T(s, k) = H \frac{\prod_{i=1}^{m}(s - z_i)}{\prod_{j=1}^{n}(s - p_j)}$$

where H, z_i, and p_j are implicit functions of k. Taking the logarithm of both sides of Equation 5.69 and assuming simple poles and zeros we have

$$\ln T = \ln H + \sum_{i=1}^{m} \ln(s - z_i) - \sum_{j=1}^{n} \ln(s - p_j)$$

Hence,

$$\frac{dT}{T} = dk\left[\frac{1}{H}\frac{\partial H}{\partial k} - \sum_{i=1}^{m}\frac{\partial z_i/\partial k}{(s - z_i)} + \sum_{j=1}^{n}\frac{\partial p_j/\partial k}{(s - p_j)}\right]$$

from which we obtain the desired expression[9]

(5.70) $$S_k^{T(s)} = S_k^H - \sum_{i=1}^{m}\frac{S_k^{z_i}}{(s - z_i)} + \sum_{j=1}^{n}\frac{S_k^{p_j}}{(s - p_j)}$$

Expression 5.70 indicates the validity of our intuitive feeling that the contribution of the pole (or zero) sensitivity to the overall sensitivity function is important in the vicinity of that pole (or zero). The formulas for computing the pole and zero sensitivities, Equations 5.66 and 5.68, can also be derived from expression 5.70.

A simple expression for the sum of root sensitivities of a polynomial $Q(s, k)$ is next derived. We first express the polynomial as

(5.71) $$Q(s, k) = \sum_{j=0}^{n} Q_j s^j = \prod_{j=1}^{n}(s - q_j)$$
$$= (A_n + kB_n)s^n + (A_{n-1} + kB_{n-1})s^{n-1} + \cdots$$

Therefore,

$$\sum \text{zeros of } Q(s, k) = \sum_{j=1}^{n} q_j = -\frac{A_{n-1} + kB_{n-1}}{A_n + kB_n}$$

---

[9] H. Ur, "Root locus properties and sensitivity relations in control systems," *IRE Trans. Automatic Control*, **AC-5**, 57–65 (January 1960).

Differentiation of both sides of the previous expression with respect to k yields

$$\sum_{j=1}^{n} \frac{dq_j}{dk} = -\frac{B_{n-1}(A_n + kB_n) - B_n(A_{n-1} + kB_{n-1})}{(A_n + kB_n)^2}$$

from which we obtain[10]

(5.72) $$\sum_{j=1}^{n} S_k^{q_j} = \frac{k(B_n A_{n-1} - B_{n-1} A_n)}{(A_n + kB_n)^2}$$

Two special cases of interest are given below:

(5.73a) $$B_n = 0 \Rightarrow \sum_{j=1}^{n} S_k^{q_j} = -\frac{kB_{n-1}}{A_n}$$

(5.73b) $$A_n = 0 \Rightarrow \sum_{j=1}^{n} S_k^{q_j} = \frac{A_{n-1}}{kB_n}$$

Equations 5.72 and 5.73 can be used to check sensitivity calculations. Some of the ideas discussed above are best illustrated by means of an example.

Example 5-4 We plan to compare the root sensitivity of each of the following possible decompositions of the denominator polynomial of a second-order Bessel filter:

(5.74a) $\quad Q(s, k) = (s^2 + 3s + 3) = (s + 4)(s + 1) - k(2s + 1)$

(5.74b) $\quad\quad\quad\quad\quad\quad\quad\quad\quad = (s + 1.732)^2 - k(0.464s)$

(5.74c) $\quad\quad\quad\quad\quad\quad\quad\quad\quad = (s + 1.5)^2 + k(0.75)$

In each of the above decompositions, the nominal value of k is unity. The first two have been obtained using $RC{:}{-}RC$ type decomposition, and the last one using $RC{:}RL$ type decomposition of $Q(s, k)$.

The original pole positions are at

$$p_1 = -\frac{3}{2} + j\frac{\sqrt{3}}{2}; \quad p_2 = -\frac{3}{2} - j\frac{\sqrt{3}}{2}$$

Consider the decomposition (5.74a) first. The return difference is given as

$$F_k(s) = \frac{s^2 + 3s + 3}{(s + 4)(s + 1)}$$

---

[10] F. F. Kuo, "Pole-zero sensitivity in network functions," *IRE Trans. on Circuit Theory*, **CT-5**, 372–373 (December 1958).

and hence

(5.75a)
$$S_k^{p_1} = \frac{\left(s + \frac{3}{2} - j\frac{\sqrt{3}}{2}\right)(s+4)(s+1)}{s^2 + 3s + 3}\bigg|_{s=-3/2+j(\sqrt{3}/2)}$$

$$= \frac{(s+4)(s+1)}{s + \frac{3}{2} + j\frac{\sqrt{3}}{2}}\bigg|_{s=-3/2+j(\sqrt{3}/2)}$$

$$= \frac{2s+1}{s + \frac{3}{2} + j\frac{\sqrt{3}}{2}}\bigg|_{s=-3/2+j(\sqrt{3}/2)}$$

$$= 1 + j\frac{2}{\sqrt{3}} = 1.527\underline{/49°8'}$$

In a similar manner, we obtain the pole sensitivities of the other two decompositions:

(5.75b)
$$S_k^{p_1} = \frac{0.464s}{s + \frac{3}{2} + j\frac{\sqrt{3}}{2}}\bigg|_{s=-3/2+j(\sqrt{3}/2)} = 0.232 + j0.402 = 0.464\underline{/60°}$$

(5.75c)
$$S_k^{p_1} = \frac{-\frac{3}{4}}{s + \frac{3}{2} + j\frac{\sqrt{3}}{2}}\bigg|_{s=-3/2+j(\sqrt{3}/2)} = j\frac{\sqrt{3}}{4} = 0.4305\underline{/90°}$$

Several interesting observations can be made from the results given in (5.75). We note that RC:$-RC$ type decomposition as given by the second decomposition (5.74b) yields a lower root sensitivity than the one given in (5.74a). But compared to the previous two, the RC:RL type decomposition given in (5.74c) yields the lowest pole sensitivity. Actually, these are in general true. "Optimum" decompositions will be discussed later.

To check our result, we can use Equations 5.72 and 5.73. We illustrate this approach for the first decomposition only.

It follows from Equations 5.67 and 5.75 that

$$S_k^{p_2} = (S_k^{p_1})^* = 1 - j\frac{2}{\sqrt{3}}$$

which implies

$$\sum_{j=1}^{2} S_k^{p_j} = 2$$

Again, comparison of Equations 5.74a and 5.71 indicates

$$A_2 = 1, \quad B_2 = 0, \quad A_1 = 5, \quad B_1 = -2$$

Use of above information in Equation 5.73a yields

$$\hat{S}_k^{p_1} + \hat{S}_k^{p_2} = -\frac{1 \times (-2)}{1} = 2$$

as expected.

Calculation of Coefficient Sensitivity

If the decomposition of the polynomial, say $Q(s, k)$, is given in the bilinear form

$$Q(s, k) = Q_1(s) + kQ_2(s)$$

and the factored form of $Q(s, k)$ for the nominal value of k is known, then the calculation of the root sensitivity presents no problem. In some cases, the pertinent polynomial may be given in the form of Equation 5.71. In this situation, the coefficient sensitivities are easier to compute:

$$\hat{S}_k^{Q_j} = \frac{k}{Q_j} \cdot B_j \tag{5.76}$$

Some interesting relations exist between the sum of the root sensitivities and the coefficient sensitivities. It follows from Equation 5.71:

$$\sum_{j=1}^{n} q_j = [\text{sum of zeros of } Q(s, k)] = -\frac{Q_{n-1}}{Q_n} \tag{5.77}$$

from which the following result is obtained:

$$\sum_{j=1}^{n} \hat{S}_k^{q_j} = (\hat{S}_k^{Q_n} - \hat{S}_k^{Q_{n-1}}) \cdot \frac{Q_{n-1}}{Q_n} \tag{5.78}$$

Similarly, from the relation

$$\prod_{j=1}^{n} q_j = (-1)^n \frac{Q_0}{Q_n} \tag{5.79}$$

the following expression can be derived:

$$\sum_{j=1}^{n} \frac{\hat{S}_k^{p_j}}{p_j} = (-1)^n [\hat{S}_k^{Q_0} - \hat{S}_k^{Q_n}] \tag{5.80}$$

Example 5-5 The decomposition of (5.74a) can be rewritten as

$$Q(s, k) = s^2 + 3s + 3 = s^2 + (5 - 2k)s + (4 - k) \tag{5.81}$$

where k is nominally unity. Use of Equation 5.76 results in

(5.82a) $\qquad \hat{S}_k^{Q_2} = 0, \qquad \hat{S}_k^{Q_1} = -\tfrac{2}{3}, \qquad \hat{S}_k^{Q_0} = -\tfrac{1}{3}$

In an analogous manner, for the second decomposition (5.74b), one can show

(5.82b) $\qquad \hat{S}_k^{Q_2} = 0, \qquad \hat{S}_k^{Q_1} = -0.1547, \qquad \hat{S}_k^{Q_0} = 0$

For the last decomposition, Equation 5.74c, we have

(5.82c) $\qquad \hat{S}_k^{Q_2} = 0, \qquad \hat{S}_k^{Q_1} = 0, \qquad \hat{S}_k^{Q_0} = 0.25$

We observe that the second RC:–RC type decomposition (5.74b) has lower coefficient sensitivities in comparison to the first one (Equation 5.74a).

A simple relation between the root sensitivity and the coefficient sensitivities is next derived.[11] The derivation is based on the method outlined earlier in Section 5-1. Let $s = q_j$ be a simple zero of $Q(s)$ that moves to a new location at $s = q'_j$ when each coefficient of $Q(s)$ increases by an incremental amount, i.e.,

(5.83)
$$Q(s) = \sum_{i=0}^{n} Q_i s^i = 0 \quad \text{at} \quad s = q_j$$
$$Q(s) + \sum_{i=0}^{n} (\delta Q_i) s^i = 0 \quad \text{at} \quad s = q'_j$$

If $s = q'_j$ is close to $s = q_j$, it follows from Equation 5.10,

(5.84) $\qquad dq_j = q'_j - q_j = -L_1^{(j)} \left[\sum_{i=0}^{n} (\delta Q_i)(q_j)^i \right]$

where

(5.85) $\qquad L_1^{(j)} = \dfrac{(s - q_j)}{Q(s)} \bigg|_{s=q_j} = \dfrac{1}{\dfrac{dQ(s)}{ds}} \bigg|_{s=q_j}$

Equation 5.85 follows from Equation 5.8. If the change in the coefficients has been caused by an incremental change of the parameter k, then we obtain from Equation 5.84 the desired expression

(5.86) $\qquad S_k^{q_j} = -L_1^{(j)} \left[\sum_{i=0}^{n} Q_i \hat{S}_k^{Q_i} \cdot (q_j)^i \right]$

[11] G. Martinelli, "On the matrix analysis of network sensitivities," *Proc. IEEE*, **54**, 72 (January 1966).

One important conclusion can be made from the above expression: for a given type of polynomial decomposition, e.g., RC:–RC type decomposition, the root sensitivity can be minimized by minimizing the coefficient sensitivities (which are of the same sign). This is illustrated by comparing Equations 5.82a and 5.82b, which show that the second RC:–RC type decomposition (Equation 5.74b) yields both lower root sensitivity and coefficient sensitivity.

5-4 SENSITIVITY FUNCTION FOR LARGE VARIATION OF A PARAMETER[12]

In this section, we develop an expression for calculating the sensitivity function for finite variation of a single network parameter only. This is easily obtained from expression 5.20. When k changes from its nominal value k_0 to $k_0 + \Delta k$, the network function $T(s, k_0)$ will change to $T(s, k_0 + \Delta k)$. Denoting

$$T(s, k_0 + \Delta k) = T(s, k_0) + \Delta T$$

we note

(5.87) $$T(s, k_0) + \Delta T = T_0(s) + \frac{(k_0 + \Delta k)A(s)B(s)}{1 - (k_0 + \Delta k)C(s)}$$

where

(5.88) $$T(s, k_0) = T_0(s) + \frac{k_0 A(s)B(s)}{1 - k_0 C(s)}$$

From Equations 5.87 and 5.88, it follows that

(5.89) $$\Delta T = \frac{A(s)B(s)\,\Delta k}{[1 - (k_0 + \Delta k)C(s)][1 - k_0 C(s)]}$$

Since

$$T(s, k_0) - T_0(s) = \frac{A(s)B(s)k_0}{1 - k_0 C(s)}$$

we obtain from Equation 5.89 the final expression:

(5.90) $$\frac{\Delta T/[T(s, k_0) - T_0(s)]}{\Delta k/k_0} = \frac{1}{1 - (k_0 + \Delta k)C(s)} = \frac{1}{F_{(k_0 + \Delta k)}}$$

where $F_{(k_0 + \Delta k)}$ is the return difference evaluated at the modified value $k_0 + \Delta k$ of the parameter k.

Expression 5.90 can be considered as defining the sensitivity of $T(s, k)$ to large variation of the parameter k. Note that in case of networks for which the leakage transmission is negligible, the left-hand side of Equation

---

[12] Truxal and Horowitz, loc. cit.

5.90 is identical in form to the expression for sensitivity to incremental variation (see Equation 5.27).

5-5 SOME GENERAL REMARKS ON SENSITIVITY MINIMIZATION[13]

We have thus far introduced various sensitivity definitions and the methods to calculate them. Out of the definitions given, the sensitivity function, the gain sensitivity, the phase sensitivity, and the pole (zero) sensitivity are the only ones which are physically measurable quantities and can be experimentally determined. The sensitivity function $S_k^{T(j\omega)}$, the gain sensitivity $S_k^{\alpha(\omega)}$, and the phase sensitivity $S_k^{\beta(\omega)}$ are useful sensitivity measures for computing the effect of incremental variation of a network parameter on the frequency responses over a band of interest. The sensitivity function or the gain sensitivity can also be used in case of pure resistive networks. Examples of networks, where these three sensitivity measures are more suitable to evaluate the network performance, are wideband filters, equalizers, attenuators, etc. Polynomial sensitivity can be used in case of networks whose frequency response is determined mainly by either the numerator or the denominator polynomial over some band of frequencies. In the case of networks having one or more dominant critical frequencies near the $j\omega$-axis (e.g., highly selective band-pass filters, notch filters), the zero (pole) sensitivity is the most useful sensitivity measure. In addition, since the pole (zero) sensitivity contributes most to the sensitivity function in the neighborhood of the corresponding pole (zero), this type of sensitivity can also be used to get a rough estimate of the general sensitivity picture.

At this point, it would be appropriate to examine and interpret the various sensitivity concepts. How good are these definitions in estimating the performance of the network in practice? To answer this question let us reexamine the parent expressions, Equation 5.27, 5.38, 5.39, 5.49, and 5.51. Observe that all of these definitions involve only the first derivative of a function with respect to the parameter k. Hence, the sensitivity function, pole sensitivity, etc., are quantitive measures of the functional performance of the network only for an infinitesimal change in the parameter k. Thus, for example, designing a network to have the gain sensitivity $S_k^{\alpha(\omega)}$ equal to zero for some nominal value of $k = k_0$ and at a frequency $\omega = \omega_c$ only means that the slope of the plot of $\alpha(\omega_c, k)$ versus k is zero at $k = k_0$. It should be especially noted that since there is no information on the relative size of higher derivatives of $\alpha(\omega_c, k)$ with

---

[13] The major portion of the discussion in this section is based on Truxal and Horowitz, *loc. cit.*

respect to k, in general for large changes of k, the corresponding change in α may not be small.

With the above considerations in mind, let us now investigate the possibilities of minimizing the sensitivities. First we consider minimization of the sensitivity function $S_k^{T(s)}$.

Examination of Equation 5.30 reveals that there are basically three ways by which $S_k^{T(s)}$ can be made small. A simple approach would be to design the network so that the leakage transmission is almost equal to the overall transmission over the frequency band of interest. In active RC network synthesis, complex poles are generally provided by means of feedback; thus to make $T_0(s)$ close to $T(s, k)$ poses a difficult problem. A second scheme would be to choose a suitable function as the return difference, $F_k(s)$. For example, $S_k^{T(j\omega)}$ can be made equal to zero at preselected points on the $j\omega$-axis by making $F_k(s)$ to have poles at these points. This is difficult to achieve if the circuit of interest is an active RC network. Then the poles of $F_k(s)$, being the poles of the RC network functions $T_0(s)$, $A(s)$, $B(s)$, and $C(s)$, can be only negative real.

The third approach is to make $F_k(j\omega)$ very large in the frequency band of interest, which in effect will make $S_k^{T(j\omega)}$ small. From expression 5.23, we note that this can be achieved by making the loop gain $kC(j\omega)$ very large over the frequency band. However, for stability reasons, the loop gain must cut off at a slow rate (less than 12 db/octave).[14] Thus large values of $kC(j\omega)$ would require a wideband network, which in some cases may be uneconomical. Of the three approaches, the last approach is the only one that is practically feasible provided that the operating range of frequency is within dc to several kHz. This is the basis of much of the circuits realizations using operational amplifiers, where the operational amplifiers are built with very large gain (typically 80 db) to make the loop gain large.

An alternate sensitivity minimization scheme was proposed by Herbst,[15] which does not require large values of loop gain. This scheme is applicable to cases, where the denominator (pole) polynomial $D(s, k)$ contributes significantly to the network response. In this case, we can make use of the definition of polynomial sensitivity as given by Equation 5.50. We observe from Equation 5.19 that

$$D(s, k) = D_1(s) + k D_2(s)$$

from which we easily obtain

(5.91) $$S_k^{D(j\omega, k)} = \frac{k D_2(j\omega)}{D(j\omega, k)}$$

[14] This is proved in Section 11-2.
[15] N. M. Herbst, *loc. cit.*

This indicates that the pole polynomial sensitivity can be minimized over the frequency band of interest by making $D_2(s)$ to have $j\omega$-axis zeros over the band. Since zeros of $D_2(s)$ are the transmission zeros of the feedback path (Equation 5.21b), this can be achieved in the case of active RC networks by properly designing the feedback path.

It is interesting to point out here two cases for which the sensitivity function and the pole polynomial sensitivity are related. In the case of active networks having zero leakage transmission, i.e., $N_1(s)$ equal to zero, we note by comparing Equations 5.54 and 5.91 that

$$S_k^{T(j\omega)} = 1 - S_k^{D(j\omega)}$$

If on the other hand, $N_2(s)$ is equal to zero, then

$$S_k^{T(j\omega)} = -S_k^{D(j\omega)}$$

For the second case, minimizing the pole polynomial sensitivity also minimizes the sensitivity function.

Now consider the problem of minimizing the pole sensitivity. Equation 5.66a indicates that the pole sensitivity can be made small by minimizing the residue of $F_k(s)^{-1}$ at each zero $s = p_j$ of $F_k(s)$. It is clear from Equation 5.66b that the residue would be small if $D_1(s)$ also has a zero in the neighborhood of $s = p_j$. In active RC network configurations, $D_1(s)$ usually has only negative real roots whereas, in general, $D(s, k)$ will be required to have complex zeros. Thus the pole sensitivity at $s = p_j$ can be minimized in the above way, provided that the pertinent pole is not too critical in determining the network response. An alternate scheme would be to make k, and hence the loop gain, sufficiently large; this, however, would lead to a stability problem, as discussed earlier.

At this point, it would be profitable to examine the relationship between the root sensitivity and the locus of the pertinent root for values of variable parameter k in the neighborhood of the nominal value. Let $s = p_m$ be a root of the polynomial $Q(s, k)$ when the variable parameter k takes the nominal value. For an incremental change δk of the parameter k, the root at $s = p_m$ will move by an incremental amount $\delta p_m = \delta \sigma_m + j\delta \omega_m$. The root sensitivity $S_k^{p_m}$ is given as

$$S_k^{p_m} = \frac{\delta p_m}{\delta k / k} = \frac{\delta \sigma_m + j\delta \omega_m}{\delta k / k}$$

If the parameter k is real, then

$$\arg S_k^{p_m} = \tan^{-1} \frac{\delta \omega_m}{\delta \sigma_m}$$

Note from Figure 5-7, that the angle of the root locus of $Q(s, k)$ at $s = p_m$ with respect to the σ-axis is also given by $\tan^{-1}(\delta\omega_m/\delta\sigma_m)$.[16] This is a very useful relation. For example, if $s = p_m$ is a pole that is very close to the $j\omega$-axis, for stability reasons it is not only desirable to make $|S_k^{p_m}|$ small but also $\arg S_k^{p_m}$ must be made close to $\pm 90°$. Let us now compare the angle of the root sensitivities of the three decompositions of $(s^2 + 3s + 3)$ as computed earlier in Example 5-4. From expressions 5.75a, 5.75b, and 5.75c, we observe the root sensitivity given by expression

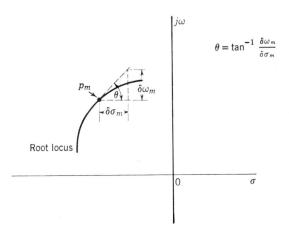

Figure 5-7 To illustrate the relation between the root sensitivity and the root-locus plot.

5.75c has an angle of 90° and also has the smallest magnitude. Thus we conclude, the third decomposition (Equation 5.74c) is the best in terms of sensitivity and stability both.

We have shown before that a practical way of minimizing either the sensitivity function or the pole sensitivity is by making the loop gain very large. But this approach leads to stability problems in many cases unless we restrict our attention to a small band in the low-frequency range. Minimum sensitivity can be achieved without increasing the loop gain if we relax the strict requirement that the network be optimal for all possible input functions. In quite a few applications, the input signals are known. For sinusoidal inputs, one is more interested either in minimizing the gain sensitivity $S_k^{\alpha(\omega)}$ or the phase sensitivity $S_k^{\beta(\omega)}$. We have shown before that if the leakage transmission can be neglected, $S_k^{\alpha(\omega)}$ is given by Equation 5.48, which indicates that the sensitivity can be minimized by making

---
[16] Pointed out by G. S. Moschytz.

$\mathcal{R}(\omega, k)$ close to zero over the band of interest. Since

$$\mathcal{R}(\omega, k) = \text{Re} \{F_k(j\omega)\} = 1 - k \, \text{Re} \{C(j\omega)\}$$

$\mathcal{R}(\omega, k)$ can be made small by making the real part of the loop gain $kC(j\omega)$ close to unity. This last approach, which usually would not lead to instability, does not appear to have been followed in optimum active network design.[17]

Very little is known on the minimization of sensitivity due to large parameter variation. Here we would like to point out a special case where the sensitivities due to incremental variation and to large finite variation are related. This is the case of active network configurations, for which the leakage transmission is negligible over the band of interest. In such cases, $S_k^{T(s)}$ for the nominal value of k is simply the reciprocal of $F_{k_0}(s)$ and the sensitivity to finite variation of k is $1/F_{(k_0+\Delta k)}(s)$. Thus, once $F_k(s)$ is determined, $F_{(k+\Delta k)}(s)$ is also determined automatically.

5-6 INCREMENTAL VARIATIONS OF SEVERAL PARAMETERS

Our interest now focuses on variations of more than one parameter in a network. Even though active elements are usually considered to have a significant variation in their characteristics, passive elements like the resistors and the capacitors generally show some amount of variation in their element values. We have also noted in Example 5-3 that minimization of sensitivity with respect to one parameter maximized the sensitivities with respect to other parameters. Hence, in general, the network desensitization must be viewed as a problem of minimizing the sensitivity with respect to several parameters at the same time. Or, in other words, it is a multiparameter sensitivity problem.

It is thus desirable to have some means of measuring the sensitivity of a network when more than one parameter changes incrementally.

Multiparameter Sensitivity Function

Multiparameter sensitivity has been investigated by many authors. The concept of return difference matrix has been introduced by Sandberg[18] for a network having a single input and single output and a multi-input and output variable subnetwork. This concept was further generalized by Kuh,[19] who used the return difference and null-return difference matrix

---

[17] An example of this approach is given in Problem 5.13.

[18] I. W. Sandberg, "Linear multi-loop feedback systems," *Bell System Tech. J.*, **42**, 355–382 (March 1963).

[19] E. S. Kuh, "Some results in linear multiple loop feedback systems," *Proc. 1st Allerton Conference in Circuit and System Theory*, Univ. of Illinois, pp. 471–487 (1963).

to define a sensitivity matrix. The development parallels the one described in Section 5-2. Many other definitions will be found in the references listed at the end of this chapter.

Here we consider the definition advanced by Kuo and Goldstein.[20] Let the vector

(5.92) $$\mathbf{k} = [k_1, k_2, \ldots, k_m]$$

define the set of m variable elements. For an incremental change in $\mathbf{k}$, we can write the fractional change in the network function $T(s, k)$ as

(5.93) $$\frac{\Delta T}{T} = d\{\ln T\} = \sum_{j=1}^{m} \frac{\partial(\ln T)}{\partial(\ln k_j)} d(\ln k_j)$$

assuming the elements k_j to be independent of each other. The multiparameter sensitivity function $\mathbf{S}_\mathbf{k}^T$ can thus be defined as the vector

(5.94) $$\mathbf{S}_\mathbf{k}^{T(s)} = \left[\frac{\partial(\ln T)}{\partial(\ln k_1)}, \frac{\partial(\ln T)}{\partial(\ln k_2)}, \ldots, \frac{\partial(\ln T)}{\partial(\ln k_m)} \right]$$
$$= [S_{k_1}^{T(s)}, S_{k_2}^{T(s)}, \ldots, S_{k_m}^{T(s)}]$$

From Equation 5.94, the multiparameter gain and phase sensitivities can be derived by taking the real part and the imaginary part of the vector $\mathbf{S}_\mathbf{k}^{T(j\omega)}$.

A reasonable criterion of designing an optimum network would be to minimize the magnitude of $\mathbf{S}_\mathbf{k}^{T(j\omega)}$, which is given by

(5.95a) $$\left| \mathbf{S}_\mathbf{k}^{T(j\omega)} \right| = \left\{ \sum_{j=1}^{m} \left| \frac{\partial(\ln T)}{\partial(\ln k_j)} \right|^2 \right\}^{1/2} = \{[\mathbf{S}_\mathbf{k}^{T(j\omega)}]^* \cdot [\mathbf{S}_\mathbf{k}^{T(j\omega)}]^t\}^{1/2}$$

where $[\mathbf{S}_\mathbf{k}^{T(j\omega)}]^t$ denotes the transpose of the vector $\mathbf{S}_\mathbf{k}^{T(j\omega)}$. If some parameters are more critical than the others, then the following factor

(5.95b) $$\left\{ \sum_{j=1}^{m} \xi_j(\omega) \left| \frac{\partial(\ln T)}{\partial(\ln k_j)} \right|^2 \right\}^{1/2}$$

could be minimized instead of $|\mathbf{S}_\mathbf{k}^{T(j\omega)}|$. In the expression 5.95b, $\xi_j(\omega)$ determines the weight of different single parameter sensitivities.

There exists no direct and simple method of designing a network with the smallest multiparameter sensitivity. The best approaches at present are the digital computer-based methods which make use of some iterative optimization techniques. Roughly, in these methods, an active network having the desired transmission characteristic is first designed. In this

---

[20] F. F. Kuo and A. J. Goldstein, "Multi-parameter sensitivity," *IEEE Trans. on Circuit Theory*, **CT-8**, 177–178 (June 1961).

respect, any of the design techniques outlined in this book can be followed. Then, many equivalent networks are generated by the computer until one with the lowest multiparameter sensitivity is found. Many variations of the above approach will be found in the literature.[21]

Some interesting relations can be formulated between the single-parameter sensitivities of $T(s)$ with respect to various network parameters.[22] Consider an active RLC network containing the four types of controlled sources. $T(s)$ then will be a function of each network element and can be written as $T(R_i, L_i, C_i, \mu_i, \alpha_i, r_i, g_i, s)$ where R_i, L_i, and C_i denote the individual passive elements, and μ_i, α_i, r_i, and g_i denote transfer constraints of the VVT, CCT, CVT, and VCT, respectively. If we now scale the impedance level of the network by an amount a, it follows that

(5.96) $\quad T\left(aR_i, aL_i, \dfrac{C_i}{a}, \mu_i, \alpha_i, ar_i, \dfrac{g_i}{a}, s\right) = H T(R_i, L_i, C_i, \mu_i, \alpha_i, r_i, g_i, s)$

where

$\quad H = a \quad$ if $T(s)$ is an impedance function,

$\quad H = \dfrac{1}{a} \quad$ if $T(s)$ is an admittance function,

and

$\quad H = 1 \quad$ if $T(s)$ is a transfer voltage or current ratio.

Differentiating both sides of Equation 5.96 with respect to a and setting $a = 1$, and then dividing both sides of the resulting expression by $T(s)$, we arrive at the following relation:

(5.97) $\displaystyle\sum_{i=1}^{n_R} S_{R_i}^T + \sum_{i=1}^{n_L} S_{L_i}^T - \sum_{i=1}^{n_C} S_{C_i}^T + \sum_{i=1}^{n_r} S_{r_i}^T - \sum_{i=1}^{n_g} S_{g_i}^T$

$= \begin{cases} 1 & \text{if } T(s) \text{ is an impedance function} \\ 0 & \text{if } T(s) \text{ is a transfer voltage or current ratio} \\ -1 & \text{if } T(s) \text{ is an admittance function.} \end{cases}$

[21] J. D. Schoeffler, "The synthesis of minimum sensitivity networks," *IEEE Trans. on Circuit Theory*, **CT-11**, 270–276 (June 1964).

D. A. Calahan, "Computer generation of equivalent networks," *IEEE International Conv. Rec.*, **17** (Part 1) 330–337 (1964).

D. A. Calahan, "A numerical algorithm for the minimization of sensitivity," *Proc. 3rd Allerton Conf. on Circuit and System Theory*, Univ. of Illinois, pp. 394–406 (1965).

[22] M. L. Blostein, "Some bounds on sensitivity in RLC networks," *Proc. 1st Allerton Conf. on Circuit and System Theory*, Univ. of Illinois, pp. 488–501 (1963).

In Equation 5.97, n_R denotes the total number of resistors, n_L the total number of inductors, and so on.

Example 5-6 For the inverting amplifier of Figure 5-5, all the sensitivities were calculated in Example 5-3. From Equation 5.61, it is seen that

$$S_{R_1}^T + S_{R_2}^T = 0$$

as expected from expression 5.97.

Next, we consider the effect of frequency scaling. Note that only the inductors and the capacitors are frequency dependent elements. It therefore follows that

(5.98) $\quad T(R_i, aL_i, aC_i, \mu_i, \alpha_i, r_i, g_i, s) = T(R_i, L_i, C_i, \mu_i, \alpha_i, r_i, g_i, as)$

As before, we differentiate both sides of Equation 5.98 with respect to a, set $a = 1$, and then divide both sides of the resulting expression by $T(s)$. This yields

(5.99) $\quad \displaystyle\sum_{i=1}^{n_L} S_{L_i}^T + \sum_{i=1}^{n_C} S_{C_i}^T = \dfrac{d(\ln T)}{d(\ln s)}$

On the $j\omega$-axis, Equation 5.99 reduces to

(5.100) $\quad \text{Re}\left\{\displaystyle\sum_{i=1}^{n_L} S_{L_i}^{T(j\omega)} + \sum_{i=1}^{n_C} S_{C_i}^{T(j\omega)}\right\} = \dfrac{d\alpha(\omega)}{d(\ln \omega)}$

and

(5.101) $\quad \text{Im}\left\{\displaystyle\sum_{i=1}^{n_L} S_{L_i}^{T(j\omega)} + \sum_{i=1}^{n_C} S_{C_i}^{T(j\omega)}\right\} = \omega\dfrac{d\beta(\omega)}{d\omega}$

The above results can be applied to derive some bounds on the worst-case deviation. Note from Equation 5.93:

(5.102) $\quad \dfrac{\Delta |T(j\omega)|}{|T(j\omega)|} \doteq \displaystyle\sum_{j=1}^{n} \text{Re } [S_{k_j}^{T(j\omega)}] \dfrac{dk_j}{k_j}$

and

(5.103) $\quad \Delta \arg T(j\omega) = \Delta\beta(\omega) \doteq \displaystyle\sum_{j=1}^{n} \text{Im } [S_{k_j}^{T(j\omega)}] \dfrac{dk_j}{k_j}$

Let us assume that each of the n elements can change at most by $\pm 100\varepsilon$ percent. We then conclude from Equation 5.102 that deviation of $|T(j\omega)|$ from its nominal value will be maximum when all the elements k_i in the network take their extremal values. Thus maximum deviation $M(\omega)$ is given approximately by

(5.104) $\quad M(\omega) \doteq \varepsilon \displaystyle\sum_{j=1}^{n} |\text{Re } S_{k_j}^{T(j\omega)}|$

$M(\omega)$ can be considered as the worst-case deviation.

Example 5-7 We are interested in obtaining the lower bound on the worst-case deviation of the transfer voltage ratio t_v of an RC two-port. In this case, Equation 5.97 simplifies to

$$(5.105) \qquad \sum_{i=1}^{n_R} S_{R_i}^{t_v} - \sum_{i=1}^{n_C} S_{C_i}^{t_v} = 0$$

and Equation 5.100 reduces to

$$(5.106) \qquad \text{Re}\left\{\sum_{i=1}^{n_C} S_{C_i}^{t_v(j\omega)}\right\} = \omega \frac{d\alpha(\omega)}{d\omega}$$

Equations 5.104 to 5.106 jointly yield the lower bound on $M(\omega)$,

$$M(\omega) \geq 2\varepsilon\omega \frac{d\alpha(\omega)}{d\omega}$$

which is independent of the actual network configuration.

Multiparameter Pole (Zero) Sensitivity

An incremental change in the root at $s = q_j$ can be expressed as

$$(5.107) \qquad dq_j = \sum_{i=1}^{m} \frac{\partial q_j}{\partial(\ln k_i)} d(\ln k_i)$$

and we define the vector

$$(5.108) \qquad \mathbf{S}_\mathbf{k}^{q_j} \triangleq \left[\frac{\partial q_j}{\partial(\ln k_1)}, \frac{\partial q_j}{\partial(\ln k_2)}, \ldots, \frac{\partial q_j}{\partial(\ln k_m)}\right]$$

as the multiparameter root (pole or zero) sensitivity with respect to the parameter vector $\mathbf{k}$.

The sensitivity vector $\mathbf{S}_\mathbf{k}^{q_j}$ can be calculated according to a method outlined by Martinelli.[23] Martinelli's method, which is a modified version of Huelsman's approach,[24] is given below. Our starting point is Equation 5.84, which we rewrite below in matrix notation as

$$(5.109) \qquad dq_j = \mathbf{U}_j \cdot \mathbf{C}^t$$

where

$$(5.110) \qquad \begin{aligned} \mathbf{U}_j &= [-L_1^{(j)}, -L_1^{(j)}q_j, -L_1^{(j)}q_j^2, \ldots, -L_1^{(j)}(q_j)^n] \\ \mathbf{C} &= [\delta Q_0, \delta Q_1, \delta Q_2, \ldots, \delta Q_n] \end{aligned}$$

Since the coefficients are functions of the parameter vector $\mathbf{k}$,

$$Q_i = Q_i(k_1, k_2, \ldots, k_m)$$

---

[23] G. Martinelli, *loc. cit.*
[24] L. P. Huelsman, "Matrix analysis of network sensitivities," *Proc. NEC*, **19**, 1–5 (1963).

we can write

(5.111) $$\delta Q_i = \sum_{r=1}^{m} \frac{\partial Q_i}{\partial(\ln k_r)} d(\ln k_r)$$

Equations 5.109 and 5.111 can be combined as

(5.112) $$dq_j = \mathbf{U}_j \cdot \mathbf{W} \cdot \mathbf{K}^t$$

where

(5.113) $$\mathbf{W} = \begin{bmatrix} \frac{\partial Q_0}{\partial(\ln k_1)} & \frac{\partial Q_0}{\partial(\ln k_2)} & \cdots & \frac{\partial Q_0}{\partial(\ln k_m)} \\ \frac{\partial Q_1}{\partial(\ln k_1)} & \frac{\partial Q_1}{\partial(\ln k_2)} & \cdots & \frac{\partial Q_1}{\partial(\ln k_m)} \\ \cdot & \cdot & \cdots & \cdot \\ \cdot & \cdot & \cdots & \cdot \\ \frac{\partial Q_n}{\partial(\ln k_1)} & \frac{\partial Q_n}{\partial(\ln k_2)} & \cdots & \frac{\partial Q_n}{\partial(\ln k_m)} \end{bmatrix}$$

(5.114) $$\mathbf{K} = [d(\ln k_1), d(\ln k_2), \ldots, d(\ln k_m)]$$

this implies

(5.115) $$\mathbf{S}_\mathbf{k}^{q_j} = \mathbf{U}_j \cdot \mathbf{W}$$

Several comments are in order: $\mathbf{U}_j$ depends only on the polynomial $Q(s)$ and is thus independent of the network configuration. On the other hand, the matrix $\mathbf{W}$ is governed by the network configuration. For simple networks, $\mathbf{W}$ can be computed by analysis; the entries, $\partial Q_i/\partial(\ln k_r)$, will all be real because the network elements are real. For complex networks, the calculation can be performed on a digital computer using topological network analysis programs. If the displacement of some other root at $s = q_l$ is desired, all we need to do is compute the $\mathbf{U}_l$ vector for that root. The following example will illustrate the method.

Example 5-8 The characteristic polynomial of the *RC* two port of Figure 5-8 is given as

(5.116) $$Q(s) = C_1 C_2 s^2 + (G_1 C_2 + G_1 C_1 + C_1 G_2)s + G_1 G_2$$

For the following nominal values of the elements,

$$C_1 = 1, \quad C_2 = 4, \quad G_1 = 2, \quad G_2 = 6$$

$$Q(s) = 4(s^2 + 4s + 3)$$

and the nominal pole positions are at $s = p_1 = -1$ and $s = p_2 = -3$.

INCREMENTAL VARIATIONS OF SEVERAL PARAMETERS / 193

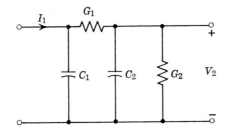

Figure 5-8 An RC two-port.

For the pole at $s = p_1$, we have

$$L_1^{(1)} = \frac{(s+1)}{4(s+1)(s+3)}\bigg|_{s=-1} = \frac{1}{8}$$

and for the second pole at $s = p_2$ we have

$$L_1^{(2)} = \frac{(s+3)}{4(s+1)(s+3)}\bigg|_{s=-3} = -\frac{1}{8}$$

Therefore,

$$\mathbf{U}_1 = [-\tfrac{1}{8} \quad -\tfrac{1}{8}(-1) \quad -\tfrac{1}{8}(-1)^2] = [-\tfrac{1}{8} \quad \tfrac{1}{8} \quad -\tfrac{1}{8}]$$

and

$$\mathbf{U}_2 = [\tfrac{1}{8} \quad \tfrac{1}{8}(-3) \quad \tfrac{1}{8}(-3)^2] = [\tfrac{1}{8} \quad -\tfrac{3}{8} \quad \tfrac{9}{8}]$$

Next step is to determine **W**. Note from Equation 5.116

$$Q_0 = G_1 G_2, \qquad Q_1 = G_1 C_2 + G_1 C_1 + C_1 G_2, \qquad Q_2 = C_1 C_2$$

Thus,

$$C_1 \frac{\partial Q_0}{\partial C_1} = 0, \qquad C_2 \frac{\partial Q_0}{\partial C_2} = 0$$

$$C_1 \frac{\partial Q_1}{\partial C_1} = C_1(G_1 + G_2), \qquad C_2 \frac{\partial Q_1}{\partial C_2} = G_1 C_2$$

$$C_1 \frac{\partial Q_2}{\partial C_1} = C_1 C_2, \qquad C_2 \frac{\partial Q_2}{\partial C_2} = C_1 C_2$$

$$G_1 \frac{\partial Q_0}{\partial G_1} = G_1 G_2, \qquad G_2 \frac{\partial Q_0}{\partial G_2} = G_1 G_2$$

$$G_1 \frac{\partial Q_1}{\partial G_1} = G_1(C_1 + C_2), \qquad G_2 \frac{\partial Q_1}{\partial G_2} = C_1 G_2$$

$$G_1 \frac{\partial Q_2}{\partial G_1} = 0, \qquad G_2 \frac{\partial Q_2}{\partial G_2} = 0$$

Substitution of the nominal values of the elements in the above equations yields

$$W = \begin{bmatrix} 0 & 0 & 12 & 12 \\ 8 & 8 & 10 & 6 \\ 4 & 4 & 0 & 0 \end{bmatrix}$$

Therefore,

(5.117) $\quad dp_1 = [-\tfrac{1}{8} \ \tfrac{1}{8} \ -\tfrac{1}{8}] \begin{bmatrix} 0 & 0 & 12 & 12 \\ 8 & 8 & 10 & 6 \\ 4 & 4 & 0 & 0 \end{bmatrix} \begin{bmatrix} dC_1/C_1 \\ dC_2/C_2 \\ dG_1/G_1 \\ dG_2/G_2 \end{bmatrix}$

$\qquad = \dfrac{1}{2}\dfrac{dC_1}{C_1} + \dfrac{1}{2}\dfrac{dC_2}{C_2} - \dfrac{1}{4}\dfrac{dG_1}{G_1} - \dfrac{3}{4}\dfrac{dG_2}{G_2}$

Similarly,

(5.118) $\quad dp_2 = \dfrac{3}{2}\dfrac{dC_1}{C_1} + \dfrac{3}{2}\dfrac{dC_2}{C_2} - \dfrac{9}{4}\dfrac{dG_1}{G_1} - \dfrac{3}{4}\dfrac{dG_2}{G_2}$

Note from Equations 5.117 and 5.118 that if the percentage change in the values of all the elements is equal, the resultant displacement of the poles is zero. This will be rigorously proved later. The worst-case deviation of the roots for 1-percent change in element values is given as

$$dp_1 = 2 \times 0.01 = 0.02$$

$$dp_2 = 6 \times 0.01 = 0.06$$

Two simple relations exist between the single-parameter root sensitivities with respect to various network elements. The development parallels the derivation of Equations 5.97 and 5.99. The relations are:[25]

(5.119) $\quad \displaystyle\sum_{i=1}^{n_L} S_{L_i}^{q_j} + \sum_{i=1}^{n_R} S_{R_i}^{q_j} - \sum_{i=1}^{n_C} S_{C_i}^{q_j} + \sum_{i=1}^{n_r} S_{r_i}^{q_j} - \sum_{i=1}^{n_g} S_{g_i}^{q_j} = 0$

(5.120) $\quad \displaystyle\sum_{i=1}^{n_L} S_{L_i}^{q_j} + \sum_{i=1}^{n_C} S_{C_i}^{q_j} = q_j$

In the above equations, L_i, R_i, and C_i denote the respective passive elements; r_i and g_i are the gains of typical CVT and VCT, respectively.

Applications of the two Equations 5.119 and 5.120 will be discussed in the following section.

---

[25] W. E. Newell, "Pole-zero sensitivity relationships," *Proc. IRE*, **49**, 1959 (December 1961).

5-7 DESIGN CONSIDERATIONS

It is clear that without going into the actual realization techniques and the resulting configurations, it is difficult to comment on the sensitivity of active networks realizing a specified transmission characteristic. However, it is possible to derive some useful results based on simple arguments. These results are general in nature and are helpful in making decisions on the design of "optimum" networks from the point of view of sensitivity and stability requirements.

The Quality Factor

Let

$$(5.121) \quad T(s) = \frac{a_1 s}{s^2 + 2\sigma s + (\sigma^2 + \omega_c^2)} = \frac{a_1 s}{s^2 + 2\sigma s + \omega_n^2}$$

be a transfer function exhibiting resonance. The sharpness of the peak of the response is defined as the ratio of resonant frequency ω_c to the bandwidth, where the bandwidth is taken as the difference of the two half-power frequencies. For very small bandwidths, this ratio is defined as the *quality factor* Q. Thus, if

$$\sigma \ll \omega_c,$$

$$(5.122) \quad Q = \frac{\omega_c}{2\sigma} \approx \frac{\omega_n}{2\sigma}$$

Roughly, the Q factor can be used to get an estimate of the positions of a pair of complex conjugate poles. Thus, a high Q pole-pair will imply that these poles are located very close to the $j\omega$-axis. As a result, from the stability point of view, the movement of these poles should not only be very small, but preferably in a direction parallel to the $j\omega$-axis.

In the case of frequency-selective filters, the Q of the pole-pair also determines the shape of the resonance. In these cases, we are also interested in the Q-sensitivity that gives the effect of element change on the resonance and also determines the stability margin.

The Q-Sensitivity

DEFINITION 5-7 *The Q-sensitivity is defined as*

$$(5.123) \quad S_{e_i}^Q \triangleq \frac{d(\ln Q)}{d(\ln e_i)} = \frac{dQ/Q}{de_i/e_i}$$

where e_i *is a network parameter.*

The Q-sensitivity and the pole-sensitivity of a complex conjugate pole-pair are related. Let $s = -\sigma_m \pm j\omega_m$ be a pole-pair and let us assume $\sigma_m \ll \omega_m$. Then, the Q of the pole-pair is given as

$$Q = \frac{\omega_m}{2\sigma_m}$$

from which we easily obtain

(5.124) $$S_{e_i}^Q = \frac{\sigma_m \, d\omega_m - \omega_m \, d\sigma_m}{\omega_m \sigma_m} \frac{e_i}{de_i}$$

Now, the pole-sensitivity of the pole at $s = p_m = -\sigma_m + j\omega_m$ is

$$S_{e_i}^{p_m} = \frac{dp_m}{de_i/e_i} = \frac{-d\sigma_m + j d\omega_m}{de_i/e_i}$$

It is easy to show that for high Q pole-pair ($\sigma_m \ll \omega_m$):

(5.125) $$\operatorname{Im}\left\{\frac{S_{e_i}^{p_m}}{p_m}\right\} = \frac{\omega_m \, d\sigma_m - \sigma_m \, d\omega_m}{\sigma_m^2 + \omega_m^2} \frac{e_i}{de_i}$$

$$\doteq \frac{\omega_m \, d\sigma_m - \sigma_m \, d\omega_m}{\omega_m^2} \frac{e_i}{de_i}$$

Substituting expression 5.125 in 5.124 we obtain the desired result:[26]

(5.126) $$S_{e_i}^Q = -2Q \operatorname{Im}\left\{\frac{S_{e_i}^{p_m}}{p_m}\right\}$$

Equation 5.126 indicates that for high Q pole-pair, the imaginary part of $S_{e_i}^{p_m}/p_m$ must be made very small to make the Q-sensitivity small.

As indicated earlier, the sensitivity of a network is a multiparameter problem. Hence, for high Q-poles a suitable design criteria for minimization could be the sum[26]

(5.127) $$\Sigma_Q = \sum_{j=1}^{n} |S_{e_j}^Q|$$

which is the sum of the magnitudes of Q-sensitivities with respect to each variable network parameter.

Example 5-9[27] Consider the *LCR* tank circuit of Figure 5-9. The resonant frequencies are given by the roots of the equation:

$$s^2 + \frac{R}{L}s + \frac{1}{LC} = 0$$

---

[26] W. E. Newell, "Selectivity and sensitivity in functional blocks," *Proc. IRE*, **50** 2517 (December 1962).

[27] W. E. Newell, "The frustrating problem of inductors in integrated circuits," *Electronics*, **37**, 50–52 (March 13, 1964).

If $R \ll \sqrt{L/C}$, the Q of the pole-pair is given by

(5.128) $$Q = \frac{1}{R}\sqrt{\frac{L}{C}}$$

It can be shown that
$$S_R^Q = -1, \quad S_L^Q = \tfrac{1}{2}, \quad S_C^Q = -\tfrac{1}{2}$$

Hence

(5.129) $$\Sigma_Q = 2$$

Note that for the LC tank circuit of Figure 5-9, Σ_Q is independent of the actual element values and thus is independent of Q. Equation 5.128 implies that to keep the Q variation within 5 percent of its nominal value, the element tolerances should be within 7.5 percent (assuming identical tolerances for each element). So we can conclude that with passive RLC high-selectivity filters, tighter element tolerances are not required.

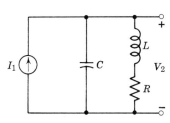

Figure 5-9 An RLC tuned circuit.

On the other hand, many commonly used active RC filters used for realizing high Q responses have a Σ_Q, which are directly proportional to Q. To show this, consider the case of active RC synthesis using the negative impedance converter or a noninverting type VVT. Here the denominator polynomial $D(s)$ of the transfer function is expressed as the difference of two polynomials. Thus, for a second-order realization, we have

$$D(s) = s^2 + 2\sigma s + \omega_n^2 = (s^2 + \gamma s + \omega_n^2) - k\eta s$$

where k is the conversion factor of the NIC or the gain of the VVT. From above we obtain

$$2\sigma = \gamma - k\eta$$

Now

$$Q \cong \frac{\omega_n}{2\sigma} = \frac{\omega_n}{\gamma - k\eta}$$

from which we have

$$\frac{dQ}{dk} = Q \cdot \frac{\eta}{\gamma - k\eta} = \frac{Q\eta}{2\sigma}$$

Therefore

$$S_k^Q = \frac{dQ/Q}{dk/k} = \frac{\gamma - 2\sigma}{2\sigma} = \frac{\gamma}{2\sigma} - 1$$

Now for a high Q pole-pair $\sigma \ll \omega_n$. Again, the requirement of RC companion network implies that the polynomial $(s^2 + \gamma s + \omega_n^2)$ has only distinct negative real roots, which implies that $\gamma > 2\omega_n$. As a result,[28]

$$S_k^Q \gg \frac{2\omega_n}{2\sigma} - 1$$

$$\gg 2Q - 1$$

Hence Σ_Q is always greater than $2Q$. Assuming S_k^Q equal to $2Q$ implies that for a Q of 25, the gain must be kept within 0.1 percent to keep the percentage variation of Q within 5 percent. As a result, active filters in general will require a tighter element tolerance for high Q responses.

The above discussion merely indicates the type of problems we face in the design of high Q active RC filters. A complete understanding of all the advantages and disadvantages of the active RC filter and the passive RLC filter will help the circuit designer in deciding which type of filter to use for a given application.

The Cascade Approach

Before a suitable synthesis procedure has been selected, the network designer would probably like to know the "best" way to realize a higher-order function: should the specified transmission function be realized as a whole in one complicated network or should it be realized as a cascade of simpler networks with proper isolation, if necessary?

For convenience in answering the above question, let $T(s)$ be a fourth-order transfer function having two pairs of complex conjugate poles:[29]

$$(5.130) \quad T(s) = \frac{N_1(s)}{s^2 + 2\sigma_1 s + \omega_1^2} \cdot \frac{N_2(s)}{s^2 + 2\sigma_2 s + \omega_2^2}$$

$$= \frac{N_1(s)}{(s - p_1)(s - p_1^*)} \cdot \frac{N_2(s)}{(s - p_2)(s - p_2^*)}$$

To simplify our calculations, we assume

$$\sigma_1 \approx 0, \quad \sigma_2 \approx 0$$

i.e., Qs of the pole-pairs are very high. If $T(s)$ is realized using a single active element of gain k, the sensitivity of the pole at $s = p_1$ will be given by $S_k^{p_1}$. On the other hand, $T(s)$ can be realized as a cascade of two

---

[28] G. C. Bown, "Sensitivity in active RC filters," *Electronics Letters*, 3, 298–299 (July 1967).

[29] Our discussion here is based on J. M. Sipress, "Synthesis of active RC networks," Doctoral Dissertation, Polytech. Inst. of Brooklyn, N.Y., June 1960.

DESIGN CONSIDERATIONS / 199

transfer functions:

(5.131a) $$T_1(s) = \frac{N_1(s)}{s^2 + 2\sigma_1 s + \omega_1^2}$$

(5.131b) $$T_2(s) = \frac{N_2(s)}{s^2 + 2\sigma_2 s + \omega_2^2}$$

If the gain of the active element in the realization of $T_1(s)$ be k_1, then the sensitivity of the pole at $s = p_1$ will be given by $S_{k_1}^{p_1}$. By considering a particular type of decomposition, we shall show that

(5.132) $$S_{k_1}^{p_1} \ll S_k^{p_1}$$

Let us assume the following forms of decomposition for the denominator polynomials of $T_1(s)$ and $T(s)$, respectively:

(5.133a) $$s^2 + 2\sigma_1 s + \omega_1^2 = (s + A_1)^2 - k_1 B_0 s$$

(5.133b) $$(s^2 + 2\sigma_1 s + \omega_1^2)(s^2 + 2\sigma_2 s + \omega_2^2)$$
$$= (s + A_2)^2(s + A_3)^2 - k B_1 s(s + B_2)^2$$

It was indicated in Example 5-4 that the above type of decompositions minimizes the root sensitivities. Since the pole polynomials are of very high Q ($\sigma_1 \approx 0$, $\sigma_2 \approx 0$), by comparing like coefficients the following identifications are obtained from Equation 5.133:

(5.134)
$$A_1 = \omega_1$$
$$k_1 B_0 = 2(\omega_1 - \sigma_1)$$
$$B_2 = \sqrt{\omega_1 \omega_2}$$
$$k B_1 = 2[\omega_1 + \omega_2 + 2\sqrt{\omega_1 \omega_2}]$$

We thus obtain for $p_1 = -\sigma_1 - j\omega_1$,

(5.135a) $$S_{k_1}^{p_1} = \frac{k_1 B_0 s(s + \sigma_1 + j\omega_1)}{s^2 + 2\sigma_1 s + \omega_1^2}\bigg|_{s=-\sigma_1-j\omega_1} \cong \omega_1 - \sigma_1 - j\sigma_1$$

(5.135b) $$S_k^{p_1} = \frac{k B_1 s(s + B_2)^2}{(s^2 + 2\sigma_2 s + \omega_2^2)(s + \sigma_1 - j\omega_1)}\bigg|_{s=-\sigma_1-j\omega_1}$$
$$\approx \frac{\omega_1[\omega_1 + \omega_2 + 2\sqrt{\omega_1 \omega_2}][\omega_2 - \omega_1 - 2j\sqrt{\omega_1 \omega_2}]}{\omega_2^2 - \omega_1^2 + 2j\omega_1(\sigma_1 - \sigma_2)}$$

For a more meaningful comparison of the above results, we further assume that $\omega_1 \approx \omega_2$, i.e., the poles are very close to each other. Then the above results simplify to:

(5.136)
$$\left|\frac{S_{k_1}^{p_1}}{\sigma_1}\right| \approx 2Q_1 - 1$$
$$\left|\frac{S_{k}^{p_1}}{\sigma_1}\right| \approx 4Q_1 \left|\left(\frac{\omega_2 + \omega_1}{\omega_2 - \omega_1}\right)\right|$$

where
$$Q_1 = \omega_1/2\sigma_1$$

We can conclude from the above results that *from the sensitivity point of view it is definitely better to realize a higher-order function by cascading second-order networks*. Of course, the price for getting better sensitivity is the use of more active devices.

Multiparameter Sensitivity Minimization

We now focus our attention to the design of "optimum" network with the aim of minimizing the multiparameter sensitivity. Even though our discussion will again be based on the realization of transfer functions with high Q pole-pairs, some of the results are general enough to be of use in other cases.

Relation between Sensitivity and Pole-Zero Locations.[30] Obviously the first question we shall ask ourselves at this time is the following: are there any preferable pole-zero locations, consistent with the approximation of the specified transmission characteristic, which will have lower sensitivity? To answer this question, let us consider a second-order transfer function,

(5.137) $$T(s) = H \frac{(s - z_1)(s - z_2)}{(s - p_1)(s - p_1^*)} = H \frac{s^2 + N_1 s + N_0}{s^2 + 2\sigma_1 s + (\omega_1^2 + \sigma_1^2)}$$

where $2\sigma_1 \ll \omega_1$. If e_i denotes an arbitrary network parameter in the realization of $T(s)$, it follows from (5.137) and (5.70):

(5.138) $$S_{e_i}^{T(s)} = S_{e_i}^H - \frac{S_{e_i}^{z_1}}{s - z_1} - \frac{S_{e_i}^{z_2}}{s - z_2} + \frac{S_{e_i}^{p_1}}{s - p_1} + \frac{S_{e_i}^{p_1^*}}{s - p_1^*}$$

which can be alternately written as

(5.139) $$S_{e_i}^{T(s)} = \frac{A(s)}{(s^2 + N_1 s + N_0)(s^2 + 2\sigma_1 s + \omega_1^2 + \sigma_1^2)} + S_{e_i}^H$$

---

[30] S. C. Lee, "Sensitivity minimization in active *RC* integrated circuit design," *Proc. 4th Allerton Conf. on Circuit and System Theory*, Univ. of Illinois, pp. 269–281 (1966).

Now, for a high selectivity filter, changes in the gain level, $S_{e_i}^H$ can be neglected. Since the frequency response in the neighborhood of $s = p_1 = -\sigma_1 + j\omega_1 \doteq j\omega_1$ is of utmost concern to us, as a first approximation we can rewrite (5.138) as

(5.140) $$S_{e_i}^{T(s)} \simeq \frac{S_{e_i}^{p_1}}{s - p_i}$$

It also follows from Equation 5.139:

(5.141) $$S_{e_i}^{p_1} = (s - p_1)S_{e_i}^{T(s)}\Big|_{s=p_1} = \frac{A(p_1)}{(p_1 - z_1)(p_1 - z_2)2j\omega_1}$$

Therefore, for $s = j\omega_1$

$$S_{e_i}^{T(j\omega_1)} \simeq \frac{A(p_1)}{(p_1 - z_1)(p_1 - z_2)2j\omega_1} \cdot \frac{1}{\sigma_1}$$

or

(5.142) $$|S_{e_i}^{T(j\omega_1)}| \simeq Q \cdot \frac{|A(p_1)|}{\omega_1^2 |(p_1 - z_1)(p_1 - z_2)|}$$

Note that the above expression is independent of the actual configuration and the type of network parameter. We can make the following observations from (5.142): (1) the sensitivity function is proportional to Q, hence the aim of approximation should be to obtain a transfer function having poles with the lowest possible Q, (2) sensitivity function tends to be higher in low frequency applications, (3) for a given pole location, the transmission zeros should be chosen to make them as far as possible from $s = p_1$, the pole in the upper left-half s-plane. The last point definitely should be made use of when higher order functions are realized as a cascade of second-order transfer functions.

It is interesting to note that use of low Q poles in the approximation procedure has also been suggested by Vlach and Bendík,[31] who designed a 4th degree and a 6th degree active filter for pulse transmission using the Linvill's method (see Section 9-3). They initially selected the Thompson approximation for maximally flat group delay, which is characterized by low Q pole-pairs. Then the approximative network function was modified by adding complex transmission zeros to improve the amplitude response without seriously influencing the phase response. They have shown experimentally that the variation of finite changes in filter elements (of the order of 10 percent) has very little effect on the amplitude responses of the realized active filters, even though the pole movements are significant.

[31] J. Vlach and J. Bendík, "Active filters with low sensitivity to element changes," *The Radio and Electronic Engineer*, **33**, 305–316 (May 1967).

Moreover, because of the locations of the pole further from the $j\omega$-axis, an instability problem did not arise until the active parameter changed by almost 18 percent.

Use of higher-order approximating functions with lower Q pole-pairs has been investigated by Gorski-Popiel.[32] Even though there is a reduction in network sensitivity, an apparent drawback of his method is the increase in the number of components of the active filter and of extra adjustments.

Multiparameter Root Sensitivity. In many cases, the circuit designer is faced with the problem of minimizing the displacement of one pole (or zero) when the network contains more than one variable parameter. For example, in a notch filter, one may be interested in keeping the notch frequency invariant. With very careful design, it may be possible to make root sensitivity very small.

To illustrate the possible approach, consider an active RC network containing a single VVT of gain μ. From Equation 5.107 we have

$$(5.143) \qquad dq_j = \frac{dR}{R} \sum_{i=1}^{n_R} S_{R_i}^{q_j} + \frac{dC}{C} \sum_{i=1}^{n_C} S_{C_i}^{q_j} + \frac{d\mu}{\mu} S_\mu^{q_j}$$

where $s = q_j$ is the root of interest. In deriving Equation 5.143, we have assumed that the percentage variations of all resistors are equal and also the percentage variations of all capacitors are equal to one another. This is not difficult to achieve if batch-fabrication techniques are used to obtain the elements. It follows from Equations 5.119 and 5.120 that

$$(5.144) \qquad \sum_{i=1}^{n_C} S_{C_i}^{q_j} = \sum_{i=1}^{n_R} S_{R_i}^{q_j} = q_j$$

which when used in Equation 5.143 yields

$$(5.145) \qquad dq_j = q_j \left(\frac{dR}{R} + \frac{dC}{C} \right) + S_\mu^{q_j} \frac{d\mu}{\mu}$$

We can make several observations on Equation 5.145:

1. If the network is a pure RC network and

$$\frac{dC}{C} = -\frac{dR}{R} = \frac{dG}{G}$$

then the root displacement is zero[33] (see Example 5-8);

---

[32] J. Gorski-Popiel, "Reduction of network sensitivity through the use of higher-order approximating functions," *Electronics Letters*, **3**, 365–366 (August 1967).

[33] W. E. Newell, *loc. cit.*

2. If the network is a pure RC network, then it follows from (5.145):

$$(5.146)^{33} \qquad \frac{dq_j}{q_j} = \frac{dR}{R} + \frac{dC}{C}$$

Since the right-hand side of (5.146) is real, the root at $s = q_j$ moves in a radial direction and this means for a pair of complex conjugate roots at $s = q_j$ and $s = q_j^*$, the Q remains unaltered.

3. If the amplifier is present, then dq_j can be made equal to zero by choosing a suitable network configuration for which

$$(5.147) \qquad S_\mu^{q_j} = \frac{-q_j \left(\frac{dR}{R} + \frac{dC}{C} \right)}{d\mu/\mu}$$

This approach was suggested recently by Gaash, Pepper, and Pederson.[34]

4. The root displacement can be made negligible by designing the amplifier to have very small percentage change and choosing RC components so that

$$\frac{dC}{C} = -\frac{dR}{R}$$

Compared to the Gaash approach, which is valid only at a single frequency, this approach is attractive for keeping the root sensitivity small over a large band of frequencies.[35]

Additional References

Angelo, E. J., Jr., "Design of feedback systems," Rept. No. R-449-55, MRI, Polytech. Inst. of Brooklyn, N.Y., 1955.
Belove, C., "Tolerance coefficients for R-C networks," *J. Appl. Phys.*, **24**, 745–747 (June 1953).
Belove, C., "Sensitivity sums for homogeneous function," *IEEE Trans. on Circuit Theory*, **CT-11**, 171 (March 1964).
Biswas, R. N., "A new definition of multi-parameter sensitivity," Proc. 4th Allerton Conf. on Circuit and System Theory, Univ. of Illinois, p. 282 (1966).
Biswas, R. N., and E. S. Kuh, "Multi-parameter sensitivity analysis for linear systems," Proc. 3rd Allerton Conf. on Circuit and System Theory, Univ. of Illinois, pp. 384–393 (1965).

[34] A. Gaash, R. Pepper, and D. Pederson, "Design of integrable desensitized frequency selective amplifiers," *Digest of Int'l Solid State Circuits Conf.*, Philadelphia, Pennsylvania, pp. 34–35, February 1966.
[35] G. S. Moschytz, unpublished report.

Broome, P. W., and F. V. Young, "The selection of circuit components for optimum circuit reproducibility," *IRE Trans. on Circuit Theory*, **CT-9**, 18–23 (March 1962).

Calahan, D. A., "Computer solution of the networks realization problem," Proc. 2nd Allerton Conf. on Circuit and System Theory, Univ. of Illinois, pp. 175–193 (1964).

Calahan, D. A., "Computer design of linear frequency selective networks," *Proc. IEEE*, **53**, 1701–1706 (November 1965).

Calahan, D. A., *Modern network synthesis*, Hayden Book Co., Inc., New York, **2**, Ch. 4 (1964).

Cartianu, G., and D. Poenaru, "Variation of transfer functions with the modification of pole location," *IRE Trans. on Circuit Theory*, **CT-9**, 98–99 (March 1962).

Fu, K. S., "Sensitivity of a linear system with variations of one or several parameters," *IRE Trans. on Circuit Theory*, **CT-7**, 348–349 (September 1960).

Géher, K., "The tolerances of linear networks and systems in the time, frequency and complex frequency domains," *Proc. Telecomm. Research Inst. (Hungary)*, **X**, 105–117 (1965).

Gorski-Popiel, J., "Classical sensitivity—a collection of formulas," *IEEE Trans. on Circuit Theory*, **CT-10**, 300–302 (June 1963).

Hakimi, S. L., and J. B. Cruz, Jr., "Measures of sensitivity for linear systems with large multiple parameter variation," *IRE WESCON Conv. Rec.*, **4** (Part 2), 109–115 (August 1960).

Huang, R. Y., "The sensitivity of the poles of linear closed-loop systems," *Trans. AIEE (Applications and Industry)*, **77**, 182–187 (September 1958).

Jones, H. E., and B. A. Shenoi, "Worst case gain sensitivity with zero phase sensitivity in active RC network synthesis," Proc. 4th Allerton Conf. on Circuit and System Theory, Univ. of Illinois, pp. 259–268 (1966).

Kuo, F. F., "Sensitivity of transmission zeros in RC network synthesis," *IRE Int'l Conv. Rec.*, **7** (Part 2), pp. 18–27 (March 1959).

Kuo, F. F., "A sensitivity theorem," *IRE Trans. on Circuit Theory*, **CT-6**, 131 (March 1959).

Lee, S. C., "On multi-parameter sensitivity," Proc. 3rd Allerton Conf. on Circuit and System Theory, Univ. of Illinois, pp. 407–420 (1965).

Leeds, J. V., Jr., "Transient and steady-state sensitivity analysis," *IEEE Trans. on Circuit Theory*, **CT-13**, 288–289 (September 1966).

Leeds, J. V., Jr., and G. I. Ugran, "Simplified multiple parameter sensitivity calculation and continuously equivalent networks," *IEEE Trans. on Circuit Theory*, **CT-14**, 188–191 (June 1967).

Lynch, W. A., "A formulation of the sensitivity function," *IRE Trans. on Circuit Theory*, **CT-4**, 289 (September 1957).

Mazer, W. M., "Specification of the linear feedback system sensitivity function," *IRE Trans. on Automatic Control*, **AC-5**, 85–93 (June 1960).

Mikulski, J. J., "A correlation between classical and pole-zero sensitivity," Tech. Note No. 5, Circuit Theory Group, Elec. Eng. Res. Lab, Univ. of Illinois, Urbana, Illinois, February 1959.

Mulligan, J. R., "The effect of pole and zero locations on the transient response of linear dynamic systems," *Proc. IRE*, **37**, 516–529 (May 1949).

Nagaraja, N. S., "Effect of component tolerances in low frequency selective amplifiers," *J. Indian Inst. Sci.*, **37** (Sec. B), 324–337 (October 1955); and **38** (Sec. B), 81–92 (April 1956).

Nagoryy, L. Y., "Dependence of the basic network parameters in one of its elements," *Telecommunications*, No. 6, pp. 43–54 (June 1961).

Papoulis, A., "Perturbations of the natural frequencies and eigen vectors of a network," *IEEE Trans. on Circuit Theory*, **CT-13**, 188–195 (June 1966).

Parker, S. R., E. Peskin, and P. M. Chirlian, "Application of bilinear theorem to network sensitivity," *IEEE Trans. on Circuit Theory*, **CT-12**, 448–450 (September 1965).

Saito, M., "Pole sensitivity in active networks," *J. Inst. Elec. Comm. Engg.* (*Japan*), **44**, 218–223 (May 1961).

Truxal, J. G., *Automatic feedback control system synthesis*, McGraw-Hill, New York, pp. 120–127 (1955).

Problems

5.1 Using the techniques outlined in Section 5-1, compute the approximate pole displacement of the input impedance $Z(s)$ of the network shown in Figure 5-10 for each of the following cases:

(a) The inductor L changes to $(L + \delta L)$.
(b) A small resistor r is added in series with the inductor L.

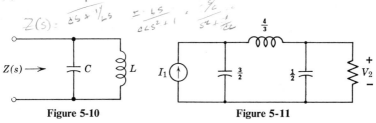

Figure 5-10   Figure 5-11

5.2 The low-pass filter of Figure 5-11 has a maximally flat magnitude characteristic. Determine the displacements of the natural frequencies for a 1 percent change of the value of each element.

5.3 Obtain a signal flow graph representation of the voltage transfer ratio V_2/V_1 of the $RC:NIC$ two-port of Figure 9-29. The parameter of interest is the conversion factor k of the CNIC.

5.4 Derive Equation 5.25.

5.5 (a) Figure 5-12 shows an attenuator along with nominal values of

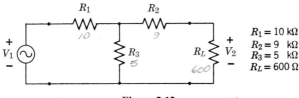

Figure 5-12

206 / SENSITIVITY AND RELATED TOPICS

the resistors. Compute the sensitivity of the transfer voltage ratio t_v with respect to each resistor.

(b) If it is desired to keep the worst-case variation of t_v within 5 percent, what should be the tolerance limits of each resistor.

5.6[36] The two feedback amplifiers of Figure 5-13 realize identical voltage

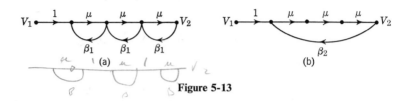

Figure 5-13

transfer ratios (by proper choice of β_1 and β_2). Show that for $\mu\beta_1 \gg 1$, the second amplifier (Figure 5-13b) is less sensitive with respect to the variation of the open loop gain μ of the amplifier.

5.7 Expression 5.148 is often referred to as *Blackman's formula*[37] and is useful in computing the input impedance of an active network:

(5.148) $$Z = (Z)_{k=0} \frac{F_k \text{ (with terminals shorted)}}{F_k \text{ (with terminals open)}}$$

where $(Z)_{k=0}$ is the input impedance Z when a specified element k of the system is zero, F_k is the return difference with reference to k with the input terminals shorted, and F_k is the return difference with reference to k with the terminals open. Derive expression 5.148.

5.8 Using Blackman's formula, calculate the input impedance of Figure 9-29 seen at port 1.

5.9 Two possible $RC:RL$ decompositions of the polynomial $(s^2 + 2s + 9)$ are given below:

$$(s^2 + 2s + 9) = (s + 1)^2 + k(2\sqrt{2})^2$$

$$= (\tfrac{1}{3}s + 3)^2 + k\left(\frac{2\sqrt{2}}{3}s\right)^2$$

where the nominal value of k is unity. Calculate the root sensitivity $S_k^{p_1}$ for both cases and show that they are equal in magnitude.

5.10 Derive the sensitivity expressions given in Equation 5.135.

5.11 Compute the pole-sensitivities of the active network of Figure 1-7 with respect to the gain of the VVT.

---

[36] J. G. Truxal, *loc. cit.*
[37] H. W. Bode, *loc. cit.*, p. 68.

5.12[38] Prove that for any network

(5.149) $$F_{k_1} = F_{k_2} \frac{(F_{k_1})_{k_2=0}}{(F_{k_2})_{k_1=0}}$$

where F_{k_1} and F_{k_2} are the return differences with respect to parameters k_1 and k_2, respectively.

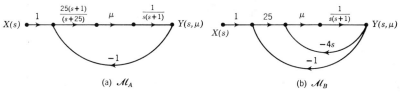

(a) $\mathcal{M}_A$ (b) $\mathcal{M}_B$

Figure 5-14

5.13[39] The two signal-flow graphs of Figure 5-14 realize the transfer function

$$T(s) = \frac{25}{s^2 + 5s + 25}$$

for unity nominal value of the gain μ of the amplifier. If the input signals are sinusoids, show that the network $\mathcal{M}_B$ has a lower gain sensitivity with respect to μ.

5.14 Derive Equations 5.119 and 5.120.

[38] J. G. Truxal, *loc. cit.*
[39] J. G. Truxal and I. M. Horowitz, *loc. cit.*

6 / *Realizability Conditions*

The synthesis of a network begins with a description of the network in terms of its network function. The network function probably was obtained by using some approximation techniques from the prescribed characteristic of the network, either in the frequency or in the time domain. Alternatively, if an equivalent network is desired, the network function may have been obtained by analyzing the given circuit.

An understanding of the realization techniques of the various types of active networks is closely associated with the properties (and hence limitations) of their network functions. Thus, a knowledge of the properties of the active network functions also helps the designer to choose the right kind of active network to satisfy his needs.

The most basic network function is the driving-point immittance function. We shall note later that properties of parameters of multiport networks can be derived from the properties of driving-point functions, thus leading to the properties of the more useful network function—the transfer function.

The derivation of the basic properties of passive and active network functions is based on the definitions of activity and passivity of a network stated in Section 6-1. Sections 6-2 and 6-3 are concerned with the derivation of the general properties of passive structures. Various types of active networks are treated in the following five sections. In Section 6-9, the potential instability conditions are derived. Finally, the main results of this chapter are summarized in the last section.

6-1 ACTIVITY AND PASSIVITY OF A NETWORK

There are many ways by which the activity and passivity of a network can be defined. From our previous acquaintance with network theory, we know that a network composed of resistances, capacitances, and inductances is a passive network. In a similar fashion, we say that a network is active if it contains such elements as transistors, vacuum tubes, and other solid-state devices.

A more precise definition of passivity and activity is necessary to get a better understanding of passive and active networks and also to formulate the realizability conditions of linear active networks. The following simple example will illustrate our point. Consider the one-port network of Figure 6-1a. Summing the currents at the input node we have

$$I_1 = V_1 + (V_1 - 2V_1)\left(\frac{s}{s+1}\right) = V_1\left(1 - \frac{s}{s+1}\right)$$

or

$$Y(s) = \frac{I_1}{V_1} = \frac{1}{s+1}$$

The input admittance of this network and the passive network shown in Figure 6-1b are identical. Thus, as far as the input port is concerned, the network of Figure 6-1a behaves "exactly" as a passive RL one-port.

Our aim is thus to define the passivity or the activity of network in terms of the terminal variables, without going into the actual construction of the network. For convenience, let us first define the passivity of a one-port. Let $v(t)$ and $i(t)$ be the voltage and current variables[1] defining

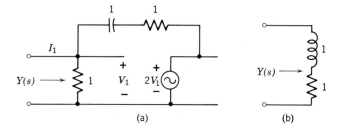

Figure 6-1 An active RC one-port and its equivalent representation.

---

[1] It is tacitly assumed that if the excitation is a real function of time, the response is also a real function of time.

the one-port $\mathcal{N}$ (Figure 6-2), where $v(t)$ may be the excitation and $i(t)$ may be the response or vice-versa. The instantaneous power delivered to $\mathcal{N}$ is

(6.1) $$p(t) = v(t)i(t)$$

and the total energy delivered to $\mathcal{N}$ at $t = t_1$ is

(6.2) $$\mathcal{E}(t_1) = \int_0^{t_1} v(\tau)i(\tau)\, d\tau$$

In defining $\mathcal{E}(t_1)$ as given in Equation 6.2, we have assumed that there is

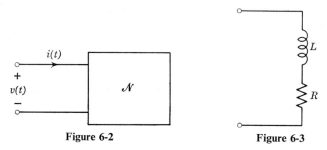

Figure 6-2 Figure 6-3

no initial energy in $\mathcal{N}$ before the application of an excitation at $t = 0$. We now make the following:[2]

DEFINITION 6-1 *A linear one-port is* passive *if with zero initial stored energy the total energy input $\mathcal{E}(t_1)$ is nonnegative for every t_1 and for all possible voltage or current excitations.*

Consider the following example.

Example 6-1 The total energy delivered to the one-port network of Figure 6-3 by a current source is given as

(6.3) $$\mathcal{E}(t_1) = \int_0^{t_1} [Li\, di + Ri^2]\, d\tau$$

$$= Li^2(t_1) + R\int_0^{t_1} i^2(\tau)\, d\tau$$

The integral on the right-hand side is always nonnegative. Thus, if the inductance and the resistance values are nonnegative, the total energy input $\mathcal{E}(t_1)$ to the one-port is always nonnegative for all values of t_1 and all possible current excitations. Hence, the network is passive.

[2] G. Raisbeck, "A definition of passive linear networks in terms of time and energy," *J. Applied Phys.*, **25**, 1510–1514 (December 1954).

Now, let us assume that the series resistor is a negative resistance, i.e., $R = -R_n$ where R_n is positive. We shall show that, in this case, $\mathcal{E}(t_1)$ can be made negative by choosing a suitable excitation and a suitable time t_1. Choosing $i(t) = \sin t$, we obtain the total energy input as:

$$\text{(6.4)} \qquad \mathcal{E}(t_1) = L \sin^2 t_1 - R_n \left[\frac{t_1}{2} - \frac{1}{4} \sin 2t \right]$$

It is clear from (6.4) that $\mathcal{E}(t_1) = -R_n t_1/2 < 0$ at $t_1 = \pi$. So we can conclude that a series connection of a positive inductance and a negative resistance is an active one-port.

An important property of a linear passive network is *causality*. By causality, we mean that the network cannot "anticipate" the response. More precisely,

DEFINITION 6-2 *A one-port is* causal, *if no voltage or current appears across the input terminals before the application of an excitation.*

It is easy to show that a linear passive network is also causal.[3] Let $v_1(t)$, $v_2(t)$, and $v_3(t)$ denote the responses of the one-port due to the excitations $i_1(t)$, $i_2(t)$, and $i_3(t)$, respectively, applied at $t = 0$. We assume that

$$\text{(6.5)} \qquad \begin{aligned} i_1(t) &\neq 0 && \text{for any } t \geq 0 \\ i_2(t) &= 0 && \text{for } 0 \leq t \leq t_2 \\ i_3(t) &= a i_1(t) + b i_2(t) \end{aligned}$$

where a and b are arbitrary real nonzero constants. Because of linearity, it follows that

$$\text{(6.6)} \qquad v_3(t) = a v_1(t) + b v_2(t)$$

Again, passivity implies

$$\text{(6.7)} \qquad \mathcal{E}(t_1) = \int_0^{t_1} v_3(\tau) i_3(\tau) \, d\tau \geq 0$$

i.e.,

$$\text{(6.8)} \qquad \int_0^{t_1} [a v_1(\tau) + b v_2(\tau)][a i_1(\tau) + b i_2(\tau)] \, d\tau \geq 0$$

The above expression is also true for $t_1 \leq t_2$. Since $i_2(t) = 0$ for $t \leq t_2$,

---

[3] H. J. Carlin and A. B. Giordano, *Network Theory*, Prentice-Hall, Englewood Cliffs, N.J., pp. 6–7 (1964).

212 / REALIZABILITY CONDITIONS

Equation 6.8 reduces to

(6.9) $$\mathcal{E}(t_1) = \int_0^{t_1} [av_1(\tau) + bv_2(\tau)] ai_1(\tau)\, d\tau$$
$$= a^2 \int_0^{t_1} v_1(\tau) i_1(\tau)\, d\tau + ab \int_0^{t_1} v_2(\tau) i_1(\tau)\, d\tau$$

for $t_1 \leq t_2$. The first integral on the right-hand side of Equation 6.9 is nonnegative because of the passivity constraint. Since a and b are arbitrary, it appears from the above equation that $\mathcal{E}(t_1)$ can be made negative by suitably choosing these constants. This being a contradiction, the second integral on the right-hand side of Equation 6.9 must be zero for all $t_1 \leq t_2$, i.e.,

(6.10) $$v_2(t) = 0 \quad \text{for} \quad t \leq t_2$$

This proves our assertion.

The activity of a linear network can be simply defined as:

DEFINITION 6-3 *A linear one-port is* active, *if it is not passive.*

It follows that a linear noncausal network is definitely an active network. Note that if a single voltage or current excitation of any waveform or a single time t_1 can be found for which the total energy input is negative, then the network is active (see Example 6-1). On the other hand, to test passivity, the total energy input $\mathcal{E}(t_1)$ must be computed for all allowable excitations and for all time t_1.

The concepts and definitions outlined above are easily extended to multiport networks. For example, for a two-port network (Figure 6-4), we have the following definition for passivity:[4]

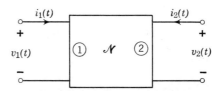

Figure 6-4 A two-port network.

DEFINITION 6-4 *A linear two-port network is passive if for zero initial stored energy, the total energy input,*

(6.11) $$\mathcal{E}(t_1) = \int_0^{t_1} \{v_1(\tau) i_1(\tau) + v_2(\tau) i_2(\tau)\}\, d\tau \geq 0$$

[4] G. Raisbeck, *loc. cit.*

While the above definitions for passivity can be used to test the network, they are difficult to apply to complicated circuits.

Since our main purpose is to synthesize a network from a specified network function, it is desirable to have a set of conditions which could be easily applied to test a specified network function and then determine the type of network that could be realized from the given function. This is the purpose of the following seven sections.

An important property of the *LLF* (linear, lumped, and finite) network is that its network functions are real rational functions of the complex frequency variable s, that is, any network function $T(s)$ can be expressed as the ratio of two polynomials in s having real coefficients only. The preceding assertion follows from the fact that the network elements comprising *LLF* networks are characterized by real parameters only. This implies that the network equations in Laplace transform will have coefficients which are real rational functions. Now, any network function can be expressed as ratio of determinants (using Cramer's rule), where the elements of the determinants will be real rational functions. As a result the determinant value also will be of the same form. This proves our assertion.

6-2 LINEAR PASSIVE ONE-PORT

Consider a passive *LLF* one-port containing a finite number of components. We wish to derive the conditions satisfied by its driving-point function. A consequence of passivity is that the natural frequencies of the one-port must be located in the left-half s plane. Natural frequencies on the $j\omega$-axis if present must be simple. To prove the validity of the above statement, we can assume the presence of a right-half s plane pole (or a multiple $j\omega$-axis pole) and show that the energy input can be made negative by choosing a proper excitation (see Problem 6.1).

To derive the necessary conditions satisfied by the driving-point impedance (admittance) function, we shall follow Brune's approach.[5] Let us compute the energy input due to an excitation of the form

(6.12) $$v(t) = Ve^{\sigma_0 t} \cos \omega_0 t = \operatorname{Re}[Ve^{s_0 t}]$$

where $s_0 = \sigma_0 + j\omega_0$, with σ_0 a nonnegative number. If we further assume that the driving-point admittance $Y(s)$ has no $j\omega$-axis poles, it can be shown that an appropriate set of initial conditions can be selected, so that the response $i(t)$ due to the excitation of Equation 6.12 will consist

---

[5] O. Brune, "Synthesis of a finite two-terminal network whose driving-point impedance is a prescribed function of frequency," *J. Math. Phys.*, **10**, 191–236 (August 1931).

214 / REALIZABILITY CONDITIONS

only of the steady-state terms,[6] i.e.,

(6.13) $$i(t) = |Y(\sigma_0 + j\omega_0)| V e^{\sigma_0 t} \cos(\omega_0 t + \theta)$$

where

(6.14) $$Y(s_0) = Y(\sigma_0 + j\omega_0) = |Y(\sigma_0 + j\omega_0)| e^{j\theta}$$

Hence the instantaneous power going into the one-port for $t \geq 0$ is

(6.15) $$p(t) = v(t)i(t) = V^2 |Y(\sigma_0 + j\omega_0)| e^{2\sigma_0 t} \cos \omega_0 t \cos(\omega_0 t + \theta)$$

There are three cases to consider.

Case 1. $\qquad \sigma_0 > 0, \qquad \omega_0 = 0$

(6.16) $$p(t) = V^2 |Y(\sigma_0)| e^{2\sigma_0 t} \cos \theta$$

Hence

(6.17) $$\mathcal{E}(t_1) = \mathcal{E}(0) + \int_0^{t_1} v(\tau)i(\tau) \, d\tau$$
$$= \mathcal{E}(0) + \int_0^{t_1} V^2 |Y(\sigma_0)| e^{2\sigma_0 \tau} \cos \theta \, d\tau$$

where $\mathcal{E}(0)$ is the initial stored energy at $t = 0$ due to the presence of initial conditions. Upon completion of the integration, Equation 6.17 becomes

6.18 $$\mathcal{E}(t_1) = \mathcal{E}(0) + \frac{V^2 |Y(\sigma_0)| \cos \theta}{2\sigma_0} [e^{2\sigma_0 t_1} - 1]$$

The second quantity on the right-hand side of Equation 6.18 is the contribution of the source; and for large values of t_1, the dominant factor in the expression for energy input will be

$$V^2 \frac{|Y(\sigma_0)| \cos \theta}{2\sigma_0} e^{2\sigma_0 t_1}$$

If $|Y(\sigma_0)| \cos \theta = \text{Re}[Y(\sigma_0)]$ is negative, the total energy input will become negative eventually due to the presence of $e^{2\sigma_0 t_1}$. Thus a necessary condition for passivity is

(6.19) $$\text{Re}[Y(\sigma_0)] \geq 0 \qquad \text{for } \sigma_0 > 0$$

Case 2. $\qquad \sigma_0 = 0, \qquad \omega_0 \neq 0$

$$p(t) = V^2 |Y(j\omega_0)| \cos \omega_0 t \cdot \cos(\omega_0 t + \theta)$$

---

[6] See Problem 6.4.

Hence,

(6.20) $\quad \mathcal{E}(t_1) = \mathcal{E}(0) + \int_0^{t_1} p(\tau) \, d\tau$

$= \mathcal{E}(0) + V^2 |Y(j\omega_0)| \, t_1 \cdot \dfrac{\cos \theta}{2}$

$+ V^2 \dfrac{|Y(j\omega_0)|}{2\omega_0} \{\sin(2\omega_0 t_1 + \theta) - \sin \theta\}$

The dominant term in the expression for $\mathcal{E}(t_1)$ for large values of t_1 is

$$V^2 |Y(j\omega_0)| \, t_1 \cdot \dfrac{\cos \theta}{2}$$

Thus if $|Y(j\omega_0)| \cos \theta = \mathrm{Re}\,[Y(j\omega_0)]$ is negative, the total energy input will be negative after some finite large t_1. This implies that a necessary condition for passivity is

(6.21) $\quad \mathrm{Re}\,[Y(j\omega_0)] = |Y(j\omega_0)| \cos \theta \geq 0 \quad \text{for all } \omega_0$

Case 3. $\qquad\qquad\qquad \sigma_0 > 0, \quad \omega_0 \neq 0$

For the general case, we have

(6.22) $\quad \mathcal{E}(t_1) = \mathcal{E}(0) + \dfrac{V^2 |Y(\sigma_0 + j\omega_0)|}{2\sigma_0}$

$\times e^{2\sigma_0 t}\{\cos \theta + \cos \psi \cos(2\omega_0 t_1 + \theta - \psi)\}$

$+ \dfrac{V^2 |Y(\sigma_0 + j\omega_0)|}{2\sigma_0} \{\cos \psi \cos(\theta - \psi) - \cos \theta\}$

where $\psi = \arg(\sigma_0 + j\omega_0)$. Thus, for large values of t_1, the second term on the right-hand side of Equation 6.22 becomes the dominant term; this term will stay positive provided

(6.23a) $\qquad \cos \theta \geq \cos \psi \qquad$ for all $\sigma_0 > 0$ and for all ω_0

i.e.,

(6.23b) $\qquad \dfrac{\mathrm{Re}\,|Y(s_0)|}{|Y(s_0)|} \geq \dfrac{\mathrm{Re}\,(s_0)}{|s_0|} \qquad$ for $\mathrm{Re}\,(s_0) > 0$

The seemingly different conditions (6.19), (6.21), and (6.23) can be expressed as one condition:

(6.24) $\qquad\qquad \mathrm{Re}\,[Y(s)] \geq 0 \qquad$ for $\mathrm{Re}\,(s) \geq 0$

Because of our assumption on lumped, linear, and finiteness structure of the one-port, we have an additional condition on $Y(s)$:

(6.25) $\qquad\qquad Y(s)$ is real for real values of s

A function $Y(s)$ that satisfied the two conditions (6.24) and (6.25) is called a *positive real function*. Summarizing our results:

Theorem 6-1 *A 2-terminal LLF network is passive if, and only if, its driving-point admittance $Y(s)$ is a positive real function.*

An important consequence of the above theorem is that if a given real rational function $Y(s)$ is positive real, then $Y(s)$ can be realized as a one-port network containing only passive network elements. This first was proved by Brune.[7] An alternate realization scheme is outlined in Chapter 10.

We have proved the necessity part of the above theorem. To prove that positive realness of $Y(s)$ implies the passivity, let us compute the total energy in the frequency domain

$$\mathcal{E}(\infty) = \frac{1}{2\pi} \int_{-\infty}^{\infty} Y(j\omega) |\mathcal{F}\{v(t)\}|^2 \, d\omega = \frac{1}{2\pi} \int_{-\infty}^{\infty} \operatorname{Re}\,[Y(j\omega)] |\mathcal{F}\{v(t)\}|^2 \, d\omega$$

where $\mathcal{F}\{v(t)\}$ is the Fourier transform of $v(t)$. Note that the integrand being nonnegative, the total energy delivered to the one-port is nonnegative.

Since the reciprocal of a positive real (p.r.) function is also positive real, we have the corollary to Theorem 6-1:

Corollary to Theorem 6-1 *The driving-point impedance of a passive LLF one-port is positive real.*

Determination of the passivity of an *LLF* one-port reduces to the problem of establishing the p.r. character of the driving-point function. Detail discussions on p.r. functions are available in books on passive network theory.[8] We shall briefly review some aspects of p.r. functions for completeness.

To simplify the testing of p.r. functions, we have the following theorem:

Theorem 6-2 *A real rational function, $Z(s)$, is p.r. if, and only if:*

(i) *$Z(s)$ is analytic in the right-half s-plane.*
(ii) *Poles of $Z(s)$ on the $j\omega$-axis are simple and the residues of $Z(s)$ at these poles are real and positive.*
(iii) *$\operatorname{Re}\,[Z(j\omega)]$ is nonnegative for all values of ω.*

---

[7] O. Brune, *loc. cit.*
[8] See for example, M. E. Van Valkenburg, *Modern network synthesis*, John Wiley, New York, Ch. 4.

An alternate theorem which avoids the computation of residues of $Z(s)$ at the $j\omega$-axis poles is due to Talbot.[9] The theorem is as follows:

Theorem 6-3 *A real rational function $Z(s) = N(s)/D(s)$ is p.r. if:*

(i) $N(s) + D(s)$ is a Hurwitz polynomial.
(ii) Re $Z(j\omega) \geq 0$ for all values of ω.

Condition (i) of Theorem 6-3 can be tested by the Routh-Hurwitz test, which requires that the continued fraction expansion of

$$\text{Ev } [N(s) + D(s)]/\text{Od } [N(s) + D(s)]$$

around infinity yields real and positive coefficients, without ending prematurely. Condition (ii) is satisfied if the numerator of the real part of $Z(j\omega)$, $A(\omega^2)$, has no zeros on the $j\omega$-axis of odd multiplicity. If we express $Z(s)$ in the usual fashion as

(6.26) $$Z(s) = \frac{N(s)}{D(s)} = \frac{m_1 + n_1}{m_2 + n_2}$$

where

(6.27)
$$m_1 = \text{Ev } N(s)$$
$$n_1 = \text{Od } N(s)$$
$$m_2 = \text{Ev } D(s)$$
$$n_2 = \text{Od } D(s)$$

then

(6.28) $$\text{Re } Z(j\omega) = \text{Ev } Z(s)\big|_{s=j\omega} = \frac{m_1 m_2 - n_1 n_2}{m_2^2 - n_2^2}\bigg|_{s=j\omega}$$

Thus $A(\omega^2)$ can alternately be expressed as

(6.29) $$A(\omega^2) = (m_1 m_2 - n_1 n_2)\big|_{s=j\omega}$$

To test $A(\omega^2)$ for zeros on the $j\omega$-axis, Sturm's test can be followed. The steps to be followed to test the positive realness of a real rational function are summarized in Table 6-1.

It follows that if the driving-point function is not p.r., then the one-port must be active. Let us illustrate the passivity test by means of several examples.

Example 6-2 The driving-point impedance of a one-port is given as

(6.30) $$Z(s) = \frac{2s^2 + 2s + 2}{s^2 + s + 2}$$

---

[9] A. Talbot, "A new method of synthesis of reactance networks," *Proc. IEE (London)*, **101** (Part C), 73–90 (February 1954).

Table 6-1 A SUMMARY OF POSITIVE REAL TESTS[10]

$$Z(s) = \frac{N(s)}{D(s)} = \frac{N_n s^n + N_{n-1} s^{n-1} + \cdots + N_1 s + N_0}{D_m s^m + D_{m-1} s^{m-1} + \cdots + D_1 s + D_0}\ [11]$$

A. *Inspection Tests:* $Z(s)$ must satisfy the following conditions:
    (i) All coefficients, N_i, D_j, be real and positive;
    (ii) $|n - m| \leq 1$;
    (iii) If $N_0 = 0$, $N_1 \neq 0$;
    (iv) If $D_0 = 0$, $D_1 \neq 0$;
    (v) $j\omega$-axis zeros of $N(s)$ and $D(s)$, if any, are simple;
    (vi) Unless all even or all odd terms are missing, $N(s)$ and $D(s)$ do not have any missing terms.

B. *Necessary and Sufficient Conditions:* If $Z(s)$ satisfies the six inspection tests, then test the following:
    (i) $N(s) + D(s)$ is a Hurwitz polynomial;
    (ii) The numerator of the real part of $Z(j\omega)$, $A(\omega^2)$, does not have any $j\omega$-axis zeros of odd multiplicity.

---

[10] Adapted from M. E. Van Valkenburg, *loc. cit.*
[11] It is assumed here that $N(s)$ and $D(s)$ do not have any common factors.

$Z(s)$ satisfies the six inspection tests of Table 6-1. For this example, it is not necessary to perform the Hurwitz test to satisfy condition (i) of Theorem 6-3. This is because the sum of the numerator and denominator polynomial of Equation 6.30 is a second-degree polynomial with all positive coefficients, and hence it is Hurwitz. To test $A(\omega^2)$, we note

$$A(\omega^2) = (2s^2 + 2)(s^2 + 2) - 2s^2 \big|_{s=j\omega} = 2(\omega^4 - 2\omega^2 + 2)$$
$$= 2(\omega^2 - 1 + j)(\omega^2 - 1 - j)$$

which is nonnegative for all ω. Therefore, $Z(s)$ is p.r., indicating that from the input terminals the network behaves like a passive network.

Example 6-3 Consider

(6.31) $$Z(s) = \frac{-s - 3}{2s^2 + 2s + 4}$$

Presence of negative coefficients in the numerator indicates that $Z(s)$ is not p.r. The real part of $Z(j\omega)$ is

$$\operatorname{Re} Z(j\omega) = \frac{\omega^2 - 3}{(2 - \omega^2)^2 + \omega^2}$$

which becomes negative as ω approaches zero, proving again that $Z(s)$ cannot be realized by a passive network.

A simple method of realization of the above impedance function is indicated next.[12] Let R_{min} denote the minimum value of Re $Z(j\omega)$. Then it is clear that

$$Z'(s) = Z(s) + R_1$$

is p.r. if $R_1 \geq |R_{min}|$. For the example on hand, we observe

$$R_{min} = -1$$

We can choose $R_1 = 1$. This implies that

(6.32) $$Z'(s) = 1 - \frac{s+3}{2s^2 + 2s + 4} = \frac{2s^2 + s + 1}{2s^2 + 2s + 4}$$

which is p.r. Thus a possible realization of $Z(s)$ of Equation 6.31 would be to realize $Z'(s)$ as given by Equation 6.32 by a passive one-port and then to connect in series a resistance $-R_1 = -1$ ohm (Figure 6-5).

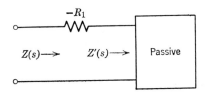

Figure 6-5 Realization of a nonpositive real driving-point function.

Example 6-4 Consider a one-port whose driving-point admittance is given by

(6.33) $$Y(s) = \frac{2s^3 + 2s^2 + s + 3}{s^3 + s^2 + s + 1}$$

We form

$$N(s) + D(s) = 3s^3 + 3s^2 + 2s + 4 = (3s^2 + 4) + (3s^3 + 2s)$$

where

$$\text{Ev } [N(s) + D(s)] = 3s^2 + 4$$
$$\text{Od } [N(s) + D(s)] = 3s^3 + 2s$$

Now perform the Hurwitz test by making a continual fraction expansion

---

[12] H. W. Bode, *Network analysis and feedback amplifier design*, Van Nostrand, New York, p. 190 (1945).

around infinity:

$$
\begin{array}{c|c|c}
3s^2 + 4 & 3s^3 + 2s & s \\
 & 3s^3 + 4s & \\
\hline
-2s & 3s^2 + 4 & -\tfrac{3}{2}s \\
 & 3s^2 & \\
\hline
 & 4 &
\end{array}
$$

The test fails because of negative coefficient, indicating that the one-port is active.

6-3 LINEAR PASSIVE TWO-PORT

The properties of the network functions of a linear passive two-port can be derived following the same reasoning outlined in the previous section. Such an approach is left as an exercise. Instead of the same approach, we shall follow an alternate approach that is illuminating in many respects.[13]

First, let us construct a one-port network out of the two-port *LLFP* network using two ideal complex transformers as shown in Figure 6-6. Assuming the existence of the open-circuit parameters of the two-port, the driving-point impedance $Z(s)$ of the new one-port can be shown to be given as

(6.34) $\quad Z(s) = u_1 u_1^* z_{11}(s) + u_1 u_2^* z_{12}(s) + u_1^* u_2 z_{21}(s) + u_2 u_2^* z_{22}(s)$

where u_1, u_2 are the complex turns ratios and $u_1^*(u_2^*)$ is the complex conjugate of $u_1(u_2)$. Since the ideal complex transformer is a lossless device, the one-port structure of Figure 6-6 is passive. As a result, $Z(s)$ given by Equation 6.34 must be p.r. for all values of u_1 and u_2. Let us denote

(6.35) $\quad z_{mn}(\sigma + j\omega) = R_{mn}(\sigma, \omega) + jX_{mn}(\sigma, \omega)$

Then

(6.36)

$$\operatorname{Re} Z(s) = \tfrac{1}{2}[Z(\sigma + j\omega) + Z^*(\sigma + j\omega)]$$
$$= u_1 u_1^* R_{11}(\sigma, \omega)$$
$$+ u_1 u_2^* \left[\frac{R_{12}(\sigma, \omega) + R_{21}(\sigma, \omega)}{2} + j\frac{X_{12}(\sigma, \omega) - X_{21}(\sigma, \omega)}{2}\right]$$
$$+ u_1^* u_2 \left[\frac{R_{12}(\sigma, \omega) + R_{21}(\sigma, \omega)}{2} - j\frac{X_{12}(\sigma, \omega) - X_{21}(\sigma, \omega)}{2}\right]$$
$$+ u_2 u_2^* R_{22}(\sigma, \omega)$$

---

[13] P. Bello, "Extension of Brune's energy function approach to the study of *LLF* networks," *IRE Trans. on Circuit Theory*, **CT-7**, 270–280 (September 1960).

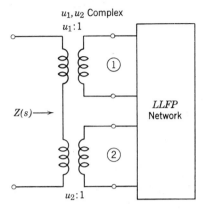

Figure 6-6 Construction of an *LLFP* one-port from an *LLFP* two-port.

If we write

(6.37) $$\Phi_{mn}(s) \triangleq \tfrac{1}{2}[z_{mn}(\sigma + j\omega) + z_{nm}{}^*(\sigma + j\omega)]$$

then Equation 6.36 can be compactly written as

(6.38) $$\operatorname{Re} Z(s) = u_1 u_1{}^* \Phi_{11} + u_1{}^* u_2 \Phi_{21} + u_1 u_2{}^* \Phi_{12} + u_2 u_2{}^* \Phi_{22}$$

By definition (6.37), Φ_{11} and Φ_{22} are real and $\Phi_{12}{}^* = \Phi_{21}$. For our purpose we rewrite Equation 6.38 as[14]

(6.39)
$$\operatorname{Re} Z(s) = \left(u_1 + \frac{\Phi_{21}}{\Phi_{11}} u_2\right)\left(u_1{}^* + \frac{\Phi_{21}{}^*}{\Phi_{11}{}^*} u_2{}^*\right) \Phi_{11} + u_2 u_2{}^* \left(\Phi_{22} - \frac{\Phi_{12} \Phi_{21}}{\Phi_{11}}\right)$$

The second condition of positive real character of $Z(s)$ requires that

$$\operatorname{Re} Z(s) \geq 0 \quad \text{for } \operatorname{Re} s \geq 0$$

which means that the right-hand side of (6.39) be nonnegative for $\operatorname{Re} s \geq 0$. Now, $\Phi_{11}(s) = \operatorname{Re} z_{11}(s)$ and hence[15]

(6.40a) $$\Phi_{11}(s) \geq 0 \quad \text{for } \operatorname{Re} s \geq 0$$

As a result, the first factor in the right-hand side of (6.39) is always nonnegative. Thus, in order to satisfy the real-part condition of $Z(s)$, we must also have

(6.40b) $$\Phi_{22} - \frac{\Phi_{12} \Phi_{21}}{\Phi_{11}} \geq 0 \quad \text{for } \operatorname{Re} s \geq 0$$

---

[14] G. Raisbeck, *loc. cit.*
[15] Because $z_{11}(s)$ is the driving-point impedance of a passive *LLF* network and by Theorem 6-1 it is p.r.

Conditions (6.40a) and (6.40b) jointly imply

(6.40c) $\quad\quad\quad \Phi_{22}(s) \geq 0 \quad$ for $\operatorname{Re} s \geq 0$

which must be true since $\Phi_{22}(s) = \operatorname{Re} z_{22}(s)$, where $z_{22}(s)$ is also p.r. We thus have the following:

Theorem 6-4[16] *The open-circuit parameters $z_{mn}(s)$ of an LLFP two-port satisfy the following conditions:*

(i) *$z_{mn}(s)$ is real for real s.*
(ii) *$\Phi_{mn}(s)$ as defined by (6.37) satisfy the inequality (6.40).*

Observe that condition (i) of Theorem 6-4 follows from the *LLF* character of the network. Theorem 6-4 can be simplified to make it easier to test the passivity of an *LLF* two-port. The simplification is based on the fact that the input impedance of the one-port of Figure 6-6 also satisfies the three conditions of Theorem 6-2. From these properties of $Z(s)$, properties of $z_{mn}(s)$ can be derived. We have noted that $z_{11}(s)$ and $z_{22}(s)$, being driving-point impedances of an *LLFP* network, are p.r. functions.

Since poles of $z_{mn}(s)$ are also poles of $Z(s)$, it is clear from Equation 6.34 that $z_{mn}(s)$ can have no poles in the right-half plane and that poles on the $j\omega$-axis, if any, must be simple. At any $j\omega$-axis pole of $Z(s)$, e.g., $s = j\omega_0$, let the residue of $Z(s)$ be k_0. So as $s \to j\omega_0$, we can write Equation 6.34 as

$$(6.40) \quad \frac{k_0}{s - j\omega_0} = u_1 u_1^* \frac{k_{11}}{s - j\omega_0} + u_1 u_2^* \frac{k_{12}}{s - j\omega_0} + u_1^* u_2 \frac{k_{21}}{s - j\omega_0} + u_2 u_2^* \frac{k_{22}}{s - j\omega_0}$$

where k_{mn} is the residue of $z_{mn}(s)$ at a possible pole at $s = j\omega_0$. Equation 6.40 reduces to

(6.41) $\quad k_0 = u_1 u_1^* k_{11} + u_2 u_2^* k_{22} + (u_1 u_2^* k_{12} + u_1^* u_2 k_{21})$

Now, k_0 is real and nonnegative for all values of u_1 and u_2. Positive realness of z_{11} and z_{22} implies that k_{11} and k_{22} be real and nonnegative. Therefore, the factor $(u_1 u_2^* k_{12} + u_1^* u_2 k_{21})$ in Equation 6.41 must be real, which will be true provided

(6.42) $\quad\quad\quad\quad\quad\quad k_{12} = k_{21}^*$

Observe that Equation 6.41 is identical in form to Equation 6.38. Hence,

[16] G. Raisbeck, *loc. cit.*

following similar arguments, we have the following conditions on the residues of $z_{mn}(s)$ at a possible simple pole on the $j\omega$-axis:

(6.43)
$$k_{11} \geq 0$$
$$k_{22} \geq 0$$
$$k_{11}k_{22} - k_{12}k_{21} \geq 0$$
$$k_{11}, k_{22} \text{ real and } k_{12} = k_{21}*$$

Condition 6.43 will be called the *residue condition* for a nonreciprocal passive *LLF* two-port.

The condition Re $Z(j\omega) \geq 0$ for all values of ω is equivalent to the condition 6.40 on $\Phi_{mn}(s)$ with $s = j\omega$. If we use the notation

(6.44) $$R_{mn}(0, \omega) + jX_{mn}(0, \omega) = r_{mn} + jx_{mn}$$

then we obtain from condition 6.40 the following:

(6.45)
$$r_{11} \geq 0$$
$$r_{22} \geq 0$$
$$r_{11}r_{22} - \left(\frac{r_{12} + r_{21}}{2}\right)^2 - \left(\frac{x_{12} - x_{21}}{2}\right)^2 \geq 0$$

for all values of ω. Summarizing the previous results, we have:

Theorem 6-5[17,18] *The open-circuit impedance matrix $z_{mn}(s)$ of a linear, lumped, finite, and passive two-port satisfies the following conditions:*

(i) *$z_{mn}(s)$ is analytic in the right-half s plane.*
(ii) *Poles of $z_{mn}(s)$ on the $j\omega$-axis, if any, are simple and the residues at each $j\omega$-axis pole satisfy condition 6.43.*
(iii) *The real and imaginary parts of $z_{mn}(j\omega)$ satisfy condition 6.45.*

Example 6-5 Consider the two-port network of Figure 6-7. The impedance matrix of this two-port is found to be

(6.46)
$$\begin{bmatrix} \dfrac{5s}{s^2 + 9} & \dfrac{s - 9}{s^2 + 9} \\ \dfrac{s + 9}{s^2 + 9} & \dfrac{2s}{s^2 + 9} \end{bmatrix}$$

[17] E. M. McMillan, "Violation of the reciprocity theorem in linear passive electromechanical system," *J. Accoust. Soc. Amer.*, **18**, 344–347 (October 1946).
[18] B. D. H. Tellegen, "The synthesis of passive resistanceless four-poles that may violate the reciprocity relation," *Phillips Res. Rept.*, **3**, 321–327 (October 1946).

224 / REALIZABILITY CONDITIONS

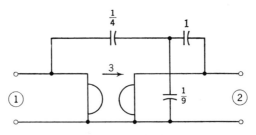

Figure 6-7 A passive nonreciprocal two-port.

The residues of z_{ij} at the $j\omega$-axis pole at $s = j3$ are

$$k_{11} = \tfrac{5}{2}; \quad k_{22} = 1; \quad k_{12} = \tfrac{1}{2} + \tfrac{3}{2}j; \quad k_{21} = \tfrac{1}{2} - \tfrac{3}{2}j$$

It is seen that the residue condition (6.43) is satisfied. Furthermore, we also note that

$$r_{11} = r_{22} = 0$$

$$r_{12} = -\frac{9}{9 - \omega^2} = -r_{21}$$

$$x_{12} = \frac{\omega}{9 - \omega^2} = x_{21}$$

which indicates that condition 6.45 is also satisfied. Therefore, the two-port of Figure 6-7 is passive.

Example 6-6 A two-port network has the following open-circuit parameters:

(6.47)
$$z_{11} = \frac{s^3 + s^2 + 5s + 2}{s^2 + s + 2}; \quad z_{12} = \frac{(s + 2)^2}{s^2 + s + 2}$$

$$z_{21} = \frac{-s(s - 2)}{s^2 + s + 2}; \quad z_{22} = \frac{3s + 2}{s^2 + s + 2}$$

It is desired to test the passivity of the two-port.

Note that the given z_{mn} are analytic in the right-half plane including the $j\omega$-axis, except for the fact that z_{11} has a private pole at infinity with positive residue. The required real and imaginary parts of z_{mn} are easily computed as:

$$r_{11} = \frac{\omega^2 + 4}{\omega^4 - 3\omega^2 + 4}; \quad r_{22} = \frac{\omega^2 + 4}{\omega^4 - 3\omega^2 + 4}$$

$$r_{12} = \frac{\omega^4 - 2\omega^2 + 8}{\omega^4 - 3\omega^2 + 4}; \quad x_{12} = \frac{-3\omega^3 + 4\omega}{\omega^4 - 3\omega^2 + 4}$$

$$r_{21} = \frac{-\omega^4 + 4\omega^2}{\omega^4 - 3\omega^2 + 4}; \quad x_{21} = \frac{-3\omega^3 + 4\omega}{\omega^4 - 3\omega^2 + 4}$$

First observe that $r_{11} > 0$, $r_{22} > 0$ for all ω. Next, substituting the above in the last equation of condition 6.45 and clearing the denominator, we obtain the following polynomial:

$$\psi(\omega^2) = (\omega^2 + 4)(\omega^2 + 4) - \left(\frac{\omega^4 - 2\omega^2 + 8 - \omega^4 + 4\omega^2}{2}\right)^2$$

$$- \left(\frac{-3\omega^3 + 4\omega + 3\omega^3 - 4\omega}{2}\right)^2$$

$$= (\omega^2 + 4)^2 - \tfrac{1}{4}(2\omega^2 + 8)^2 = 0$$

Condition 6.45 is thus satisfied. Therefore, the two-port is passive.

This example illustrates the fact that if the denominators of the z parameters are identical, then condition 6.45 becomes:

(6.48)[19] $(\text{Num } r_{11}) \geq 0$

$$\psi(\omega^2) = (\text{Num } r_{11})(\text{Num } r_{22}) - \left(\frac{\text{Num } r_{12} + \text{Num } r_{21}}{2}\right)^2$$

$$- \left(\frac{\text{Num } x_{12} - \text{Num } x_{21}}{2}\right)^2 \geq 0$$

Several comments are here in order. For a reciprocal two-port $z_{12} = z_{21}$; hence, the residues at a $j\omega$-axis pole of the transfer impedance are now equal, i.e.,

$$k_{12} = k_{21}$$

which also requires that they must be real. In this case, condition 6.43 on residues reduces to the following:

(6.49) k_{11}, k_{22}, k_{12} real
$k_{11} \geq 0$
$k_{11}k_{22} - k_{12}^2 \geq 0$

Condition 6.49 is the more familiar *residue condition* for a reciprocal two-port. Furthermore, we note

$$r_{12} = r_{21}, \quad x_{12} = x_{21}$$

Hence, condition 6.45 becomes

(6.50) $r_{11} \geq 0$
$r_{11}r_{22} - r_{12}^2 \geq 0$

---

[19] Numerator polynomial of a real rational function $R(\omega)$ will be denoted by Num $R(\omega)$. Thus, if $R(\omega) = U(\omega)/V(\omega)$, then Num $R(\omega) = U(\omega)$.

The above condition is called the *real-part condition* for a reciprocal two-port. Summarizing, we have:

Theorem 6-6 *A linear reciprocal lumped two port is passive if, and only if, the open-circuit parameters z_{mn} satisfy the following three conditions:*

   (i) Poles of $z_{mn}(s)$ are in the left-half s-plane, and $j\omega$-axis poles, if any, are simple.
   (ii) The residues at the $j\omega$-axis pole satisfy the residue condition 6.49.
   (iii) The real part of the impedances for $s = j\omega$ satisfy the real-part condition 6.50.

The conditions on the short-circuit admittance parameters, $y_{mn}(s)$, can be derived in an analogous manner. In this case we shall have:

Theorem 6-7 *The short-circuit admittances, $y_{mn}(s)$, of a LLFP two-port network satisfy the following necessary and sufficient conditions:*

   (i) Poles of $y_{mn}(s)$ are in the left-half s-plane and $j\omega$-axis poles, if any, are simple.
   (ii) The residues at each $j\omega$-axis pole satisfy condition 6.43.
   (iii) If $y_{mn}(j\omega) = g_{mn} + jb_{mn}$, then

$$g_{11} \geq 0, g_{22} \geq 0$$
$$4g_{11}g_{22} - (g_{12} + g_{21})^2 - (b_{12} - b_{21})^2 \geq 0$$

(6.51)

6-4 ACTIVE NETWORKS

In the previous two sections, we derived the properties of network functions of passive (*LLFP*) networks from energy considerations. Theorems 6-1, 6-2, 6-3, 6-5, and 6-7 can be used to test the passivity (or activity) of a network.

In the remaining part of this book, we shall be concerned with more specific types of active networks. To help understand the usefulness and limitations of these *LLF* networks, it is desirable to formulate the physical realizability conditions for the pertinent cases.

For the purpose of simpler identification and to facilitate our later discussions, the following notations, among others, for the *LLF* networks will be used:

   (i) *RC:–R* network is a network composed of resistances, capacitances, and negative resistances.
   (ii) *RC:NIC* network consists of resistances, capacitances, and negative impedance converters. (This includes the class of *RC:–R,–C* networks.)

(iii) $LC{:}{-}R$ network is composed of inductances, capacitances, and negative resistances.
(iv) $RLC{:}{-}R$ network contains resistances, capacitances, inductances, and negative resistances.
(v) $RC{:}{-}C$ network has negative capacitance as the active elements and (positive) resistances, and capacitances as the passive elements.
(vi) $RC{:}t_d$ network contains resistances, capacitances, and N-type negative resistances (tunnel diodes).[20]
(vii) $RC{:}VCT$ circuit contains resistances, capacitances, and voltage controlled current sources.

Thus a general notation of an LLF network will be "passive components: active components." Sometimes, a superscript may be used over the active part to indicate the number of active elements in the circuit. For example, an RC network in which N tunnel diodes have been embedded, will be designated as an $RC{:}t_d^{(N)}$ network.

Derivation of the physical realizability conditions of the LLF networks can be achieved by the energy-functions approach originally formulated by Brune[21] for the study of $LLFPB$ networks. Extension of this method to the general class of LLF networks has been made by Bello,[22] and we refer the interested reader to the original paper.

The properties of the type of active networks considered in this text can be derived alternately. Our approach will be based on the assumption that the properties of the passive "companion" network are known. In Section 6-5, we consider networks containing negative resistances as the active element. Networks containing gyrators, NIC's, and controlled sources are covered in the subsequent sections.

6-5 NETWORKS CONTAINING NEGATIVE RESISTANCES

In this section, LLF networks containing negative resistances will be studied. First, we shall derive the properties of network functions of $RC{:}{-}R$ networks from a knowledge of the properties of RC networks. Next, $RC{:}$tunnel-diode networks will be treated. We shall conclude this section with a discussion of $LC{:}{-}R$ networks, followed by a treatment of

---

[20] It is mentioned in Section 7-1 that a simple equivalent circuit of a tunnel diode is identical to that of an N-type negative resistance. This is why we shall use the symbol t_d to represent all N-type negative resistances.
[21] O. Brune, *loc. cit.*
[22] P. Bello, *loc. cit.*

LC: tunnel-diode networks. Throughout this section we shall be concerned with the necessary properties of the network functions. Sufficiency conditions will be treated in the next chapter on synthesis.

RC:$-R$ Network[23]

Consider an RC:$-R^{(1)}$ one-port as shown in Figure 6-8 where, without any loss of generality, the negative resistance has been taken as -1 ohm. The input admittance $Y(s)$ is given as

$$(6.52) \qquad Y(s) = y_{11}(s) - \frac{y_{12}^{2}(s)}{y_{22}(s) - 1}$$

where $y_{ij}(s)$ is the short-circuit parameter of the RC two-port of Figure 6-8. Observe that the poles of $Y(s)$ are either the same as poles of $y_{ij}(s)$ or are due to a zero of $[y_{22}(s) - 1]$. Since $y_{22}(s)$ is a monotonically

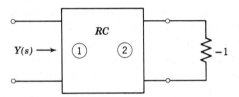

Figure 6-8 Representation of an RC:$-R^{(1)}$ one-port.

increasing function for nonnegative values of σ, $Y(s)$ can have at most one pole on the positive real axis. The remaining poles are on the negative real axis, including origin and infinity.

Let $s = \rho$ be a finite pole of $Y(s)$ and its corresponding residue of $Y(s)$ be denoted by k_ρ. Then

$$(6.53) \quad k_\rho = Y(s)(s - \rho)\big|_{s=\rho} = y_{11}(s)(s - \rho)\big|_{s=\rho} - \frac{y_{12}^{2}(s)(s - \rho)}{y_{22}(s) - 1}\bigg|_{s=\rho}$$

If $s = \rho$ is a pole of the y parameters, Equation 6.53 reduces to

$$(6.54) \qquad k_\rho = k_{11\rho} - \frac{k_{12\rho}^{2}}{k_{22\rho}}$$

where $k_{ij\rho}$ is the residue of $y_{ij}(s)$ at the pole at $s = \rho$. Because of the residue condition satisfied by the passive RC two-port,

$$(6.55) \qquad k_{11\rho}k_{22\rho} - k_{12\rho}^{2} \leq 0$$

it is seen that k_ρ will be real and nonpositive. Note that if $s = \rho$ is a

---

[23] The treatment of Section 6-5 follows S. K. Mitra, "The realizability of tunnel-diode-RC networks," *J. Franklin Inst.*, 275, 205–216 (March 1963).

NETWORKS CONTAINING NEGATIVE RESISTANCES / 229

private pole of $y_{11}(s)$, k_ρ will be still real and nonpositive. Finally, consider the case when $(s - \rho)$ is a factor of $[y_{22}(s) - 1]$. Then Equation 6.53 reduces to

$$(6.56) \quad k_\rho = -\left.\frac{y_{12}^2(s)}{[y_{22}(s) - 1]/(s - \rho)}\right|_{s=\rho} = -\left.\frac{y_{12}^2(\rho)}{\dfrac{d}{d\sigma}[y_{22}(\sigma) - 1]}\right|_{\sigma=\rho}$$

$y_{22}(s)$ being an RC driving-point admittance, $d[y_{22}(\sigma) - 1]/d\sigma$, which is the slope of $y_{22}(\sigma) - 1$, is always positive. This implies that k_ρ is real and nonpositive also in this case.

Possible pole at infinity can occur two ways. It is either a pole of $y_{ij}(s)$, or $y_{22}(\infty)$ is equal to unity. In the former case, the residue of $Y(s)$ at $s = \infty$ is

$$(6.57) \quad k_\infty = k_{11\infty} - \frac{k_{12\infty}^2}{k_{22\infty}}$$

where $k_{ij\infty}$ denotes the residue of $y_{ij}(s)$ at a pole at infinity. Since

$$(6.58) \quad k_{11\infty}k_{22\infty} - k_{12\infty}^2 \geq 0$$

k_∞ must be real and nonnegative. In the second case, it can be shown that k_∞ will still be real and nonnegative. To prove this, let us express $y_{ij}(s) = N_{ij}(s)/D(s)$. We rewrite Equation 6.52 as

$$(6.59) \quad Y(s) = \frac{N_{11}(s)}{D(s)} - \frac{N_{12}^2(s)}{D(s)[N_{22}(s) - D(s)]}$$

If $y_{22}(\infty)$ is equal to unity, we can write

$$(6.60) \quad \begin{aligned} N_{22}(s) &= s^m + A_{m-1}s^{m-1} + \cdots \\ D(s) &= s^m + B_{m-1}s^{m-1} + \cdots \\ N_{12}(s) &= C_m s^m + C_{m-1}s^{m-1} + \cdots \end{aligned}$$

In the neighborhood of infinity,

$$(6.61) \quad Y(s) \to -\frac{C_m^2 s}{[A_{m-1} - B_{m-1}]} \quad \text{as} \quad s \to \infty$$

But

$$[A_{m-1} - B_{m-1}] = [\Sigma \text{ zeros of } y_{22}(s) - \Sigma \text{ poles of } y_{22}(s)] < 0$$

[This follows from the fact that $y_{22}(s)$ is an RC admittance.] Thus we conclude from expression 6.61 that a possible pole at infinity due to the second reason will also be associated with a real and positive residue for $Y(s)$.

Finally, we note from Equation 6.52 that if $Y(s)$ has no finite pole on the positive real axis (including the origin), the second term on the right-hand side can have any real value at $s = 0$. This indicates that $Y(0)$ may have any real value. On the other hand, if there is a finite pole, $[y_{22}(0) - 1]$ is negative, implying that $Y(0)$ must then be real and positive. Summarizing the above results, we have the following theorem:

Theorem 6-8 *The driving-point admittance $Y(s)$ of an $RC{:}{-}R^{(1)}$ network satisfies the following necessary and sufficient conditions:*

(i) *Poles of $Y(s)$ are real and simple, and there can be at most one pole on the positive real axis.*
(ii) *At each finite pole, the residues of $Y(s)$ are real and negative. The possible pole at infinity must have a real and positive residue.*
(iii) *If $Y(s)$ has a finite pole on the positive real axis, then $Y(0)$ is real and nonnegative. Otherwise, it can have any value.*

Sufficiency of Theorem 6-8 will be shown in the next chapter.

Let us now derive the properties of an $RC{:}{-}R^{(1)}$ two-port. Connecting ideal (real) transformers as shown in Figure 6-9, we first construct

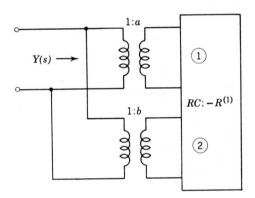

Figure 6-9 Construction of an $RC-R^{(1)}$ one-port from an $RC{:}{-}R^{(1)}$ two-port.

a one-port $RC{:}{-}R^{(1)}$ network. The driving-point admittance of this one port can be expressed as:

(6.62) $$Y(s) = a^2 y_{11}(s) + 2ab y_{12}(s) + b^2 y_{22}(s)$$

where $y_{ij}(s)$ represent the y parameters of the $RC{:}{-}R^{(1)}$ two-port. It is clear that $Y(s)$ must satisfy the conditions of Theorem 6-8. Pursuing the arguments of Section 6-2, we arrive at the following theorem:

Theorem 6-9 *The necessary conditions satisfied by the y parameters, $y_{ij}(s)$, of an $RC{:}{-}R^{(1)}$ two-port are:*

(i) *Poles of $y_{ij}(s)$ are real and simple, with at most one pole on the positive real axis.*

(ii) *At a finite pole at $s = \rho$, the residue $k_{ij\rho}$ of $y_{ij}(s)$ is real and satisfies the residue condition,*

$$k_{11\rho}k_{22\rho} - k_{12\rho}^2 \leq 0$$

(iii) *At a possible pole at infinity, the residue $k_{ij\infty}$ of $y_{ij}(s)$ is real and satisfies the condition*

$$k_{11\infty}k_{22\infty} - (k_{12\infty})^2 \geq 0$$

(iv) *If $y_{ij}(s)$ has a finite pole on the positive real axis, then the matrix $[y_{ij}(0)]$ is real and positive semidefinite. Otherwise, it can have any value.*

A consequence of Theorem 6-8 is that on the real axis, the driving-point admittance of an $RC{:}{-}R^{(1)}$ network is a monotonically increasing function. This fact, along with the results of Theorem 6-9, enables us to extend the results to the case of $RC{:}{-}R^{(2)}$ networks. The approach is similar to the one outlined previously, i.e., we consider an $RC{:}{-}R^{(2)}$ one-port as an $RC{:}{-}R^{(1)}$ two-port terminated at one port by a negative resistance. In this way, the results of Theorem 6.8 can be generalized to the case of an $RC{:}{-}R^{(N)}$ driving-point admittance; we leave this as an exercise to the reader. Instead, we mention below similar conditions for the driving-point impedance of an $RC{:}{-}R^{(N)}$ network to point out the major differences.

Theorem 6-10 *The necessary and sufficient conditions satisfied by an $RC{:}{-}R^{(N)}$ driving-point impedance $Z(s)$ are:*

(i) *All poles are simple and lie on the real axis. There can be at most N poles on the positive real axis including infinity.*

(ii) *The residue at a finite pole is real and positive, whereas at a possible pole at infinity the residue must be real and negative.*

(iii) *$Z(\infty)$ is real and nonnegative if $Z(s)$ has N finite positive real poles. $Z(\infty)$ can have any real value, if $Z(s)$ has less than N positive real poles including the point at infinity.*

The sufficiency proof is outlined in Section 7-2.

$RC{:}t_d$ Network

As mentioned earlier, an $RC{:}t_d$ network represents the general class of active network composed of resistances, capacitances, and N-type negative resistances. It is obvious that $RC:t_d$ networks must satisfy the necessary conditions of Theorem 6-10. The purpose of the following discussion is to obtain some additional restrictions that arise due to the presence of the shunt capacitor across the negative resistances. The reciprocal of product $C_d R_d$ for a tunnel diode (Figure 6-10) will be called the dissipation factor, δ_d, of the diode.

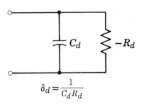

Figure 6-10 Equivalent circuit of an N-type negative resistance (or tunnel diode).

First we note that at $s = \infty$, the negative resistances are effectively short-circuited by the shunt capacitances. Hence at infinity, $RC:t_d$ networks behave as RC networks. Next, we consider a network with a single tunnel diode. An $RC:t_d^{(1)}$ one-port can be represented as shown in Figure 6-11 where $\mathcal{M}$ is a passive RC two-port. The natural frequencies of such a network are then given by the roots of the equation

(6.63) $$Y_0(s) = Y_{22}(s) + C_1(s - \delta_1) = 0$$

where δ_1 is the dissipation factor of the diode. $Y_{22}(s)$ in Equation 6.63 is a passive RC driving-point admittance, and its poles and zeros alternate with the critical frequency near the origin being a zero. That is,

(6.64) $$Y_{22}(s) = k \frac{\Pi(s + z_i)}{\Pi(s + p_j)}$$

where $z_1 < p_1 < z_2 < p_2 \ldots$. From Equations 6.63 and 6.64, we observe that the natural frequencies are given by the zeros of $A(s) + B(s)$, where

(6.65) $$A(s) = C_1(s - \delta_1)\Pi(s + p_j)$$
$$B(s) = k\Pi(s + z_i)$$

It is seen from a plot of $A(\sigma)$ and $-B(\sigma)$ (Figure 6-12) that there can be at most one real and positive natural frequency at $s = \alpha$ where

$$\alpha \leq \delta_1$$

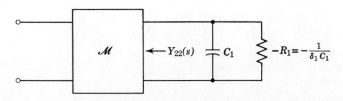

Figure 6-11 Representation of an $RC:t_d$ one-port.

NETWORKS CONTAINING NEGATIVE RESISTANCES / 233

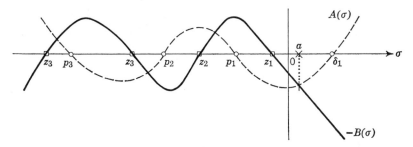

Figure 6-12 To determine the natural frequency locations of an $RC:t_d^{(1)}$ one-port.

The above results lead to the following theorem:

Theorem 6-11 *The driving-point impedance $Z(s)$ of an $RC:t_d^{(1)}$ network has the following properties:*

(i) All poles of $Z(s)$ are simple and real with real and positive residues.

(ii) $Z(s)$ can have at most one pole at $s = \alpha$ and at most one zero at $s = a$ on the positive real axis satisfying the condition:

$$0 \leq a \leq \alpha \leq \delta_1$$

(iii) $Z(\infty)$ is a nonnegative real constant.

Sufficiency proof of Theorem 6-11 will be taken up in the next chapter. It is fairly obvious that $Z(s + \delta_1)$ is a passive RC driving-point impedance.

Now, let us generalize the results of Theorem 6-11 to the case of two tunnel diodes. We can represent an $RC:t_d^{(2)}$ one-port as shown in Figure 6-11, where $\mathcal{M}$ is now an $RC:t_d^{(1)}$ two-port. For convenience, let us assume that the tunnel diode inside the two-port has the smaller dissipation factor δ_2. We look at the zeros of $Y_0(s)$ (see Figure 6-11) to determine the location of the natural frequencies. $Y_{22}(s)$ in Equation 6.63 is the driving-point admittance of an $RC:t_d^{(1)}$ network and hence $1/Y_{22}(s)$ has the properties indicated in Theorem 6-11. Two possible cases may arise:

1. $Y_{22}(s)$ has only a positive real zero;
2. $Y_{22}(s)$ has a zero and a pole on the positive real axis.

As before, the natural frequencies are given by the zeros of the polynomial $A(s) + B(s)$. In Figures 6-13a and 6-13b, the two cases are illustrated, from which we conclude that the driving-point impedance of an $RC:t_d^{(2)}$ network can have either:

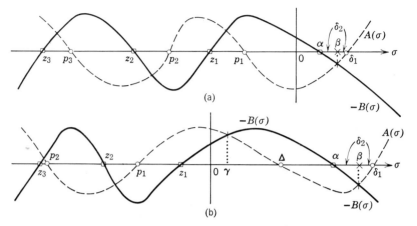

Figure 6-13 To determine the natural frequency locations of an $RC{:}t_d^{(2)}$ one-port.

1. At most one pole in the interval $(\delta_2, \delta_1]$ and at most one pole in the interval $(0, \delta_2]$; or
2. At most two poles in the interval $(0, \delta_2]$.

The above reasoning can be repeated for more than two tunnel diodes. In the case of an $RC{:}t_d^{(N)}$ network, let the dissipation factors of the tunnel diodes satisfy the relation

$$\delta_1 > \delta_2 > \delta_3 \cdots > \delta_N$$

Denoting the origin as δ_{N+1}, we have the following theorem:

Theorem 6-12 *The driving-point impedance $Z(s)$ of an $RC{:}t_d^{(N)}$ network satisfies the following necessary and sufficient conditions:*

(i) *All poles of $Z(s)$ are simple and lie on the real axis.*
(ii) *The residues of $Z(s)$ at the poles are real and positive.*
(iii) *There can be at most j poles on the positive real axis in the interval $(\delta_{j+1}, \delta_1]$ where $j = 1, 2, \ldots, N$.*
(iv) $Z(\infty)$ *is a nonnegative constant.*

The necessity of the above conditions has been shown. Sufficiency will be shown in the next chapter. Extension of the above theorem to the multiport case is straightforward, and the reader is urged to work it out as an exercise. Although the results presented above are on an impedance basis, analogous results on the admittance basis can be easily formulated.

LC:−R Network

We first consider a network containing a single negative resistance. An $LC{:}{-}R^{(1)}$ one-port can be represented as shown in Figure 6-14a, where

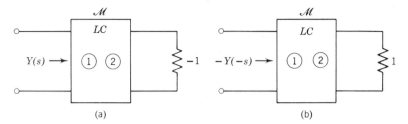

Figure 6-14 (a) An $LC{:}{-}R^{(1)}$ one-port, (b) An RLC one-port.

the negative resistance has been taken as -1 ohm without any loss of generality. In terms of the y parameters of the two port $\mathcal{M}$, y_{mn}, the driving-point admittance $Y(s)$ of the one-port is given by

$$(6.66) \qquad Y(s) = y_{11}(s) - \frac{y_{12}^2(s)}{y_{22}(s) - 1}$$

From Equation 6.66 we note that

$$(6.67) \qquad -Y(-s) = -y_{11}(-s) + \frac{y_{12}^2(-s)}{y_{22}(-s) - 1} = y_{11}(s) - \frac{y_{12}^2(s)}{y_{22}(s) + 1}$$

because $y_{mn}(s)$, being the short-circuit admittances of a lossless reciprocal network, are odd functions, $y_{mn}(-s) = -y_{mn}(s)$. The expression on the right-hand side of Equation 6.67 represents the driving-point admittance of the two-port $\mathcal{M}$ of Figure 6-14a when terminated at port 2 by a positive resistance of value one (Figure 6-14b). Hence we have:

Theorem 6-13 *If $Y(s)$ is an $LC{:}{-}R^{(1)}$ driving-point admittance, then $-Y(-s)$ is a positive real function.*

The above condition is also sufficient. It can be proved by realizing $-Y(-s)$ by the Darlington method[24] in the form of Figure 6-14b and replacing the positive resistance by a negative resistance of equal value. Of course, in the realization of the lossless two-port, ideal transformers may be needed. It is a simple exercise to verify that the driving-point function of an $LC{:}{-}R^{(N)}$ network has the same property as that of an $LC{:}{-}R^{(1)}$ network. Next, we shall consider the constraint imposed by a shunt capacitor across the negative resistance.

$LC{:}t_d$ Network[25]

As before, we restrict our attention to an $LC{:}t_d^{(1)}$ one-port as shown by Figure 6-15, where we have normalized the negative resistance to -1.

---

[24] See, for example, M. E. Van Valkenburg, *loc. cit.*
[25] Our development closely parallels B. K. Kinariwala, "The Esaki diode as a network element," *IRE Trans. on Circuit Theory*, **CT-8**, 389–397 (December 1961).

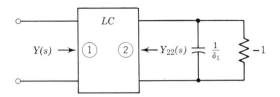

Figure 6-15 An $LC:t_d^{(1)}$ one-port.

The natural frequencies are given by the roots of the equation:

(6.68) $$Y_0(s) = \frac{1}{\delta_1}s - 1 + Y_{22}(s) = 0$$

where $Y_{22}(s)$ is a lossless driving-point admittance. Let

(6.69) $$\frac{1}{\delta_1}s + Y_{22}(s) = \frac{A_n s^n + A_{n-2}s^{n-2} + \cdots}{B_{n-1}s^{n-1} + B_{n-3}s^{n-3} + \cdots}$$

We observe from Equation 6.69 that

$$\frac{A_n}{B_{n-1}} \geq \frac{1}{\delta_1}$$

Furthermore, $(1/\delta_1)s + Y_{22}(s)$ being a lossless driving-point function, zeros of $Y_0(s)$ as given by Equation 6.68 are in the right-half s-plane. From Equations 6.68 and 6.69, we have

(6.70) $$Y_0(s) = \frac{A_n s^n - B_{n-1}s^{n-1} + A_{n-2}s^{n-2} - \cdots}{B_{n-1}s^{n-1} + B_{n-3}s^{n-3} + \cdots}$$

If we denote the zeros of $Y_0(s)$ by z_i, it is clear from Equation 6.70 that

$$\frac{B_{n-1}}{A_n} = \sum_i z_i \leq \delta_1$$

We now state the realizability conditions:

Theorem 6-14 *The driving-point admittance $Y(s)$ of an $LC:t_d^{(1)}$ one-port has the following necessary properties:*

   (i) $-Y(-s)$ *is positive real.*
   (ii) $Y(s)$ *has a pole or a zero at infinity.*
   (iii) Num [Ev $Y(s)$] *is a square of a polynomial with real coefficients.*
   (iv) *The sum of the zeros (poles) of $Y(s)$ is less than or equal to the dissipation factor of the tunnel diode.*

Condition (i) follows from Theorem 6-13. Condition (ii) results from the lossless behavior of the one-port at $s = \infty$. Condition (iii) will be obvious if Ev $Y(s)$ is computed in terms of the short-circuit parameters of the lossless two-port of Figure 6-15. The conditions given in Theorem 6-14 are also sufficient. We shall prove the sufficiency in Section 10.2 for the case when nonreciprocal lossless elements (gyrators) are used for the realization of the lossless two-port. It will be shown that condition (iii) is not necessary in this latter case.

Multiport $LC:t^{(1)}$ network has been investigated by Kinariwala[26] and Sandberg.[27]

$RLC:-R$ Network

It is obvious that the driving-point impedance of an $RLC:-R$ network is a real rational function. This is also a sufficient condition. In fact, we shall show in the following chapter that any real rational function can be realized by an $RLC:-R^{(2)}$ one-port. In Chapter 12 it will be shown that any real rational function is realizable by embedding a single negative resistance in an RLC network.

$RLC:t_d$ networks have been studied by Shenoi.[28] In addition to deriving the properties of such networks, he has presented several synthesis procedures.

6-6 NETWORKS CONTAINING GYRATORS

It was pointed out that gyrators are lossless devices; hence, RLC networks containing gyrators are passive networks. Therefore, the driving-point functions of such networks must be positive real functions. Similarly, for an RLC:gyrator two-port, the open-circuit and short-circuit parameters satisfy the conditions of Theorems 6-5 and 6-7, respectively. Since inductors can be realized by terminating an ideal gyrator by a capacitor, RC:gyrator network functions have the same properties as the corresponding functions of RLC:gyrator networks.

From the results of Theorems 6-5 and 6-7, it is fairly evident that the transfer functions of RLC:gyrator networks have lesser restrictions than the driving-point functions. Obviously, the poles of transfer functions

---

[26] B. K. Kinariwala, *loc. cit.*

[27] I. W. Sandberg, "The realizability of multi-port structures obtained by imbedding a tunnel diode in a lossless reciprocal network," *Bell System Tech. J.*, **41**, 857–876 (May 1962).

[28] B. A. Shenoi, "A general synthesis of tunnel diode networks and sensitivity minimization," *Proc. Nat. Elect. Conf.*, **18**, 114–126 (November 1962).

must lie in the left-half plane and $j\omega$-axis poles must be simple. The zeros of the transfer function can be anywhere on the s-plane.

In the chapter on synthesis using gyrators (Chapter 10), it will be shown that the positive real property is sufficient to guarantee the realization of a driving-point function by means of an RLC:gyrator network. Conditions of Theorems 6-5 and 6-7 are also sufficient, the proof of which is beyond the scope of this book.

There are further restrictions on the natural frequencies of an RC:gyrator network containing a single gyrator. It can be shown that the natural frequencies must satisfy an angle criterion, details of which will be provided in Chapter 10.

6-7 NETWORKS CONTAINING NEGATIVE IMPEDANCE CONVERTERS

When negative impedance converters are embedded in an *LLFPB* network, the network functions are definitely real rational functions. It will be shown in Chapter 9 that any real rational function can be realized as the driving-point function of a network containing resistances, capacitances, and a single negative impedance converter or a generalized impedance converter. Sufficiency of a single negative impedance converter embedded in an *RC* network in realizing any real rational function as the transfer function of a two-port will also be shown in Chapter 9.

In the case of multiport networks, Sandberg has proved the following theorem:[29]

Theorem 6-15 *An arbitrary symmetric $(n \times n)$ matrix of real rational functions of s can always be realized as the impedance (admittance) matrix of an n-port composed of resistances, capacitances, and at most n negative impedance converters.*

In addition, Sandberg has advanced a procedure for the realization of an arbitrary $(n \times n)$ short-circuit admittance matrix in the form of an unbalanced transformerless active *RC* network containing at most n negative-impedance converters.

6-8 OTHER *LLF* NETWORKS

The concept of generalized inverse and dual networks can be used effectively to derive the properties of various types of *LLF* networks. In

---

[29] I. W. Sandberg, "Synthesis of transformerless active *n*-port networks," *Bell System Tech. J.*, **40,** 761–784 (May 1961).

this section, we indicate how this can be achieved for two cases. Results of many other cases can similarly be derived. For a detailed discussion on properties of *LLF* networks, the reader is referred to Bello.[30]

Networks Containing Controlled Sources

In our discussion on equivalent circuits in Section 2-5, we noted that all of the proposed active devices can be represented by means of controlled sources. It seems natural to assume that any arbitrary real rational function can be synthesized by means of networks containing resistances, capacitances, and a single controlled source. This is, indeed, the case as we shall point out in a later part of this book. Similarly, the transfer functions do not have any restrictions, except that they be real rational functions.

Regarding multiport networks, we state below, without proof, a theorem due to Sandberg:[31]

Theorem 6-16 *Any arbitrary matrix of order n, whose elements are real rational functions can be realized as the short-circuit admittance matrix of a transformerless n-port network containing resistances, capacitances, and n controlled sources. In general, such a matrix cannot be realized with less than n controlled sources.*

***RC*:–*C* Networks**

An *RC*:–*C* network can be obtained by taking the capacitive dual of an *RC*:–*R* network. Making use of the concept of capacitive inverse on the results of Theorem 6-12, we arrive at the following theorem:

Theorem 6-17 *An RC:–$C^{(N)}$ driving-point admittance $Y_0(s)$ has the following properties:*

  (i) *All poles of $Y_0(s)$ are real and simple, except possibly a double order pole at infinity, and there can be at most N poles on the positive real axis.*
  (ii) *The residues of $Y_0(s)$ at the negative real poles are real and negative, whereas the residues at the poles on the positive real axis must be real and positive.*
  (iii) *The Laurent expansion around infinity is of the form*

$$+k'_\infty s^2 + k''_\infty s + k'''_\infty + \cdots$$

  where k'_∞ is real and nonpositive, k''_∞ can have any real value, and k'''_∞ is real and nonnegative.

[30] P. Bello, *loc. cit.*
[31] I. W. Sandberg, "Synthesis of *n*-port active *RC* networks," *Bell System Tech. J.*, **40**, 329–347 (January 1961).

The conditions given in Theorem 6-17 will be evident if it is noted that $Y_0(s)/s$ has the same properties as the driving-point impedance of an $RC\!:\!-R^{(N)}$ network. The properties of the driving-point impedance of an $RC\!:\!-C$ network can easily be established. By making use of a transformer network, the results of Theorem 6-17 can be extended to multiport cases.

RC: Capacitive-Gyrator Networks

In the section on ideal active elements in Chapter 2, we defined the reactive gyrator as a two-port having the z parameters, $z_{11} = z_{22} = 0$, $z_{12} = -z_{21} = \mp \alpha(s)$. If $\alpha(s)$ is chosen as K/s, we call the complex gyrator a *capacitive gyrator*. Noting that RC:capacitive-gyrator networks are capacitive duals of RC:gyrator networks, we have the following result:

Theorem 6-18 *The driving-point impedance $Z_0(s)$ of an RC:capacitive-gyrator network has the property that $1/s\,Z_0(s)$ is a positive real function.*

Consequently the natural frequencies of such networks must lie in the left-half plane. Sufficiency of the above theorem can be shown by realizing $1/sZ_0(s)$ as the input impedance of an RC:gyrator network and then forming the capacitive dual of the realized network.

6-9 STABILITY CONDITIONS

The concepts of potential instability and unconditional stability of a network with respect to passive terminations were introduced in the Chapter 3. In this section, we shall derive some inequalities that can be used to determine whether a network is potentially unstable or always stable irrespective of the terminations used.

Terminated One-Port

For a passive one-port, the poles and zeros of the driving-point function are restricted to the left-half s plane with possibly simple $j\omega$-axis critical frequencies. Although, theoretically, a passive network can be marginally stable, in practice it is always strictly stable. This is due to the unavoidable losses in physical inductors and capacitors. Hence, a passive one-port is always open-circuit and short-circuit stable. Next, we observe that since the sum of two positive real functions is positive real, a passive one-port is also unconditionally stable.

Consider now an active two-terminal network described by a driving-point impedance $Z_A(s)$. If $Z_A(s)$ and its reciprocal are strictly stable network functions, then $\text{Re}\,Z_A(j\omega)$ must be negative at one or more

STABILITY CONDITIONS / 241

points on the $j\omega$-axis.[32] This is precisely what makes it potentially unstable. The following example will illustrate our observation.

Example 6-7 The network of Figure 6-16a has an input impedance

$$Z_A(s) = \frac{s^2 + 3s + 8}{s^2 + s + 1}$$

Note that $Z_A(s)$ is both open-circuit and short-circuit stable. Hence, $Z_A(s)$ will be positive real (i.e., unconditionally stable) if Re $Z_A(j\omega)$ is nonnegative for all values of ω. To check the real part, we compute the

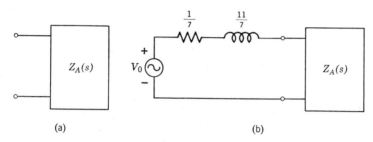

Figure 6-16

numerator of the real part:

$$A(\omega^2) = (s^2 + 8)(s^2 + 1) - 3s^2 \big|_{s=j\omega}$$
$$= \omega^4 - 6\omega^2 + 8 = (\omega^2 - 2)(\omega^2 - 4)$$

Re $Z_A(j\omega)$ has zeros of odd multiplicity for real positive values of ω, and it is negative for $\sqrt{2} < \omega < 2$. Therefore, $Z_A(s)$ represents an active one-port, which is open-circuit and short-circuit stable. For $s = j\sqrt{3}$,

$$Z_A(j\sqrt{3}) = -\frac{1}{7} - j\frac{11\sqrt{3}}{7}$$

If this network is excited by a voltage source of internal impedance $Z_p(s)$ (Figure 6-16b) given as

$$Z_p(s) = \frac{1}{7} + \frac{11}{7}s$$

then the natural frequencies of the overall system are given by the roots

---

[32] If Re $Z_A(j\omega) \geq 0$ for all values of ω, then $Z_A(s)$ is realizable by a passive one-port.

242 / REALIZABILITY CONDITIONS

of $Z_p(s) + Z_A(s)$, which is zero at $s = \pm j\sqrt{3}$:

$$Z_p(+j\sqrt{3}) + Z_A(+j\sqrt{3}) = \frac{1}{7} + j\frac{11\sqrt{3}}{7} - \frac{1}{7} - j\frac{11\sqrt{3}}{7} = 0$$

This indicates that $Z_A(s)$ is potentially unstable.

Terminated Two-Port

Examination of a terminated two-port (Figure 6-17) for instability is slightly more complicated. This is because, in general, four parameters are needed for a complete description of the network.

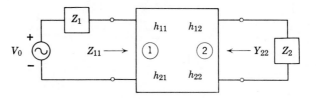

Figure 6-17

The potential instability conditions can be developed in terms of any of the following two-port parameter sets: z_{ij}, y_{ij}, g_{ij}, or h_{ij}. We shall outline the method using the h parameters. Parallel development follows for the other three.

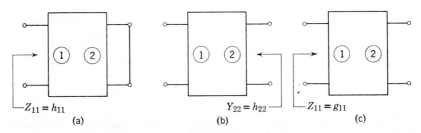

Figure 6-18

By assumption, Z_1 and Z_2 are positive real functions. Now h_{11} is the input impedance seen at port 1 when port 2 is short-circuited (Figure 6-18a), and h_{22} is the output admittance seen at port 2 with port 1 open-circuited (Figure 6-18b). Thus a necessary condition for unconditional stability would be that h_{11} and h_{22} be positive real.

To derive the exact condition, we shall follow a generalization of

Gewertz's approach[33] as outlined by Bolinder.[34] The input impedance of the terminated network is given as

$$\text{(6.70)} \qquad Z_{11} = h_{11} - \frac{h_{12}h_{21}}{h_{22} + Y_2} = h_{11}\frac{g_{22} + Y_2}{h_{22} + Y_2}$$

Note that g_{22} is the output admittance with port 1 shorted ($Z_1 = 0$), hence g_{22} must be positive real. Again, $g_{22} + Y_2$ and $h_{22} + Y_2$ are also positive real functions. Therefore, it follows that Z_{11} is both open-circuit and short-circuit stable. Hence, the system of Figure 6-17 would be unconditionally stable if, and only if, Re $Z_{11}(j\omega)$ is nonnegative for all values of ω. For convenience, we introduce the following notations:

$$\text{(6.71)} \qquad \begin{aligned} h_{mn}(j\omega) &= a_{mn} + jb_{mn} \\ h_{12}(j\omega)h_{21}(j\omega) &= \mathcal{A} + j\mathcal{B} \\ Y_2(j\omega) &= G_2 + jB_2 \end{aligned}$$

Substituting Equation 6.71 in Equation 6.70, we obtain after some algebra

$$\text{(6.72)} \qquad \text{Re}\,[Z_{11}(j\omega)] = a_{11} - \frac{\mathcal{A}(a_{22} + G_2) + \mathcal{B}(b_{22} + B_2)}{(a_{22} + G_2)^2 + (b_{22} + B_2)^2}$$

or

$$\text{(6.73)} \qquad \text{Re}\,[Z_{11}(j\omega)] = \frac{N(\omega, G_2, B_2)}{D(\omega, G_2, B_2)}$$
$$= \frac{a_{11}(a_{22} + G_2)^2 + a_{11}(b_{22} + B_2)^2 - \mathcal{A}(a_{22} + G_2) - \mathcal{B}(b_{22} + B_2)}{(a_{22} + G_2)^2 + (b_{22} + B_2)^2}$$

Examination of expression 6.73 reveals that Re $[Z_{11}(j\omega)]$ will be nonnegative if the numerator of Re $[Z_{11}(j\omega)]$, $N(\omega, G_2, B_2)$ is nonnegative. It is profitable to rewrite $N(\omega, G_2, B_2)$ as

$$\text{(6.74)} \quad N(\omega, G_2, B_2) = a_{11}G_2^2 + (2a_{11}a_{22} - \mathcal{A})G_2 + a_{11}B_2^2$$
$$+ (2a_{11}b_{22} - \mathcal{B})B_2 + (a_{11}a_{22}^2 + a_{11}b_{22}^2 - \mathcal{A}a_{22} - \mathcal{B}b_{22})$$

which is reexpressed as

$$\text{(6.75)} \qquad \frac{N(\omega, G_2, B_2)}{a_{11}} = \left[G_2 + a_{22} - \frac{\mathcal{A}}{2a_{11}}\right]^2$$
$$+ \left[B_2 + b_{22} - \frac{\mathcal{B}}{2a_{11}}\right]^2 - \frac{\mathcal{A}^2 + \mathcal{B}^2}{4a_{11}^2}$$

[33] C. M. Gewertz, "Synthesis of a finite, four-terminal network from its prescribed driving-point functions and transfer functions," *J. Math. Phys.*, **12**, 1–257 (January 1933).

[34] E. F. Bolinder, "Survey of some properties of linear networks," *IRE Trans. on Circuit Theory*, **CT-4**, 70–78 (September 1957).

244 / REALIZABILITY CONDITIONS

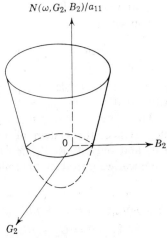

Figure 6-19

Observe that for all values of ω, G_2, and B_2, Equation 6.75 describes a paraboloid of revolution (Figure 6-19). The portion of the paraboloid above the G_2, B_2 plane corresponds to positive values of $N(\omega, G_2, B_2)/a_{11}$ and hence positive values of $N(\omega, G_2, B_2)$ since $a_{11} \geq 0$. Now, the intersection of the paraboloid with the G_2, B_2 plane is a circle (Figure 6-20) which represents $N(\omega, G_2, B_2)$ equal to zero. The radius of the

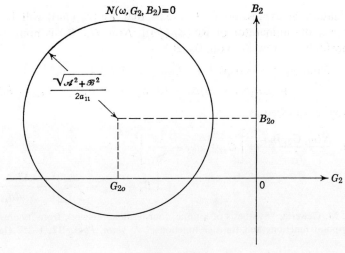

Figure 6-20

circle is

$$\frac{\sqrt{\mathcal{A}^2 + \mathcal{B}^2}}{2a_{11}}$$

and the center is at

(6.76)
$$G_2 = G_{2_0} = \frac{\mathcal{A}}{2a_{11}} - a_{22}$$

$$B_2 = B_{2_0} = \frac{\mathcal{B}}{2a_{11}} - b_{22}$$

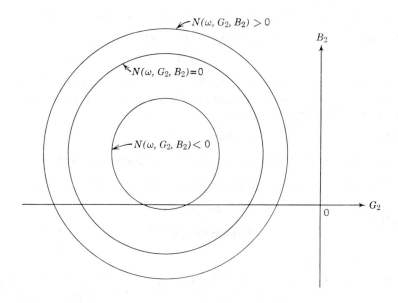

Figure 6-21

Note that the loci of constant $N(\omega, G_2, B_2)/a_{11}$ are circles concentric with the previous circle and are in planes parallel to the G_2, B_2 plane. A projection of a few of these circles on the G_2, B_2 plane are shown in Figure 6-21. The circles which are outside the circle corresponding to $N(\omega, G_2, B_2) = 0$ represent nonnegative values of $N(\omega, G_2, B_2)$. It is now clear that in order that $N(\omega, G_2, B_2)$ be nonnegative for $G_2 \geq 0$ (because of assumed passivity of Z_2), the circle corresponding to $N(\omega, G_2, B_2) = 0$ must be in the left half of the G_2, B_2 plane. This will be satisfied provided that the radius of the circle is less than or equal to $|G_{2_0}|$ and G_{2_0} is negative, i.e.,[35]

$$\text{(6.77)} \qquad \frac{\sqrt{\mathcal{A}^2 + \mathcal{B}^2}}{2a_{11}} \leq \left| \frac{\mathcal{A}}{2a_{11}} - a_{22} \right|$$

$$\text{(6.78)} \qquad \frac{\mathcal{A}}{2a_{11}} - a_{22} < 0$$

for all values of ω. A little thought will convince the reader that conditions 6.77 and 6.78 can be combined as one condition:

$$\text{(6.79)} \qquad \frac{\sqrt{\mathcal{A}^2 + \mathcal{B}^2}}{2a_{11}} \leq a_{22} - \frac{\mathcal{A}}{2a_{11}}$$

Equation 6.79 is the desired result. This can be alternately written as:

$$\text{(6.80a)} \qquad 4a_{11}a_{22}(a_{11}a_{22} - \mathcal{A}) - \mathcal{B}^2 \geq 0$$

$$\text{(6.80b)} \qquad a_{11} \geq 0$$

The inequalities given above, if satisfied, imply the unconditional stability of the two-port. Conditions 6.80 can be compactly written by making use of the following notation:

$$\text{(6.81)} \qquad h_{121} = \sqrt{h_{12}h_{21}} = a_{121} + jb_{121}$$

Summarizing the above results we arrive at the following theorem:

Theorem 6-19 *A linear two-port is unconditionally stable if, and only if,*

$$\text{(6.82)} \qquad \begin{array}{c} a_{11} \geq 0 \\ a_{11}a_{22} - (a_{121})^2 \geq 0 \end{array}$$

for all values of ω.

It follows that conditions 6.82 imply $a_{22} \geq 0$. Conditions 6.82 are known as *Llewellyn's stability conditions*.[36] Since for a reciprocal network $a_{12} = a_{21} = a_{121}$, Gewertz's conditions[37] for a reciprocal network follow from condition 6.82.

Invariant Stability Factor

Use of the relation

$$\text{(6.83)} \qquad a_{121}^2 = \tfrac{1}{2}\{|h_{12}h_{21}| + \text{Re}\,(h_{12}h_{21})\}$$

---

[35] L. DePian, *Linear active network theory*, Prentice-Hall, Englewood Cliffs, N.J., Ch. 7, 1962.

[36] F. B. Llewellyn, "Some fundamental properties of transmission systems," *Proc. IRE*, **40**, 271–283 (March 1952).

[37] C. M. Gewertz, *loc. cit.*

yields an alternate form of the stability conditions 6.82:

(6.84) $\quad 2\,\text{Re}\,(h_{11})\,\text{Re}\,(h_{22}) \geq |h_{12}h_{21}| + \text{Re}\,(h_{12}h_{21})$

where it is tacitly assumed that $\text{Re}\,(h_{11}) \geq 0$ and $\text{Re}\,(h_{22}) \geq 0$.

Let us define

(6.85a) $$\eta \triangleq \frac{2\,\text{Re}\,(h_{11})\,\text{Re}\,(h_{22}) - \text{Re}\,(h_{12}h_{21})}{|h_{12}h_{21}|}$$

The quantity η, originally defined by Rollett,[38] is invariant under parameter substitution, i.e.,

(6.85b) $$\eta = \frac{2\,\text{Re}\,(z_{11})\,\text{Re}\,(z_{22}) - \text{Re}\,(z_{12}z_{21})}{|z_{12}z_{21}|}$$

(6.85c) $$= \frac{2\,\text{Re}\,(y_{11})\,\text{Re}\,(y_{22}) - \text{Re}\,(y_{12}y_{21})}{|y_{12}y_{21}|}$$

(6.85d) $$= \frac{2\,\text{Re}\,(g_{11})\,\text{Re}\,(g_{22}) - \text{Re}\,(g_{12}g_{21})}{|g_{12}g_{21}|}$$

η is known as the *stability invariant factor*. In each of the above equations, the additional constraint $\text{Re}\,(z_{11}) \geq 0$, $\text{Re}\,(y_{11}) \geq 0$, $\text{Re}\,(g_{11}) \geq 0$ is assumed to hold. From expressions 6.85, it immediately follows that for unconditional stability

(6.86) $$\eta \geq 1$$

for all values of ω. Potential instability condition is determined by setting either $\text{Re}\,(h_{11}) = 0$ or $\text{Re}\,(h_{22}) = 0$. Thus, if

(6.87) $$-1 \leq \eta < 1$$

for all ω, the two-port is potentially unstable.

It is a simple exercise to show that a passive *LLF* two-port is unconditionally stable.

Example 6-8 A negative impedance converter is described by the following h parameters:

$$h_{11} = h_{22} = 0$$
$$h_{12} = h_{21} = k$$

This implies

$$\eta = -1$$

or, in other words, the NIC is a potentially unstable two-port.

---

[38] J. M. Rollett, "Stability and power gain invariants of linear two-ports," *IRE Trans. on Circuit Theory*, **CT-9**, 29–32 (March 1962).

6-10 SUMMARY

Properties of the network function, in particular driving-point functions, of various classes of active and passive networks have been investigated in this chapter. We now plan to summarize some of the significant results of the chapter.

One important and useful result of this chapter is that the driving-point function of a linear, lumped, finite, and passive (LLFP) network is a positive real function (Theorem 6-1). This result can be used to test the passivity of an *LLF* one-port. Procedures for testing the positive real character of a real rational function are outlined in Table 6-1.

In the case of an *LLFP* two-port, we observe that the open-circuit impedance matrix and the short-circuit admittance matrix have similar restrictions (Theorems 6-5 and 6-7). For example, the open-circuit parameters, $z_{ij}(s)$, are analytic in the right-half s-plane. Poles of $z_{ij}(s)$ on the $j\omega$-axis are simple and the corresponding residues satisfy the residue condition (6.43). In addition, the real and imaginary parts of $z_{ij}(j\omega)$ satisfy condition 6.45. From these properties, some general properties of transfer functions are easily obtained. The transfer function must have all its poles in the left-half s-plane and $j\omega$-axis poles (if any) must be simple. For a transfer voltage or current ratio, there can be no pole at $s = 0$ or at $s = \infty$. There are no restrictions on the multiplicity and location of the transmission zeros. It should be noted that the *LLFP* network class includes the RLC:gyrator and RC:gyrator type networks.

The next significant result derived is the necessary conditions satisfied by the driving-point function of a RC:$-R$ network (Theorem 6-10) and RC:tunnel-diode network (Theorem 6-12). It is shown that these type of active networks have properties similar to that of passive RC networks with the following exceptions. The driving-point function of an RC:$-R$ network can have poles on the positive real axis. Even though the properties of two-port parameters of such networks were not considered, some obvious conclusions on the transfer functions can be made at this point. Poles of the transfer function are restricted to be located on the real axis, whereas there is no restriction on the location of the transmission zeros. In fact, we shall show in Chapter 7 that positive real transmission zero can be realized by a grounded RC:$-R$ two-port. On the other hand, a passive RC grounded two-port without transformers cannot realize positive real transmission zeros.

We have observed that the network functions of *LLF* networks containing negative impedance converter (NIC) or controlled source are necessarily real rational functions. This condition is also sufficient. In later chapters, it will be shown that any real rational function can be realized as the driving-point (transfer) function of a network composed of resistors, capacitors, and a single NIC (or a single controlled source).

Finally, we note that the conditional stability of a linear two-port can be determined from the value of the invariant stability factor η as defined in (6.85). If $\eta \geq 1$, the two-port is unconditionally stable with respect to any passive terminations. On the other hand, if $-1 \leq \eta < 1$, the two-port is potentially unstable.

Additional References

Bahrs, G. S., "Stable amplifiers employing potentially unstable transistors," *IRE WESCON Conv. Rec.*, **1** (Part 2), 185–189 (1957).

Brodie, J. H., "Passivity conditions for three-terminal networks," *IRE Trans. on Circuit Theory*, **CT-9**, 284 (September 1963).

Carlin, H. J., "On the physical realizability of linear non-reciprocal networks," *Proc. IRE*, **43**, 608–616 (May 1955).

Cederbaum, I., "On the physical realizability of linear non-reciprocal networks," *IRE Trans. on Circuit Theory*, **CT-3**, 155 (June 1956).

DePian, L., *Linear active network theory*, Prentice-Hall, Englewood Cliffs, N.J., 1962.

DePian, L., "Passivity conditions for three-terminal networks," *IRE Trans. on Circuit Theory*, **CT-8**, 360–361 (September 1961).

Desoer, C. A., and E. S. Kuh, "Bounds on natural frequencies of linear active networks," Proc. Symp. Active Networks and Feedback Systems, MRI Symp. Series, Vol. X, Polytech. Inst. of Brooklyn, pp. 415–436 (April 1960).

Fjällbrant, T., "Activity and stability of linear networks," *IEEE Trans. on Circuit Theory*, **CT-12**, 12–17 (March 1965).

Karni, S., "A note on n-port networks terminated with n gyrators," *IRE Trans. on Circuit Theory*, **CT-10**, 526–527 (December 1963).

Kinariwala, B. K., "Necessary and sufficient conditions for the existence of $\pm R, C$ networks," *IRE Trans. on Circuit Theory*, **CT-7**, 330–335 (September 1960).

Kuh, E. S., "Regenerative modes in active networks," *IRE Trans. on Circuit Theory*, **CT-7**, 62–63 (March 1960).

Kuh, E. S. and R. A. Rohrer, *Theory of Linear Active Network*, Holden-Day, San Francisco, Calif., 1967.

Mason, S. J. "Docile behavior of feedback amplifiers," *Proc. IRE*, **44**, 781–787 (June 1956).

McMillan, B., "Introduction to formal realizability theory—I and II," *Bell System Tech. J.*, **31** (1) 217–279, **31**(2), 541–600 (1952).

Newcomb, R. W., "On causality, passivity and single-valueness," *IRE Trans. on Circuit Theory*, **CT-9**, 87–89 (March 1962).

Newcomb, R. W., "On network realizability conditions," *Proc. IRE*, **50**, 1995 (September 1962).

Oono, Y., "Formal realizability of linear networks," Proc. Symp. Active Networks and Feedback Systems, MRI Symp. Series, Vol. X, Polytech. Inst. of Brooklyn, pp. 475–486 (1960).

Page, D. F., and A. R. Boothroyd, "Instability in two-port active networks," *IRE Trans. on Circuit Theory*, **CT-5**, 133–139 (June 1958).

Resh, J. A., "A note concerning the *n*-port passivity condition," *IEEE Trans. on Circuit Theory*, **CT-13**, 238–239 (June 1966).

Roberts, S., "Conjugate-image impedances," *Proc. IRE*, **34**, 198–204 (April 1946).

Sandberg, I. W., "The realizability of transformerless multi-port networks containing positive resistors, and positive and negative reactive elements," *IRE Trans. on Circuit Theory*, **CT-9**, 377–384 (December 1962).

Sharpe, G. E., J. L. Smith, and J. R. Smith, "A power theorem on absolutely stable two-ports," *IRE Trans. on Circuit Theory*, **CT-6**, 159–163 (June 1959).

Shekel, J., "Reciprocity relations in active 3-terminal elements," *Proc. IRE*, **42**, 1268–1271 (August 1954).

Sipress, J. M., "Necessary and sufficient conditions for $+R$, $+L$, $+C$, $-C$ networks," *IRE Trans. on Circuit Theory*, **CT-9**, 95–97 (March 1962).

Stern, A. P., "Considerations on the stability of active elements and applications to transistors," *IRE Nat'l Conv. Rec.*, **4** (Part 2), 46–52 (1956).

Su, K. L., *Active network synthesis*, McGraw-Hill, New York, Ch. 3, 1965.

Thornton, R. D., "Active *RC* networks," *IRE Trans. on Circuit Theory*, **CT-4**, 70–78 (September 1957).

Youla, D., L. Castriota, and H. Carlin, "Bounded scattering matrices and the foundations of linear passive network theory," *IRE Trans. on Circuit Theory*, **CT-6**, 102–124 (March 1959).

Youla, D., "A note on the stability of linear, non-reciprocal *n*-ports," *Proc. IRE*, **48**, 121–122 (January 1960).

Youla, D., "Physical realizability criteria," *IRE Trans. on Circuit Theory*, **CT-7** (Special Supplement), 50–68 (August 1960).

Problems

6.1 Show that if the driving-point function of a one-port has poles in the right-half *s*-plane and/or multiple $j\omega$-axis poles, then the one-port is not passive.

6.2 Prove the following theorem:[39]

Theorem 6-20 *Let an n-port network be composed of M two-terminal elements (Figure 6-22), where $\{v_k, i_k\}$ represent the port variables and $\{e_l, j_l\}$ represent the voltage and current defining the two-terminal elements. Then*

$$\sum_{k=1}^{n} v_k \cdot i_k = \sum_{l=1}^{M} e_l \cdot j_l$$

[39] B. D. H. Tellegen, "A general network theorem, with applications," *Philips Research Reports*, **7**, 259–269 (1952).

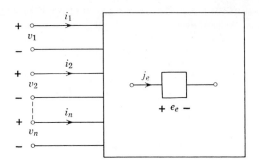

Figure 6-22

6.3 Using Tellegen's theorem (Theorem 6-20), show that an *RLC* network is passive.

6.4 Show that for the one-port of Figure 6-23, if the initial current

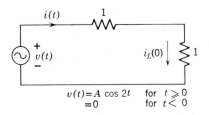

Figure 6-23

through the inductor at $t = 0$ is $A/5$ amp., the response current $i(t)$, due to a voltage excitation $v(t)$ of the form shown, will contain only steady-state terms, for all $t \geqslant 0$.

6.5 Test the following driving-point functions for passivity.

(a) $\dfrac{4s^2 + 3s + 3}{s^2 + s + 4}$ \qquad (b) $\dfrac{(s + 1)(s + 3)}{(s + 2)(s + 4)}$

(c) $\dfrac{s^2 + 3s + 3}{(s + 1)(s + 2)}$ \qquad (d) $\dfrac{3}{s + 5} + 4 + \dfrac{6s}{s^2 + 1} + \dfrac{3s}{s + 2}$

(e) $\dfrac{\sqrt{s} + 3}{s + 1}$ \qquad (f) $\left(\dfrac{s + 1}{s + 2}\right)^2$

6.6 Prove that if $Z(s)$ and $W(s)$ are positive real functions, then the following functions are also positive real:

(a) $Z\left(\dfrac{1}{s}\right)$, \quad (b) $\dfrac{1}{Z(s)}$, \quad (c) $Z[W(s)]$, \quad (d) $Z(s) + W(s)$

6.7 A minimum reactive[40] positive real function $Z(s)$ can be written as a sum of two functions $Z_1(s)$ and $Z_2(s)$, where

$$Z_1(s) = k\,\frac{kZ(s) - sZ(k)}{k^2 - s^2}$$

$$Z_2(s) = s\,\frac{kZ(k) - sZ(s)}{k^2 - s^2}$$

Show that $Z_1(s)$ and $Z_2(s)$ are p.r. functions.[41]

6.8 Show that the numerator polynomial of the even part of a p.r. function has a quadrantal symmetry.

6.9 If $Z(s)$ is p.r., can you make $[Z(s) - K]$ p.r. by choosing proper real value of K?

6.10 Show that the ratio of two RC driving-point impedances is always a p.r. function.

6.11 Test the passivity of a two-port described by the following z matrix:

$$\frac{1}{4(s+1)(s^3 + s^2 + 3s + 1)}$$

$$\times \begin{bmatrix} (3s^4 + 2s^3 + 4s^2 + 2s + 1) & -(s^2 - 1)(s^2 + 1)(s + 2) \\ s(s^2 + 1)(s^2 - 1) & (3s^4 + 2s^3 + 4s^2 + 2s + 1) \end{bmatrix}$$

6.12 A two-port is characterized by the following y matrix:

$$\begin{bmatrix} \dfrac{9s^2 + 1}{s} & -\dfrac{6s^2 + 5s + 1}{s} \\ -\dfrac{6s^2 - 5s + 1}{s} & \dfrac{4s^2 + 1}{s} \end{bmatrix}$$

Is it passive? Is it lossless?

6.13 Prove Theorem 6-7.

6.14 If $Z(s)$ represents an RC:$-R$ driving-point impedance show that

$$\frac{dZ(\sigma)}{d\sigma} < 0$$

6.15 Show that the poles and zeros of an RC:$-R$ driving-point impedance alternate with each other on the real axis.

6.16 Derive the necessary and sufficient conditions satisfied by an RC:$-R$ driving-point impedance function.

---

[40] If a positive real function does not have any $j\omega$-axis poles, then it is defined as a minimum reactive positive real function.

[41] D. Hazony, "An alternate approach to the Bott-Duffin cycle," *IRE Trans. on Circuit Theory*, **CT-9**, 363 (September 1961).

6.17 Determine which of the following are $RC:-R$ driving-point impedance or admittance functions:

(a) $\dfrac{(s-2)(s+1)}{(s-1)(s+3)}$

(b) $\dfrac{(s+1)(s+3)}{(s-1)(s+6)}$

(c) $\dfrac{2s^2+5s+1}{2s^2+s-1}$

(d) $\dfrac{s^2+6s+8}{s^2+4s+3}$

6.18 Is any of the driving-point functions in Problem 6.17 realizable by an $RC:t_d$ one-port? If so, what are the maximum value of dissipation factors of the tunnel diodes?

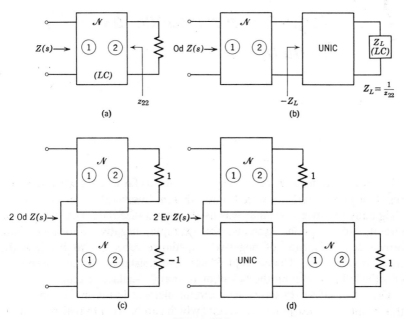

Figure 6-24

6.19 Let $Z(s)$ be a p.r. impedance realized by the network of Figure 6-24a. Show that the input impedances of the one-ports of Figures 6-24b, c, and d are, respectively, Od $Z(s)$, 2[Od $Z(s)$] and 2[Ev $Z(s)$].[42]

6.20 Prove Theorem 6-17.

6.21 Prove Theorem 6-18.

6.22 Derive the physical realizability conditions of the following LLF one-ports using the generalized duality concept.

(a) $LC:-C$, (b) $LC:-L$, (c) LC:Inductive-Gyrator

[42] V. Cimagalli, "Active synthesis of one-port and two-port network with zero attenuation or zero phase," *IEEE Trans. on Circuit Theory*, **CT-12**, 440–441 (September 1965).

7 / *Negative Resistance as a Circuit Element*

The concept of negative resistance dates back as far as 1911 when negative resistance characteristics were found inherent in several electronic devices. Originally negative resistance was used to neutralize the losses in a passive circuit. In telephone systems, for example, negative impedances are introduced by means of negative impedance repeaters to increase the transmission gain. The concept of negative resistance also has been used effectively in examining the behavior of many oscillator circuits.

Use of negative resistance as a circuit element is the main concern of this chapter. A few practical circuits which can be used to realize negative resistances are presented in the first section. Use of negative resistances in compensating the losses in inductors is discussed in the next section. In the remaining two sections, driving-point and transfer function realizations are discussed. Even though some synthesis techniques using ideal negative resistances are presented, emphasis is on the use of N-type nonideal negative resistances. Main results are summarized in the last section.

7-1 PRACTICAL CONSIDERATIONS

It was pointed out in Section 2-2 that a physical circuit exhibiting negative resistance must have a nonlinear v-i characteristic. Depending on the shape of their v-i characteristics, circuits are classified into two groups: N-type or voltage-controlled, and S-type or current-controlled (see Figure 2-13). Thus, an N-type negative resistance device is *short-circuit stable* and an S-type is *open-circuit stable*.

A short-circuit stable negative impedance Z_n is also called a *shunt-type* negative impedance. The reason is as follows. If an impedance $Z_p (Z_p < Z_n)$ is found so that when connected across the negative impedance it results in a stable system, then any impedance can be added across Z_p guaranteeing stability. Analogously, we observe that an impedance Z_p ($Z_p > Z_n$) in series with an open-circuit stable impedance

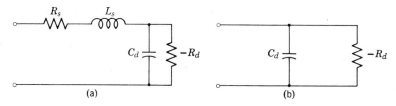

Figure 7-1 Small-signal equivalent circuit of a biased tunnel diode.

Z_n ensures stable operation. Thus, any additional impedance in series does not impair stability. This has led to the name *series-type* negative impedance for the S-type devices.

Various electronic devices exhibit negative resistance characteristics under some operating conditions. One of the recent additions is the *tunnel diode*.[1] The tunnel diode or the Esaki diode (as it is called after its inventor) is an extremely simple device having a single p-n junction between two heavily doped semiconductor regions. A complete small-signal linear approximation of a tunnel diode is shown in Figure 7-1a. The intrinsic diode is an N-type negative resistance device represented by the parallel combination of a negative resistor $(-R_d)$ and a positive capacitor (C_d). The series resistor (R_s) and the inductor (L_s) represent semiconductor body losses and connecting leads, and for many purposes can be neglected. In this book, we shall refer to any N-type negative resistance as the tunnel diode and represent it by the equivalent circuit of Figure 7-1b.

A complete discussion of all the available negative resistance devices is beyond the scope of this book. A representative survey can be found elsewhere.[2] A few typical electronic circuits, which have been found to

---

[1] L. Esaki, "New phenomenon in narrow Ge p-n junction," *Phys. Rev.*, **109**, 603–604 (January 1958).
[2] K. L. Su, *Active network synthesis*, McGraw-Hill, New York, pp. 9–26 (1965).

be useful in producing negative resistances, will be described below. Even though any of these circuits can be used to produce negative resistances of arbitrary value, usually negative resistances of moderate values (several kilohms) have better stability.

Positive Feedback Arrangement

Two simple circuits that employ noninverting type VVT's were introduced by Crisson.[3] Consider the arrangement of Figure 7-2a first; nodal

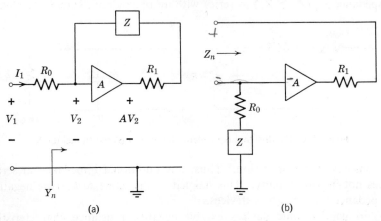

Figure 7-2 Positive feedback arrangement of producing negative resistances. (a) N-type (b) S-type.

analysis yields for this circuit:

$$(7.1) \quad I_1 = (V_1 - V_2)\frac{1}{R_0} = \frac{(1-A)V_2}{Z+R_1}$$

from which we obtain the expression for the input admittance as

$$(7.2) \quad Y_n = \frac{I_1}{V_1} = \frac{-(A-1)}{Z+R_1-(A-1)R_0} = -\frac{A-1}{Z}$$

if $R_1 = (A-1)R_0$. To make Y_n a negative admittance A the gain of the amplifier must be greater than unity. The input impedance of the amplifier has been assumed to be infinite. Finite output impedance of the amplifier can be considered as R_1. If R_1 is zero, then R_0 is not needed.

[3] E. L. Gintzon, "Stabilized negative impedances," *Electronics*, **18**, 140–150 (July 1945).

A similar analysis for the second circuit (Figure 7-2b) yields the following expression for the input impedance:

(7.3) $\quad Z_n = -Z(A - 1) + R_1 + (1 - A)R_0 = -Z(A - 1)$

if $R_1 = (A - 1)R_0$. Again, we observe that the gain of the voltage amplifier must be greater than one to produce negative impedance. As before, finite output impedance of the amplifier can be compensated.

A careful examination of Equations 7.2 and 7.3 reveals the type of the negative resistance produced by each circuit. The gain of a practical amplifier drops off at higher frequencies. For the circuit of Figure 7-2a, the input impedance approaches infinite value when A approaches unity. This indicates that the first circuit is a voltage-controlled or short-circuit stable (N-type) device. On the other hand, as $A \to 1$, the input admittance of the one-port of Figure 7-2b approaches infinity, indicating that the second circuit is a current-controlled or open-circuit stable device.

Gintzon[4] discusses in detail various aspects of negative impedances obtained by positive feedback arrangement. Following his approach, we can compute the sensitivity of the two circuits. In general, the amplifier gain, A, will be stabilized by means of negative feedback. If μ is the gain of the amplifier without feedback and β is the feedback factor, then

(7.4) $$A = \frac{\mu}{1 - \mu\beta}$$

For the N-type circuit of Figure 7-2a, the input impedance can be reexpressed as

(7.5) $$Z_n = \frac{Z(1 - \mu\beta)}{1 - \mu - \mu\beta}$$

using Equation 7.4. From above it follows that

(7.6) $$S_\mu^{Z_n} = \frac{\partial Z_n/Z_n}{\partial \mu/\mu} = \frac{\mu}{(1 - \mu\beta)(1 - \mu - \mu\beta)}$$

Note that the sensitivity factor $S_\mu^{Z_n}$ is independent of Z. The optimum value of β can be found from above by setting $dS_\mu^{Z_n}/d\beta$ equal to zero, which results in

(7.7)
$$\beta_{\text{opt}} = \frac{1}{2}\left(\frac{2}{\mu} - 1\right)$$

$$(S_\mu^{Z_n})_{\text{opt}} = -\frac{4}{\mu}$$

[4] E. L. Gintzon, *loc. cit.*

Equation 7.7 indicates that for large values of μ, $\beta = -\frac{1}{2}$ yields the best sensitivity factor.

Similar analysis can be carried out for the S-type negative resistance circuit of Figure 7-2b. We write Equation 7.3 as

$$(7.8) \qquad Y_n = \frac{Y}{1-A} = \frac{Y(1-\mu\beta)}{1-\mu-\mu\beta}$$

where $Y = 1/Z$. Equation 7.8 being identical in form to Equation 7.5,

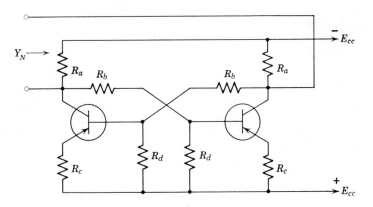

Figure 7-3 A bipolar N-type negative-resistance circuit.

it immediately follows that $S_\mu^{Y_n}$ is also given by Equation 7.6 and the optimum value of β is again the same as before.

Realizations of noninverting type VVT is considered in the next chapter.

A Bipolar N-type Negative Resistance Circuit

A bipolar N-type negative resistance circuit was proposed by Indiresan.[5] The proposed configuration, shown in Figure 7-3, is essentially a modification of Waldman-Bierri's circuit.[6]

To analyze its operation, let us first assume the transistors to be ideal $(r_e \to 0, r_b \to 0, r_c \to \infty, \alpha \to 1)$ and

$$R_a = R_d \to \infty, \qquad R_b = 0$$

---

[5] P. V. Indiresan, "A negative resistance for d.c. computers," *J. Brit. IRE*, **19**, 401–410 (July 1959).

[6] L. Waldmann and R. Bieri, "A negative resistance for direct and alternating current," *Z. Naturforsch*, **10A**, 814–820 (1955).

With these assumptions, the circuit reduces to that of Figure 7-4a, where each transistor has been replaced by its nullator-norator equivalent circuit. It is clear from the figure that

$$V = -2R_c I$$

Thus the input admittance of the idealized circuit is

(7.9) $$Y_n = -\frac{1}{2R_c}$$

If the transistor α's are not exactly equal to one, then we can use the

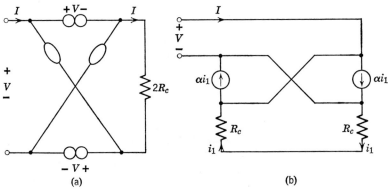

Figure 7-4 Equivalent representations of the N-type negative-resistance circuit of Figure 7-3.

circuit of Figure 7-4b to compute the input admittance. From the figure the following relations are easily established:

$$I = \alpha i_1 - i_1 + \alpha i_1 = (2\alpha - 1)i_1$$
$$V = -2R_c i_1$$

which implies

(7.10) $$Y_n = -\frac{2\alpha - 1}{2R_c}$$

Since α is frequency dependent, Y_n approaches zero value as α approaches $\frac{1}{2}$. This indicates that the circuit is short-circuit stable or N-type negative resistance.

A more accurate analysis of the circuit of Figure 7-3 yields

(7.11) $$Y_n \cong \frac{1}{2}\left[\frac{1}{R_a} - \frac{\dfrac{\alpha}{R_c + r_e} - \dfrac{1}{R_d}}{\dfrac{R_b}{R_d} + 1}\right]$$

assuming identical transistors and if $(1 - \alpha)R_b \ll (R_c + r_e)$ and $r_c \to \infty$.

The suggested circuit, which is suitable for d-c operation, has a high degree of stability. A nomogram for evaluating the component values for a specified value of the negative resistance will be found in the original paper.

An S-type Circuit with No Internal Bias Supplies

Recently, Nagata[7] reported an S-type negative impedance circuit that does not require any internal bias supplies. The circuit (shown in Figure 7-5) uses two transistors and is d-c coupled. For an approximate analysis,

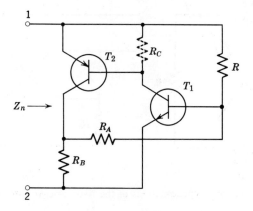

Figure 7-5 An S-type negative-resistance circuit requiring no internal bias supplies.

let us assume the transistors to be ideal ($r_e = r_b = 0$, $r_c \to \infty$, $\alpha = 1$). Replacing the transistors by their nullator-norator equivalent circuit, we arrive at the representation of Figure 7-6a, which simplifies to that of Figure 7-6b. The current through R, I_1, must go through R_A since the current through the nullator must be zero. Again, the nullator being a short-circuit for voltage, the following equations are obtained by inspection:

$$V_B = -R_A I_1$$
$$V = I_1 R$$
$$I_2 = \frac{V_B}{R_B} - I_1 = -\left(\frac{R_A}{R_B} + 1\right)I_1$$

---

[7] M. Nagata, "A simple negative impedance circuit with no internal bias supplies and good linearity," *IEEE Trans. on Circuit Theory*, **CT-12**, 433–434 (September 1965).

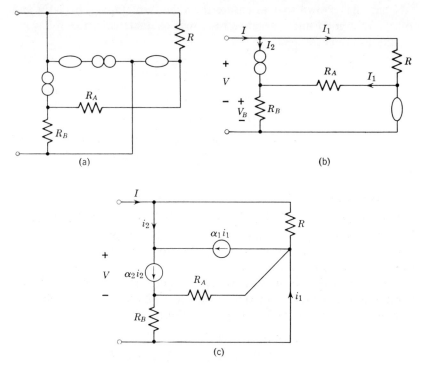

Figure 7-6 Equivalent representations of the S-type negative-resistance circuit of Figure 7-5.

Therefore, the input current is

$$I = I_1 + I_2 = -\frac{R_A}{R_B} I_1 = -\frac{R_A}{R_B} \cdot \frac{V}{R}$$

or the input resistance is

(7.12) $$Z_n = \frac{V}{I} = -\frac{R_B}{R_A} R$$

If the transistor α's are not exactly equal to unity, we can use the equivalent circuit of Figure 7.6c for analysis. In this case we have

(7.13) $$Z_n = \frac{V}{I} = \frac{(1-\alpha_1)(1-\alpha_2)(R_A+R_B) - \alpha_1\alpha_2 R_B}{(1-\alpha_1)(1-\alpha_2)(R_A+R_B) + \alpha_1\alpha_2 R_A} \cdot R$$

It is seen from (7.13) that Z_n goes to zero as α decreases with frequency, which implies that the circuit of Figure 7-5 is an open-circuit stable (S-type) negative resistance.

Nagata has shown that to ensure the validity of Equation (7-12) and to achieve a good linear negative resistance characteristic, the following conditions are necessary:

(i) $\beta_1\beta_2 \gg \left(\dfrac{R_A}{R_B} + 1\right)$

(ii) $R \gg (R_A + R_B)$

(iii) $R_A < 24R_B$

where β_1 and β_2 are the common emitter current gains of the transistors. Nagata also suggests adding another resistance R_C as shown to improve linearity.

An *N*-type Circuit with No Internal Bias Supplies

A negative resistance circuit having an *N*-type (short-circuit stable) *i-v* characteristic and requiring no internal bias supplies is shown in Figure 7-7a. The circuit proposed by Ramanan and Varshney[8] employs two *n-p-n* transistors and five resistors.

For an approximate analysis, we replace each transistor by its nullator-norator equivalent representation as shown in Figure 7-7b. The following equations are easily established by inspection:

$$I = I_1 + I_2 + I_3$$
$$V = (R_A + R_B)I_1 = R_C I_2 + R_E I_3$$
$$I_1 R_B = I_2 R_D$$

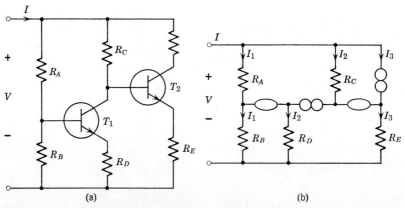

Figure 7-7 An *N*-type negative-resistance circuit requiring no internal bias supplies and its nullator-norator equivalent representation.

[8] K. V. Ramanan and R. C. Varshney, "New short-circuit-stable negative-resistance circuit with no internal bias supplies," *Electronics Letters*, **3**, 186–188 (May 1967).

which can be solved to yield

(7.14) $$Y_n = \frac{I}{V} = \frac{R_D(R_E + R_A + R_B) + R_B(R_E - R_C)}{(R_A + R_B)R_D R_E}$$

If R_E is chosen such that

$$R_E \ll (R_A + R_B)$$
$$R_E \ll R_C$$

then Equation 7.14 reduces to

(7.15) $$Y_n \cong -\frac{R_B R_C - R_D(R_A + R_B)}{(R_A + R_B)R_D R_E}$$

In order that Y_n be negative, it is seen that the following condition must be satisfied:

$$R_B R_C > R_D(R_A + R_B)$$

Performing a more accurate analysis, it can be shown that the circuit of Figure 7-7a is a short-circuit stable negative resistance.

It should be noted that use of only *n-p-n* transistors permits the realization of this circuit in integrated form and also enables a higher frequency range of operation. Effect of variations of R_E, R_B, and R_D will be found in the original paper.

An alternate realization of the *N*-type negative resistance requiring no internal bias supplies has been reported by Martinelli.[9] However, Martinelli's circuit uses a Zener diode which restricts the lower limit of the peak voltage of the *i-v* characteristic to be greater than the breakdown voltage of the Zener diode. In addition, large values of peak-to-valley current ratios are not easily obtainable.

Negative Resistance Using Negative-Impedance Converter

If one of the ports of a negative-impedance converter (NIC) is terminated by a positive resistance, the input resistance seen at the other port is negative. Thus, an NIC can also be used to realize negative resistance. It will be shown in Chapter 9 that one of the ports of a negative-impedance converter is open-circuit stable and the other one is short-circuit stable. As a result, both types of negative resistance can be generated by a single NIC. Realizations of the NIC is considered in Section 9-1.

---

[9] G. Martinelli, "Realization of a short-circuit-stable negative-resistance circuit with no internal bias supplies," *Electronics Letters*, **2**, 308 (1966).

Negative Resistance Using Negative-Impedance Inverter

In a similar manner, a negative-impedance inverter is also suitable for realizing a negative resistance. However, for a negative-impedance inverter, either both ports are open-circuit stable or both are short-circuit stable. Hence, only one type of negative resistance can be obtained from a given negative-impedance inverter.

Recently, Su[10] advanced a four-transistor circuit for realizing a four-terminal negative-impedance inverter. Many other circuits for realizing the three-terminal (input and output have a common terminal) and the four-terminal negative-impedance inverters will be found elsewhere.[11]

Negative Resistance Using Operational Amplifier

Both types of negative resistances can be readily realized using the operational amplifier which is now commercially available in monolithic integrated circuit form. One method of realization is based on the positive feedback approach of Section 7-1, p. 256 (see Problem 7.3). Another approach is by constructing a negative-impedance converter by means of an operational amplifier and then using the negative-impedance converter to realize the negative resistance. Details on both approaches will be found in Section 11-3.

7-2 COMPENSATION OF LOSSES IN REACTIVE ELEMENTS

One simple but effective use of negative resistances is in the cancellation of dissipations in reactive elements, in particular the inductors.[12] There are two possible ways the compensation can be achieved.

We first observe that for most practical purposes a lossy inductor can be represented by a series combination of an inductance and a resistance (Figure 7-8a). The series resistance is usually of the order of a few ohms. Thus, one approach to cancel the inductor dissipation will be to connect a negative resistance in series with the lossy inductor (Figure 7-8b). For stability reasons, the negative resistance must be of the series type, i.e., open-circuit stable. However, the value of the negative resistance is required to be a few ohms, and this poses a difficult practical problem.

---

[10] K. L. Su, "A method for realizing the negative-impedance inverter," *IEEE J. Solid-State Circuits*, **SC-2**, 22–25 (March 1967).

[11] S. K. Mitra, "Non-reciprocal negative-impedance inverter," *Electronics Letters*, **3**, 388 (August 1967).

S. K. Mitra, "Alternate realizations of the 3-terminal and 4-terminal negative-impedance inverters," *Proc. IEEE*, **56**, 368 (March 1968).

[12] J. T. Bangert, "The transistor as a network element," *Bell System Tech. J.*, **33**, 329–352 (March 1954).

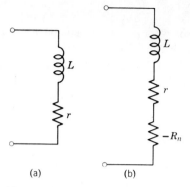

Figure 7-8 (a) Representation of a lossy inductor. (b) Series compensation arrangement.

An alternate approach would be to connect a negative resistance across the lossy inductor (Figure 7-9). For a stable operation, the negative resistance in this case must be of the shunt type, i.e., short-circuit stable. The effective input impedance of the combination can be shown to be equal to

(7.16) $$Z(j\omega) \cong r - \frac{\omega^2 L^2}{R_n} + j\omega L$$

if $\omega L \ll R_n$. Hence for an exact compensation at a given frequency ω, the negative resistance R_n must have a value

(7.17) $$R_n = \frac{\omega^2 L^2}{r} = \omega L Q$$

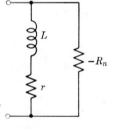

Figure 7-9 Shunt-compensation arrangement of a lossy inductor.

where Q represents the Q-factor of the inductor. Thus an inductor of value 10 mh and a Q of 300 will require a negative resistance of value approximately equal to 1.9 mΩ at 100 kHz. As a result, the second approach requires negative resistances of very large values.

7-3 SYNTHESIS OF DRIVING-POINT FUNCTIONS

In this section we are concerned with the realization of a specified driving-point function using negative resistances. More precisely, we shall prove the sufficiency of some of the theorems of Section 6-5.

RC:-R Network

The sufficiency of Theorem 6–10 can be proved by making a partial fraction expansion of the specified driving-point impedance and then realizing each term of the expansion. A possible pole at infinity is realized by terminating a reciprocal negative-impedance inverter by a capacitor of proper value.[13] Let us illustrate the technique by means of an example.

Example 7-1 Realize

$$Z(s) = \frac{-s^3 - 2s^2 + 3s + 3}{s^2 + s - 2} \tag{7.18}$$

A partial fraction expansion of the above function yields

$$Z(s) = -s - 1 + \frac{1}{s+2} + \frac{1}{s-1} \tag{7.19}$$

From the above, it is clear that the specified $Z(s)$ satisfies the conditions of Theorem 6-10. Realization of Equation 7.19 as given by Figure 7-10a should be self-evident. The total number of negative resistances in

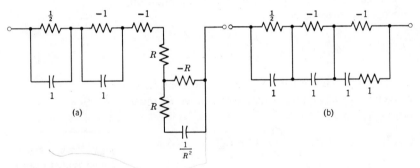

Figure 7-10 Realization of the RC:-R driving-point impedance of Example 7-1.

Figure 7-10a can be minimized by choosing $R = 1$. The final realization is shown in Figure 7-10b. Alternately $Z(s)$ can be realized by a continued fraction expansion (Problem 7.7).

RC: Tunnel-Diode Network

Sufficiency of Theorem 6-12 follows in a similar manner and will be illustrated by means of an example.

---

[13] P. Bello, "Extension of Brune's energy function approach to the study of *LLF* networks," *IRE Transactions on Circuit Theory*, **CT-7**, 270–280 (September 1960).

Example 7-2 Let us realize

(7.20) $$Z(s) = \frac{4s^2 - 4s - 2}{s^3 - 2s^2 - s + 2}$$

using two tunnel diodes of dissipation factors 2 and 3, respectively. First we make a partial fraction expansion of the specified $Z(s)$:

$$Z(s) = \frac{1}{s+1} + \frac{1}{s-1} + \frac{2}{s-2}$$

From the above we observe that the necessary conditions of Theorem 6-12

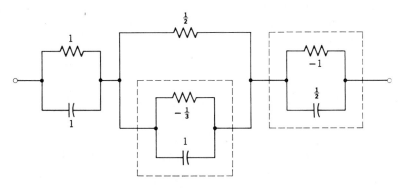

Figure 7-11 Realization of the $RC{:}{-}t_d$ driving-point impedance of Example 7-2.

are satisfied. Final realization is shown in Figure 7-11, where the desired tunnel diodes are shown by means of dotted boxes.

LC:–R Network

It is fairly obvious from Theorem 6-13 that if $Z(s)$ represents the driving-point impedance of an $LC{:}{-}R$ network, then $-Z(-s)$ is p.r. From a consideration of Figure 6-14, the realization of the driving-point function of a $LC{:}{-}R^{(1)}$ network should be evident. We illustrate the idea by means of the following example:

Example 7-3 Synthesize

(7.21) $$Z(s) = \frac{s^2 - s + 1}{s^3 - s^2 + 2s - 1}$$

From Equation 7.21 we obtain

$$-Z(-s) = \frac{s^2 + s + 1}{s^3 + s^2 + 2s + 1}$$

The denominator of $-Z(-s)$ is Hurwitz. The numerator of the even part of $-Z(-s)$ is unity, implying that $-Z(-s)$ is p.r. Realization of $-Z(-s)$ can be achieved by Miyata technique as shown in Figure 7-12a. By replacing the positive resistor by a negative resistor of equal value, we obtain the realization of the above $Z(s)$ as indicated in Figure 7-12b.

Note that since a p.r. function can always be realized by a lossless two-port terminated by a positive resistor following Darlington's technique, sufficiency of Theorem 6-13 is guaranteed. Of course, ideal transformers may be needed in the realization of the lossless two-port.

LC: Tunnel-Diode Network

To prove the sufficiency of Theorem 6-14, it is enough to show that in the Darlington realization of $-Y(-s)$ a capacitor of value at least equal to the desired value of time constant of the tunnel diode is produced

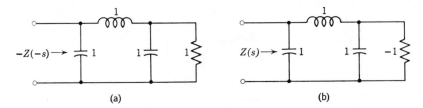

Figure 7-12 Realization of the LC:$-R$ driving-point impedance of Example 7-3.

across the resistance at the output port of the lossless two-port. In Section 10-2, the sufficiency of Theorem 6-14 will be proved. It can be mentioned here that the necessary and sufficient conditions for the realizability of an n port LC:$t_d^{(1)}$ network have been independently derived by Kinariwala[14] and Sandberg.[15]

RLC:–*R* Network

Several schemes are available for the realization of an arbitrary real rational function as the driving-point impedance of an RLC:$-R$ network. A very simple synthesis method advanced by Boesch and Wohlers[16] will be presented next. The final realization consists of lossy inductors and capacitors and, at most, two negative resistors.

---

[14] B. K. Kinariwala, "The Esaki diode as a network element," *IRE Trans. on Circuit Theory*, **CT-8**, 389–397 (December 1961).

[15] I. W. Sandberg, "The realizability of multi-port structures obtained by imbedding a tunnel diode in a lossless reciprocal network," *Bell System Tech. J.*, **41**, 857–876 (May 1962).

[16] F. T. Boesch and M. R. Wohlers, "On network synthesis with negative resistances," *Proc. IRE*, **48**, 1656–1657 (September 1960).

Let $Y(s) = N(s)/D(s)$ be the specified driving-point admittance, where $N(s)$ and $D(s)$ are polynomials with real coefficients. Choose a polynomial $Q(s)$ with distinct negative real roots such that

$$Q(s)^0 = \max\,[N(s)^0,\ D(s)^0]$$

By Theorem 3-5, we can write $N(s)/Q(s)$ as

(7.22) $$\frac{N(s)}{Q(s)} = \frac{N_p(s)}{Q_p(s)} - \frac{N_n(s)}{Q_n(s)} = \frac{N_p(s)Q_n(s) - Q_p(s)N_n(s)}{Q_p(s)Q_n(s)}$$

where $N_p(s)/Q_p(s)$ and $N_n(s)/Q_n(s)$ are RC driving-point impedances.

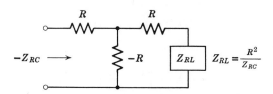

Figure 7-13 RLC:-$R^{(1)}$ realization of a negative RC driving-point impedance.

From Equation 7.22 we have

(7.23) $$N(s) = N_p(s)Q_n(s) - Q_p(s)N_n(s) = N_1(s) - N_2(s)$$

Note that $Q(s)$ can always be selected so that $N_1(s) = N_p(s)Q_n(s)$, and $N_2(s) = Q_p(s)N_n(s)$ have all distinct finite negative real roots. This implies that we can write

(7.24) $$Y(s) = \frac{N(s)}{D(s)} = \frac{N_1(s)}{D(s)} + \frac{-N_2(s)}{D(s)} = \frac{1}{Z_1(s)} + \frac{1}{Z_2(s)}$$

where $Z_1(s) = D(s)/N_1(s)$, and $Z_2(s) = D(s)/-N_2(s)$. Observe that $N_1(s)^0$ and $N_2(s)^0$ are equal to or one greater than $D(s)^0$. Consider the realization of $Z_1(s)$ first. Again, by Theorem 3-5 we can express

(7.25) $$Z_1(s) = Z_{RC}^{(1)} - Z_{RC}^{(3)}$$

A simple method of realization of a negative RC impedance $-Z_{RC}$ is to terminate an NIV by a positive RL impedance $Z_{RL} = R^2/Z_{RC}$, as shown in Figure 7-13. An alternate scheme is as follows. In general, $-Z_{RC}^{(3)}$ will have the form

(7.26) $$-Z_{RC}^{(3)} = -K_\infty - \sum_i \frac{K_i}{s + \sigma_i}$$

where K_∞ and K_i are real and positive numbers. We now choose a real positive number R_A such that

$$R_A \geq \sum_i K_i/\sigma_i$$

and rewrite Equation 7.26 as

(7.27) $\quad -Z_{RC}^{(3)} = (-K_\infty - R_A) + \left(R_A - \sum_i \frac{K_i}{s + \sigma_i}\right) = -R_1 + Z_{RL}^{(1)}$

where

$$Z_{RL}^{(1)} = R_A - \sum_i \frac{K_i}{s + \sigma_i}$$

is guaranteed to be an RL impedance. Using (7.27) in (7.25) we obtain

(7.28) $\quad\quad\quad Z_1(s) = Z_{RC}^{(1)} + Z_{RL}^{(1)} - R_1$

In a similar manner, we can express

(7.29) $\quad\quad\quad Z_2(s) = Z_{RC}^{(2)} + Z_{RL}^{(2)} - R_2$

The final realization of $Y(s)$ is obtained by paralleling Z_1 and Z_2 (Figure 7-14a). In some cases, the realization of $Z_{RC}^{(i)}$ may have a large positive resistance which will swamp out $-R_i$, thus decreasing the number of negative resistances. $Z_{RC}^{(i)}$ and $Z_{RL}^{(i)}$ can be realized in Foster's form to make the reactive elements lossy. A similar technique can be followed on an impedance basis; the following example will illustrate the method.

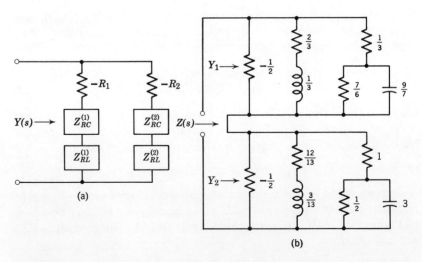

Figure 7-14 (a) An $RLC:-R^{(2)}$ configuration for driving-point admittance realization. (b) Realization of the driving-point impedance of Example 7-4.

SYNTHESIS OF TRANSFER FUNCTIONS / 271

Example 7-4 Realize
$$Z(s) = \frac{N(s)}{D(s)} = \frac{2}{s^2 + s + 1}$$

Choosing $Q(s) = (s+1)(s+3)$ we obtain

$$\frac{N(s)}{(s+1)(s+3)} = \frac{1}{s+1} - \frac{1}{s+3} = \left(\frac{1}{s+1} + 1\right) - \left(\frac{1}{s+3} + 1\right)$$
$$= \frac{(s+2)(s+3) - (s+1)(s+4)}{(s+1)(s+3)}$$

The above implies that we can write

$$Z(s) = \frac{(s+2)(s+3)}{s^2+s+1} + \frac{-(s+1)(s+4)}{s^2+s+1} = \frac{1}{Y_1} + \frac{1}{Y_2}$$

where

$$Y_1 = \frac{s^2+s+1}{(s+2)(s+3)} = 1 + \frac{3}{s+2} - \frac{7}{s+3}$$
$$= \frac{3}{s+2} + \frac{3s+2}{s+3} - 2$$

and

$$Y_2 = -\frac{s^2+s+1}{(s+1)(s+4)} = \frac{\frac{13}{3}}{s+4} - \frac{\frac{1}{3}}{s+1} - 1$$
$$= \frac{\frac{13}{3}}{s+4} + \frac{s+\frac{2}{3}}{s+1} - 2$$

The final realization of $Z(s)$ is obtained by making a series connection of Y_1 and Y_2 as shown in Figure 7-14b.

It should be observed from this example that because of the nature of the partial fraction expansion, constants can be added and subtracted from the expansion to facilitate the synthesis of the impedance function.

In Chapter 10, an approach to the realization of an impedance function by means of an $RLC:-R^{(1)}$ network will be outlined.

7-4 SYNTHESIS OF TRANSFER FUNCTIONS

Several transfer function realization schemes will be outlined in this section. Some of the techniques are simple modifications of known passive *RC* synthesis methods; the remaining techniques are novel methods having no analogous methods in passive network theory. In this section, our primary interest will be in the realization of stable transfer functions using *N*-type negative resistances.

$RC:-R$ Network

There are two cases of interest: real transmission zeros and complex transmission zeros.

Realization of Real Transmission Zeros. The well-known zero-shifting technique[17] can be easily extended to $RC:-R$ ladder networks in order to realize transmission zeros on the real axis. If the zeros are on the negative real axis, the technique is identical to the passive RC case. A zero on the positive real axis can be realized in two ways:

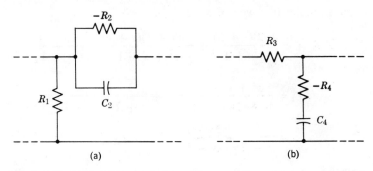

Figure 7-15 Zero-shifting technique using negative resistances as applied to the realization of positive real transmission zeros.

1. Zero shifting may be done by a shunt resistance (R_1) and the transmission zero realized by a parallel combination of a positive capacitance and a negative resistance in the series branch (Figure 7-15a) or

2. The zero shifting may be done by a positive resistor (R_3) in the series branch and the transmission zero realized by a series $-R, C$ circuit in the shunt arm (Figure 7-15b).

Consider the following example:

Example 7-5 Let us realize

$$z_{21}(s) = \frac{H(s-1)^2}{s(s+2)}$$

A compatible passive RC driving-point impedance z_{11} would be as follows:

$$z_{11}(s) = \frac{(s+1)(s+3)}{s(s+2)}$$

---

[17] See for example, M. E. Van Valkenburg, *Modern network synthesis*, John Wiley, New York, pp. 276–288 (1962).

SYNTHESIS OF TRANSFER FUNCTIONS / 273

Note that
$$z_{11}(1) = \tfrac{8}{3}$$

Hence, to produce a transmission zero at $s = 1$, we remove from $1/z_{11}(s)$ a resistance $R_1 = \tfrac{8}{3}$ in shunt (Figure 7-16). The remainder admittance function is

$$Y_a(s) = \frac{s(s+2)}{(s+1)(s+2)} - \frac{3}{8} = \frac{(s-1)(5s+9)}{8(s+1)(s+3)}$$

$Y_a(s)$ has a zero at $s = 1$ as expected. The pole of $1/Y_a$ at $s = 1$ is next

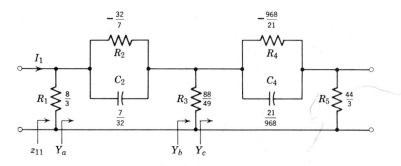

Figure 7-16 Realization of the open-circuit transfer impedance of Example 7-5 using zero shifting technique.

removed as R_2 and C_2 in parallel as a series branch (Figure 7-16):

$$\frac{1}{Y_a} = \frac{\tfrac{32}{7}}{s-1} + \frac{56s+120}{7(5s+9)}$$

which implies $R_2 = -\tfrac{32}{7}$ and $C_2 = \tfrac{7}{32}$. The admittance function present after extraction is

$$Y_b = \frac{35s + 63}{56s + 120}$$

From the above, a conductance $1/R_3 = \tfrac{49}{88}$ can be removed to produce the second transmission zero at $s = 1$:

$$Y_c = \frac{35s + 63}{56s + 120} - \frac{49}{88} = \frac{21(s-1)}{11(28s+60)}$$

To remove the pole of $1/Y_c$ we write

$$\frac{1}{Y_c} = \frac{968}{21(s-1)} + \frac{44}{3}$$

This indicates that R_4 and C_4 in the series branch (Figure 7-16) are given as: $R_4 = -\frac{968}{21}$ and $C_4 = \frac{21}{968}$. The final shunt resistance R_5 is $\frac{44}{3}$. The complete network is shown in Figure 7-16.

Realization of Complex Transmission Zeros. A voltage transfer ratio t_v having complex zeros and negative real poles can be realized by means of an $RC:-R^{(1)}$ two-port. The proposed configuration,[18] shown in Figure 7-17, is characterized by

$$(7.30) \qquad t_v = \left.\frac{V_2}{V_1}\right|_{I_2=0} = -\frac{y_{21} - 1}{y_{22} - 1}$$

where y_{21} and y_{22} are the short-circuit parameters of the RC two-port. A specified voltage transfer ratio $t_v = N(s)/D(s)$ can be put in the form of Equation 7.30 as follows.

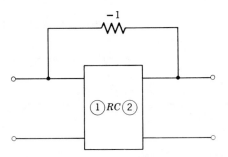

Figure 7-17 An $RC:-R^{(1)}$ two-port suitable for the realization of complex transmission zeros.

Select a polynomial $Q(s)$ so that $D(s)/Q(s)$ is an RC driving-point admittance function. Since $Q(s)$ has only distinct negative real roots, we can express $N(s)/Q(s)$ as

$$(7.31) \qquad \frac{N(s)}{Q(s)} = \frac{N_1(s)}{Q_1(s)} - \frac{N_2(s)}{Q_2(s)} = \frac{N_1(s) + KQ_1(s)}{Q_1(s)} - \frac{N_2(s) + KQ_2(s)}{Q_2(s)}$$

where $N_1(s)/Q_1(s)$ and $N_2(s)/Q_2(s)$ are RC driving-point admittances. This implies that $N(s)$ and $D(s)$ must satisfy the following degree requirement:

$$(7.32) \qquad N(s)^0 \leq D(s)^0$$

[18] I. M. Horowitz, "Active network synthesis," *IRE Conv. Record*, 4 (Part 2), 38–45 (March 1956).

Hence, a possible decomposition of $N(s)$ would be

(7.33) $\quad N(s) = \{N_1(s) + KQ_1(s)\}Q_2(s) - \{N_2(s) + KQ_2(s)\}Q_1(s)$

where K is an arbitrary real positive constant. Using Equation 7.33, we can then write

(7.34) $\quad \dfrac{N(s)}{D(s)} = \dfrac{\dfrac{\{N_1(s) + KQ_1(s)\}Q_2(s)}{\{N_2(s) + KQ_2(s)\}Q_1(s)} - 1}{\dfrac{D(s)}{\{N_2(s) + KQ_2(s)\}Q_1(s)}}$

Comparing Equations 7.30 and 7.34, we obtain

(7.35)
$$-y_{21} = \dfrac{\{N_1(s) + KQ_1(s)\}Q_2(s)}{\{N_2(s) + KQ_2(s)\}Q_1(s)}$$

$$y_{22} = \dfrac{D(s)}{\{N_2(s) + KQ_2(s)\}Q_1(s)} + 1$$

For suitable values of K, the zeros of $N_2(s) + KQ_2(s)$ will move closer to those of $Q_2(s)$ and the zeros of $\{N_2(s) + KQ_2(s)\}Q_1(s)$ can be made to alternate with the zeros of $D(s)$. This will make y_{22} RC realizable. Furthermore, the zeros of y_{21} as given by Equation 7.35 are always negative real. Hence, $-y_{21}$ and y_{22} can be developed into an unbalanced RC network, provided Fialkow-Gerst conditions[19] are satisfied. In some cases, the two-port can be realized as a ladder network.

Example 7-6 Consider the realization of

$$t_v = \dfrac{N(s)}{D(s)} = \dfrac{s-2}{(s+1)(s+3)}$$

Let $Q(s) = (s+2)(s+4)$. This guarantees $D(s)/Q(s)$ to be an RC admittance as required.

$$\dfrac{N(s)}{Q(s)} = \dfrac{s-2}{(s+2)(s+4)} = \dfrac{3}{s+4} - \dfrac{2}{s+2}$$

which means that $N_1(s) = 3$, $Q_1(s) = (s+4)$, $N_2(s) = 2$, and $Q_2(s) = (s+2)$. From Equation 7.35, we have

$$y_{22} = \dfrac{(s+1)(s+3)}{[2 + K(s+2)](s+4)} + 1$$

---

[19] See Appendix D.

which indicates that $K = 4$ is a suitable value. Hence,

(7.36)
$$y_{22} = \frac{5s^2 + 30s + 43}{(4s + 10)(s + 4)}$$

$$-y_{21} = \frac{4s^2 + 17s + 18}{(4s + 10)(s + 4)}$$

As given by Equation 7.36, y_{22} and $-y_{21}$ satisfy the Fialkow-Gerst conditions for unbalanced networks. Final realization is left as an exercise.

Note that a slight modification of the procedure will guarantee a capacitance across the negative resistance in Figure 7-17, allowing us to use N-type negative resistance.[20] But it should be evident that, for such a realization, the transfer voltage ratio must be a constant at infinity $[N(s)^0 = D(s)^0]$.

Realization of Short-Circuit Admittance Matrix. The necessary conditions for the realizability of an RC:$-R^{(1)}$ two-port, as given by Theorem 6-9, are easily extended to an RC:$-R^{(N)}$ two-port. These conditions are also sufficient, provided that ideal transformers are used in the realization. Of course, ideal transformers can be replaced by their $\pm R$ equivalent circuit. An alternate approach to this problem has been suggested;[21] the method begins with a partial fraction expansion of the short-circuit parameters which, in general, can be expressed as

$$y_{ij} = g_{ij} + k_{ij}^\infty s + \frac{k_{ij}^0}{s} + \sum_m \frac{k_{ij}^{(m)} s}{s + \sigma_m}$$

Each set of terms in the expansion $\{g_{ij}\}$, $\{k_{ij}^\infty s\}$, etc., is realized separately in the form of the network of Figure 7-18. A parallel connection of these simpler networks yields the complete realization of $\{y_{ij}\}$. For details, the reader is referred to the original paper.

RC: Tunnel-Diode Network

Several typical synthesis schemes of RC:t_d networks using a minimum number of tunnel diodes, i.e., N-type negative resistances, will be presented in this section.

---

[20] G. J. Herskowitz and M. S. Ghausi, "Transfer function synthesis employing one tunnel diode and a passive RC ladder network," *IRE Wescon Conv. Rec.*, 1962.

[21] C. L. Phillips and K. L. Su, "Synthesis of $\pm R$, C networks," *IEEE Trans. on Circuit Theory*, **CT-11**, 80–82 (March 1964).

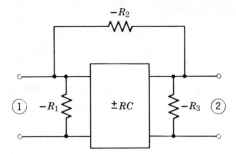

Figure 7-18 The basic RC:$-R$ configuration for the realization of a short-circuit admittance matrix.

The Reverse Predistortion Method.[22] A very simple technique to realize transfer functions using passive elements and negative resistors is the *reverse predistortion method*. The method is as follows. Let the specified transfer function be

(7.37) $$T(s) = K \frac{N(s)}{D(s)}$$

In uniform reverse predistortion, a modified transfer function $T_f(s)$ is first obtained by letting

(7.38) $$s \rightarrow s + d$$

giving

(7.39) $$T_f(s) = K \frac{N(s+d)}{D(s+d)}$$

$T_f(s)$ is then realized by usual passive synthesis techniques. Finally, the effect of reverse predistortion is removed by replacing each inductor L_v by a series combination of L_v and a negative resistance $-R_v = -L_v d$, and by replacing each capacitor C_μ by a parallel combination of C_μ and a negative resistor $-R_\mu = -1/C_\mu d$. The minimum value of d is determined by the condition that $T_f(s)$ be passively realizable. If $T(s)$ is already passively realizable, the value of d then can be determined by the required value of the gain constant K.

If $T_f(s)$ is RC realizable, it is evident that the final realization of $T(s)$ is in the form of an RC:tunnel-diode network. Let us illustrate the technique by means of a simple example.

[22] L. Weinberg, "Synthesis using tunnel diodes and masers," *IRE Trans. on Circuit Theory*, **CT-8**, 66–75 (March 1961).

Example 7-7 Let us realize

$$t_v = \frac{N(s)}{D(s)} = \frac{H}{s^2 + 5s + 5}$$

Note that the given t_v is RC realizable and that the maximum value of H obtained in that case is 5. To obtain an RC:tunnel-diode realization, we can pick d arbitrarily. Assuming $d = 2$, then

(7.40) $$t_{vf} = \frac{H_f}{(s+2)^2 + 5(s+2) + 5} = \frac{H_f}{s^2 + 9s + 19} = \frac{z_{21f}}{z_{11f}}$$

The poles of z_{21f} and z_{11f} can be chosen to make z_{11f} an RC realizable driving-point impedance. One solution will be

$$z_{11f} = \frac{s^2 + 9s + 19}{s(s+4)}$$

Since all the transmission zeros are at infinity, z_{11f} can be realized by means of a continued fraction expansion:

$$s^2 + 4s \overline{\smash{\big)}\, s^2 + 9s + 19} \quad 1$$
$$\underline{s^2 + 4s}$$
$$5s + 19 \overline{\smash{\big)}\, s^2 + 4s} \quad \tfrac{1}{5}s$$
$$\underline{s^2 + \tfrac{19}{5}s}$$
$$\tfrac{1}{5}s \overline{\smash{\big)}\, 5s + 19} \quad 25$$
$$\underline{5s}$$
$$19 \overline{\smash{\big)}\, \tfrac{1}{5}s} \quad \tfrac{1}{95}s$$
$$\underline{\tfrac{1}{5}s}$$
$$0$$

The resulting network is shown in Figure 7-19a; removing the effect of the reverse predistortion, we obtain the final realization as shown in Figure 7-19b. The value of the multiplying constant is now equal to 19 (Equation 7.40).

The reverse predistortion technique can be applied to realize any type of transfer function. The above technique, although simple, has a major drawback: the number of tunnel diodes is equal to the number of capacitances in the realization of $T_f(s)$. Minimization of the number of tunnel diodes is desirable. We shall now discuss two methods of transfer-function realization that employ a single tunnel diode.

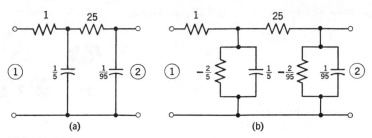

Figure 7-19 Illustration of the reverse predistortion method of Example 7-7.

Realization of Transfer Function with No Positive Real Transmission Zeros.[23] A transfer voltage ratio whose poles are restricted to negative real axes and zeros anywhere in the complex plane (except the positive real axis) can be realized in the form of Figure 7-20, provided that there is at least one transmission zero at infinity. The method of realization is as follows: let $t_v = N(s)/D(s)$ be the specified transfer function. t_v must be expressed as

$$(7.41) \qquad t_v = \frac{V_2}{V_1} = \frac{-y_{21}}{y_{22} - 1}$$

where y_{21} and y_{22} are the short-circuit admittances of the RC two-port. Assuming that $D(s)$ can be decomposed as

$$(7.42) \qquad D(s) = D_a(s) - D_b(s)$$

where $D_a(s)/D_b(s)$ is an RC driving-point admittance, it is clear that we can write

$$(7.43) \qquad t_v = H\frac{N(s)}{D(s)} = \frac{H\dfrac{N(s)}{D_b(s)}}{\dfrac{D_a(s)}{D_b(s)} - 1}$$

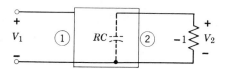

Figure 7-20 An RC:$-R^{(1)}$ configuration for realizing a complex transmission zero (excluding positive real zeros).

[23] D. L. Losee and S. K. Mitra, "Transfer function synthesis of RC tunnel-diode networks," *IEEE Trans. on Circuit Theory*, **CT-11**, 357–362 (September 1964).

280 / NEGATIVE RESISTANCE AS A CIRCUIT ELEMENT

Comparing Equations 7.41 and 7.43, we identify

(7.44)
$$-y_{21} = H \frac{N(s)}{D_b(s)}$$

$$y_{22} = \frac{D_a(s)}{D_b(s)}$$

The RC two-port can be realized by conventional RC two-port synthesis techniques. Note that the Fialkow-Gerst conditions for unbalanced network realization can be satisfied by properly choosing H. If the decomposition (Equation 7.42) is such that $D_a(s)$ is one degree higher than that of $D_b(s)$, y_{22} will have a pole at $s = \infty$, which can be removed as a shunt capacitor across port 2. The parallel combination of the positive capacitor and the negative resistor can finally be replaced by an N-type negative resistance. The structure of Fig. 7-20 requires that there be at least one transmission zero at infinity. We shall now consider the choice of $D_b(s)$. Let us select a polynomial $D_b(s)$ so that $D(s)/D_b(s)$ is an RC driving-point admittance with a pole at infinity, i.e., $D_b(s)^0 = D(s)^0 - 1$. Therefore,

(7.45)
$$\frac{D(s)}{D_b(s)} + 1 = \frac{D(s) + D_b(s)}{D_b(s)} = \frac{D_a(s)}{D_b(s)}$$

is also guaranteed to be an RC driving-point admittance with a pole at infinity.

The following example will illustrate the technique.

Example 7-8 Consider the transfer voltage ratio of Example 7-7.

$$t_v = \frac{N(s)}{D(s)} = \frac{H}{s^2 + 5s + 5}$$

Let us choose $D_b(s) = dD(s)/ds$. Note that $D(s)/[dD/ds]$ is guaranteed to have the required properties. We thus have

$$D_a(s) = (s^2 + 5s + 5) + \frac{d}{ds}(s^2 + 5s + 5)$$

$$= s^2 + 7s + 10$$

which yields

$$y_{22} = \frac{s^2 + 7s + 10}{2s + 5}; \quad -y_{21} = \frac{H}{2s + 5}$$

The complete realization of the given transfer ratio is obtained by making a continued fraction expansion of y_{22} (Figure 7-21). The realized value of the gain constant can be shown to be equal to 10.

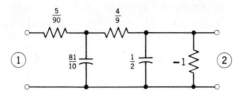

Figure 7-21 Realization of the transfer voltage ratio of Example 7-8.

Note that by choosing $D_b(s) = k[dD(s)/ds]$ and adjusting the value of the positive real constant k, almost any value of the initial capacitor can be obtained.

Unrestricted Transmission Zeros.[24] By employing a parallel ladder structure (Figure 7-22), stable transfer functions having transmission zeros anywhere in the complex plane can be realized. From Figure 7-22, we note that

(7.46) $$t_v = -\frac{y_{21}}{y_{22}} = -\frac{y'_{21} + y''_{21}}{y'_{22} + y''_{22}}$$

where y'_{ij} and y''_{ij} are the short-circuit parameters of the networks $\mathcal{N}'$ and $\mathcal{N}''$, respectively. A given transfer voltage ratio $N(s)/D(s)$ can be put into the form of Equation 7.46:

(7.47) $$\frac{N(s)}{D(s)} = \frac{a'\dfrac{(s-\alpha)N'(s)}{Q(s)} + a''\dfrac{N''(s)}{Q(s)}}{K'\dfrac{D(s)}{Q(s)} + K''\dfrac{D(s)}{Q(s)}}$$

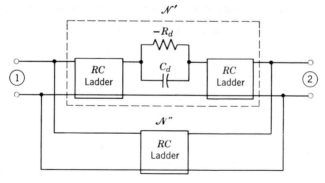

Figure 7-22 An RC:$-R^{(1)}$ configuration for realizing unrestricted transmission zeros.

[24] D. L. Losee and S. K. Mitra, *loc. cit.*

where $Q(s)$ is selected to make $D(s)/Q(s)$ an *RC* driving-point admittance and $K' + K'' = 1$. Equation 7.47 suggests that $N(s)$ must be decomposed in the form

(7.48) $$N(s) = a'(s - \alpha)N'(s) + a''N''(s)$$

where, of course, $N'(s)$ [$N''(s)$] is a polynomial with negative real roots and unity leading coefficients. It can be shown[25] that any polynomial with real coefficients and positive leading coefficient can be decomposed in the form of Equation 7.48.

Now, realize $y_{22}^{(1)} = D(s)/Q(s)$ and $y_{21}^{(1)} = (s - \alpha)N'(s)/Q(s)$, and designate the two-port as $\mathcal{N}_1$. Similarly, by the zero-shifting technique, realize $y_{22}^{(2)} = D(s)/Q(s)$ and $y_{21}^{(2)} = N''(s)/Q(s)$, and let the realized two-port be called $\mathcal{N}_2$. It is clear that since there is no control over the constant multipliers, the respective gain constants of $\mathcal{N}_1$ and $\mathcal{N}_2$ will be, for example, B_1 and B_2. Change the admittance level of $\mathcal{N}_1$ by Ka'/B_1, and of $\mathcal{N}_2$ by Ka''/B_2, where K will be determined shortly. Now if the two modified $\mathcal{N}_1$ and $\mathcal{N}_2$, i.e., $\mathcal{N}'$ and $\mathcal{N}''$, are connected in parallel, the overall y parameters are given by:

(7.49)
$$\frac{Ka'}{B_1} y_{21}^{(1)} + \frac{Ka''}{B_2} y_{21}^{(2)} = K y_{21} \text{ (prescribed)}$$
$$\frac{Ka'}{B_1} y_{22}^{(1)} + \frac{Ka''}{B_2} y_{22}^{(2)} = K\left(\frac{a'}{B_1} + \frac{a''}{B_2}\right) y_{22} \text{ (prescribed)}$$

It is clear from the above that the following condition must hold:

(7.50) $$K\left(\frac{a'}{B_1} + \frac{a''}{B_2}\right) = 1$$

which determines the value of K. Consider the following example.

Example 7-9 Let the prescribed functions be

$$y_{22} = \frac{(s + 2)(s + 4)}{(s + 3)}$$

$$-y_{21} = \frac{s^2 - s + 2}{s + 3}$$

The numerator of $-y_{21}$ can be decomposed in the form of Equation 7.48 in several ways:

$$s^2 - s + 2 = s(s - 1) + 2$$
$$= (s - 2)(s + 1) + 4$$
$$= s(s - 2) + 4(s + 2)$$

[25] D. L. Losee and S. K. Mitra, *loc. cit.*

From above, we select

$$-y_{21}^{(1)} = \frac{s(s-1)}{s+3} \; ; \quad -y_{21}^{(2)} = \frac{1}{s+3}$$

Note that $a' = 1$ and $a'' = 2$. Note that $y_{21}^{(1)}$ has a positive real axis

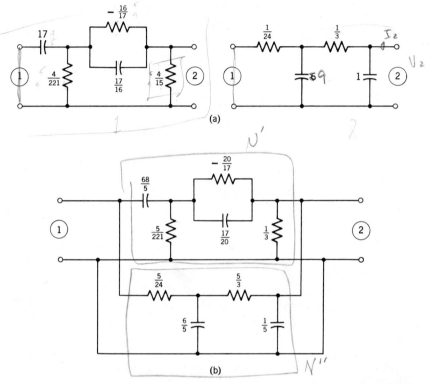

Figure 7-23 Realization of the transfer voltage ratio of Example 7-9.

transmission zero. Hence the two-port $\mathcal{N}_1$ is realized using the zero-shifting technique outlined earlier. On the other hand, $y_{21}^{(2)}$ has its only transmission zero at infinity. Therefore, $\mathcal{N}_2$ is realized by simply making a continued fraction expansion of y_{22}. The realized networks are shown in Figure 7-23a. By analysis, we obtain $B_1 = 1$ and $B_2 = 8$. From Equation 7.50, we have

$$K = \frac{1}{\dfrac{a'}{B_1} + \dfrac{a''}{B_2}} = \frac{4}{5}$$

The admittance level of $\mathcal{N}_1$ is now changed by $Ka'/B_1 = \frac{4}{5}$ and that of $\mathcal{N}_2$ by $Ka''/B_2 = \frac{1}{5}$. Final realization is shown in Figure 7-23, which realizes the transfer function with a gain constant of 5.

Several comments are here in order. Although the realization techniques have been demonstrated for transfer voltage ratios, they can be easily modified for other types of transfer functions. If it is desired to have resistive termination at either or both ends, methods analogous to passive RC two-port synthesis can be followed. Finally, although the last method proves the sufficiency of a single tunnel diode in realizing transfer functions with arbitrary zero pattern, it has a major disadvantage: the transmission zeros of the overall network are critically dependent on the element values and in practice it may be difficult to align the network properly. An answer to this problem is the cascade development,[26] where a pair of complex transmission zeros or a pair of real zeros can be realized by means of "zero-section." It can be shown that a number of tunnel diodes required for cascade development is almost equal to half of the total number of right-half plane transmission zeros.

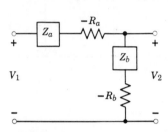

Figure 7-24 An $RLC:-R^{(2)}$ configuration for the realization of transfer voltage ratio.

$RLC:-R$ Network

As pointed out in the previous chapter, $RLC:-R$ network functions do not have limitations other than they be only real rational functions. In fact, we shall presently show that any real rational function can be realized as the transfer voltage ratio of an $RLC:-R^{(2)}$ two-port.

Realization of Voltage Transfer Ratio.[27] Consider the structure of Figure 7-24, for which

$$(7.51) \qquad t_v = \left.\frac{V_2}{V_1}\right|_{I_2=0} = \frac{Z_b - R_b}{Z_b - R_b + Z_a - R_a} = \frac{1}{1 + \dfrac{Z_a - R_a}{Z_b - R_b}}$$

A specified transfer function can be expressed in the form of Equation 7.51 as follows:

$$(7.52) \qquad t_v = \frac{N(s)}{D(s)} = \frac{1}{1 + \dfrac{D(s) - N(s)}{N(s)}}$$

[26] D. L. Losee and S. K. Mitra, *loc. cit.*
[27] F. T. Boesch and M. R. Wohlers, *loc. cit.*

where $N(s)$ and $D(s)$ are arbitrary polynomials with real coefficients. Choose a polynomial $Q(s)$ with negative real roots and of degree greater than or equal to max (D^0, N^0). After proper grouping of terms, a partial fraction expansion of $(D-N)/Q$ and N/Q will yield

(7.53)
$$\frac{D(s) - N(s)}{Q(s)} = Z'_{RC} + Z'_{RL} - R'$$

$$\frac{N(s)}{Q(s)} = Z''_{RC} + Z''_{RL} - R''$$

In Equation 7.53, Z'_{RC} and Z''_{RC} are RC impedances, and Z'_{RL} and Z''_{RL} are RL impedances. Note that R' and R'' are real constants. From Equations 7.51 through 7.53, we identify

$$Z_a = Z'_{RC} + Z'_{RL}; \quad R_a = R'$$
$$Z_b = Z'_{RC} + Z'_{RL}; \quad R_b = R''$$

Synthesis is completed by realizing the respective RC (RL) impedances and connecting the one-ports in the form of Figure 7-24.

Example 7-10   Let us realize

$$t_v = \frac{N(s)}{D(s)} = \frac{4}{s^2 + s + 1}$$

Hence $D(s) - N(s) = (s^2 + s - 3)$ and $N(s) = 4$. Let $Q(s) = (s+1) \times (s+2)$. As a result we have

$$\frac{D - N}{Q} = \frac{s^2 + s - 3}{(s+1)(s+2)} = 1 - \frac{3}{s+1} + \frac{1}{s+2}$$

$$= 3 - \frac{3}{s+1} + \frac{1}{s+2} - 2$$

and

$$\frac{N}{Q} = \frac{4}{(s+1)(s+2)} = \frac{4}{s+1} - \frac{4}{s+2} = \frac{4}{s+1} + 2 - \frac{4}{s+2} - 2$$

which results in

$$Z'_{RC} = 1/(s+2); \quad Z'_{RL} = 3s/(s+1); \quad R_a = 2$$
$$Z''_{RC} = 4/(s+1); \quad Z''_{RL} = 2s/(s+2); \quad R_b = 2$$

Final realization is shown in Figure 7-25. Several interesting observations can be made from the above example. The gain constant can be chosen arbitrarily and, as a result, can be made as large as desired. In some cases,

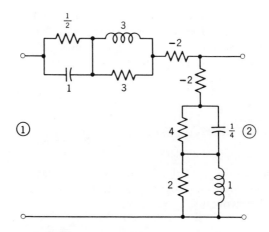

Figure 7-25 Realization of the transfer voltage ratio of Example 7-10.

with a proper selection of $Q(s)$, $R'(R'')$ can be made equal to zero, thus reducing the number of negative resistances.

An interesting variation of the configuration of Figure 7-24 was advanced by Martinelli.[28] His method is based on the admittance identification and with proper manipulation results in an $RLC:-R^{(1)}$ network.

7-5 SUMMARY

Several typical schemes for realizing the N-type (voltage-controlled) and the S-type (current-controlled) negative resistances are provided in Section 7-1.

A simple application of the negative resistance is considered in Section 7-2, which outlines two methods of cancellation of incidental dissapation of inductors.

Section 7-3 is concerned with the synthesis of driving-point functions employing negative resistances as the active elements. More specifically, realization methods of $RC:-R$, RC:tunnel-diode, LC:tunnel-diode, $RLC:-R$ one-ports are outlined.

Remaining portion of the chapter deals primarily with the synthesis of networks using the N-type negative resistances. Several synthesis schemes for realizing transfer functions, having all poles on the negative

---

[28] G. Martinelli, "Voltage transfer-function synthesis using one negative resistance," *Electronics Letters*, **2**, 87 (March 1966).

real axis, are presented. An important result shown here is that a stable transfer voltage ratio, having all negative real distinct poles and unrestricted transmission zeros, can always be realized by a grounded RC two-port containing a single N-type negative resistance.

Additional References

Brunetti, C., and L. Greenhough, "Some characteristics of a stable negative resistance," *Proc. IRE*, **30**, 542–546 (December 1942).

Carlin, H. J., "General n-port synthesis with negative resistors," *Proc. IRE*, **48**, 1174–1175 (June 1960).

Carlin, H. J., and D. C. Youla, "Network synthesis with negative resistors," *Proc. IRE*, **49**, 907–920 (May 1961).

Gafni, H., and I. Cederbaum, "Synthesis with negative resistors," *Proc. IEE (London)*, **113**, 783–787 (May 1966).

Gammie, J., and J. L. Merrill, Jr., "Stability of negative impedance elements in short transmission lines," *Bell System Tech. J.*, **34**, 333–360 (March 1955).

Harris, H. E., "Simplified Q multiplier," *Electronics*, **24**, 130–134 (May 1951).

Herold, E. W., "Negative resistance and devices for obtaining it," *Proc. IRE*, **23**, 1201–1223 (October 1935).

Howard, W. G., Jr., and D. O. Pederson, "Integrated voltage-controlled oscillators," *Proc. Nat. Elect. Conf.*, **23**, 279–284 (1967).

Huang, J. S. T., and A. A. Pandiscio, "Reactance associated with a class of negative resistances," *J. Electronics and Control*, **12**, 265–271 (April 1962).

Indiresan, P. V., and Sashilbala Sataindra, "A series type d.c. negative resistance for analogue computers," *J. Brit. IRE*, **23**, 417–420 (November 1963).

Kawakami, M., T. Yanagisawa, and H. Shibayama, "Highly selective band-pass filters using negative resistances," Proc. Symp. Active Networks and Feedback Systems, MRI Symposia Series, Vol. X, Polytech. Inst. of Brooklyn, pp. 369–378 (April 1960).

Kinariwala, B. K., "Necessary and sufficient conditions for the existence of $\pm R, C$ Networks," *IRE Trans. on Circuit Theory*, **CT-7**, 330–335 (September 1960).

Linvill, J. G., "The synthesis of active filters," Proc. Symp. Modern Network Synthesis, MRI Symposia Series, Vol. I, Polytech. Inst. of Brooklyn, pp. 453–476 (April 1955).

Lundry, W. R., "Negative impedance circuits: some basic relations and limitations," *IRE Trans. on Circuit Theory*, **CT-4**, 132–139 (September 1957).

Merrill, J. L., "A negative impedance repeater," *Trans. AIEE*, **70**, 49–54 (1951).

Merrill, J. L., Jr., A. F. Rose, and J. Q. Smethurst, "Negative impedance telephone repeaters," *Bell System Tech. J.*, **33**, 1055–1092 (March 1955).

Paul, J. C., "Design of a series tuned negative resistance," *Semiconductor Products*, **5**, 29–35 (1962).

Phillips, C. L., "Synthesis of three-terminal $\pm R, C$ networks," Ph.D. Thesis, Georgia Inst. of Tech., 1962.

Reich, H. J., "Circuits for producing high negative conductance," *Proc. IRE*, **43**, 228 (February 1955).

Reich, H. J., "More about negative resistance circuits," *Proc. IEEE*, **52**, 1058 (1964).

Scanlan, J. O., *Analysis and synthesis of tunnel diode circuits*, John Wiley, New York, 1966.

Shenoi, B. A., "A general method of designing tunnel diode amplifiers," *SCP and Solid State Technology*, pp. 23–29 (February 1964).

Shenoi, B. A., "A general synthesis of tunnel diode networks and sensitivity minimization," *Proc. Nat. Elect. Conf.*, **18**, 114–126, November 1962.

Su, K. L., "Cascade synthesis of *RC* networks using negative resistances," *IRE Trans. on Circuit Theory*, **CT-9**, 423–425 (December 1962).

Su, K. L., "Field-effect-transistor negative resistance circuit," *Electronics Letters*, **2**, 420, November 1966.

Tillman, J. R., "A note on electronic negative resistors," *Wireless Engineer*, **22**, 17–24 (January 1945).

Tombs, D. M., "Negative and positive resistance," *Wireless Engineer*, **19**, 341–346 (August 1942).

Todd, C. D., "Transistor tunnel diode produces an *N*-type negative resistance characteristic," *Communication and Electronics*, p. 284 (1962).

Yasuda, Y., T. Kasami, H. Ozaki, and H. Watanabe, "Synthesis of $\pm R$, L, C ladder networks," *J. Inst. Elec. Comm. Engrs. (Japan)*, **45**, 622–628 (May 1962).

Problems

7.1 Figure 7-26 shows the Colpitts type negative resistance circuit which is fairly stable against the variation of power supply voltage.[29] Calculate the input impedance of the circuit and show that it has an *N*-type negative input resistance.

7.2 A transistorized version of the circuit of Figure 7-26 was proposed recently and is shown in Figure 7-27.[30] Determine its input impedance.

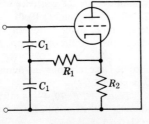
Figure 7-26

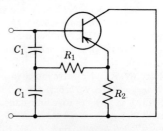
Figure 7-27

7.3 Figure 7-28 shows two operational amplifier circuits realizing finite gain inverting and noninverting type VVTs. Using the VVTs and

[29] H. E. Harris, *loc. cit.*
[30] M. Kawakami, T. Yanagisawa, and H. Shibayama, *loc. cit.*

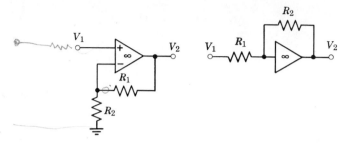

Figure 7-28

the positive feedback arrangement outlined in Section 7-1, p. 256, develop circuits to realize N-type and S-type negative resistances.

7.4 Realize using a minimum number of negative resistors, the functions of Problem 6.17, which are $RC:-R$ realizable driving-point functions.

7.5 Show that any real rational function having simple poles on $j\omega$-axis with real and positive residue can always be realized as the driving-point impedance of an $RLC:-R^{(1)}$ one-port.[31]

7.6 Using the previous result and the concept of predistortion technique, show that any real rational function is realizable as the driving-point function of an $RLC:-R^{(1)}$ one-port.[32]

7.7 Realize the driving-point impedance of Example 7-1 by a continued fraction expansion.

7.8 Complete the realization of the transfer voltage ratio of Example 7-6.

7.9 Realize $t_v = H/(s^2 + 3s + 1)$ in the form of Figure 7-20.

7.10 Realize

$$t_v = \frac{H(s-2)(s-4)}{(s+2)(s+4)}$$

in the form of Figure 7-22.

7.11 Synthesize the transfer voltage ratio of Problem 7.10 by means of an $RLC:-R^{(2)}$ two-port.

---

[31] H. Carlin and D. C. Youla, *loc. cit.*
[32] H. Gafni and I. Cederbaum, *loc. cit.*

8 / *Synthesis Using Controlled Sources*

In the last chapter, we were concerned with the use of the negative resistance as a network element. Although, the use of the negative resistance has many advantages, a major problem arises upon the exclusion of inductors from the realization process. In this latter situation, realizability is restricted to network functions having real poles only, and for stability reasons the poles must be situated on the negative real axis. This problem is avoided if instead we use two-port active elements like the controlled sources, the negative-impedance converter, etc.

The main purpose of this chapter is to develop synthesis techniques which hinge upon the use of the controlled source as the active network element. Later chapters deal with the use of the negative-impedance converter, gyrator, and the operational amplifier. In Section 8-1, a systematic approach of controlled source realizations using transistors is introduced. The next two sections are concerned with some typical synthesis techniques of driving-point and transfer functions employing the controlled sources as the active elements. Several active RC network design schemes based on coefficient matching techniques are introduced in Section 8-4. Section 8-5 is concerned with the sensitivity problem. Main results of the chapter are summarized in the final section.

The major step in almost all active RC synthesis methods is the decomposition and partitioning of the specified network functions into simpler subfunctions. The next step is the manipulation of these subfunctions to yield the identification of the parameters characterizing the "companion" *RC* network. From these parameters the passive *RC* portion is synthesized following known *RC* synthesis procedures. In most cases, the *RC* network is either a collection of one-ports or two-ports or both.[1] For an *RC* one-port, the network parameter is a driving-point function. It can be realized by a partial fraction expansion (Foster's method) or a continued fraction expansion (Cauer's method) or a combination of both. For an *RC* two-port, the network parameters are usually a driving-point function and a transfer impedance (admittance) function. Depending on the location of the transmission zeros, there are different methods of synthesis. If the transmission zeros are all at the origin or at infinity, network realization is obtained by a continued fraction expansion of the driving-point function. If the transmission zeros are on the negative real axis, then the network is realized by the zero-shifting technique. However, if the transmission zeros are complex, then the realization technique is a little complicated. There are three possible techniques to handle this situation: Dasher's method, Fialkow-Gersts' method, Guillemin's method.[2]

8-1 PRACTICAL CONSIDERATIONS

We now present a systematic approach of transistor-circuit realizations of the controlled sources. In general, it is not possible to design an ideal controlled source. However, if properly designed, a transistorized circuit realization can be made to approach their idealized counterpart for small-signal applications.

Nullator-Norator Representation

The equivalent representation of controlled sources using the two degenerate network elements, the nullator and the norator, was introduced earlier in Chapter 2. The four basic nullator-norator circuits and other equivalent representations of the controlled sources derived from the basic circuits are shown again in Figure 8-1 for convenience.

Transistorized realizations of various controlled sources are easily obtained from the nullator-norator equivalent circuits. The realization scheme as mentioned earlier in Section 2-7, p. 74, is based on the equivalence of an idealized transistor ($r_e \to 0$, $r_b \to 0$, $r_c \to \infty$, $\alpha \to 1$) and the

---

[1] Physical realizability conditions of passive networks are summarized in Appendices B and C.
[2] For a review of *RC* realization techniques, see any textbook listed at the end of Appendices B and C.

nullator-norator circuit of Figure 8-2. Thus, in order to convert a nullator-norator circuit into a transistor realization, a nullator and a norator must occur in a pair with a common junction. With reference to Figure 8-2, we observe that the common junction of the nullator-norator pair is the emitter terminal, the other terminals of the nullator and the norator being, respectively, the base and the collector terminals of the transistor.

It can be seen from Figure 8-1 that in many equivalent representations, the nullator and the norator do not appear with a common terminal. In order to convert these circuits into transistorized realizations, the identity depicted in Figure 8-3 can be used to advantage. This indicates that in a given nullator-norator equivalent representation, between any two nodes, a nullator-norator series combination can be added without modifying circuit operation. Such an addition of superfluous elements often leads to circuits that are more suitable for a transistorized conversion.

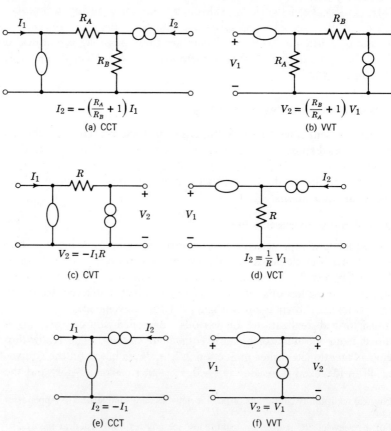

Figure 8-1 Nullator-norator representations of the controlled sources.

PRACTICAL CONSIDERATIONS / 293

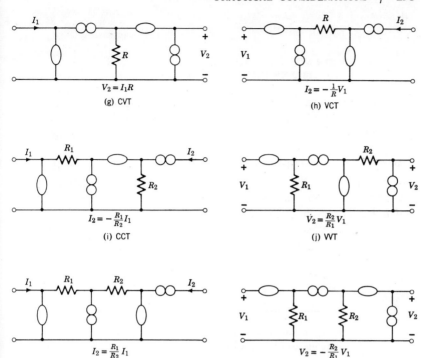

Figure 8-1 (Continued)

Transistor Realizations

We consider first the conversion of the VVT circuits of Figure 8-1 into transistor circuits. It follows from Figures 8-1f and 8-2 that a unity gain noninverting type[3] VVT can be realized by a common-collector transistor as shown in Figure 8-4.

The performance of the unity gain VVT of Figure 8-4 (also called an

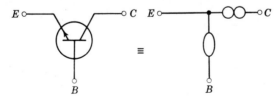

Figure 8-2 The nullator-norator equivalent representation of an ideal transistor.

---

[3] Output polarity is same as the input polarity.

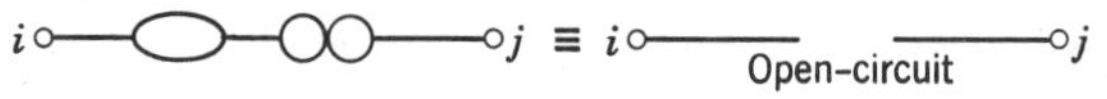

294 / SYNTHESIS USING CONTROLLED SOURCES

$$i \circ\!\!-\!\!\bigcirc\!\!-\!\!\infty\!\!-\!\!\circ j \equiv i \circ\!\!-\!\!\underset{\text{Open-circuit}}{\rule{2cm}{0.4pt}}\!\!-\!\!\circ j$$

Figure 8-3 Formation of an open-circuit condition by a series connection of a nullator and a norator.

emitter follower) can be improved to a certain extent by cascading two such circuits. While such a cascade leads to high input impedance and low output impedance, the phase-shift problem worsens due to charge storage in the transistors used.

Making use of the identity shown in Figure 8-3, we arrive at the alternate realization of the unity gain VVT as indicated in Figure 8-5. The transistor circuit shown does not include the biasing arrangement. This circuit has a very high input impedance even with moderate β (of the

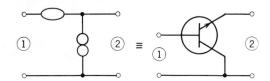

Figure 8-4 Transistorized realization of a unity gain noninverting type VVT.

order of 50) transistors. Gile[4] suggests the use of a diode D_1 as shown in the figure primarily for compensating the base-emitter offset voltage of T_1 and for temperature compensation.

For constructing noninverting type VVT with gain larger than unity, the nullator-norator circuit of Figure 8-1b can be used. Since a direct conversion to a transistor circuit is not possible, use of the identity shown in Figure 8-3 can be made. Out of four possible modifications (with two

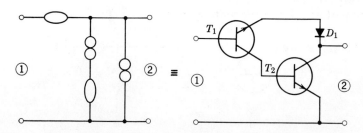

Figure 8-5 A modified unity gain noninverting-type VVT.

[4] W. Gile, "Solid-state low-frequency filter," *Electro-Technology*, **74,** 34–37 (September 1964).

transistors), only two lead to stable operations; they are shown in Figure 8-6 (the circuit of Figure 8-6a is more well known as the *series-shunt feedback pair*).

Figure 8-7a shows the series-shunt feedback pair with its biasing resistors and supplies. An improved version of this noninverting type VVT

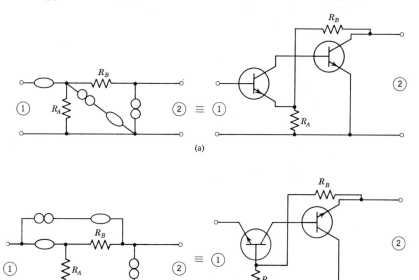

Figure 8-6 Transistorized realizations of noninverting-type VVT having a voltage gain greater than unity.

(Figure 8-6a) is indicated in Figure 8-7b along with its biasing arrangement. In a recent paper Kerwin[5] discusses in detail the design principles of this voltage amplifier.

Another possible realization of the noninverting type VVT is obtained by converting the nullator-norator circuit of Figure 8-1j. The corresponding transistor circuit is sketched in Figure 8-8.

An inverting type VVT (output polarity opposite to that of the input) can be realized by making use of the nullator-norator circuit of Figure 8-1l. It's transistor realization is shown in Figure 8-9.

We now consider the transistor realization of the current-controlled current source (CCT). It is seen from Figures 8-2 and 8-1e that a unity gain CCT can be realized by a common base transistor amplifier.

[5] W. J. Kerwin, "An active *RC* elliptic function filter," *1966 IEEE Region VI Conference Record*, Tucson, Arizona, pp. 647–654 (April 1966).

296 / SYNTHESIS USING CONTROLLED SOURCES

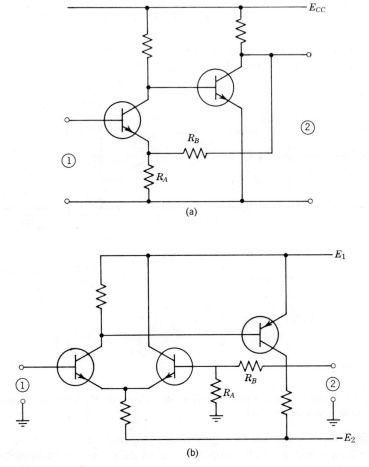

Figure 8-7 (*a*) The VVT circuit of Figure 8-6*a* with appropriate biasing arrangement. (*b*) An improved version of the same circuit.

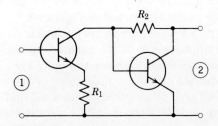

Figure 8-8 Transistorized realization of a noninverting-type VVT.

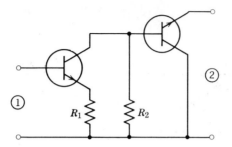

Figure 8-9 Transistorized realization of an inverting-type VVT.

To obtain CCT having current gain of more than unity, the nullator-norator circuit of Figure 8-1a can be used. Two possible stable realizations are developed in Figures 8-10a and 8-10b. The current amplifier of Figure 8-10a is more commonly known as *shunt-series feedback pair*. It is redrawn in Figure 8-10c showing the possible biasing arrangement. An alternate transistor realization of the same type of CCT can be obtained by converting the nullator-norator circuit of Figure 8-1i. The resultant circuit is sketched in Figure 8-11.

Figure 8-1k represents a CCT of opposite type than those discussed above. Its transistorized conversion is indicated in Figure 8-12.

Next, we introduce the realizations of the voltage-controlled current source (VCT). The basic VCT representation of Figure 8-1d can be easily converted to a transistor circuit as depicted in Figure 8-13a. An improved version of this circuit shown in Figure 8-13b is obtained by making use of the identity of Figure 8-3.

Conversion of the nullator-norator circuit of Figure 8-1h results in a transistor realization of the VCT of negative gain. The desired circuit is shown in Figure 8-14.

Lastly, we consider the realizations of the current-controlled voltage source (CVT). The transistor realization of the basic CVT nullator-norator representation of Figure 8-1c and two of its modifications are sketched in Figure 8-15. A CVT of positive gain can be obtained from the circuit of Figure 8-1g as indicated in Figure 8-16.

Following the methods outlined here, realizations of controlled sources using more than two transistors can be easily constructed.

Analysis of Some Typical Circuits

For a better appraisal of the circuits of Figures 8-4 through 8-16, it is necessary to perform an analysis of these circuits by replacing the transistor

298 / SYNTHESIS USING CONTROLLED SOURCES

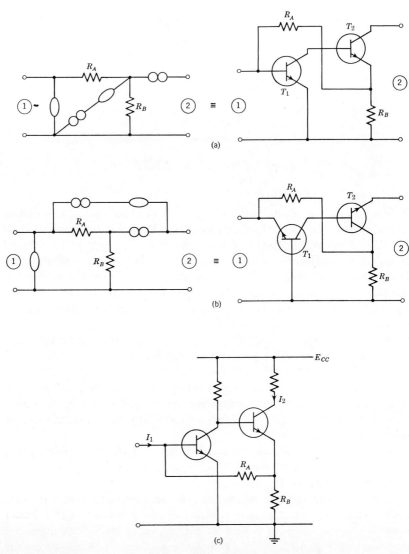

Figure 8-10 Transistorized realizations of noninverting-type CCTs having a gain greater than unity.

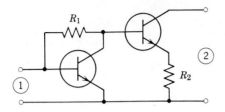

Figure 8-11 Realization of a noninverting-type CCT.

by its T equivalent circuit (Figure 2-45b). We present below the results of such an analysis of a few typical circuits as reported by Lewis.[6]

For the VVT of Figure 8-4, one easily obtains the following g-parameters:

(8.1)
$$g_{11} = \frac{1}{r_c} \cong 0$$
$$g_{12} = -(1 - \alpha)$$
$$g_{21} = 1$$
$$g_{22} = r_e + (1 - \alpha)r_b$$

provided $r_b \ll r_c$. Analysis of the noninverting voltage amplifier of Figure 8-6a yields the following g-parameters:

(8.2)
$$g_{11} \cong \frac{(1 - \alpha_1)(1 - \alpha_2)}{\alpha_1 R_A} + \frac{1}{\alpha_1 r_{c1}}\left(1 + \frac{1}{R_A r_{c2}}\right) + \frac{(1 - \alpha_1)}{\alpha_1 r_{c2}}\left(1 + \frac{R_B}{R_A}\right)$$
$$g_{12} \cong -\frac{(1 - \alpha_1)(1 - \alpha_2)}{\alpha_1} - \frac{1}{\alpha_1 r_{c1} r_{c2}}$$
$$g_{21} \cong 1 + \alpha_2 \frac{R_B}{R_A}$$
$$g_{22} \cong \frac{(1 - \alpha_2)}{\alpha_1 r_{c1}}\left(1 + \frac{R_B}{R_A}\right) + R_A \frac{(1 - \alpha_2)}{\alpha_1}$$

The derivation of Equation 8.2 is based on the following assumptions:
$$r_{e1} \ll (1 - \alpha_1)r_{c1}, \qquad r_{e2} \ll (1 - \alpha_2)r_{c2}$$
$$r_{e1}(1 - \alpha_2) \ll R_A \ll r_{c1}$$
$$R_B \ll r_{c2}$$
$$(1 - \alpha_1)(1 - \alpha_2) \ll 1$$
$$(R_A + R_B)r_{e1} \ll r_{c2} R_A$$
$$r_{b1} \approx r_{b2} \approx 0$$

[6] P. H. Lewis, "The design of quasi-ideal negative impedance circuits," Proc. 6th Midwest Symp. on Circuit Theory, Univ. of Wisconsin, Madison, Wis., pp. C01–C24 (May 1963).

300 / SYNTHESIS USING CONTROLLED SOURCES

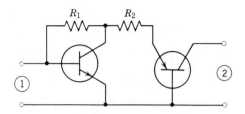

Figure 8-12 Transistorized realization of an inverting-type CCT.

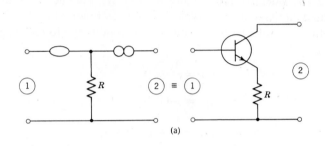

(a)

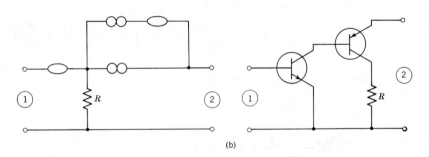

(b)

Figure 8-13 Transistorized realizations of a VCT of positive gain.

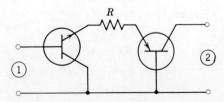

Figure 8-14 Realization of a VCT of negative gain.

PRACTICAL CONSIDERATIONS / 301

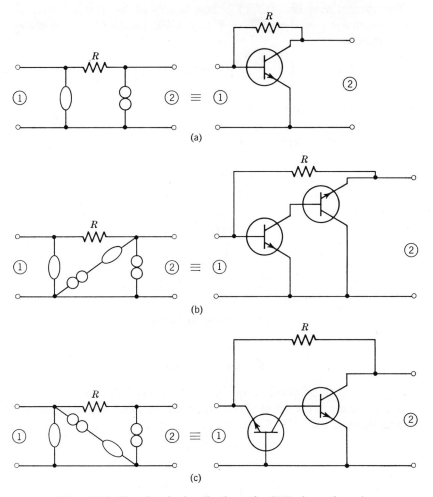

Figure 8-15 Transistorized realizations of a CVT of negative gain.

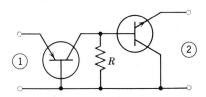

Figure 8-16 Realizations of a CVT of positive gain.

The current amplifier (CCT) is best described by h-parameters. A rigorous analysis of the CCT of Figure 8-10a results in the following h-parameters:

$$h_{11} \cong \frac{r_{e1}}{\alpha_1}\left[\left(\frac{R_A + R_B}{R_B}\right)(1 - \alpha_2) + R_A\left(\frac{1}{r_{c1}} + \frac{1}{r_{c2}}\right)\right]$$

$$h_{12} \cong r_{e1}/\alpha_1 r_{c2}$$

(8.3) $$h_{21} \cong -\alpha_2\left(1 + \frac{R_A}{R_B}\right)$$

$$h_{22} \cong \frac{1}{r_{c2}}\left[1 + \frac{r_{e1}}{\alpha_1 R_B} + \frac{R_A + R_B}{R_B} \cdot \frac{(1 - \alpha_1)}{\alpha_1}\right]$$

provided
$$r_{b1} = r_{b2} \cong 0$$

$$r_{e1} \ll (1 - \alpha_1)r_{c1}; \quad r_{e2} \ll (1 - \alpha_2)r_{c2}$$

$$\frac{R_A + R_B}{R_B} \ll \frac{\alpha_1 r_{c2}}{r_{c2}(1 - \alpha_1)(1 - \alpha_2) + r_{e2}}$$

$$r_{e1}(1 - \alpha_2) \ll R_B \ll r_{c2}$$

$$R_A \ll r_{c1}$$

$$R_A \ll r_{c2}/(1 - \alpha_1)$$

For a voltage-controlled current source (VCT), a suitable set of two-port parameters is the set of short-circuit parameters. For example, the y-parameters of the VCT of Figure 8-13a are given as:

$$y_{11} \cong \frac{1 - \alpha}{R} + \frac{1}{r_c}$$

$$y_{12} \cong \frac{1}{r_c} + \frac{r_b}{Rr_c}(1 - \alpha)$$

(8.4) $$y_{21} \cong \frac{\alpha}{R}$$

$$y_{22} \cong \frac{1}{r_c} - \frac{\alpha r_b}{Rr_c}$$

under the assumptions
$$r_b \ll r_c$$
$$r_e + (1 - \alpha)r_b \ll R \ll r_c$$

The modified two-transistor VCT circuit of Figure 8-13b can be similarly analyzed, which yields

(8.5)
$$y_{11} \cong \frac{(1-\alpha_1)(1-\alpha_2)r_{c1} + r_{e2}}{\alpha_2 R r_{e1}} + \frac{(1-\alpha_1)}{\alpha_2 r_{c2}}$$

$$y_{12} = -\frac{(1-\alpha_1)}{\alpha_2 r_{c2}}$$

$$y_{21} \cong \frac{1}{R}$$

$$y_{22} \cong \frac{1}{r_{c2}}\left[\frac{(r_{e1}+r_{e2})}{\alpha_2 R} + \frac{(1-\alpha_1)}{\alpha_2}\right]$$

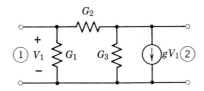

Figure 8-17 Nonideal representation of the VCT of Figure 8-13a.

Derivation of Equation 8.5 is based on the following assumptions:

$$r_{b1} = r_{b2} = 0$$
$$r_{e1} \ll (1-\alpha_1)r_{c1}; \quad r_{e2} \ll (1-\alpha_2)r_{c2}$$
$$(1-\alpha_1)(1-\alpha_2) \ll 1$$
$$R \ll r_{c2}/(1-\alpha_1)$$
$$R \gg r_{e1}(1-\alpha_2)$$
$$R \gg r_{e2}$$

A popular model of a nonideal VCT often used in synthesis is shown in Figure 8-17. For example, if the active two-port of Figure 8-13a is used, the pertinent parameters can be easily computed from Equation

8.4. We have for this case

(8.6)
$$G_1 \cong \frac{(1-\alpha)}{R}\left[1 + \frac{r_b}{r_c}\right]$$
$$G_2 \cong \frac{1}{r_c}\left[1 - \frac{r_b}{R}(1-\alpha)\right]$$
$$G_3 \cong \frac{r_b}{Rr_c}(1-2\alpha)$$
$$g \cong \frac{\alpha}{R} + \frac{1}{r_c} - \frac{r_b}{Rr_c}(1-\alpha)$$

Finally, we consider the analysis of two CVT circuits, which are usually described by their open-circuit parameters. The z-parameters of the CVT circuit of Figure 8-15a can be shown to be given as:

(8.7)
$$z_{11} \cong r_e + (1-\alpha)r_b + \frac{\alpha r_b R}{r_c}$$
$$z_{12} \cong r_e + (1-\alpha)r_b$$
$$z_{21} \cong -\alpha R$$
$$z_{22} \cong r_e + (1-\alpha)(R + r_b)$$

where we have assumed
$$r_b \ll r_e$$
$$r_e + (1-\alpha)r_b \ll R \ll r_c$$

In a similar manner, routine analysis of the circuit of Figure 8-15b yields:

(8.8)
$$z_{11} \cong r_{e1}\left[\frac{1-\alpha_2}{\alpha_1} + \frac{(r_{c1}+r_{c2})R}{\alpha_1 r_{c1} r_{c2}}\right]$$
$$z_{12} \cong \frac{(1-\alpha_2)r_{e1}}{\alpha_1}$$
$$z_{21} \cong R$$
$$z_{22} \cong \frac{R}{\alpha_1}\left[(1-\alpha_1)(1-\alpha_2) + \frac{r_{e2}}{r_{c1}}\right] + \frac{r_{e1}}{\alpha_1}(1-\alpha_2)$$

provided
$$r_{b1} = r_{b2} = 0$$
$$r_{e1} \ll (1-\alpha_1)r_{c1}, \quad r_{e2} \ll (1-\alpha_2)r_{c2}$$
$$r_{e1}(1-\alpha_2) \ll R$$
$$R \ll r_{c1}$$
$$R \ll r_{c2}/(1-\alpha_1)$$

The performance of many of the transistor circuit realizations of controlled sources introduced earlier can be improved by using a Darlington compound pair of compound transistors.[7] For example, the operation of the unity gain VVT of Figure 8-4 can be improved by using a Darlington compound pair in place of the single transistor, thereby making the overall α almost close to unity.

8-2 SYNTHESIS OF DRIVING-POINT FUNCTIONS

Realization of driving-point impedances is a basic problem. It offers insight into the general area of synthesis. In addition, as noted in Section 3-4, a transfer function realization problem can be easily converted into that of one-port synthesis. Since any active element can have a controlled source representation, synthesis techniques outlined in the succeeding chapters are easily adaptable for use of the controlled source as the active element.

The following section outlines several techniques that are centered around the controlled source as the active device.

Realization Using *RC* One-Ports and Two Controlled Sources

A novel structure consisting of four *RC* one-ports and two ideal voltage-controlled voltage sources (VVT) was proposed by Sandberg.[8] The proposed configuration is shown in Figure 8-18a. The input impedance of this active one-port is given as

(8.9) $$Z_{in} = \frac{V_1}{I_1} = 1 + \frac{y_4 - y_3}{y_2 - y_1} \Rightarrow \text{Fig 8-18a}$$

A prescribed rational function $N(s)/D(s)$ can be cast into the form of Equation 8.9 by writing it as

$$Z_{in} = \frac{N(s)}{D(s)} = 1 + \frac{N(s) - D(s)}{D(s)}$$

(8.10) $$= 1 + \frac{\dfrac{N(s) - D(s)}{Q(s)}}{\dfrac{D(s)}{Q(s)}} \quad \begin{array}{l} y_4 - y_3 \\ \\ y_2 - y_1 \end{array}$$

---

[7] J. G. Linvill and J. F. Gibbons, *Transistors and Active Circuits*, McGraw-Hill, New York, 1961.

[8] I. W. Sandberg, "Active *RC* networks," MRI Res. Rept. R-662-58, PIB-590, Polytech. Inst. of Brooklyn, 1958.

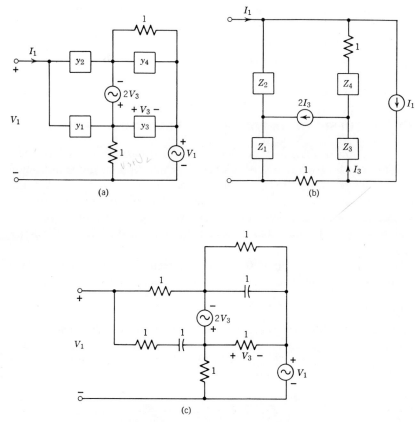

Figure 8-18 Active *RC* realization of a driving-point function. (*a*) Proposed configuration using two VVTs. (*b*) Proposed configuration using two CCTs. (*c*) Realization of the driving-point impedance of Example 8-1.

where $Q(s)$ is a polynomial with distinct negative real roots and of degree governed by

(8.11) $$Q(s)^0 + 1 \geq \max\,[N(s)^0,\, D(s)^0]$$

It follows from Theorem 3-5 that we can write

(8.12)
$$\frac{N(s) - D(s)}{Q(s)} = Y_{RC}{}^a - Y_{RC}{}^b$$

$$\frac{D(s)}{Q(s)} = Y_{RC}{}^c - Y_{RC}{}^d$$

SYNTHESIS OF DRIVING-POINT FUNCTIONS / 307

From Equations 8.9, 8.10, and 8.12, we identify

(8.13) $\quad y_1 = Y_{RC}{}^d, \qquad y_2 = Y_{RC}{}^c, \qquad y_3 = Y_{RC}{}^b, \qquad y_4 = Y_{RC}{}^a$

It should be clear that the identifications of y_1 and y_2 (y_3 and y_4) can be interchanged.

Note that the proposed synthesis technique imposes no restriction on the driving-point impedance to be realized. The dual structure of Figure 8-18b is characterized by an input admittance:

(8.14) $\qquad Y_{in} = \dfrac{I_1}{V_1} = 1 + \dfrac{Z_3 - Z_4}{Z_1 - Z_2} \quad \Rightarrow \text{Fig 8-18} b$

Realization steps are similar to the previous one and will be omitted.

Example 8-1 Consider the realization of $Z_{in} = N(s)/D(s) = s^2$. Note that the specified driving-point function is not p.r. Choose $Q(s) = s + 1$. A partial fraction expansion of $[N(s) - D(s)]/sQ(s)$ and $D(s)/sQ(s)$ is next carried out:

$$\dfrac{N(s) - D(s)}{sQ(s)} = \dfrac{s^2 - 1}{s(s+1)} = 1 - \dfrac{1}{s}$$

$$\dfrac{D(s)}{sQ(s)} = \dfrac{1}{s(s+1)} = \dfrac{1}{s} - \dfrac{1}{s+1}$$

This implies

$$\dfrac{N(s) - D(s)}{Q(s)} = \dfrac{s^2 - 1}{s+1} = s - 1$$

$$\dfrac{D(s)}{Q(s)} = \dfrac{1}{s+1} = 1 - \dfrac{s}{s+1}$$

from which we easily identify

$$y_4 = s, \qquad y_3 = 1$$

$$y_2 = 1, \qquad y_1 = \dfrac{s}{s+1}$$

The complete realization is shown in Figure 8-18c.

Several comments are here in order. Since the order[9] of the impedance function is two, any active realization must have at least two capacitors. In this respect, the network of Figure 8-18c is optimal. $Q(s)$ was purposely chosen as $(s + 1)$ to cancel out with the factor $(s + 1)$ in $N(s) - D(s)$.

---

[9] Order $\triangleq \max [N(s)^0, D(s)^0]$

308 / SYNTHESIS USING CONTROLLED SOURCES

Any other choice would thus have led to a larger number of passive R,C elements. However, $Q(s)$ can also be chosen to minimize the pole or zero sensitivity.

Other Methods

Most of the earlier works on the active RC synthesis of driving-point function concentrated on using a single controlled source of unrestricted gain. The main aim of every author was to prove the following basic theorem on active RC one-port synthesis:

Theorem 8-1 *Any real rational function of the complex frequency variable s can be realized as the driving-point impedance of a one-port obtained by embedding a single controlled source of unrestricted gain in a transformerless RC network.*

The proof of the basic theorem (Theorem 8-1) was first advanced by Sandberg.[10] A variation of his synthesis method will be discussed in the next chapter.

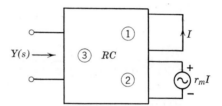

Figure 8-19 A general active RC configuration containing a single CVT.

An alternate proof of sufficiency of a single controlled source to realize any real rational driving-point function was shown by Kinariwala.[11] We discuss briefly here his approach. The active RC one-port configuration proposed by Kinariwala is shown in Figure 8-19. The input admittance of the active one-port is given as

(8.15) $$Y(s) = \frac{y_{33} - r_m(y_{13}y_{23} - y_{12}y_{33})}{1 + r_m y_{12}}$$

where $\{y_{ij}\}$ represent the short-circuit parameters of the RC three-port. From a specified driving-point admittance function, y_{12}, y_{23}, y_{33} can be

---

[10] I. W. Sandberg, *loc. cit.*
[11] B. K. Kinariwala, "Synthesis of active RC networks," *Bell System Tech. J.*, **38**, 1269–1316 (September 1959).

identified by suitable decomposition and partitioning. The remaining parameters are arbitrary; they can be chosen so that the *RC* three-port can be realized without transformers. Since multiport synthesis is beyond the scope of this book, we refer the reader to the original paper.

Another driving-point synthesis method is due to DeClaris.[12] He makes use of a structure containing two *RC* two-ports and a four-terminal voltage-controlled current source (VCT). Realization steps are based on a successive application of the root locus approach. DeClaris' method is restricted to a class of driving-point function. An excellent exposition of his method will be found in Su.[13]

8-3 SYNTHESIS OF TRANSFER FUNCTIONS

In most cases, one is interested in the realization of a specified transfer function. Some useful network specifications are summarized in Table A-3 of Appendix A.

Several elegant and practical methods are available for the realization of transfer functions by means of an active *RC* network where the active device is some type of controlled source. In particular, some of the methods do account for the nonidealness of a practical controlled source.

Two Basic Structures[14]

A transfer impedance $Z_{21}(s)$ having complex poles anywhere in the s plane and negative real transmission zeros can be realized in the configuration shown in Figure 8-20a. Analysis reveals

$$(8.16) \qquad Z_{21}(s) = \left.\frac{V_2}{I_1}\right|_{I_2=0} = \frac{\mu}{Y_1 - (\mu - 1)Y_2}$$

where μ is the gain of the noninverting voltage amplifier. A specified $Z_{21}(s) = N(s)/D(s)$ can be realized as follows: From the results of Theorem 3-5 it is clear that if

$$N(s)^0 \geq D(s)^0 - 1$$

we can write

$$(8.17) \qquad \frac{D(s)}{N(s)} = Y_{RC}{}^A - Y_{RC}{}^B$$

---

[12] N. DeClaris, "Synthesis of active networks: driving-point functions," *IRE Natl. Conv. Rec.*, 7 (Part 2), 23–39 (1959).

[13] K. L. Su, *Active network synthesis*, McGraw-Hill, New York, pp. 116–121 (1965).

[14] Section 8-3, p. 309 through p. 327, is based on S. K. Mitra, "Transfer function realization using *RC* one-ports and two grounded voltage amplifiers," Proc. First Annual Princeton Conference on Information Sciences and Systems, Princeton Univ., N.J., pp. 18–23 (March 1967).

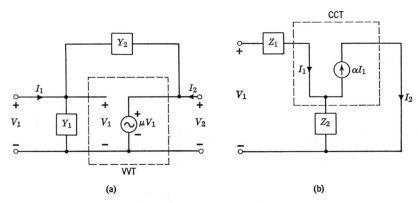

Figure 8-20 Basic "pole sections."

where $Y_{RC}{}^A$ and $Y_{RC}{}^B$ are passive *RC* driving-point admittances. If $\mu > 1$, we identify

(8.18) $\qquad Y_1 = Y_{RC}{}^A, \qquad Y_2 = Y_{RC}{}^B/(\mu - 1)$

The dual structure shown in Figure 8-20b employs a current amplifier and is characterized by a transfer admittance

(8.19) $\qquad Y_{21} = \dfrac{I_2}{V_1} = \dfrac{\alpha}{Z_1 - (\alpha - 1)Z_2}$

A major disadvantage in using the structure is that the current amplifier (CCT) requires a "floating" power supply. In both configurations of Figure 8-20, the pole locations can be controlled by adjusting the gain of the amplifiers. The circuits of Figure 8-20 will be called the basic "pole section."

If an inverting voltage amplifier is used, then a transfer admittance with complex zeros and negative real poles can be synthesized. The proposed configuration is indicated in Figure 8-21a for which

(8.20) $\qquad Y_{21} = \dfrac{-I_2}{V_1}\bigg|_{V_2=0} = Y_3 - \mu Y_4$

where $-\mu$ is the gain of the amplifier. Using the results of Theorem 3-5, a specified transfer admittance $P(s)/Q(s)$ can be realized in the form of Figure 8-21a with Y_3 and Y_4 as *RC* admittances provided

$$Q(s)^0 \geq P(s)^0 - 1$$

The dual configuration is indicated in Figure 8-21b, which uses a current amplifier. For practical reasons, the dual structure is not convenient to use. It should be noted that use of the active two-ports of

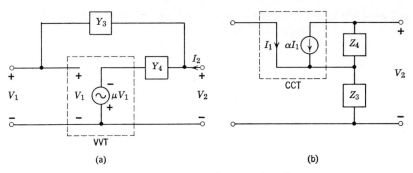

Figure 8-21 Basic "zero sections."

Figure 8-21 has one definite advantage: the transmission zeros are easily controlled by adjusting the gain of the amplifiers. We shall refer the circuits of Figure 8-21 as the basic "zero sections."

The circuits of Figure 8-20 and 8-21 form two basic types of circuits, which can be used to construct more general types of active *RC* configurations. The development of such configurations follows.

Kuh's Method

The structure proposed by Kuh[15] is obtained by cascading an *RC* two-port with the basic "pole section" of Figure 8-20a as shown in Figure 8-22. The voltage transfer ratio of this circuit is given as

$$(8.21) \qquad \frac{V_o}{V_{in}} = \frac{-y_{21}}{y_{22} + Y_1 - (\mu - 1)Y_2}$$

where y_{21}, y_{22} represent the short-circuit admittances of the *RC* two-port, and Y_1, Y_2 are required to be *RC* one-ports. To account for the input

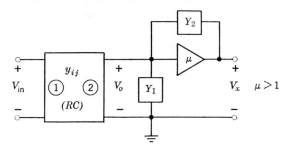

Figure 8-22 Kuh's active *RC* configuration.

[15] E. S. Kuh, "Transfer function synthesis of active *RC* networks," *IRE Trans. on Circuit Theory*, **CT-7** (Special Supplement) 3–7 (August 1960).

312 / SYNTHESIS USING CONTROLLED SOURCES

impedance of the amplifier, $Y_1(0)$ must be a nonzero constant. The feedback and output impedances can be included in Y_2 provided $Y_2(0)$ and $Y_2(\infty)$ are finite.

The synthesis procedure is as follows. Without any loss of generality, assume $Y_1 = 1$. Let $N(s)/D(s)$ be the specified voltage transfer ratio. Choose a polynomial $Q(s)$ with distinct negative real roots and satisfying the following degree requirement:

$$Q(s)^0 + 1 \geq \max [N(s)^0, D(s)^0]$$

We now make a partial fraction expansion of $D(s)/sQ(s)$ and rewrite it as

(8.22) $$\frac{D(s)}{Q(s)} = k_0 + k_\infty s + \sum_i \frac{k_i s}{s + \sigma_i} - \sum_j \frac{h_j s}{s + \delta_j}$$

where k_i and h_j are positive and nonnegative. Comparing Equation 8.22 with the denominator of expression 8.21, we can make the following identifications:

(8.23)
$$y_{22} = k_0' + k_\infty s + \sum_i \frac{k_i s}{s + \sigma_i} + \sum_j \frac{h_j' s}{s + \delta_j}$$

$$(\mu - 1)Y_2 = k_0'' + \sum_j \frac{h_j'' s}{s + \delta_j}$$

where $k_0 = k_0' - k_0'' + 1$, $h_j = h_j'' - h_j' > 0$. Because of the suggested decomposition, $Y_2(0)$ and $Y_2(\infty)$ will be finite and positive. y_{21} is identified as

(8.24) $$-y_{21} = \frac{N(s)}{Q(s)}$$

The RC two-port is realized following standard passive RC two-port synthesis techniques. There are several drawbacks of this method which, in most cases, are of no importance. First, since there is no control over the multiplier constant in the development of the RC two-port, the transfer function is always realized within a multiplicative constant. Second, if a common ground is desired, this method cannot realize positive real transmission zeros. The primary attraction of this approach would be the possible use of a nonideal controlled source, having finite input, output, and feedback impedances, as the active device.

Let us illustrate the method by means of two examples.

Example 8-2 Consider the realization of a second order Bessel filter,

$$t_v = \frac{N(s)}{D(s)} = \frac{H}{s^2 + 3s + 3}$$

SYNTHESIS OF TRANSFER FUNCTIONS / 313

Let $\mu = 2$ be the gain of the voltage amplifier. Choose $Q(s) = (s + 1)$. We shall have to express $D(s)/Q(s)$ as the difference of two passive RC driving-point admittances. To this end, we first make a partial fraction expansion of $D(s)/sQ(s)$.

$$\frac{D(s)}{sQ(s)} = \frac{s^2 + 3s + 3}{s(s + 1)} = 1 + \frac{3}{s} - \frac{1}{s + 1}$$

which implies

$$\frac{D(s)}{Q(s)} = (s + 3) - \left(\frac{s}{s + 1}\right)$$

Selecting $k'_0 = \frac{3}{1}$ and $h'_j = 1$, we have $h''_j = h_j + h'_j = 2$. Thus

$$(\mu - 1)Y_2 = Y_2 = 1 + \frac{2s}{s + 1} = \frac{3s + 1}{s + 1}$$

$$y_{22} = s + 3 + \frac{s}{s + 1} = \frac{s^2 + 5s + 3}{s + 1}$$

$$-y_{21} = \frac{H}{s + 1}$$

Since y_{21} as given above has its only transmission zero at $s = \infty$, the RC two-port can be realized by making a continued fraction expansion of y_{22} around infinity:

$$s + 1 \overline{\smash{)}\, s^2 + 5s + 3} \, \big| \, s$$
$$\underline{s^2 + s}$$
$$4s + 3 \overline{\smash{)}\, s + 1} \, \big| \, \tfrac{1}{4}$$
$$\underline{s + \tfrac{3}{4}}$$
$$\tfrac{1}{4} \overline{\smash{)}\, 4s + 3} \, \big| \, 16s$$
$$\underline{4s}$$
$$3 \, \big| \, \tfrac{1}{4} \, \big| \, \tfrac{1}{12}$$
$$\underline{\tfrac{1}{4}}$$
$$0$$

The corresponding realization of the RC two-port is shown in Figure 8-23a. The value of the multiplier constant is found by computing y_{21} at $s = 0$ which yields a value of 3 for H. To account for the output impedance

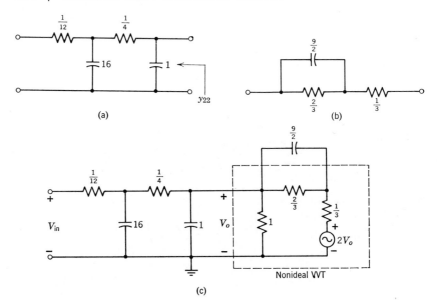

Figure 8-23 Realization of the transfer voltage ratio of Example 8-2, using Kuh's method.

of the VVT, a series resistance must be extracted from Y_2:

$$\frac{1}{Y_2} = \frac{s+1}{3s+1} = \frac{1}{3} + \frac{\frac{2}{3}}{3s+1} = \frac{1}{3} + \frac{1}{\frac{9}{2}s + \frac{3}{2}}$$

Realization of Y_2 is indicated in Figure 8.23b. The complete realization of the transfer voltage ratio is shown in Figure 8.23c, where the dotted box represents a non-ideal VVT.

If the output impedance of the amplifier is small, then V_x can be taken as the output and the network of Figure 8-22 can be cascaded without additional buffer amplifiers. Such a realization of the function of Example 8-2 is shown in Figure 8-24.

It should be noted that both the networks of Figures 8-23c and 8-24 allow for compensation of generator impedance.

The stability of the realized network is of interest. Because Y_1 is chosen to be unity, the characteristic equation is given as

$$y_{22} + 1 - (\mu - 1)Y_2 = 0$$

For the example on hand, the above equation, after some algebra, becomes

$$(s^2 + 9s + 5) - \mu(3s + 1) = 0$$

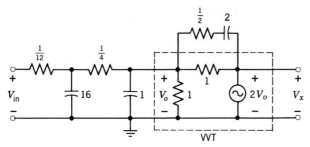

Figure 8-24 An alternate realization of the transfer voltage ratio of Example 8-2.

which indicates that for strict left-half s-plane zeros, i.e., strict stability, the amplifier gain must satisfy the relation

$$\mu < 3$$

Example 8-3 Consider the realization of

$$t_v = \frac{N(s)}{D(s)} = \frac{H(s^2 + 2)}{s^2 + s + 1}$$

This type of function arises in the realization of elliptic-function filter. Choosing $Q(s) = (s + 1)$ and $\mu = 2$ we obtain

$$Y_2 = 1 + \frac{2s}{s+1} = \frac{3s}{s+1}$$

(8.25) $$y_{22} = s + 1 + \frac{s}{s+1} = \frac{s^2 + 3s + 1}{s+1}$$

$$-y_{12} = \frac{H(s^2 + 2)}{s+1}$$

Note that Y_2 is identical to that obtained for in the preceding example. Its realization is thus given by Figure 8-23b.

Realization of the RC two-port poses a problem because of the presence of complex transmission zero. We shall illustrate the realization of the RC two-port by means of Guillemin's technique. In this method, the RC two-port is realized by a parallel connection of several RC ladder networks. For our example we need two such ladders, N_A and N_B. The initial identification of the parameters characterizing N_A and N_B are obtained from Equation 8.25 as:

(8.26a) $$y'_{22A} = \frac{s^2 + 3s + 1}{s+1}, \quad -y'_{12A} = \frac{s^2}{s+1}$$

(8.26b) $$y'_{22B} = \frac{s^2 + 3s + 1}{s+1}, \quad -y'_{12B} = \frac{2}{s+1}$$

316 / SYNTHESIS USING CONTROLLED SOURCES

A two-port is first realized from the set of parameters given by Equation 8.26a. Let us designate this two-port as $\mathcal{N}'_A$. Realization of $\mathcal{N}'_A$ is simply achieved by making a continued fraction expansion of y'_{22A} around origin, since the transmission zeros are all at the origin,

$$\begin{array}{r|l|l}
1+s & 1+3s+s^2 & 1 \\
& \underline{1+s} & \\
& 2s+s^2 & 1+s & \dfrac{1}{2s} \\
& & \underline{1+\dfrac{s}{2}} & \\
& & \dfrac{s}{2} & 2s+s^2 & 4 \\
& & & \underline{2s} & \\
& & & s^2 & \dfrac{s}{2} & \dfrac{1}{2s} \\
& & & & \underline{\dfrac{s}{2}} & \\
& & & & 0 &
\end{array}$$

The realized network $\mathcal{N}'_A$ is shown in Figure 8-25a.

Next, the two-port (which we shall designate as $\mathcal{N}'_B$) defined by the set (8.26b) is realized. Note that in this case the transmission zero is at infinity; hence, $\mathcal{N}'_B$ can be realized by a continued fraction expansion of y'_{22B} around infinity:

$$\begin{array}{r|l|l}
s+1 & s^2+3s+1 & s \\
& \underline{s^2+\ s} & \\
& 2s+1 & s+1 & \tfrac{1}{2} \\
& & \underline{s+\tfrac{1}{2}} & \\
& & \tfrac{1}{2} & 2s+1 & 4s \\
& & & \underline{2s} & \\
& & & 1 & \tfrac{1}{2} & \tfrac{1}{2} \\
& & & & \underline{\tfrac{1}{2}} & \\
& & & & 0 &
\end{array}$$

The realized network $\mathcal{N}'_B$ is sketched in Figure 8-25b.

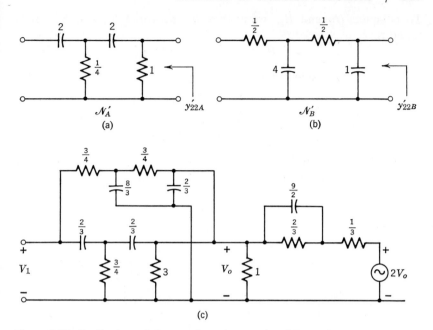

Figure 8-25 Realization of the transfer voltage ratio of Example 8-3, using Kuh's method.

Before we connect N'_A and N'_B in parallel, we shall have to adjust the admittance levels of both networks. Let the multiplier constants of the realized networks be H_A and H_B. We change the admittance level of N'_A by $K\alpha_A/H_A$, where α_A is the desired multiplier constant of N'_A. Similarly, we change the admittance level of N'_B by $K\alpha_B/H_B$ where α_B is the desired multiplier constant of N'_B. Now if the modified networks are connected in parallel, the over all y-parameters are given as:

$$\frac{K\alpha_A}{H_A} y'_{12A} + \frac{K\alpha_B}{H_B} y'_{12B} = Ky_{12} \text{ (prescribed)}$$

$$\frac{K\alpha_A}{H_A} y'_{22A} + \frac{K\alpha_B}{H_B} y'_{22B} = K\left(\frac{\alpha_A}{H_A} + \frac{\alpha_B}{H_B}\right) y_{22} \text{ (prescribed)}$$

It is now clear that K must be chosen to make

(8.27) $$K\left(\frac{\alpha_A}{H_A} + \frac{\alpha_B}{H_B}\right) = 1$$

For our example, it is seen from (8.26)

$$\alpha_A = 1, \quad \alpha_B = 2$$

318 / SYNTHESIS USING CONTROLLED SOURCES

To compute H_A and H_B, we analyze the networks $\mathcal{N}'_A$ and $\mathcal{N}'_B$. It is found that

$$H_A = 1, \qquad H_B = 1$$

Using this information in (8.27) we obtain

$$K = \frac{1}{\dfrac{\alpha_A}{H_A} + \dfrac{\alpha_B}{H_B}} = \frac{1}{1+2} = \frac{1}{3}$$

This implies the admittance level of $\mathcal{N}'_A$ has to be raised by $\frac{1}{3}$ and that of $\mathcal{N}'_B$ has to be increased by $\frac{2}{3}$. The realization of the specified transfer

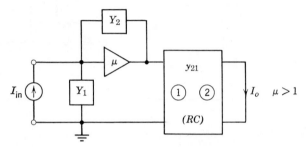

Figure 8-26 Hakim's active RC configuration.

voltage ratio is now obtained by connecting these two networks in parallel and then connecting the voltage amplifier and RC one-port Y_2 (Figure 8-23b) in the way indicated in Figure 8-22. The complete realization is shown in Figure 8-25c.

An interesting observation can be made here. Note that $\mathcal{N}'_B$ could have been obtained from $\mathcal{N}'_A$ by using $RC:CR$ transformation (Section 3-2).

Hakim's Method

The structure proposed by Hakim[16] is obtained by cascading the basic pole section of Figure 8-20a with an RC two-port as indicated in Figure 8-26. The active two-port is suitable for the realization of a current transfer ratio. By inspection, we obtain

$$(8.28) \qquad t_I = \frac{I_o}{I_{in}} = \frac{-\mu y_{21}}{Y_1 - (\mu - 1)Y_2}$$

[16] S. S. Hakim, "RC active filters using an amplifier as the active element," *Proc. IEE (London)*, **112**, 901–912 (May 1965).

SYNTHESIS OF TRANSFER FUNCTIONS / 319

The realization steps are described next. If $N(s)/D(s)$ is the prescribed transfer ratio, we choose a polynomial $Q(s)$ with distinct negative real roots and of degree $Q(s)^0$ so that

$$Q(s)^0 + 1 \geq \max [N(s)^0, D(s)^0]$$

By Theorem 3-5, we can write

(8.29) $$\frac{D(s)}{Q(s)} = Y_{RC}{}^A - Y_{RC}{}^B$$

Comparing the denominator of expression 8.28 with Equation 8.29, we obtain

(8.30) $$Y_1 = Y_{RC}{}^A$$
$$Y_2 = \frac{Y_{RC}{}^B}{\mu - 1}$$

and the RC two-port is described by

(8.31) $$-y_{21} = \frac{N(s)}{Q(s)}$$

Note that in the realization of the RC two-port, a suitable y_{11} (or y_{22}) is selected and then developed along with y_{21} as given above. If the transmission zeros are on the negative real axis, well-known zero-shifting technique can be followed.

Example 8-4 Consider the realization of the following function:

$$t_I = \frac{N(s)}{D(s)} = \frac{H}{s^2 + 3s + 3}$$

Choice of $Q(s) = (s + 1)$ yields the following expansion of $D(s)/Q(s)$;

$$\frac{D(s)}{Q(s)} = (s + 3) - \left(\frac{s}{s + 1}\right)$$

Assuming $\mu = 2$, implies that

$$Y_1 = s + 3$$
$$Y_2 = \frac{s}{s + 1}$$
$$-y_{21} = \frac{H}{s + 1}$$

A suitable y_{22} would be

$$y_{22} = \frac{2s+1}{s+1}$$

Since $y_{22}(\infty)$ is finite, a series resistance at the output can be developed to account for load resistance. The complete network is sketched in Figure 8-27. The two-port is stable for $\mu < 5$.

Note that $Y_1(0)$, being finite and nonzero, guarantees compensation of the input resistance of the amplifier. The suggested configuration does not, however, allow simultaneous compensation of feedback and output resistance of the amplifier. Compared to Kuh's method, Hakim's approach will have fewer passive elements. This is due to the additional terms usually added to y_{22} in the former case to ensure the same poles as y_{21}. An added advantage of the latter synthesis method is in the realization

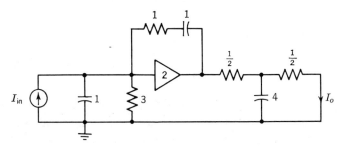

Figure 8-27 Realization of the transfer current ratio of Example 8-4, using Hakim's method.

of the *RC* two-port. Since only y_{21} is specified, for simple functions, the *RC* two-port can be selected from experience to give the desired transmission zeros. The final network can then be analyzed to determine the poles, and hence $Q(s)$.

It is seen that by proper scaling of the *RC* two-port or the *RC* one-port networks, any gain constant can be achieved in Hakim's method. Only restriction of this method is that if a common ground is desired between the input and output, then positive real axis transmission zeros cannot be realized.

The following example illustrates the use of tables in network realization.

Example 8-5 Let us realize

$$t_I = \frac{N(s)}{D(s)} = \frac{4(s^2+2)}{s^2+s+1}$$

by Hakim's method. Note that this function was realized as transfer voltage ratio in Example 8-3. Choosing $\mu = 2$ and $Q(s) = (s + 1)$ we arrive at

(8.32) $$-y_{21} = \frac{2(s^2 + 2)}{s + 1}$$

(8.33) $$Y_A - Y_B = \frac{s^2 + s + 1}{s + 1}$$

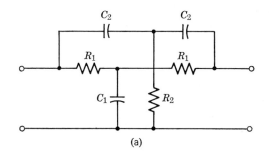

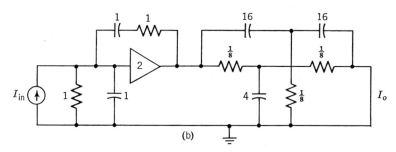

Figure 8-28 Realization of the transfer current ratio of Example 8-5, using Hakim's method.

In Table D-1 (Appendix D), RC networks realizing second-order short-circuit transfer admittances have been tabulated. We observe that the network N_7 of this table, reproduced in Figure 8-28a for convenience, has a y_{21} which is identical in form to that required:

(8.34) $$-y_{21} = \frac{s^2 + a_0}{s + p_1}$$

and the element values are given as:

$$R_1 = \frac{p_1}{2a_0}$$

$$R_2 = \frac{1}{4p_1}$$

$$C_1 = \frac{4a_0}{p_1^2}$$

$$C_2 = 2$$

Comparing (8.32) and (8.34), and noting that the admittance level has to be raised by a factor of 2, we obtain the final element values of the *RC* two-port as:

$$R_1 = \tfrac{1}{8}$$
$$R_2 = \tfrac{1}{8}$$
$$C_1 = 16$$
$$C_2 = 4$$

Y_A and Y_B are easily determined from (8.33). We first make a partial fraction expansion of $(s^2 + s + 1)/s(s + 1)$ which yields

$$\frac{s^2 + s + 1}{s(s + 1)} = 1 + \frac{1}{s} - \frac{1}{s + 1}$$

Thus

$$Y_A - Y_B = \frac{s^2 + s + 1}{s + 1} = (s + 1) - \left(\frac{s}{s + 1}\right)$$

i.e.

$$Y_A = (s + 1); \qquad Y_B = \frac{s}{s + 1}$$

The complete realization is indicated in Figure 8-28b. Due to the presence of the shunt resistor, the input source can be converted into a voltage source having a finite generator impedance.

An alternate connection of the basic circuit and an *RC* two-port, shown in Figure 8-29, permits the use of nonideal amplifiers.[17] For this network, analysis reveals

(8.35) $$\frac{I_o}{I_{in}} = \frac{-y_{21}}{y_{11} + Y_1 - (\mu - 1)Y_2}$$

The synthesis steps are identical to Kuh's method and will be omitted.

---

[17] S. K. Mitra, *loc. cit.*

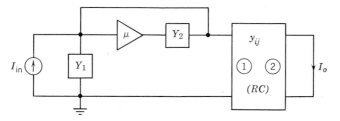

Figure 8-29 An alternate active *RC* configuration for current transfer ratio realization.

The synthesis procedure in this case, like Kuh's and Hakim's method, can realize only transfer functions that have no positive real zeros and are within a multiplicative constant. The disadvantages of the above three methods can be overcome by using an additional active device.

Realization Using Two Amplifiers and *RC* One-Ports

Cascade connection of the basic circuits of Figures 8-20*a* and 8-21*a* in the form of Figure 8-30*a* results in an active *RC* two-port suitable for the

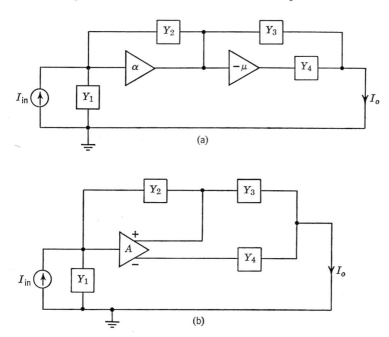

Figure 8-30 An active *RC* configuration employing two voltage amplifiers and four *RC* one-ports for current transfer ratio realization.

324 / SYNTHESIS USING CONTROLLED SOURCES

realization of a current transfer ratio. The pertinent network function is given by

(8.36) $$t_I = \frac{I_o}{I_{in}} = \frac{\alpha(Y_3 - \mu Y_4)}{Y_1 - (\alpha - 1)Y_2}$$

The two amplifiers of Figure 8-30a can be combined into a single amplifier having a differential output; the complete circuit takes the form

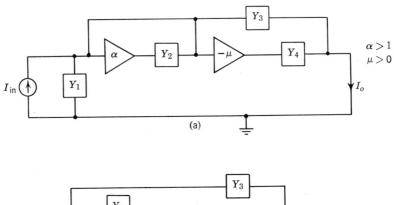

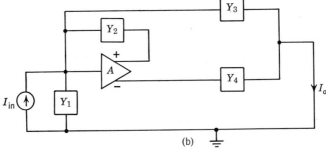

Figure 8-31 An alternate two-amplifier configuration for current transfer ratio realization.

shown in Figure 8-30b. This circuit was advanced by Bobrow[18] and has a current transfer ratio:

(8.37) $$t_I = \frac{A(Y_3 - Y_4)}{Y_1 - (A - 1)Y_2}$$

The realization method is left as an exercise.

By connecting the input terminals of the basic circuits in parallel, an alternate configuration (Figure 8-31a) is obtained,[19] which is characterized

---

[18] L. S. Bobrow, "On active RC synthesis using an operational amplifier," *Proc. IEEE*, **53**, 1648–1649 (October 1965).

[19] S. K. Mitra, *loc. cit.*

by the current transfer ratio

(8.38) $$t_I = \frac{I_o}{I_{in}} = \frac{Y_3 - \mu Y_4}{Y_1 + Y_3 - (\alpha - 1)Y_2}$$

The circuit of Figure 8-31a can be easily modified to use a differential output voltage amplifier. The modified network, indicated in Figure 8-31b, has a current transfer ratio

(8.39) $$t_I = \frac{Y_3 - AY_4}{Y_1 + Y_3 - (A - 1)Y_2}$$

To realize a specified current transfer ratio $N(s)/D(s)$ in the form of Figure 8-31a, we rely on the familiar key. We select a polynomial $Q(s)$ with distinct negative real roots and of degree $Q(s)^0$ so that

$$Q(s)^0 + 1 \geq \max \, [N(s)^0, D(s)^0]$$

By Theorem 3-5, we can express

(8.40)
$$\frac{N(s)}{Q(s)} = Y_{RC}{}^A - Y_{RC}{}^B$$
$$\frac{D(s)}{Q(s)} = Y_{RC}{}^C - Y_{RC}{}^D$$

Comparing Equations 8.38 and 8.40, we can make the following identifications:

(8.41)
$$Y_3 = Y_{RC}{}^A$$
$$Y_4 = \frac{Y_{RC}{}^B}{\mu}$$
$$Y_1 = Y_{RC}{}^C$$
$$Y_2 = \frac{1}{\alpha - 1}[Y_{RC}{}^D + Y_{RC}{}^A]$$

By choosing suitable $Q(s)$ and, if necessary, by adding proper constants to $Y_{RC}{}^A$, $Y_{RC}{}^B$ and to $Y_{RC}{}^C$, $Y_{RC}{}^D$, the amplifiers can be compensated for their parasitics.

For a voltage transfer ratio realization, the active two-port of Figure 8-32 can be used.[20] The structure has been obtained by cascading the basic

---

[20] This active *RC* configuration has been independently proposed by S. K. Mitra, *loc. cit.*; R. J. Paul, "Active network synthesis using one-port *RC* networks," *Proc. IEE (London)*, **113**, 83–88 (January 1966); R. E. Cooper and C. O. Harbourt, "Sensitivity reduction in amplifier-*RC* synthesis," *Proc. IEEE*, **54**, 1578 (November 1966).

326 / SYNTHESIS USING CONTROLLED SOURCES

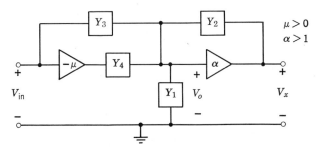

Figure 8-32 A two-amplifier configuration for voltage transfer ratio realization.

circuits in opposite order to that discussed above. The pertinent network function is given as

$$(8.42) \quad t_v = \frac{V_o}{V_{in}} = \frac{Y_3 - \mu Y_4}{Y_1 + Y_3 + Y_4 - (\alpha - 1)Y_2}$$

The synthesis steps are similar to the previous method. We shall illustrate it by means of an example.

Example 8-6 Realize

$$t_v = \frac{N(s)}{D(s)} = \frac{s^2 - s + 1}{s^2 + s + 1}$$

Choice of $Q(s) = (s + 1)$ yields

$$(8.43) \quad t_v = \frac{\dfrac{N(s)}{Q(s)}}{\dfrac{D(s)}{Q(s)}} = \frac{(s+1) - \left(\dfrac{3s}{s+1}\right)}{(s+1) - \left(\dfrac{s}{s+1}\right)}$$

Let us make $\mu = 1$ and $\alpha = 2$ without any loss of generality. Comparing expressions 8.42 and 8.43, we identify

$$Y_3 = s + 1$$

$$Y_4 = \frac{3s}{s+1}$$

$$Y_1 = 0$$

$$Y_2 = \frac{4s}{s+1}$$

The above choice does not allow compensation for input impedance of the noninverting amplifier. To enable compensation and also to account for a load impedance, let us make Y_1 equal to a constant and add the same constant to Y_2. We express

$$Y_1 = 2$$

$$Y_2 = \frac{4s}{s+1} + 2 = \frac{6s+2}{s+1}$$

The final realization is shown in Figure 8-33 where the dotted boxes

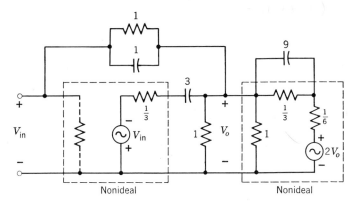

Figure 8-33 Two-amplifier realization of the voltage transfer ratio of Example 8-6.

represent nonideal amplifiers. If the input source V_{in} is a low-impedance source, the input impedance of the inverting amplifier will have no effect on the overall transfer function.

All active configurations of this section have several attractive features. The poles and zeros can be controlled independently by changing the gains of the amplifiers. The synthesis steps are simple since the "companion" RC networks are one-ports. None of the methods imposes any restriction on the transfer function. In addition, the last method is suitable for cascading without buffer stages if a noninverting voltage amplifier with very low output impedance is available.

Other Methods

There are several other synthesis procedures that employ the controlled source as the active element. Most of the methods are restricted to second-order transfer functions. Size of the book limits the discussion of all known methods. We refer the interested reader to the list of references at the end of this chapter.

In passing, we would like to mention two general methods of transfer-function synthesis. The synthesis method proposed by Khazanov[21] is based on a structure containing two RC two-ports and a single nonideal current amplifier. For a higher-order function, a computer may be needed to decompose the denominator polynomial suitably for identification of "companion" RC network parameters. Another general method is the one due to Hazony and Joseph.[22] They begin by considering a general active RC configuration employing an ideal finite gain VVT. From the specified transfer function, the transfer voltage ratio vector characterizing the three-terminal grounded RC network is first obtained and then is synthesized by following the multiterminal network synthesis procedure.

8-4 NETWORK DESIGN BY COEFFICIENT MATCHING

There are several realization procedures that are based on the coefficient matching technique. Usually these methods begin with a network of given topology, which are then analyzed to determine the type of transfer functions they can realize. Design equations are then developed by equating like coefficients.

The active element used is usually a noninverting type voltage amplifier of gain greater than unity. For convenience, in cascading without additional buffer amplifiers, the amplifier output terminal is also taken as the output of the filter. This causes the active filter to have low output impedance. Since higher-order transfer functions can be realized by cascading second-order stages, design methods in general are developed for the second-order transfer functions.

The most significant work in this direction has been done by Sallen and Key,[23] who have tabulated many such second-order networks. In this section, we shall present design equations for a low-pass, a high-pass, and a band-pass second-order active filter. In addition, we shall consider the active RC design of a second-order voltage transfer ratio having a pair of $j\omega$-axis transmission zeros. Finally, a novel method of designing an active filter having all transmission zeros either at origin or at infinity will be introduced.

---

[21] G. L. Khazanov, "Synthesis of selective amplifiers with negative feedback," *Telecomm and Radio Engg*, (Part 1), No. 12, p. 37 (1961).

[22] D. Hazony and R. D. Joseph, "Transfer matrix synthesis with one amplifier but no inductors," *Proc. IEEE*, **52**, 1748 (December 1964).

[23] R. P. Sallen and E. L. Key, "A practical method of designing RC active filters," *IRE Trans. on Circuit Theory*, **CT-2**, 74–85 (March 1955).

A Second-Order Low-Pass Filter Section

The active RC configuration of Figure 8-34a was originally proposed by Sallen and Key[24] for realizing low-pass filter section. By analysis we obtain for this structure:

(8.44) $$t_v = \frac{V_2}{V_1} = \frac{K \dfrac{G_1 G_2}{C_1 C_2}}{s^2 + \left[\dfrac{G_2}{C_2}(1-K) + \dfrac{G_1}{C_1} + \dfrac{G_2}{C_1}\right]s + \dfrac{G_1 G_2}{C_1 C_2}}$$

Let

(8.45) $$t_v = \frac{H}{s^2 + \beta s + \gamma}$$

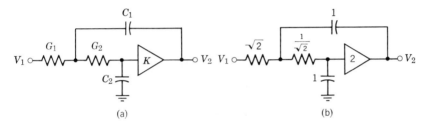

Figure 8-34 (a) A second-order low-pass active RC filter section. (b) Realization of the transfer function of Example 8-7. Element values in mhos and farads.

be the specified voltage transfer ratio. Equating like coefficients of Equations 8.44 and 8.45 we obtain:

(8.46)
$$H = K \frac{G_1 G_2}{C_1 C_2}$$
$$\beta = \frac{G_2}{C_2}(1-K) + \frac{G_1}{C_1} + \frac{G_2}{C_1}$$
$$\gamma = \frac{G_1 G_2}{C_1 C_2}$$

Note from (8.46) that the number of unknown quantities is greater than the number of equations. One simple way to solve (8.46) is to preselect some element values and then determine the remaining unknown quantities. For example, one possible choice would be

(8.47)
$$C_1 = C_2 = 1$$
$$K = 2$$

[24] R. P. Sallen and E. L. Key, *loc. cit.*

Substituting (8.47) in (8.46) we obtain easily:

(8.48)
$$G_1 = \beta$$
$$G_2 = \frac{\gamma}{\beta}$$
$$H = 2\gamma$$

The following example will illustrate the procedure.

Example 8-7 Let us realize a second-order low pass Butterworth filter.

$$t_v = \frac{H}{s^2 + \sqrt{2}s + 1}$$

This implies

$$\beta = \sqrt{2}$$
$$\gamma = 1$$

Using this information in (8.48) and (8.47) we obtain

$$G_1 = \sqrt{2}$$
$$G_2 = \frac{1}{\sqrt{2}}$$
$$H = 2$$
$$C_1 = C_2 = 1$$
$$K = 2$$

The complete realization is sketched in Figure 8-34b.

A Second-Order High-Pass Filter Section

A second-order high-pass active *RC* filter advanced by Sallen and Key[25] is indicated in Figure 8-35. A design procedure for this filter can be

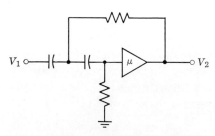

Figure 8-35 A second-order high-pass active *RC* filter section.

[25] R. P. Sallen and E. L. Key, *loc. cit.*

NETWORK DESIGN BY COEFFICIENT MATCHING / 331

easily developed in a manner similar to the one described for the low-pass section.

An alternate approach based on the use of $RC:CR$ transformation (Section 3-2) is illustrated by means of an example.

Example 8-8 Realize the high-pass transfer voltage ratio:

(8.49) $$t_v(s) = \frac{H_1 s^2}{s^2 + 2s + 3}$$

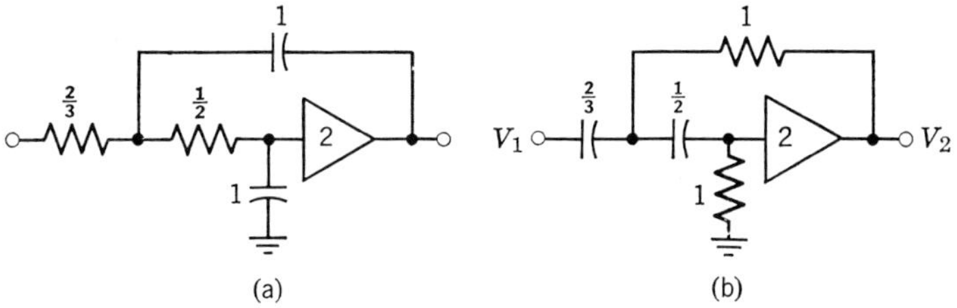

Figure 8-36 Steps in the realization of the high pass transfer function of Example 8-8. Element values in mhos and farads.

Use of high-pass to low-pass transformation

$$s \to \frac{1}{s}$$

on the function of Equation 8.49 yields the following:

(8.50) $$t'_v(s) = \frac{\frac{H_1}{3}}{s^2 + \frac{2}{3}s + \frac{1}{3}}$$

Using the design equations (8.47) and (8.48) we obtain the following element values for the realization of (8.50):

$$C_1 = C_2 = 1$$
$$K = 2$$
$$G_1 = \tfrac{2}{3}$$
$$G_2 = \tfrac{1}{2}$$
$$\frac{H_1}{3} = \frac{2}{3}$$

The realization of (8.50) is indicated in Figure 8-36a. Making an $RC:CR$ transformation on the network of Figure 8-36a yields the desired realization of the function (8.49) as shown in Figure 8.36b.

A Second-Order Band-Pass Filter Section

The second-order transfer function having a band-pass type response is of the form:

$$(8.51) \qquad t_v = \frac{Hs}{s^2 + \beta s + \gamma}$$

A possible active RC network realization of this type of function is indicated in Figure 8-37a.[26] Routine analysis yields for this two-port:

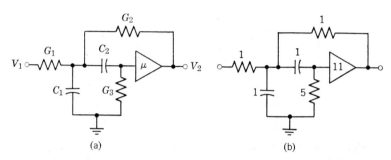

Figure 8-37 (a) A second-order band-pass active RC filter section. (b) Realization of the transfer function of Example 8-9. Element values in mhos and farads.

$$(8.52) \qquad t_v = \frac{V_2}{V_1} = \frac{\mu \dfrac{G_1}{C_1} s}{s^2 + \left(\dfrac{G_1}{C_1} + \dfrac{G_2}{C_1} + \dfrac{G_3}{C_2} + \dfrac{G_3}{C_1} - \mu \dfrac{G_2}{C_1}\right)s + \dfrac{G_3(G_1 + G_2)}{C_1 C_2}}$$

Equating like coefficients of (8.51) and (8.52) we obtain

$$H = \mu G_1 / C_1$$

$$(8.53) \qquad \beta = \frac{G_1}{C_1} + \frac{G_2}{C_1} + \frac{G_3}{C_2} + \frac{G_3}{C_1} - \mu \frac{G_2}{C_1}$$

$$\gamma = \frac{G_3(G_1 + G_2)}{C_1 C_2}$$

Since the number of unknown parameters is more than the number of equations, Equation 8.53 can be conveniently solved by preselecting a number of elements. One such choice will be

$$(8.54) \qquad \begin{aligned} G_1 &= G_2 = 1 \\ C_1 &= C_2 = 1 \end{aligned}$$

[26] W. J. Kerwin and L. P. Huelsman, "The design of high performance active RC band-pass filters," *IEEE Int'l Conv. Rec.*, **14** (Part 10), 74-80 (1960).

NETWORK DESIGN BY COEFFICIENT MATCHING / 333

which when used in Equation 8.53 yields:
$$G_3 = \gamma/2$$
(8.55)
$$\mu = 2 + \gamma - \beta$$
$$H = \mu$$

Note that for complex pole-pair $\beta^2 < 4\gamma$ and hence μ is always positive.

Example 8-9 Realize
$$t_v = \frac{Hs}{s^2 + s + 10}$$

We have for this example
$$\beta = 1$$
$$\gamma = 10$$

Therefore from Equations 8.55 and 8.54 we obtain
$$G_3 = 5$$
$$\mu = 11 = H$$
$$G_1 = G_2 = C_1 = C_2 = 1$$

The complete realization is sketched in Figure 8-37b.

A Second-Order Filter Section Having a Pair of $j\omega$-Axis Zeros

In the design of elliptic-function active RC filter, a basic filter section needed is of the form,

(8.56)
$$t_v = \frac{V_2}{V_1} = \frac{H(s^2 + \alpha)}{s^2 + \beta s + \gamma}$$

Realization of this type of transfer voltage ratio is considered here. There are three cases of interest:

(8.57)
$$\text{Case 1.} \quad \gamma > \alpha$$
$$\text{Case 2.} \quad \gamma < \alpha$$
$$\text{Case 3.} \quad \gamma = \alpha$$

We consider first Case 1. Realization of the function 8.56 with $\gamma > \alpha$ can be achieved by means of the active RC configuration shown in Figure 8-38.[27] Analysis of this active two-port yields

(8.58) $$t_v = \frac{V_2}{V_1} = \frac{\mu(s^2 + a^2)}{s^2 + (k+1)a\left(\frac{1}{R} + \frac{2-\mu}{k}\right)s + \left(1 + \frac{k+1}{R}\right)a^2}$$

[27] W. J. Kerwin, *loc. cit.*

334 / SYNTHESIS USING CONTROLLED SOURCES

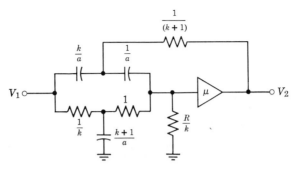

Figure 8-38 A second-order active filter section having a pair of $j\omega$-axis transmission zeros.

Comparing (8.58) and (8.56) and equating like coefficients we obtain

$$H = \mu$$
$$\alpha = a^2$$

(8.59)
$$\gamma = \left(1 + \frac{k+1}{R}\right)a^2$$

$$\beta = (k+1)a\left(\frac{1}{R} + \frac{2-\mu}{k}\right)$$

The above equations can be solved to yield the unknown parameters:

$$a = \sqrt{\alpha}$$

(8.60)
$$R = \frac{(k+1)}{\left(\dfrac{\gamma}{\alpha} - 1\right)}$$

$$\mu = 2 - k\left[\frac{\beta}{(k+1)a} - \frac{1}{R}\right]$$

$$H = \mu$$

Note that the value of k can be chosen arbitrarily to minimize either the element spread of the element value or minimize the sensitivity of the filter. A convenient design criteria would be to minimize the sensitivity function S_μ^β. For practical purposes, a compromise between the sensitivity and element spread has to be made.

We now consider the second case ($\gamma < \alpha$). An active RC configuration and its associated design procedure for this case has been proposed by Kerwin. We illustrate an alternate approach to this problem by an example.

Example 8-10 Consider the realization of

(8.61)
$$t_v = \frac{H_1(s^2 + 4)}{s^2 + 2s + 3}$$

Making a low-pass to high-pass transformation on this function we arrive at

(8.62)
$$t'_v = \frac{H_1\left(\frac{1}{s^2} + 4\right)}{\frac{1}{s^2} + \frac{2}{s} + 3} = \frac{\frac{4H_1}{3}(s^2 + \frac{1}{4})}{s^2 + \frac{2}{3}s + \frac{1}{3}}$$

Realization of the above function is next carried out following the technique outlined above. We have

(8.63)
$$H = \frac{4H_1}{3}$$

$$\alpha = \tfrac{1}{4}$$

$$\beta = \tfrac{2}{3}$$

$$\gamma = \tfrac{1}{3}$$

Using this in Equation 8.60 we obtain

$$a = \tfrac{1}{2}$$

$$R = 3(k + 1)$$

$$\mu = 2 - \frac{k}{k+1}$$

$$H_1 = \frac{3H}{4} = \frac{3\mu}{4}$$

Letting $k = 1$, we then arrive at the active RC network realization of function 8.62 as shown in Figure 8-39a. The realization of the original function (8.61) is finally obtained by applying a $RC:CR$ transformation on the network of Figure 8-39a. The resultant network is indicated in Figure 8-39b.

Development of a realization procedure for case 3 is left as an exercise (Problem 8.12).

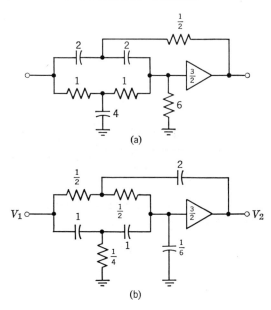

Figure 8-39 Steps in the realization of the transfer function of Example 8-10. Element values in ohms and farads.

An Active *RC*-Chain Network

Bach[28] recently suggested a novel method for the realization of a restricted class of voltage transfer ratios, using unity gain voltage amplifiers as the active device.

Figure 8-40 shows the proposed circuit for various orders of all-pole transfer functions. Generalizations of the chain-type network (Figure 8-40) to higher-order transfer functions should be evident. Simple calculation yields the following transfer functions:

For the circuit of Figure 8-40*a*:

$$\frac{V_2}{V_1} = \frac{\omega_1}{s + \omega_1} \tag{8.64}$$

For the circuit of Figure 8-40*b*:

$$\frac{V_2}{V_1} = \frac{\omega_1 \omega_2}{s^2 + \omega_1 s + \omega_1 \omega_2} \tag{8.65}$$

---

[28] R. E. Bach, "Selecting *R-C* values for active filters," *Electronics*, **33**, 82–85 (May 13, 1960).

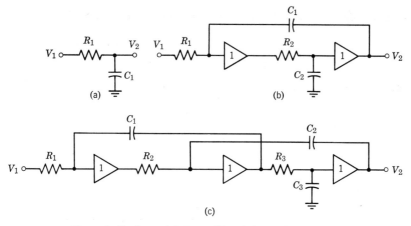

Figure 8-40 Several RC:amplifier chain configurations.

For the circuit of Figure 8-40c:

$$(8.66) \qquad \frac{V_2}{V_1} = \frac{\omega_1\omega_2\omega_3}{s^3 + \omega_1 s^2 + \omega_1\omega_2 s + \omega_1\omega_2\omega_3}$$

In Equations 8.64 through 8.66,

$$(8.67) \qquad \omega_k = \frac{1}{R_k C_k} \qquad k = 1, 2, 3$$

For the nth order chain network, the transfer voltage ratio will be given by

$$(8.68) \qquad \frac{V_2}{V_1} = \frac{\prod_{i=1}^{n} \omega_i}{s^n + \omega_1 s^{n-1} + \omega_1\omega_2 s^{n-2} + \cdots + \prod_{i=1}^{n} \omega_i}$$

An arbitrary transfer voltage ratio of the form

$$(8.69) \qquad \frac{V_2}{V_1} = \frac{H}{s^n + a_1 s^{n-1} + a_2 s^{n-2} + \cdots + a_n}$$

can be compared with Equation 8.68 to yield

$$(8.70) \qquad a_k = \prod_{i=1}^{k} \omega_i \qquad k = 1, 2, \ldots, n$$

Finally, the above equations are solved for the element values of the passive RC network. The following example will illustrate the above technique.

338 / SYNTHESIS USING CONTROLLED SOURCES

Example 8-11 Let us realize a third-order Butterworth filter:

(8.71) $$\frac{V_2}{V_1} = \frac{H}{(s+1)(s^2+s+1)} = \frac{H}{s^3+2s^2+2s+1}$$

Comparing Equations 8.71 and 8.66 we obtain:

$$\omega_1 = 2$$
$$\omega_1\omega_2 = 2$$
$$\omega_1\omega_2\omega_3 = 1$$

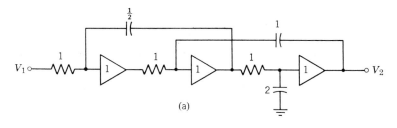

(a)

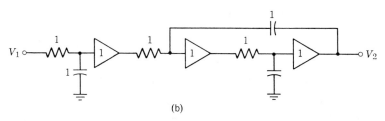

(b)

Figure 8-41 Realization of the transfer function of Example 8-11. Element values in ohms and farads.

This implies that

(8.72)
$$\omega_1 = \frac{1}{R_1C_1} = 2$$
$$\omega_2 = \frac{1}{R_2C_2} = 1$$
$$\omega_3 = \frac{1}{R_3C_3} = \frac{1}{2}$$

Note from Equation 8.72 that much freedom exists in selecting the values of the resistances and capacitances. If we desire that all the resistances have identical value of unity, then the final network will be of the form of Figure 8-41a. The spread of the capacitor values is seen to be about four to one.

The same filter can be designed using a cascade of first-order and second-order filter sections. If we cascade the networks of Figures 8-40a and 8-40b, the over-all transfer function is given by the product of Equations 8.64 and 8.65:

$$(8.73) \quad \frac{V_2}{V_1} = \frac{\omega_1' \omega_1'' \omega_2}{(s + \omega_1')(s^2 + \omega_1'' s + \omega_1'' \omega_2)}$$

Comparing Equation 8.73 with the factored form of desired transfer function as given by Equation 8.71, we obtain:

$$\omega_1' = 1$$
$$\omega_1'' = 1$$
$$\omega_1'' \omega_2 = 1$$

Setting all resistors to have the value unity, the final network is as shown in Figure 8-41b. Note that in the alternate network, all the capacitors also have identical value, which is definitely attractive to the designer.

8-5 SENSITIVITY CONSIDERATIONS

The sensitivity problem in active RC filters is a serious problem. In most active filters employing controlled sources as the active elements, the Q-sensitivity is significantly large. In fact, we have shown in Section 5.7, p. 198, that if the complex poles are generated as the difference of the two polynomials, the Q-sensitivity is usually greater than $2Q$. This high-sensitivity feature limits the use of most of the synthesis techniques outlined in this chapter to the realization of network functions having low Q poles (Qs of the order of 20 or less).

For synthesis techniques using the RC:$-RC$ type decomposition, there exists a decomposition which minimizes the pole sensitivity. Development of this decomposition is considered in the next chapter. It should be noted that the lowest Q-sensitivity obtainable is $2Q$.

The restriction of active RC network using a controlled source to low-selectivity filter realization can be eliminated to a certain extent by using multiloop active filter configuration containing more than one controlled source. Such an approach has made possible the design of high-selectivity filters (Q of the order of 300) as reported recently by Moschytz.[28] Absolutely stable multiloop active RC configuration has also been extensively studied by Lee and Wyndrum.[29]

8-6 SUMMARY

Several transistor-circuit realizations of the four types of controlled sources are developed in Section 8-1, p. 293, from their nullator-norator representations constructed in Section 8-1, p. 291.

A driving-point synthesis method employing two controlled sources is outlined in Section 8-2, p. 305.

Several transfer function synthesis techniques are developed in Section 8-3 by making use of the basic "pole-section" and the "zero section" (Figures 8-20 and 8-21).

One such method is the Kuh's method (Section 8-3, p. 311) which is based on the active RC configuration of Figure 8-22 and basically realizes a transfer voltage ratio. A major attraction of Kuh's method is the possibility of using a nonideal noninverting type VVT having finite input, output, and feedback impedances. There are essentially three drawbacks of Kuh's method: (1) realization of complex transmission zero pose a somewhat difficult problem, (2) transfer function is always realized within a multiplicative constant, (3) positive real transmission zeros cannot be realized if a common ground is desired between the input and output ports. The active RC configuration of Figure 8-29, which is suitable for the realization of current transfer ratio, has all the features of Kuh's method including the disadvantages.

Hakim's method (Section 8-3, p. 318) is primarily for current transfer ratio synthesis and is based on Figure 8-26. Hakim's method is somewhat simpler than Kuh's method but has all the drawbacks except the second one. In addition, it is not possible to compensate the VVT for the output and feedback impedances. One particular advantage this second method has is in the realization of second-order transfer functions. In this case, the RC two-port can be easily realized by using Table D-1 of Appendix D.

Most of the disadvantages are not inherent in the two-amplifier synthesis techniques outlined in Section 8-3, p. 323. Any real rational transfer function can be realized by the methods of this section, which also allows

---

[28] G. S. Moschytz, "Miniaturized filter building blocks using frequency emphasizing networks," *Proc. N.E.C.*, **23**, 364–369 (1967).

[29] S. C. Lee and R. W. Wyndrum, Jr., "Absolutely stable multi-loop active *R-C* synthesis," Proc. Asilomar Conf. on Circuits and Systems, Monterey, California, November 1967.

independent adjustment of the poles and zeros. In most cases, nonideal voltage amplifiers can be used.

Several design procedures using the coefficient matching techniques are reviewed in Section 8-4. In particular, realization of second-order low-pass, high-pass, band-pass transfer functions are discussed along with the design of second-order transfer functions having $j\omega$-axis zeros.

Additional References

Armstrong, D. B., and F. M. Reza, "Synthesis of transfer functions by active RC networks with feedback loop," *IRE Trans. on Circuit Theory*, **CT-1**, 8–17 (June 1954).

Bachman, A. E., "Transistor active filters using twin-T rejection circuits," *Proc. IEE (London)*, **106B**, 170–174 (1959).

Balabanian, N., and C. I. Clinkilic, "Expansion of an active synthesis technique," *IEEE Trans. on Circuit Theory*, **CT-10**, 290–298 (June 1963).

Balabanian, N., and B. Patel, "Active realization of complex zeros," *IEEE Trans. on Circuit Theory*, **CT-10**, 299–300 (June 1963).

Balabanian, N., "Active network synthesis with a non-ideal device," *J. Franklin Inst.*, **279**, 334–346 (May 1965).

Bangert, J. T., "The transistor as a network element," *Bell System Tech. J.*, **33**, 329–352 (March 1954).

Blecher, F. H., "Design principles for single loop transistor feedback amplifiers," *IRE Trans. on Circuit Theory*, **CT-4**, 145–157 (September 1957).

Bobrow, L. S., and L. S. Hakimi, "A note on active RC realization of voltage transfer functions," *IEEE Trans. on Circuit Theory*, **CT-11**, 493–494 (December 1964).

Bodner, H. A., and S. K. Mitra, "Active all-pass network," *Electronics Letters*, **1**, 98 (June 1965).

Bogdanov, M., R. Boite, and H. Leich, "Synthese des filteres actifs a caracteristique elliptique," *Revue Haut Frequence (Belgique)*, **6**, 245–258 (1965).

Bongiorno, J. J., "Synthesis of active RC single-tuned band-pass filters," *IRE Conv. Record*, **6** (Part 2), 30–41 (1958).

Boyce, A. H., "A theoretical and practical study of active filters," *Marconi Review*, 2nd Quarter, pp. 68–97 (1967).

Calfee, R. W., "An active network equivalent to constant-resistance lattice with delay circuit applications," *IEEE Trans. on Circuit Theory*, **CT-10**, 532 (December 1963)

Chirlian, P. M., and V. A. Marsocci, "The controlled superconductor as a linear amplifier," *IRE Trans. on Component Parts*, **CP-8**, 84–88 (June 1961).

Chirlian, P. M., and V. A. Marsocci, "A cryotron linear amplifier," *IRE Trans. on Component Parts*, **CP-10**, 144–146 (December 1963).

Cimagalli, V., "Active synthesis of one-port and two-port network functions with zero attenuation or zero phase," *IEEE Trans. on Circuit Theory*, **CT-12**, 440–441 (September 1965).

Corza, J. E., "Synthesis of RC active networks using transistor as the active element," AIEE District Conference Paper, Paper No. DP 62–991, May 15, 1962.

Cowley, P. E. A., "An active filter for the measurement of process dynamics," *Proc. 19th Annual ISA Conf.*, New York, October 1964.

Crombie, D. D., "The design of low-frequency high-pass *RC* filters," *Electronic Engg*, **28**, 254–256 (1956).

Dietzold, R. L., "Frequency discriminative electric transducer," U.S. Patent No. 2,549,065, April 1951.

Franks, L. E., and I. W. Sandberg, "An alternative approach to the realization of network transfer functions: the N-Path filter," *Bell System Tech. J.*, **39**, 1321–1350 (September 1960).

Franklin, D. P., "Active low-pass *RC* filters," *Electron. Technol.*, **38**, 278–282 (1961).

Fryer, W. D., "How to design low cost audio filters," *Electronics*, **32**, 68–70 (April 10, 1959).

Ganguly, U. S., "An inductorless all-pass phase shifter," *Proc. IEEE*, **54**, 1462–1463 (October 1966).

Ghausi, M. S., "Optimum design of the shunt-series feedback pair with a maximally flat magnitude response," *IRE Trans. on Circuit Theory*, **CT-8**, 448–453 (December 1961).

Ghausi, M. S., and D. O. Pederson, "A new design approach for feedback amplifiers," *IRE Trans. on Circuit Theory*, **CT-9**, 274–284 (September 1962).

Golembeski, J. J., M. S. Ghausi, J. H. Mulligan, Jr., and S. S. Shamis, "A class of minimum sensitivity amplifiers," *IEEE Trans. on Circuit Theory*, **CT-14**, 69–74 (March 1967).

Hakim, S. S., "Synthesis of *RC* active filters with prescribed pole sensitivity," *Proc. IEE (London)*, **112**, 2235–2242 (1965).

Hakim, S. S., "Synthesis of *RC* active filters using common base and common-collector configurations," *Proc. IEE (London)*, **113**, 788–790 (May 1966).

Hakim, S. S., "A synthesis procedure of *RC*-active filters using unity gain current and voltage amplifiers," *Int. J. of Control*, **3**, 553–564 (1966).

Hazony, D., "Grounded *RC*-unity gain amplifier transfer vector synthesis," *IEEE Trans. on Circuit Theory*, **CT-14**, 75–76 (March 1967).

Hogin, J. L., "Active *RC* networks utilizing the voltage follower," Ph.D. Dissertation, Montana State University, Bozeman, Montana, March 1966.

Holt, A. G. J., and R. Linggard, "*RC* active synthesis procedure for polynomial filters," *Proc. IEE (London)*, **113**, 777–782 (May 1966).

Holt, A. G. J., and R. Linggard "Active Chebyshev filters," *Electronics Letters*, **1**, 130–131 (1965).

Horowitz, I. M., "*RC*-transistor network synthesis," *Proc. NEC*, **12**, 818–829 (1956).

Horowitz, I. M., "Active network synthesis," *IRE Conv. Record*, **4** (Part 2), 38–45 (March 1956).

Horowitz, I. M., "Exact design of transistor *RC* band-pass filters with prescribed active parameter insensitivity," *IRE Trans. on Circuit Theory*, **CT-7**, 313–320 (September 1960).

Horowitz, I. M., "Optimum design of single stage gyrator-*RC* filters with prescribed sensitivity," *IRE Trans. on Circuit Theory*, **CT-8**, 85–94 (June 1961).

Horowitz, I. M., "Active *RC* synthesis," *IEEE Trans. on Circuit Theory*, **CT-13**, 101–102 (March 1966).

Humpherys, D. S., "Active crystal filters," Electro-technology, **78**, 43–47 (July, 1966).

Jagoda, N. H., "An active realization of the elliptic function approximation," *IRE Trans. on Circuit Theory*, **CT-9**, 423 (December 1962).

Joseph, R. D., and D. Hilberman, "Immittance matrix synthesis with active networks," *IEEE Trans. on Circuit Theory*, **CT-13**, 324 (September 1966).

MacDonald, J. R., "Active band-pass filter has sharp cutoff," *Electronics*, **31**, 84–87 (August 15, 1958).

MacDonald, J. R., "Active adjustable audio band-pass filter," *J. Acc. Soc. Am.*, **29**, 1348 (December 1957).

McCamey, R. E., and G. J. Thaler, "Design of some active compensators of feedback controls," *IEEE Trans. on Applications and Industry*, **82**, 84–89, May 1963.

McVey, P. J. W., "An active *RC* filter using cathode-followers," *Electronic Eng.*, **34**, 458 (1962).

Margolis, S. G., "On the design of active filters with Butterworth characteristics," *IRE Trans. on Circuit Theory*, **CT-3**, 202 (September 1956).

Markarian, B. K., "Network partitioning techniques applied to the synthesis of transistor amplifiers," *IRE Conv. Record*, **2** (Part 2), 130–134 (March 1954).

Maupin, J. T., "Constant resistance transistor stages," *IRE Trans on Circuit Theory*, **CT-8**, 480–481 (December 1961).

Moschytz, G. S. "Miniaturized *RC* filters using phase-locked loop," *Bell System Tech. J.*, 44, 823–870 (May–June 1965).

Moschytz, G. S., "Sallen and key filter networks with amplifier gain larger than or equal to unity," *IEEE J. Solid-State Circuits*, **SC-2**, 114–116 (September 1967).

Moschytz, G. S. "Active *RC* filter building blocks using frequency emphasizing networks," *IEEE J. Solid-state Circuits*, **SC-2**, 59–62 (June 1967).

Myers, B. R., "Transistor *RC* network synthesis," *IRE WESCON Conv. Record*, **3** (Part 2), 65–74 (August 1959).

Nai-tuing, T'eng, "Synthesis of electrical filters from cascades of *RC* sections and one amplifier with *RC* feedback," *Telecomm. and Radio Engg.*, Part 2, No. 3, p. 33 (1962).

Nai-tuing, T'eng, "Synthesis of electrical filters with two section *RC* circuit," *Telecomm. and Radio Engg.*, Part 1, No. 4, p. 47 (1963).

Piercey, R. N. G., "Synthesis of active *RC* filter networks," *A.T.E.J.*, **21**, 61–75 (April 1965).

Scott, H. H., "A new type of selective circuit and some applications," *Proc. IRE*, **26**, 226–235 (February 1938).

Scott, L., "Criteria for the design of active filters using resistance and capacitance elements in feedback circuits," *Solid-state Electronics*, **9**, 641–651 (1966).

Shorabji, N., "*RC* filters and oscillators using junction transistors," *Electronic Engg.*, **29**, 606 (1957).

Shumard, C. C., "Design of high-pass, low-pass and band-pass filters using *RC* networks and direct current amplifiers with feedback," *RCA Rev.*, **11**, 534 (1950).

Slavskii, G. N., "Active low-pass *RC* filters using transistors," *Telecomm. Radio Engg.*, **18** (Part 2) **12**, 47–56 (1963).

Smith, K. D., "An active Bessel-Butterworth filter," *IEEE Int'l. Conv. Record*, **12** (Part 8) 1–11 (March 1964).

Taylor, P. L., "Flexible design method for active *RC* two-ports," *Proc. IEE (London)*, **110**, 1607–1616 (September 1963).

Temby, A. C., "Active filter synthesis using feedback blocks," *Proc. IRE (Australia)*, **24**, 631–638 (1963).

Thiele, A. N., "The design of filters using only *RC* sections and gain stages," *Electronic Engg*, **28**, 31–36 and 80–82 (1956).

Thorp, J., "Realization of variable active networks," *IEEE Trans. on Circuit Theory*, **CT-12**, 511–514 (December 1965).

Waldhauer, F. D., "Wide-band feedback amplifiers," *IRE Trans. on Circuit Theory*, **CT-4**, 178–190 (September 1957).

Wilson, M. G., "Low-pass amplifier with adjustable bandwidth," *Electronics*, 90–91 (May 30, 1966).

Woroncow A., and J. Croney, "Inductorless band-pass I.F. amplifiers," *Radio and Electronic Engr.*, **33**, 184–186 (March 1967).

Zai Y. F., "RC–active filters using unity–gain amplifiers," *Electronics Letters* **3**, 461 (October 1967).

Problems

8.1 Realize the following driving-point impedances using Sandberg's approach outlined in Section 8-2, p. 305.

(a) s

(b) $s^2 + s + 1$

(c) $\dfrac{s^2 + s + 2}{2s^2 + s + 1}$

8.2 A variation of the basic circuit of Figure 8-21a is shown in Figure 8-42.

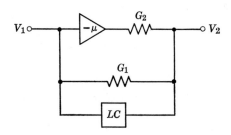

Figure 8-42

(a) Determine the condition under which the voltage transfer ratio is an all-pass function.[30]

(b) Realize the circuit using a single transistor.

(c) Develop a suitable synthesis method.

8.3 Which of the following transfer functions are all-pass functions?

---

[30] Several versions of this basic circuit have been reported in the literature. See, for example: H. Bodner and S. K. Mitra, *loc. cit.*; U. S. Ganguly, *loc. cit.*; R. W. Calfee, *loc. cit.*

Synthesize only the realizable functions in the form of Figure 8-42.

(a) $\dfrac{s^2 - s + 1}{s^2 + s + 1}$

(b) $\dfrac{s^3 - 6s + 15s - 15}{s^3 + 6s + 15s + 15}$

(c) $\dfrac{s^2 + s - 3}{s^2 + s + 3}$

8.4 The active RC structure proposed by Armstrong and Reza[31] for

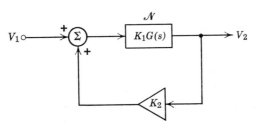

Figure 8-43

realizing voltage-transfer ratios $t_v(s)$ is shown in Figure 8-43 in block diagram form. Show that if $t_v(s)$ is a biquadratic function, then the network $\mathcal{N}$ is always RC realizable. Develop a synthesis procedure for this configuration and realize the following all-pole transfer function.

$$t_v = \dfrac{1}{s^2 + 3s + 3}$$

8.5 Apply Kuh's method (Section 8-3, p. 311) to realize the following transfer voltage ratios:

(a) $\dfrac{H}{s^3 + 2s^2 + 2s + 1}$

(b) Hs^2

(c) $\dfrac{Hs^3}{s^3 + 6s^2 + 15s + 15}$

(d) $\dfrac{H(s + 1)}{s^2 + 3s + 3}$

[31] D. B. Armstrong and F. M. Reza, *loc. cit.*

8.6 Realize the transfer functions of Problem 8.5 as transfer current ratios, using Hakim's method and with a minimum number of elements. How would you incorporate generator and load resistances in your network realization?

8.7 Are the following functions realizable using Kuh's method or Hakim's method if a common ground is desired between the input and the output?

(a) $\dfrac{6}{s^2 + 3s + 3}$

(b) $\dfrac{H(s - 1)}{s^2 + s + 1}$

(c) $\dfrac{H(s^2 + 1)}{s^2 + s + 1}$

(d) $\dfrac{s^2 - s + 2}{s^2 + s + 2}$

If some of the functions cannot be realized using Kuh's or Hakim's method, synthesize them using two amplifiers and RC one-ports.

8.8 Realize the second-order transfer functions of Problems 8.5 and 8.7 using Hakim's method and the RC network tables of Table D-1 (Appendix D).

8.9 Realize the following transfer voltage ratios by the coefficient matching technique outlined in Section 8-4, pp. 328–335.

(a) $\dfrac{H}{s^2 + \sqrt{2}s + 1}$

(b) $\dfrac{Hs^2}{s^2 + \sqrt{2}s + 1}$

(c) $\dfrac{Hs}{s^2 + 0.2s + 3}$

(d) $\dfrac{H(s^2 + 1)}{s^2 + s + 1}$

(e) $\dfrac{H}{s^3 + 2s^2 + 2s + 1}$

(f) $\dfrac{Hs}{(s^2 + s + 1)(s^2 + s + 4)}$

Use more than one stages in cascade if necessary.

8.10 Using Bach's method (Section 8-4, p. 336) realize the following transfer voltage ratios:

(a) $\dfrac{H}{s^3 + 6s^2 + 15s + 15}$

(b) $\dfrac{Hs^3}{s^3 + 2s^2 + 2s + 1}$

Obtain at least two realizations for each function.

8.11 Show that the transfer voltage ratio of the generalized Kuh's structure of Fig. 8-44 is

$$\frac{V_2}{V_1} = \frac{-\mu y_{21A}}{y_{22A} + y_{11B} - \mu(-y_{12B})}$$

Develop an appropriate synthesis procedure for this structure and

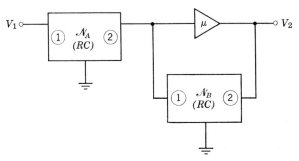

Figure 8-44

realize a second-order Butterworth filter so that the pole polynomial sensitivity has $j\omega$-axis zeros at $\pm j1$.[32]

8.12 Develop the design equations for the realization of voltage transfer ratio of the form of Equation 8.56 for the case of $\gamma = \alpha$ by modifying the active network configuration of Figure 8-38.

---

[32] N. M. Herbst, "Optimization of pole sensitivity in active RC networks," Res. Rept. EE 569, Cornell University, Ithaca, N.Y., 1963.

9 / *Negative-Impedance Converter as an Active Element*

The concept of the negative-impedance converter (NIC) was introduced by Merrill in 1950 and the NIC was initially used to compensate for losses in telephone transmission lines. In 1953, Linvill advanced a method of synthesizing arbitrary transfer functions by embedding a single NIC in an *RC* network. In a sense, Linvill was the first to initiate a systematic approach to active network synthesis. Since then, a considerable amount of interest has been focused on the area of active networks employing an impedance converter as the active device, mainly because it has simplified the synthesis procedure.

The purpose of this chapter is to introduce the negative-impedance converter as a network element and discuss some of its applications. Practical realizations of an NIC are considered in Section 9-1 from a fundamental approach. Sections 9-2 and 9-3 are concerned with the methods of synthesis of driving-point and transfer functions. The problem of minimizing the sensitivity of a network employing the NIC is considered in the next section. We conclude the chapter with a summary of main results.

9-1 PRACTICAL CONSIDERATIONS

We shall first discuss the nullator-norator representations of "ideal" negative-impedance converters. Later, transistorized versions of some of the equivalent representations will be developed. This section will be concluded with a discussion of sensitivity and stability of the NIC.

Nullator-Norator Representation[1]

In Section 2-8 it was mentioned that the nullator and the norator form a basic set of active elements, and a collection of nullator-norator circuits

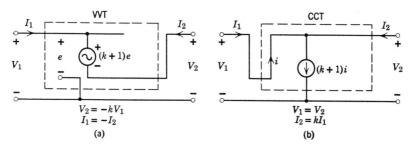

Figure 9-1 Equivalent representations of the negative-impedance converter using controlled sources.

representing controlled sources was presented at that time. Transistor realizations of the controlled sources were developed in Section 8-1 based on these nullator-norator representations. Development of transistorized NIC circuits follow similar lines. First a set of nullator-norator equivalent circuits of the NIC are generated. Then transistor NIC circuits are developed by making use of nullator-norator representation of the ideal transistor (Figure 8-2).

Construction of NIC representations is based on their equivalent circuits using the positive gain VVT and the negative gain CCT, shown again in Figure 9-1. Replacing the controlled sources in Figure 9-1 by their nullator-norator equivalent circuits, representations of the VNIC and the CNIC are easily obtained. Note from Figure 9-1 that, for negative-impedance conversion property, the VVT (and the CCT) must have a gain greater than unity. An examination of the nullator-norator VVT and CCT circuits of Figure 8-1 reveals that only circuits given in Figures 8-1a, 8-1b, 8-1i and 8-1j can have gains greater than unity by choosing proper values of the resistances. A few more suitable VVT and CCT circuits are now developed.

By cascading the CCT's of Figures 8-1a and 8-1e, the CCT circuit of Figure 9-2a is formed. Because of the presence of the nullators, nodes

---

[1] S. K. Mitra, "Nullator-norator equivalent circuits of linear active elements and their applications," *Proc. Asilomar Conf. on Circuits and Systems*, Monterey, Calif., **1**, 267–276 (November 1967).

"a" and "b" are at virtual ground. Moreover, since there is no current through the nullator, the nullator across the input port can be connected across "a" and "b" (as shown in Figure 9-2b) without changing the operation of the circuit.

In a similar manner, a VVT can be obtained by cascading the networks of Figures 8-1b and 8-1f, as indicated in Figure 9-2c. Since the potentials at "a" and "b" are fixed (they depend on V_1 only), the norator across the output port can be connected across "a" and "b" leading to a new VVT

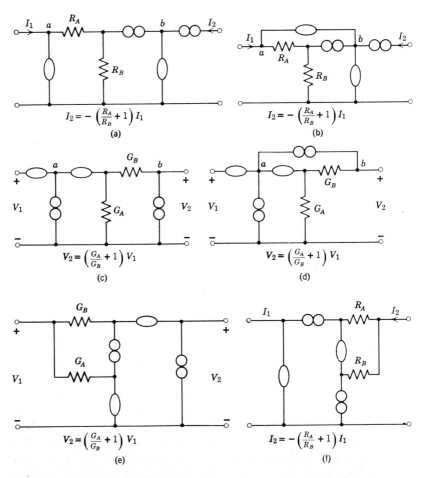

Figure 9-2 Additional nullator-norator models of the current-controlled current source and the voltage-controlled voltage source.

circuit (see Figure 9-2d). Duals of Figures 9-2b and 9-2d are shown in Figures 9-1e and 9-1f, respectively.[2]

Development of the NIC representations is now straightforward. By replacing the CCT of Figure 9-1b by the nullator-norator CCT circuits of Figures 8-1a, 8-1i, 9-2b, and 9-2f, we easily obtain the first four CNIC representations tabulated in Table 9-1. Likewise, replacement of the

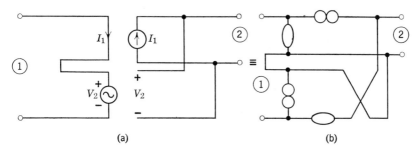

Figure 9-3 Development of the nullator-norator model of a balanced VNIC (Type V).

VVT of Figure 9-1a by the nullator-norator VVT circuits of Figures 8-1b, 8-1j, 9-2d, and 9-2e yields the first four VNIC representations cataloged in Table 9-2.

The development of the Type V negative impedance converter does not follow the above pattern. Instead, it is based on the balanced representation of the NIC indicated in Figure 2-42a. To facilitate the construction, Figure 2-42a has been redrawn (Figure 9-3a); it shows that a balanced VNIC can be established by making a series-parallel connection of a CCT and a VVT. Replacing the unity gain CCT and the VVT by their simple nullator-norator equivalent circuits (Figures 8-1e and 8-1f), the balanced model of a VNIC is obtained (see Table 9-2). The dual of this balanced VNIC is the type V balanced CNIC circuit of Table 9-1. The models of Tables 9-1 and 9-2 were cataloged by Braun[3] without any derivation.

Transistor Realizations

The nullator-norator representation of the ideal transistor (Figure 8-2) will form the basis of transistor realizations of the negative-impedance

---

[2] It should be remembered that the dual of a nullator is a nullator and the dual of a norator is a norator.

[3] J. Braun, "Equivalent NIC networks with nullator and norator," *IEEE Trans. on Circuit Theory*, **CT-12**, 441–442 (September 1965).

Table 9-1 *CNIC CIRCUITS*

Type	CNIC Model	Conversion Factor k	Parent CCT Circuit
I		$\dfrac{R_A}{R_B}$	Figure 8-1a
II		$\dfrac{R_1}{R_2} - 1$	Figure 8-1i
III		$\dfrac{R_A}{R_B}$	Figure 9-2b
IV		$\dfrac{R_A}{R_B}$	Figure 9-2f
V		1	

Table 9-2 VNIC CIRCUITS

Type	VNIC model	Conversion Factor k	Parent VVT Circuit	Dual CNIC Model
I	(circuit with R_A, R_B)	$\dfrac{R_B}{R_A}$	Figure 8-1b	I
II	(circuit with R_1, R_2)	$\dfrac{R_2}{R_1} - 1$	Figure 8-1j	II
III	(circuit with R_A, R_B)	$\dfrac{R_B}{R_A}$	Figure 9-2d	IV
IV	(circuit with R_A, R_B)	$\dfrac{R_B}{R_A}$	Figure 9-2e	III
V	(crossed circuit)	1		V

354 / NEGATIVE-IMPEDANCE CONVERTER AS AN ACTIVE ELEMENT

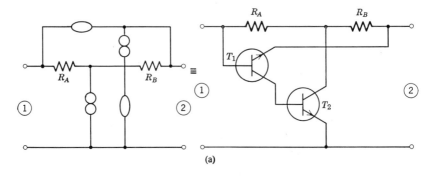

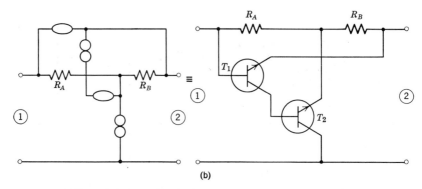

Figure 9-4 Transistorized realizations of the Type I-CNIC.

converters. As it will be noted, some of the transistor NIC circuits are well known.

Current-Inversion Type NIC. Consider first the realization of the Type I-CNIC. It is clear that a direct realization is not possible; however, the circuit operation remains invariant if a nullator-norator series combination is connected between any two nodes (see Figure 8-3). Several such modifications and their transistorized versions are indicated in Figure 9-4. The CNIC circuit of Figure 9-4b was originally advanced by Larky.[4] The other CNIC realization was advanced by Sandberg[5] as a generalized

---

[4] A. I. Larky, "Negative-impedance converters," *IRE Trans. on Circuit Theory*, **CT-4**, 124–131 (September 1957).

[5] I. W. Sandberg, "Synthesis of driving-point impedance with active *RC* networks," *Bell System Tech. J.*, **39**, 947–962 (July 1960).

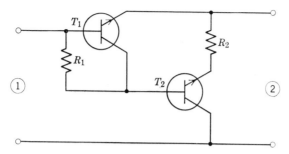

Figure 9-5 A Type II-CNIC circuit.

impedance converter by making R_A and R_B complex impedances. The development indicated in Figure 9-4 was recently pointed out by Myers.[6] An examination of Figure 7-6b reveals that Nagata's negative resistance circuit is essentially a Type I CNIC.

Realization of Type II-CNIC is shown in Figure 9-5; the transistorized version was advanced by Hakim.[7] The transistor CNIC circuit of Figure 9-6 represents the Type III model and is due to Myers.[8]

Conversion of Type IV-CNIC circuit is possible in two ways (Figure

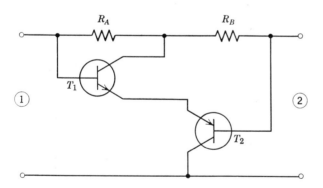

Figure 9-6 A Type III-CNIC circuit.

[6] B. R. Myers, "Nullor model of the transistor," *Proc. IEEE*, **53**, 758–759 (1965).
[7] S. S. Hakim, "Some new negative-impedance converters," *Electronics Letters*, **1**, 9–10 (March 1965).
[8] B. R. Myers, "New subclass of negative impedance converters with improved gain-product sensitivities," *Electronics Letters*, **1**, 68–70 (May 1965).

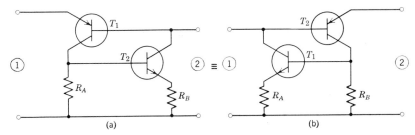

Figure 9-7 Type IV-CNIC circuits.

9-7). The circuit of Figure 9-7a was initially presented by Yanagisawa,[9] the second circuit (Figure 9-7b) will be found in Sandberg.[10]

Type V-CNIC can be similarly realized. Note that the negative resistance circuit of Figure 7-3 is essentially a terminated Type V-CNIC circuit.

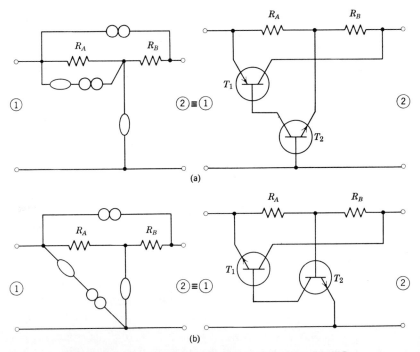

Figure 9-8 Transistorized realizations of the Type I-VNIC.

[9] T. Yanagisawa, "*RC* active networks using current inversion type negative-impedance converters," *IRE Trans. on Circuit Theory*, **CT-4**, 140–144 (September 1957).
[10] I. W. Sandberg, *loc. cit.*

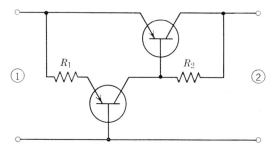

Figure 9-9 A Type II-VNIC circuit.

Voltage-Inversion Type NIC. Several Type I-VNIC circuits are developed in Figure 9-8. Both of these circuits were advanced by Myers.[11] The transistor realization shown in Figure 9-8a was independently developed by Hakim.[12]

Conversion of Type II-VNIC model is shown in Figure 9-9. This circuit was originally suggested by Hakim.[12] Transistor realization of Type III model[11] (Figure 9-10) is straightforward.

In 1953, Linvill[13] proposed a transistorized VNIC circuit that is essentially a Type IV realization. The circuit is indicated in Figure 9-11. Khazanov[14] has also presented a similar realization.

The balanced NIC models of Type V can easily be converted to circuits proposed by Linvill.[15] Realization of this is left as an exercise.

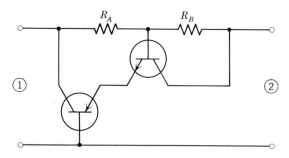

Figure 9-10 A Type III-VNIC circuit.

[11] B. R. Myers, "New subclass of negative-impedance converters with improved gain-product sensitivities," *Electronics Letters*, **1**, 68–70 (May 1965).

[12] S. S. Hakim, *loc. cit.*

[13] J. G. Linvill, "Transistor negative-impedance converters," *Proc. IRE*, **41**, 725–729 (June 1953).

[14] G. L. Khazanov, "Synthesis of *RC* active filters with prescribed transfer admittance," *Elektrosviaz*, **13**, 63–72 (March 1959).

[15] J. G. Linvill, *loc. cit.*

358 / NEGATIVE-IMPEDANCE CONVERTER AS AN ACTIVE ELEMENT

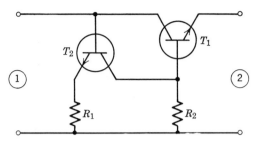

Figure 9-11 Type IV-VNIC circuits.

Analysis of Some Typical NIC Circuits

For a more accurate analysis of the previous NIC circuits, the transistor must be replaced by its T equivalent circuit (Figure 9-12). For example, Linvill's VNIC circuit (Figure 9-11) has the following h parameters:

(9.1)
$$h_{11} = \frac{\{r_{e1} + (r_{b1} + R_2)(1 - \alpha_1)\}\{R_1 + r_{e2} + r_{b2}(1 - \alpha_2)\}}{\Delta}$$

$$h_{12} = \frac{\alpha_1}{\Delta}\{R_1 + r_{e2} + r_{b2}(1 - \alpha_2)\}$$

$$h_{21} = \frac{\alpha_2 R_2}{\Delta}$$

$$h_{22} = \frac{1 - \alpha_2}{\Delta}$$

where
$$\Delta = (1 - \alpha_2)\{r_{e1} + (r_{b1} + R_2)(1 - \alpha_1)\} - \alpha_1\alpha_2 R_2$$

Note that any variation in the emitter resistance r_{e2} of T_2 will be swamped out by R_1. From Equation 9.1 we observe that for most practical

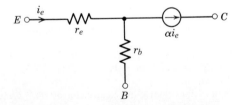

Figure 9-12 The T-equivalent circuit of a biased transistor suitable for small-signal analysis.

purposes, the *h*-parameters can be approximated as:

$$h_{11} \simeq 0$$

$$h_{12} \simeq -\frac{R_1}{\alpha_2 R_2}$$

(9.2)

$$h_{21} \simeq -\frac{1}{\alpha_1}$$

$$h_{22} \simeq 0$$

Detailed analysis of the CNIC circuits of Figures 9-4b and 9-7a and b are given elsewhere.[16] For the circuit of Figure 9-4b, we have:

(9.3)

$$h_{11} = \frac{\{r_{b1}(1-\alpha_1) + r_{e1}\}(R_A + R_B)}{\{(r_{b1} + R_A + R_B)(1-\alpha_1) + r_{e1}\} - \frac{\alpha_1 \alpha_2}{1-\alpha_2} R_B}$$

$$h_{12} = 1$$

$$h_{21} \simeq \frac{\dfrac{R_A}{R_B} + \dfrac{1-\alpha_2}{\alpha_1 \alpha_2}\left[1 + \dfrac{R_A + r_{b1}(1-\alpha_1) + r_{e1}}{R_B}\right]}{1 - \dfrac{(1-\alpha_1)(1-\alpha_2)}{\alpha_1 \alpha_2}\left[1 + \dfrac{(r_{b1} + R_A)(1-\alpha_1) + r_{e1}}{(1-\alpha_1)R_B}\right]}$$

$$h_{22} = 0$$

The CNIC of Figure 9-7a has the following *h*-parameters:

$$h_{11} = (1-\alpha_1)r_{b1} + r_{e1}$$

$$h_{12} = 1$$

(9.4)

$$h_{21} = \frac{\alpha_1 \alpha_2 \dfrac{R_A}{R_B} - (1-\alpha_1)\left[1 + \dfrac{(1-\alpha_2)(R_A + r_{b2}) + r_{e2}}{R_B}\right]}{1 + \dfrac{(1-\alpha_2)(R_A + r_{b2}) + r_{e2}}{R_B}}$$

$$h_{22} = 0$$

A similar analysis can be carried out for the remaining NIC circuits.

Sensitivity Considerations

A practical negative-impedance converter is a non-ideal device characterized, in general, by $h_{11} \neq 0$ and $h_{22} \neq 0$. If h_{11} and h_{22} are real quantities, then they can be compensated by connecting additional passive elements as indicated in Figure 2-17. It has been shown that for perfect

---

[16] I. W. Sandberg, *loc. cit.*

compensation the following relations must hold:

(9.5)
$$h_{12}h_{21} = h_{11}h_{22} + A$$
$$AB = 1$$

where

(9.6)
$$A = \frac{h_{22}}{Y_A}$$
$$B = \frac{1}{Y_B h_{11}}$$

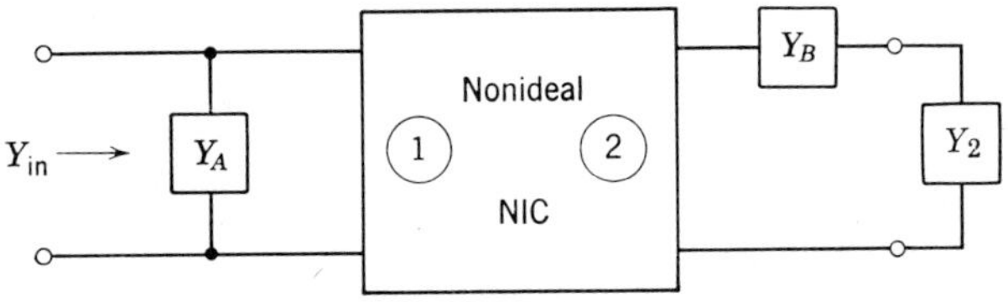

Figure 9-13 A fully-compensated nonideal NIC.

Generally, in non-ideal NIC, each of the h parameters will be a slowly variable quantity; hence, for a meaningful result, we plan to compute the sensitivity of the compensated NIC with respect to each h-parameter.[17] The input admittance of a compensated NIC terminated by an admittance Y_2 at the output port (Figure 9-13) is

(9.7)
$$Y_{in} = Y_A + \frac{h_{22}(Y_B + Y_2) + Y_B Y_2}{(h_{11}h_{22} - h_{12}h_{21})(Y_B + Y_2) + h_{11}Y_B Y_2}$$

The sensitivity of Y_{in} with respect to various h parameters of the compensated NIC is:

(9.8)
$$S_{h_{11}}^{Y_{in}} \triangleq \frac{h_{11}}{Y_{in}}\left(\frac{dY_{in}}{dh_{11}}\right) = h_{11}Y_2\left(1 + \frac{Y_A}{Y_2}\right)^2$$
$$S_{h_{22}}^{Y_{in}} = \frac{h_{22}}{Y_2}\left(1 + \frac{Y_2}{Y_B}\right)^2$$
$$S_{h_{12}h_{21}}^{Y_{in}} = -h_{11}Y_B\left(1 + \frac{Y_2}{Y_B}\right)\left(1 + \frac{Y_A}{Y_2}\right)$$

In deriving Equation 9.8, use of Equations 9.5 and 9.6 has been made. In general, h_{11} and h_{22} are small, so without any loss of generality we can

[17] A. I. Larky, *loc. cit.*

write:

(9.9)
$$Y_B > Y_2 > Y_A$$
$$A \simeq h_{12}h_{21}$$

which when used in Equation 9.8 yields the following approximate relations.

(9.10)
$$S_{h_{11}}^{Y_{in}} \simeq \left|\frac{Y_2}{Y_B}\right| < 1$$
$$S_{h_{22}}^{Y_{in}} \simeq \left|\frac{Y_A}{Y_2}\right| < 1$$
$$S_{h_{21}h_{12}}^{Y_{in}} \simeq -1$$

assuming unity conversion factor, i.e., $h_{12}h_{21} = 1$.

Expression 9.10 indicates that in the design of an NIC, the following two conditions should be satisfied: (1) $h_{12}h_{21}$ is the conversion factor and hence its variation should be small over the frequency band of interest; and (2) h_{11} and h_{22} should be made very small, which in turn minimizes the effect of their variations.

With these preliminaries, let us now examine several NIC circuits from a sensitivity point of view. Consider Linvill's VNIC circuit of Figure 9-11, for which

(9.11)
$$h_{12}h_{21} \simeq \frac{R_1}{\alpha_1\alpha_2 R_2}$$

from which we obtain

$$S_{\alpha_1}^{h_{12}h_{21}} = S_{\alpha_2}^{h_{12}h_{21}} = -1$$

Any further improvement in either sensitivity cannot be obtained. Since the conversion factor, $h_{12}h_{21}$, is dependent on α_1 and α_2, to keep the variation of $h_{12}h_{21}$ small, it is desirable to reduce the alpha variations by using a Darlington compound pair (Figure 9-14) for both transistors.[18]

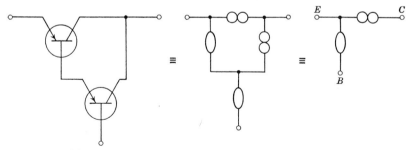

Figure 9-14 A Darlington compound pair of transistors.

[18] For a discussion on compound transistors, see J. G. Linvill and J. F. Gibbons, *Transistors and active circuits*, McGraw-Hill, New York, 1961.

In the case of the Type I-CNIC circuit of Larky (Figure 9-4b), we obtain the following approximate expression for h_{21} from Equation 9.3:

$$(9.12) \qquad h_{21} \cong \frac{R_A}{R_B} + \frac{1 - \alpha_2}{\alpha_1 \alpha_2}\left(1 + \frac{R_A}{R_B}\right)$$

from which one easily obtains

$$(9.13) \qquad S_{\alpha_1}^{h_{12}h_{21}} \cong -\frac{(1 - \alpha_2)}{\alpha_1 \alpha_2}\left(1 + \frac{R_B}{R_A}\right)$$

$$S_{\alpha_2}^{h_{12}h_{21}} \cong -\frac{1}{\alpha_1 \alpha_2}\left(1 + \frac{R_B}{R_A}\right)$$

From Equation 9.13 we conclude that $S_{\alpha_1}^{h_{12}h_{21}}$ can be minimized by replacing the transistor T_2 by a Darlington-compound pair; this in turn also minimizes $S_{\alpha_2}^{h_{12}h_{21}}$ by stabilizing variations of α_2. For this arrangement, it is not necessary to replace T_1 by a compound pair.

For the Yanagisawa-CNIC circuit (Type IV) shown in Figure 9-7a, we have the approximate expression for the conversion factor as

$$(9.14) \qquad h_{12}h_{21} = h_{21} \cong \alpha_1\alpha_2 \frac{R_A}{R_B} - (1 - \alpha_1)$$

from which one obtains

$$(9.15) \qquad S_{\alpha_1}^{h_{12}h_{21}} \cong 1 + \frac{R_B}{\alpha_2 R_A}$$

$$S_{\alpha_2}^{h_{12}h_{21}} \cong 1$$

For most practical purposes it is thus sufficient to stabilize α_2, which can be achieved by a compound pair of transistors in place of T_2.

Conversion-factor-sensitivities of the NIC circuits of Figure 9-4a, 9-4b, 9-6, 9-8a, 9-8b, and 9-10 have been tabulated by Myers.[19] He has shown that the VNIC circuits of Figure 9-8a and 9-8b have the lowest conversion-factor sensitivities.

Stability Considerations

Stability of an arbitrary two-port was investigated in Section 6-9. We can use the results of that section to determine the stability conditions of an NIC. In terms of h parameters, the stability invariant factor is

---

[19] B. R. Myers, "New subclass of negative-impedance converters with improved gain-product sensitivities," *Electronics Letters*, **1**, 68–70 (May 1965).

given as

$$\eta = \frac{2\text{Re}(h_{11})\text{Re}(h_{22}) - \text{Re}(h_{12}h_{21})}{|h_{12}h_{21}|}$$

For a potentially unstable two-port, $-1 \leq \eta < 1$ and for unconditional stability, $\eta \geq 1$.

For an NIC with real parameters, $\eta = -1$ indicating potential instability of the ideal device.

Consider an ideal NIC ($h_{12} = h_{22} = 0$) terminated at port 1 by an impedance Z_x and at port 2 by an impedance Z_y (Figure 9-15). Stability is determined by

(9.16) $$Z_x - h_{12}h_{21}Z_y = Z_x - Z_y = 0$$

This indicates that if the system is stable for some values of Z_x and Z_y,

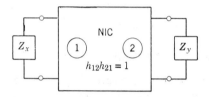

Figure 9-15 An NIC terminated at both ports.

the circuit obtained by interchanging Z_x and Z_y will still be stable. In practice, however, it is found that a NIC circuit is stable for a low impedance at one-port and a high impedance at the other port. In the reverse situation, the system is unstable.

A simple explanation of the above phenomenon, which follows next, was provided recently by Hoskins.[20] To a first approximation, we can represent a practical NIC by the following parameters:

(9.17)
$$h_{11} = h_{22} \simeq 0$$
$$h_{12}h_{21} \simeq e^{-s\tau}$$

which accounts for the finite (however small) delay experienced in any physical circuit. For simplicity, assume the terminating impedances to be pure resistances, R_x and R_y. Stability of the system is then determined by the following equation:

(9.18) $$R_x - e^{-s\tau}R_y = 0$$

---

[20] R. F. Hoskins, "Stability of negative-impedance converter," *Electronics Letters*, **2**, 341 (September 1966).

which can be reexpressed as

(9.19) $$e^{-s\tau} = e^{-\sigma\tau}(\cos \omega\tau - j \sin \omega\tau) = \frac{R_x}{R_y}$$

Since the right-hand side is real and positive, roots of the Equation 9.19 are given as:

(9.20) $$s_n = \frac{1}{\tau} \log \frac{R_y}{R_x} \pm jn\pi$$
$$n = 0, 1, 2, \ldots$$

It is seen that if $R_y < R_x$, the roots are in the left-half plane and the system is stable. If, on the other hand, $R_y > R_x$, the system is unstable because of right-half plane roots. The port which must be terminated by a low impedance is then short-circuit stable and the other port, which must face a high impedance, is thus open-circuit stable.

This property of a practical NIC has also been explained by other authors following different arguments.[21]

9-2 DRIVING-POINT FUNCTION SYNTHESIS

In this section, we shall outline several synthesis methods for realizing *RC* one-ports where the active element is some form of immittance converter. Specifically, we shall concentrate on methods advanced by Kinariwala, Sandberg, and Sipress. In a later chapter, the polynomial decomposition approach to the synthesis of such networks will be discussed.

Two Basic Structures

Figure 3-14 shows two simple *RC*:NIC configurations that can be used to realize a restricted class of driving-point function. The synthesis methods are a direct consequence of Theorem 3-5.

Consider a driving-point impedance $Z(s) = N(s)/D(s)$ having the following restrictions: (1) $D(s)$ has only negative real simple zeros; (2) $D(s)^0 \geq N(s)^0$. It follows from Theorem 3-5 that we can always express

(9.21) $$Z(s) = Z_{RC}{}^A - Z_{RC}{}^B$$

where $Z_{RC}{}^A$ and $Z_{RC}{}^B$ are *RC* driving-point impedances. Thus a suitable realization $Z(s)$ will be in the form of Figure 3-14a.

[21] J. D. Brownlie, "On the stability properties of a negative-impedance converter," *IEEE Trans. on Circuit Theory*, **CT-13**, 98 (March 1966); A. F. Schwartz, "On the stability properties of a negative-immittance converter," *IEEE Trans. on Circuit Theory*, **CT-14**, 77 (March, 1967).

If $Z(s)$ has simple negative real zeros with $N(s)^0 \geq D(s)^0 - 1$, then part (ii) of Theorem 3-5 suggests a realization of $Z(s)$ in the form of Figure 3-14b where

(9.22) $$\frac{1}{Z(s)} = Y_{RC}{}^a - Y_{RC}{}^b$$

In Equation 9.22, $Y_{RC}{}^a$ and $Y_{RC}{}^b$ are RC driving-point admittances.

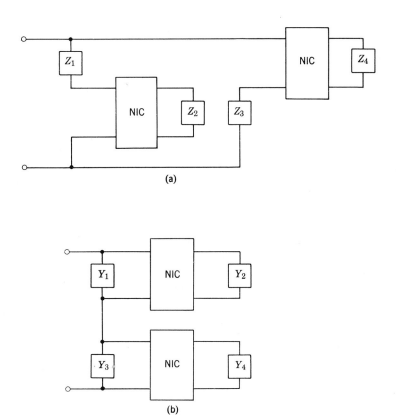

Figure 9-16 Two general RC:NIC one-port structures.

It is a simple exercise to show that a parallel connection of two of the basic networks of Figure 3-14a, as shown in Figure 9-16a, can realize any real rational function as the driving-point impedance. In a dual manner, it can be shown that the structure of Figure 9-16b can realize any driving-point function (Problem 9.2).

Kinariwala's Method

One of the negative-impedance converters is not needed if one RC network in the basic configurations is generalized into a two-port network. Such a structure, suggested by Kinariwala,[22] is shown in Figure 9-17. In terms of the open-circuit parameters of the RC two-port, the input impedance is given as

$$(9.23) \qquad Z(s) = z_{11} - \frac{z_{12}^2}{z_{22} - kZ_L}$$

where k is the conversion factor of the NIC. Without any loss of generality, let us assume k to be unity.

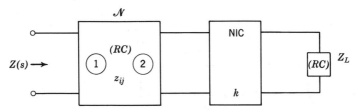

Figure 9-17 The cascade RC:NIC configuration of Kinariwala.

A specified driving-point function $N(s)/D(s)$ can be cast into the form of Equation 9.23 as follows:

Let $N(s)/D(s)$ be positive on at least one section of the negative-real axis. Choose a polynomial $Q(s)$ with distinct negative-real zeros only, lying anywhere on the section or sections of the negative-real axis where the function $N(s)/D(s)$ is finite and positive. Moreover, let

$$(9.24) \qquad Q(s)^0 \geq \max \left[N(s)^0, D(s)^0 \right]$$

Under these assumptions we can express

$$(9.25) \qquad \frac{N(s)}{D(s)} = \frac{N(s)/Q(s)}{D(s)/Q(s)} = \frac{\dfrac{N_a}{Q_a} - \dfrac{N_b}{Q_b}}{\dfrac{D_a}{Q_a} - \dfrac{D_b}{Q_b}}$$

where $Q(s) = Q_a Q_b$. The right-hand side expression of Equation 9.25 is obtained by making a partial fraction expansion of N/Q and D/Q and grouping the resulting terms accordingly to make N_a/Q_a, N_b/Q_b, D_a/Q_a, and D_b/Q_b RC impedances (Theorem 3-5).

[22] B. K. Kinariwala, "Synthesis of active RC networks," *Bell Systems Tech. J.*, **38**, 1269–1316 (September 1959).

Rewrite Equation 9.25 as

(9.26) $$\frac{N(s)}{D(s)} = \frac{N_b}{D_b} - \frac{\dfrac{N_b D_a - N_a D_b}{D_b^2}}{\dfrac{D_a}{D_b} - \dfrac{Q_a}{Q_b}}$$

Comparing Equations 9.26 and 9.23, we identify:

(9.27)
$$z_{11} = \frac{N_b}{D_b}$$
$$z_{22} = \frac{D_a}{D_b}$$
$$z_{12}^2 = \frac{N_b D_a - N_a D_b}{D_b^2}$$
$$Z_L = \frac{Q_a}{Q_b}$$

It can be shown[23] that a polynomial $Q(s)$ can always be chosen so that Q_a/Q_b and N_b/D_b are RC impedances. Simple calculation will convince the reader that the residue conditions are satisfied by the z parameters given by Equation 9.27. The main problem is thus to ensure that z_{12} can be made a rational function. The following example will illustrate some key ideas behind the method.

Example 9-1 Let us realize

(9.28) $$Z(s) = \frac{N(s)}{D(s)} = \frac{s^2 + s + 2}{s^2 + s + 4}$$

Note that the above function is positive on the entire negative real axis. Choosing $Q(s) = (s+1)(s+2)$, we obtain

(9.29) $$Z(s) = \frac{\dfrac{s^2+s+2}{(s+1)(s+2)}}{\dfrac{s^2+s+4}{(s+1)(s+2)}} = \frac{\dfrac{s+3}{s+1} - \dfrac{4}{s+2}}{\dfrac{s+5}{s+1} - \dfrac{6}{s+2}}$$

We identify $Q_a = (s+2)$ and $Q_b = (s+1)$. This is necessary to make Q_a/Q_b an RC impedance. Forming the numerator of z_{12}^2, we note that it is not a perfect square as desired. In this case, we rewrite Equation 9.29

[23] B. K. Kinariwala, *loc. cit.*

as

$$Z(s) = \frac{\dfrac{4}{s+2} - \dfrac{s+3}{s+1}}{\left(\dfrac{6}{s+2} + R\right) - \left(\dfrac{s+5}{s+1} + R\right)}$$

and recompute the numerator of z_{12}^2 to obtain

(9.30) $\quad N_b D_a - N_a D_b = (s+3)(6 + 2R + Rs) - 4(s + 5 + Rs + R)$

$\qquad\qquad\qquad = Rs^2 + (2+R)s + 2(R-1)$

The right-hand side of Equation 9.30 can be made a perfect square by making

$$7R^2 - 12R - 4 = 0$$

which implies that

$$R = 2$$

Thus the parameters characterizing the companion RC network are given by

(9.31)
$$z_{11} = \frac{s+3}{3s+7}, \qquad z_{22} = \frac{2s+10}{3s+7}$$

$$z_{12} = \frac{\sqrt{2}(s+1)}{3s+7}$$

$$Z_L = \frac{s+2}{s+1}$$

In general, the two-port network realized using the z-parameters of Equation 9.27 will contain an ideal transformer. Transformerless realization can be achieved, in some cases, by taking advantage of the impedance conversion action of the NIC. The method is as follows.[24]

Instead of realizing z_{11}, z_{12}, and z_{22} simultaneously, first realize z_{11} satisfying the zeros of z_{12}. The resulting network will be in the form of a ladder, if the zeros of z_{12} are negative real. Designate this two-port by $\mathcal{N}'$ (Figure 9-18a). In general, the open-circuit transfer impedance z'_{12} of $\mathcal{N}'$ will be different from z_{12} by a multiplicative constant α, i.e.,

$$z'_{12} = \alpha z_{12}$$

[24] S. K. Mitra, "A new approach to active RC network synthesis," *J. Franklin Inst.*, **274**, 185–197 (September 1962).

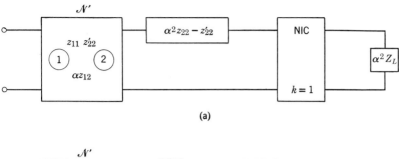

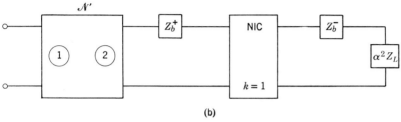

Figure 9-18 Steps illustrating the transformerless realization of a driving-point function.

Moreover, the open-circuit output impedance of $\mathcal{N}'$, z'_{22}, will be different from z_{22}. Rewrite Equation 9.23 as

$$(9.32) \qquad Z(s) = z_{11} - \frac{(\alpha z_{12})^2}{\alpha^2 z_{22} - \alpha^2 Z_L}$$

The problem now is to design a two-port $\mathcal{N}$ whose z-parameters are given by z_{11}, αz_{12}, and $\alpha^2 z_{22}$. We have already realized a two-port $\mathcal{N}'$ whose z-parameters are z_{11}, αz_{12}, and z'_{22}. This implies that the desired result is obtained by connecting an impedance $(\alpha^2 z_{22} - z'_{22})$ in series at the output as shown in Figure 9-18a. We can always write

$$(9.33) \qquad \alpha^2 z_{22} - z'_{22} = Z_b^+ - Z_b^-$$

where Z_b^+ and Z_b^- are passive RC impedances. This indicates that final form of realization of $Z(s)$ will be as shown in Figure 9-18b.

Example 9-2 Let us apply the above technique to Example 9-1. We first realize $z_{11} = (s+3)/(3s+7)$ by the zero-shifting technique to obtain a transmission zero at $s = -1$. The resulting two-port is shown in Figure 9-19a, for which we compute z'_{12} and z'_{22}:

$$z'_{12} = \frac{s+1}{3s+7} \qquad z'_{22} = \frac{s+5}{3s+7}$$

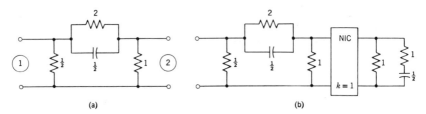

Figure 9-19 Realization of the driving-point impedance of Example 9-1, using the method shown in Figure 9-18.

Comparing z'_{12} as given above with z_{12} given by Equation 9.31 we conclude that $\alpha = 1/\sqrt{2}$. We now scale z_{22} and Z_L as given by Equation 9.31 by $\alpha^2 = \frac{1}{2}$. Note that $\alpha^2 z_{22} = z'_{22}$. Thus $Z_b^+ = Z_b^- = 0$. The final realization is given in Figure 9-19b. The two 1-ohm resistors across the input and output of the negative-impedance converter cancel each other, leaving five elements for the realization of the specified $Z(s)$. It has been shown that if an NIC is used as the active element, then the number of passive elements needed to realize a biquadratic driving-point function is five.[25]

A similar method can be followed on the admittance basis, resulting in a dual structure to Figure 9-18b.

It should be noted that in place of the negative-impedance converter, a voltage amplifier or a current amplifier can be used. The circuit then modifies to those of Figure 9-20, which are identical to those proposed by Kuh for transfer-function synthesis (see Section 8-3, p. 311).

Synthesis Using Generalized Impedance Converter

A major problem of the previous method is to make z_{12} a real rational function. In addition, realization of an RC two-port is not necessarily

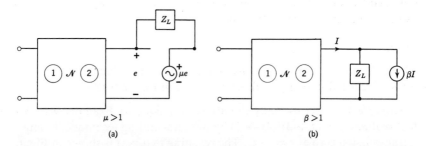

Figure 9-20 Controlled-source representations of the cascade configuration of Figure 9-17.

[25] B. R. Myers, "Ladder realization of biquadratic driving-point function," *Summaries of Papers, I.C.M.C.I.*, Part 2, Tokyo, Japan, pp. 75–76 (1964).

DRIVING-POINT FUNCTION SYNTHESIS / 371

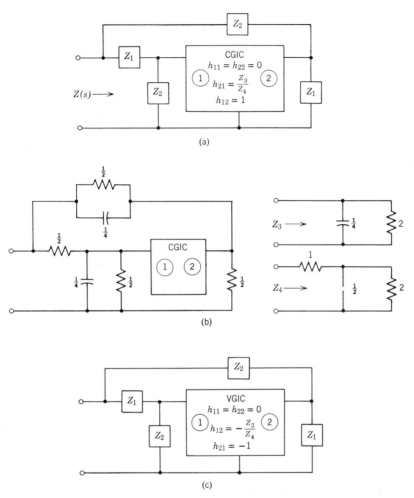

Figure 9-21 Driving-point function realization using generalized impedance converter: (a) An $RC:CGIC$ structure. (b) Realization of the driving-point function of Example 9-3. (c) An $RC:VGIC$ structure.

simple. These problems are avoided in the method of Sandberg described next. The active RC configuration (Figure 9-21a) proposed by Sandberg[26] employs a CGIC and four RC one-ports. The synthesis method is very straightforward and is as follows.

The input impedance of the one-port configuration of Figure 9-21a can

---

[26] I. W. Sandberg, *loc. cit.*

be expressed as

$$\text{(9.34)} \qquad Z(s) = \frac{Z_4 - Z_3}{\dfrac{Z_4}{Z_1} - \dfrac{Z_3}{Z_2}}$$

Let $Z(s) = N(s)/D(s)$ be the specified driving-point function which is positive at least on a section on the negative real axis. Then following the technique outlined in Section 9-2, p. 366, we can always choose a polynomial $Q(s)$ with distinct negative real roots located on the section or sections where $Z(s)$ is positive and express $Z(s)$ as:

$$\text{(9.35)} \qquad Z(s) = \frac{N(s)/Q(s)}{D(s)/Q(s)} = \frac{\dfrac{N_a}{Q_a} - \dfrac{N_b}{Q_b}}{\dfrac{D_a}{Q_a} - \dfrac{D_b}{Q_b}}$$

where $Q(s) = Q_a Q_b$ and N_a/Q_a, N_b/Q_b, D_a/Q_a, and D_b/Q_b are RC driving-point impedance functions. We rewrite Equation 9.35 as

$$\text{(9.36)} \qquad Z(s) = \frac{\dfrac{N_a}{Q_a} - \dfrac{N_b}{Q_b}}{\dfrac{D_a + \alpha Q_a}{Q_a} - \dfrac{D_b + \alpha Q_b}{Q_b}}$$

Comparing expressions 9.34 and 9.36, we identify

$$\text{(9.37)} \qquad Z_1 = \frac{N_a}{D_a + \alpha Q_a}, \qquad Z_2 = \frac{N_b}{D_b + \alpha Q_b}$$

$$Z_3 = \frac{N_b}{Q_b}, \qquad Z_4 = \frac{N_a}{Q_a}$$

Note that Z_3 and Z_4 are RC impedances. Now the degree of Q_a is as great as the degree of D_a. Since N_a/Q_a is an RC impedance, we can always find a finite, real, and positive α to make Z_1 an RC impedance. Similarly, Z_2 can be made an RC impedance function.

Example 9-3 We consider the driving-point function of Example 9-1. From Equation 9.29, we have

$$Z(s) = \frac{s^2 + s + 2}{s^2 + s + 4} = \frac{\dfrac{s+3}{s+1} - \dfrac{4}{s+2}}{\dfrac{s+5}{s+1} - \dfrac{6}{s+2}}$$

where we have chosen $Q(s) = (s + 1)(s + 2)$. Therefore

$$Z_1 = \frac{(s + 3)}{(\alpha + 1)s + (\alpha + 5)}; \qquad Z_2 = \frac{4}{\alpha s + (2\alpha + 6)}$$

$$Z_4 = \frac{(s + 3)}{(s + 1)}; \qquad Z_3 = \frac{4}{s + 2}$$

Note that Z_2 is an *RC* impedance for all positive values of α. Z_1 is *RC* realizable for $\alpha \geq 1$. Choosing $\alpha = 1$, the final realization as shown in Figure 9-21b is obtained.

The active *RC* one-port of Figure 9-21c, which employs a **VGIC** as an active device, is also suitable for synthesis of driving-point functions.

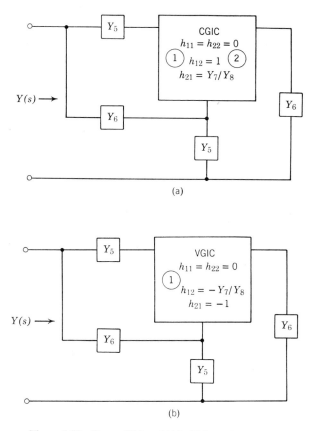

Figure 9-22 Two additional *RC*:GIC configurations.

It can be shown[27] that the input impedance of this structure is identical to that of Figure 9-21a as given by Equation 9.34.

Two alternate RC:GIC structures are available for realizing driving-point functions. The configuration shown in Figure 9-22a was proposed by Sandberg[28] and has an input admittance given as

$$(9.38) \qquad Y(s) = \frac{Y_7 - Y_8}{\dfrac{Y_7}{Y_6} - \dfrac{Y_8}{Y_5}}$$

The second structure (Figure 9-22b) was shown recently[29] to have the same input admittance. Observe that the form of the input admittance as given by Equation 9.38 is similar in form to expression 9.34. A prescribed $Y(s)$ that is positive over at least one section of the negative real axis can always be realized in the form of Figure 9-22. The realization steps are similar to those outlined earlier and will be omitted.

Note that the active elements in these latter configurations do not have a common ground with the input and hence require a "floating supply." This is a disadvantage from the practical point of view.

It is interesting to point out here that the configuration of Figure 9-22a is the capacitive dual of Figure 9-21c and that the one-port of Figure 9-22b is the capacitive dual of Figure 9-21a.

The Basic Theorem on Active RC One-Port Synthesis

Initially, one of the major objectives of many authors of active RC one-port synthesis procedures was to attempt to prove the sufficiency of the following theorem:

Theorem 9-1 *Any real rational function of the complex frequency variable can be realized as the driving-point impedance of a one-port obtained by embedding a single active two-port[30] in a transformerless RC network.*

The first proof of the above theorem was advanced by Sandberg[31] using the one-port structure of Figure 9-22a. Another proof was offered by Kinariwala. A brief outline of his approach was sketched in Section 8-2, p. 308.

[27] S. K. Mitra and N. M. Herbst, "Synthesis of active RC one-ports using generalized impedance converters," *IEEE Trans. on Circuit Theory*, **CT-10**, 532 (December 1963).
[28] I. W. Sandberg, *loc. cit.*
[29] S. K. Mitra and N. M. Herbst, *loc. cit.*
[30] By an active two-port we mean here a controlled source of unrestricted gain, or an NIC or a GIC.
[31] I. W. Sandberg, "Active RC networks," Report R-662-58, PIB-590, MRI, Polytech. Inst. of Brooklyn, May 28, 1958.

In what follows we present an alternate proof of the sufficiency of a single active element, also advanced by Sandberg.[32]

We first observe that a major restriction imposed by his synthesis technique is that the driving-point function $Z(s)$ be positive at least on a section of the negative real axis. This restricts an important class of functions including the inductance.

This restriction can be circumvented as follows. If $Z(s)$ is nonpositive on the entire negative real axis, then $Z'(s) = Z(s) - Z_{RC}(s)$ can be made positive on one section of the negative real axis by choosing a suitable passive RC impedance, $Z_{RC}(s)$. Now $Z'(s)$ can be realized following the method outlined in the previous section. Realization of $Z(s)$ is then obtained by connecting in series with $Z'(s)$ an RC impedance $Z_{RC}(s)$. A suitable $Z_{RC}(s)$ is

$$Z_{RC}(s) = \frac{a}{s+b}, \qquad a > 0, \qquad b \geq 0$$

Example 9-4 Suppose it is desired to realize an inductance of 1 henry by means of an active RC network. Note that the specified driving-point function $Z(s) = s$ is negative on the entire negative real axis. As a result, consider instead

$$Z'(s) = s - \frac{9}{s} = \frac{s^2 - 9}{s}$$

which is positive on the negative real axis between $s = -3$ and origin. Choosing $Q(s) = (s+1)(s+2)$ and following the method of Section 9-2, p. 370, the network realization of an inductance is obtained as shown in Figure 9-23. The reader is urged to go through the intermediate steps.

This approach also removes the same restriction on Kinariwala's method.

Sipress' Method

A completely different approach to the driving-point synthesis problem is taken by Sipress.[33] By applying successively a root-locus approach,[34] he has shown that any real rational function can always be realized in the form of Figure 9-24, where the RC two-ports $\mathcal{N}_a$ and $\mathcal{N}_b$ can be made to have ladder structures. In effect, his synthesis method offers another alternative proof of Theorem 9-1.

---

[32] I. W. Sandberg, "Synthesis of driving-point impedance with active RC networks," *Bell System Tech. J.*, **39**, 947–962 (July 1960).

[33] J. M. Sipress, "Synthesis of active RC networks," *IRE Trans. on Circuit Theory*, **CT-8**, 260–269 (September 1961).

[34] For a discussion on root locus techniques, see, for example, J. G. Truxal, *Control system synthesis*, McGraw-Hill, New York, Ch. 4 (1955).

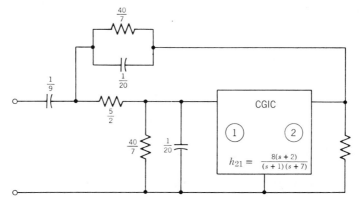

Figure 9-23 Realization of an inductance, using an RC:GIC structure (Example 9-4).

It is convenient to develop the synthesis method on the y basis. The input admittance of the cascade-feedback configuration of Figure 9-24 is given as

$$(9.39) \qquad Y(s) = (y_{11a} + y_{11b}) - \frac{(y_{12a} + y_{12b})(y_{12a} - ky_{12b})}{Y_1 + y_{22a} - k(y_{22b} + Y_2)}$$

where $\{y_{ija}\}$ and $\{y_{ijb}\}$ represent the short-circuit parameters of the two-port $\mathcal{N}_a$ and $\mathcal{N}_b$, respectively, and $1/k$ is the forward conversion factor of the VNIC. Note that if the VNIC is replaced by a CNIC described by $h_{12} = 1$ and $h_{21} = 1/k$, the expression for the input admittance (Equation 9.39) is unchanged.

Let $N(s)/D(s)$ be the specified driving-point admittance whose realization is desired, and let

$$(9.40) \qquad \delta = \max \{N(s)^0, D(s)^0\}$$

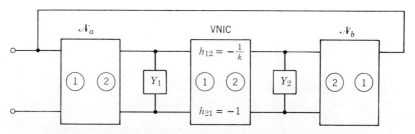

Figure 9-24 The RC:NIC cascade-feedback configuration of Sipress.

The first step of the synthesis method is to choose arbitrarily an *RC* driving-point admittance $P(s)/Q(s)$ having no zero at the origin and $P(s)^0 = Q(s)^0 = \delta$. We let

(9.41) $$(y_{11a} + y_{11b}) = K_1 \frac{P(s)}{Q(s)}$$

K_1 is to be determined shortly. From Equations 9.39 and 9.41 we obtain

(9.42) $$\frac{(y_{12a} + y_{12b})(y_{12a} - ky_{12b})}{Y_1 + y_{22a} - k(y_{22b} + Y_2)} = \frac{K_1 P(s)D(s) - N(s)Q(s)}{D(s)Q(s)} = \frac{A(s)}{B(s)}$$

where the degrees of $A(s)$ and $B(s)$ are both less than or equal to 2δ. Now $A(s)$ will always have at least $(\delta - 1)$ distinct negative real roots and by choosing K_1 properly, we can make $A(s)$ to have δ negative real roots, i.e., we can express

(9.43) $$A(s) = K_1 P(s)D(s) - N(s)Q(s) = U(s)V(s)$$

where $U(s)$ is a polynomial of degree δ having distinct negative real roots. The leading coefficient of $U(s)$ is made positive by adjusting the signs of the coefficients of $V(s)$. It is clear that $V(s)^0 \leq \delta$.

Define two polynomials $R_a(s)$ and $R_b(s)$ as:

(9.44)
$$R_a(s) \triangleq \frac{kK_2 U(s) + V(s)}{1 + k}$$

$$R_b(s) \triangleq \frac{K_2 U(s) - V(s)}{1 + k}$$

It follows from root-locus considerations that for a large enough positive value of the constant K_2, $R_a(s)$ and $R_b(s)$ can be made to have only distinct negative real roots with positive leading coefficients. In addition, $R_a(s)^0 = R_b(s)^0 = \delta$. Observe from expression 9.44 that

(9.45)
$$R_a(s) + R_b(s) = K_2 U(s)$$

$$R_a(s) - kR_b(s) = V(s)$$

We now rewrite Equation 9.42 as

(9.46) $$\frac{(y_{12a} + y_{12b})(y_{12a} - ky_{12b})}{Y_1 + y_{22a} - k(y_{22b} + Y_2)} = \frac{U(s)V(s)}{D(s)Q(s)}$$

$$= \frac{\{R_a(s) + R_b(s)\}\{R_a(s) - kR_b(s)\}}{K_2 D(s)Q(s)}$$

from which we obtain the following identifications:

$$-y_{12a} = K_3 \frac{R_a(s)}{Q(s)}$$

(9.47)
$$-y_{12b} = K_3 \frac{R_b(s)}{Q(s)}$$

$$Y_1 + y_{22a} - ky_{22b} - kY_2 = K_2 K_3^2 \frac{D(s)}{Q(s)}$$

We can arbitrarily split $K_1 P(s)/Q(s)$ to identify

$$y_{11a} = K_{1a} \frac{P_a(s)}{Q(s)}$$

(9.48)
$$y_{11b} = K_{1b} \frac{P_b(s)}{Q(s)}$$

where $K_1 P(s) = K_{1a} P_a(s) + K_{1b} P_b(s)$. Of course, the split is governed by the requirement that y_{11a} and y_{11b} as given by expression 9.48 be RC admittances with no zero at the origin. A possible solution would be to make $P_a(s) = P_b(s) = P(s)$ and $K_{1a} + K_{1b} = K_1$. K_3 is determined in a manner analogous to parallel ladder synthesis. Let the leading coefficients of $R_a(s)$ and $R_b(s)$ be h_a and h_b, respectively. Now realize $y'_{11a} = [P(s)/Q(s)]$ and $-y'_{12a} = [R_a(s)/Q(s)] = h_a[R'_a(s)/Q(s)]$ by the zero-shifting method. The short-circuit transfer admittance of this two-port, which we shall designate $\mathcal{N}''_a$, will be proportional to $-y'_{12a}$ and will be given as $H_a[R'_a(s)/Q(s)]$. A similar development of $y'_{11b} = [P(s)/Q(s)]$ and $-y'_{12b} = h_b[R'_b(s)/Q(s)]$ will result in a two-port $\mathcal{N}''_b$ having a short-circuit transfer admittance $H_b[R'_b(s)/Q(s)]$. Scale the admittance level of the two-port $\mathcal{N}''_a$ by an amount $(K_3/H_a)h_a$ and the admittance level of two-port $\mathcal{N}''_b$ by $(K_3/H_b)h_b$. K_3 is determined from the condition below:

(9.49)
$$K_3 \left(\frac{h_a}{H_a} + \frac{h_b}{H_b} \right) = K_1$$

Then y_{22a} and y_{22b} are determined from the scaled networks. Only the identification of Y_1 and Y_2 remains. We rewrite the last equation of (9.47) as

(9.50)
$$Y_1 - kY_2 = ky_{22b} - y_{22a} + K_2 K_3^2 \frac{D(s)}{Q(s)}$$

An RC:$-RC$ expansion (Theorem 3-5) of the right-hand side of Equation 9.50 yields the desired expressions for Y_1 and Y_2. The following example will illustrate the method.

Example 9-5 Let us realize $Y(s) = 1/s$ using a UNIC. Let

$$\frac{P(s)}{Q(s)} = \frac{s+2}{2s+9}$$

Choosing $K_1 = 1$, we observe from Equation 9.43 that $U(s) = (s+3)$ and $V(s) = (s-3)$. Next, selecting $K_2 = 1$ in expression 9.44 results in $R_a(s) = s$ and $R_b(s) = 3$. Therefore, from expression 9.47,

$$-y_{12a} = K_3 \frac{s}{2s+9}, \qquad -y_{12b} = K_3 \frac{3}{2s+9}$$

Note that $h_a = 1$ and $h_b = 3$. The two-port $\mathcal{N}'_a$ is first developed from the following parameters:

$$y'_{11a} = \frac{s+2}{2s+9}, \qquad -y'_{12a} = \frac{s}{2s+9}$$

The transmission zero is at the origin. Hence a continued fraction expansion of y'_{11a} around the origin is made:

$$9 + 2s \overline{\big)\, 2 + s\, \big(\, \tfrac{2}{9}}$$
$$\underline{2 + \tfrac{4}{9}s}$$
$$\tfrac{5}{9}s\, \overline{\big)\, 9 + 2s\, \big(\, \tfrac{81}{5s}}$$
$$\underline{9}$$
$$2s\, \overline{\big)\, \tfrac{5}{9}s\, \big(\, \tfrac{5}{18}}$$
$$\underline{\tfrac{5}{9}s}$$
$$0$$

The resultant network is shown in Figure 9-25a. The multiplier constant H_a of the realized two-port, is found to be $\tfrac{5}{9}$.

Next, the two-port $\mathcal{N}'_b$ is developed from the following set of parameters:

$$y'_{11b} = \frac{s+2}{2s+9}, \qquad -y'_{12b} = \frac{3}{2s+9}$$

In this case, we note that the transmission zero is at infinity. This implies that $\mathcal{N}'_b$ can be obtained by making a continued fraction expansion of

380 / NEGATIVE-IMPEDANCE CONVERTER AS AN ACTIVE ELEMENT

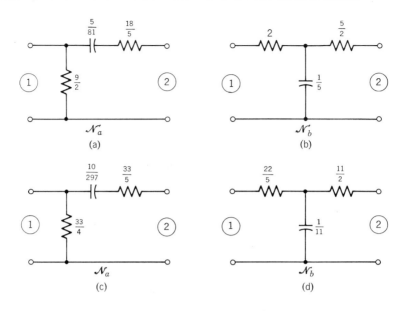

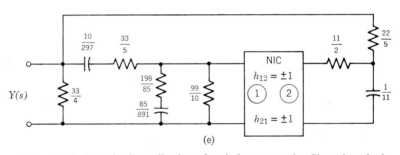

Figure 9-25 Steps in the realization of an inductance, using Sipress' method.

y'_{11b} around infinity:

$$s+2 \begin{vmatrix} 2s+9 & 2 \\ 2s+4 & \end{vmatrix}$$

$$5 \begin{vmatrix} s+2 & \dfrac{s}{5} \\ s & \end{vmatrix}$$

$$2 \begin{vmatrix} 5 & \dfrac{5}{2} \\ 5 & \\ 0 & \end{vmatrix}$$

The desired two-port N_b'' is sketched in Figure 9-25b. The realized value of H_b is 2. Using this information in Equation (9.49), we obtain

$$K_3 = \frac{K_1}{\left(\dfrac{h_a}{H_a} + \dfrac{h_b}{H_b}\right)} = \frac{1}{(\tfrac{9}{5} + \tfrac{3}{2})} = \frac{10}{33}$$

The admittance level of N_a'' is then raised by $\tfrac{6}{11}$ and that of N_b'' is raised by $\tfrac{5}{11}$. The resulting networks are shown in Figures 9-25c and 9-25d, from which by analysis we obtain

$$y_{22a} = \frac{\tfrac{10}{33}s}{2s + 9}$$

$$y_{22b} = \frac{\tfrac{2}{11}(2s + 5)}{2s + 9}$$

Now $D(s)/Q(s)$ is equal to $s/(2s + 9)$. Using all this information in Equation 9.50, we arrive at

$$Y_1 - Y_2 = \frac{\tfrac{2}{11}(2s + 5)}{2s + 9} - \frac{\tfrac{10}{33}s}{2s + 9} + \frac{s}{2s + 9} = \frac{\tfrac{35}{33}s + \tfrac{10}{11}}{2s + 9}$$

This implies that

$$Y_2 = 0 \quad \text{and} \quad Y_1 = \frac{\tfrac{35}{33}s + \tfrac{10}{11}}{2s + 9}$$

The complete realization is then as shown in Figure 9-25e.

Observe that from a practical point of view, the above method is advantageous because the two RC networks can always be realized in unbalanced forms. Compared to other synthesis methods, Sipress' approach is slightly more complicated.

9-3 SYNTHESIS OF TRANSFER FUNCTIONS

Use of a negative-impedance converter as a network element in the realization of a specified transfer function is considered next. The two most elegant transfer-function synthesis methods are due to Linvill and Yanagisawa. A systematic development of these synthesis methods are outlined in this section. It should be noted that some of the methods discussed in the previous chapter can also be considered as synthesis using NIC. Specifically note that the active RC configurations of Figures 8-22 and 8-29 are essentially RC:NIC structures.

Several Basic RC:NIC Structures

The basic RC:NIC two-port of Figure 9-26a is characterized by a transfer admittance

(9.51) $$Y_{21} = \frac{I_2}{V_1} = Y_1 - Y_2$$

Thus, if Y_1 and Y_2 are required to be RC one-ports, a specified transfer

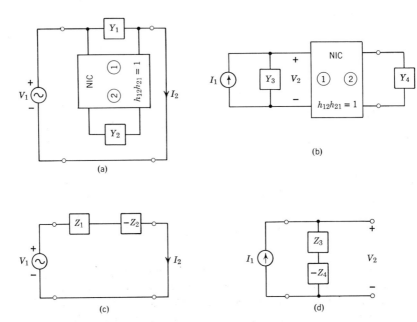

(a)

(b)

(c)

(d)

Figure 9-26 Several basic RC:NIC structures for transfer-function realizations.

admittance $N(s)/D(s)$ can be realized in the form of Figure 9-26a provided that:

(i) $D(s)$ has only distinct negative real zeros
and
(ii) $D(s)^0 + 1 \geq N(s)^0$.

The second structure shown in Figure 9-26b has a transfer impedance given by

(9.52) $$Z_{21} = \frac{V_2}{I_1} = \frac{1}{Y_3 - Y_4}$$

To make Y_3 and Y_4 RC one-port networks, a given transfer impedance $P(s)/Q(s)$ must have the following restrictions:

(i) $P(s)$ has only distinct negative real zeros

and

(ii) $P(s)^0 + 1 \geq Q(s)^0$.

Observe the similarity between the basic structures of Figures 9-26a and 9-26b and those of Figures 8-20 and 8-21.

The third basic configuration of Figure 9-26c has a transfer admittance

(9.53) $$Y_{21} = \frac{I_2}{V_1} = \frac{1}{Z_1 - Z_2}$$

and the last two-port (Figure 9-26d) is characterized by a transfer impedance of the form

(9.54) $$Z_{21} = \frac{V_2}{I_1}\bigg|_{I_2=0} = Z_3 - Z_4$$

Realization steps are similar to the ones described earlier.

Linvill's Method

If both of the RC networks of Figure 9-26b (or Figure 9-26c) are generalized into RC two-ports, the cascade RC:NIC configuration of Linvill[35] (Figure 9-27) is obtained. In terms of the z-parameters of the two ports, the transfer impedance is given as

(9.55) $$Z_{21} = \frac{V_2}{I_1}\bigg|_{I_2=0} = \frac{z_{12a}z_{12b}}{z_{22a} - kz_{11b}}$$

if the active device is a VNIC ($h_{12} = -k$, $h_{21} = -1$). If, however, the negative-impedance converter is chosen to be of the current inversion type ($h_{12} = 1$, $h_{21} = k$), then Z_{21} is given as:

(9.56) $$Z_{21} = \frac{V_2}{I_1}\bigg|_{I_2=0} = \frac{kz_{12a}z_{12b}}{kz_{11b} - z_{22a}}$$

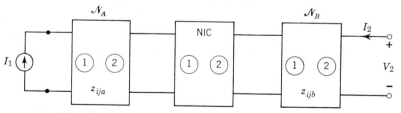

Figure 9-27 The cascade RC:NIC configuration of Linvill.

[35] J. G. Linvill, "A new RC filter employing active elements," *Proc. N.E.C.*, **9**, 342–352 (1963); "RC active filters," *Proc. IRE*, **42**, 555–564 (March 1964).

Without any loss of generality, we shall assume the active device to be a UVNIC ($k = 1$) and base our synthesis steps on expression 9.55. Similar steps can be followed for a CNIC.

Let $Z_{21}(s) = N(s)/D(s)$ be a specified real rational function. Select a polynomial $Q(s)$ having all distinct negative real zeros so that

$$Q(s)^0 \geq \max [N(s)^0, D(s)^0]$$

A partial fraction expansion of $D(s)/Q(s)$ yields (Theorem 3-5),

$$\frac{D(s)}{Q(s)} = \frac{D_a(s)}{Q_a(s)} - \frac{D_b(s)}{Q_b(s)}$$

where $D_a(s)/Q_a(s)$ is the sum of terms in the partial fraction expansion with positive residues and $D_b(s)/Q_b(s)$ is the corresponding sum with negative residues. The numerator polynomial $N(s)$ can be grouped into two factors $N_a(s)N_b(s)$. We thus write

(9.57) $$Z_{12} = \frac{N(s)}{D(s)} = \frac{\dfrac{N_a(s)}{Q_a(s)} \cdot \dfrac{N_b(s)}{Q_b(s)}}{\dfrac{D_a(s)}{Q_a(s)} - \dfrac{D_b(s)}{Q_b(s)}}$$

Comparing Equations 9.57 and 9.55, we obtain

(9.58)
$$z_{12a} = \frac{N_a(s)}{Q_a(s)}, \quad z_{22a} = \frac{D_a(s)}{Q_a(s)}$$
$$z_{12b} = \frac{N_b(s)}{Q_b(s)}, \quad z_{22b} = \frac{D_b(s)}{Q_b(s)}$$

Note that $N_a(s)$ and $N_b(s)$ are arbitrarily chosen except for the fact that $\{z_{ija}\}$ and $\{z_{ijb}\}$ as given by expression 9.58 must form RC realizable sets. Complete realization is achieved by synthesizing the networks $\mathcal{N}_A$ and $\mathcal{N}_B$ by means of standard passive RC synthesis techniques.

Two drawbacks of this method are now evident. Since the transmission zeros are realized by two RC two-ports, it is not possible for the structure of Figure 9-27 to have a common ground between input and output if there exists positive real transmission zeros. Next, we note that the maximum realizable value of the multiplier constant is limited by the maximum obtainable by means of transformerless RC two-ports.

Example 9-6 Realize a second-order maximally flat delay filter. The specified transfer function is

$$Z_{21} = \frac{H}{s^2 + 3s + 3}$$

Choose $Q(s) = s(s + 1)$. This results in the following z-parameters:

$$z_{12a} = \frac{H_a}{s}, \quad z_{22a} = \frac{s+3}{s}$$

$$z_{12b} = \frac{H_b}{s+1}, \quad z_{11b} = \frac{1}{s+1}$$

The final realization is shown in Figure 9-28. Simple calculation will show that the realized value of the multiplier constant H is 3, which is the same as that which would have been obtained by a transformerless RLC network.

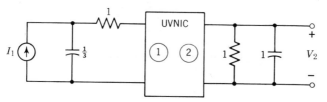

Figure 9-28 Realization of the transfer impedance of Example 9-6, using Linvill's approach.

Yanagisawa's Method

A major problem of Linvill's method is the realization of the RC two-ports. If the transmission zeros are on the negative real axis, then the realization can be accomplished by the zero-shifting method. In case of complex zeros, either Fialkow-Gerst or Dasher's synthesis techniques must be followed. In Yanagisawa's method,[36] the companion RC networks are RC one-ports and thus eliminate the above problem.

The RC:NIC configuration proposed by Yanagisawa can be developed from the basic structures indicated in Figure 9-26.[37] Cascading the two structures, we obtain the two-port of Figure 9-29a, where for simplicity we have not drawn the negative-impedance converters. Essentially, the network can be considered as a voltage divider and thus the voltage transfer ratio is given as:

$$(9.59) \quad \left.\frac{V_2}{V_1}\right|_{I_2=0} = \frac{\dfrac{1}{Y_3 - Y_4}}{\dfrac{1}{Y_1 - Y_2} + \dfrac{1}{Y_3 - Y_4}} = \frac{1}{1 + \dfrac{Y_3 - Y_4}{Y_1 - Y_2}}$$

Let us now realize the negative admittance by terminating a UCNIC by

---

[36] T. Yanagisawa, *loc. cit.*

[37] Based on an approach suggested by A. W. Keen in "Recent Developments in Network Theory", S. R. Deards, Ed., Pergamon Press, 1965.

Y_4 as indicated in Figure 9-29b. Since a UCNIC inverts the direction of current, keeping the port voltages equal to each other, the voltage transfer ratio of the two-port stays invariant if the branch $-Y_2$ is replaced by Y_2 and connected as shown in Figure 9-29c. This last configuration was originally proposed by Yanagisawa. Myers[38] and Thomas[39] have independently proposed a lattice origin of the same configuration.

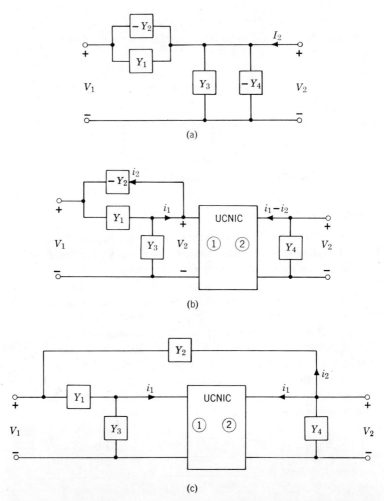

Figure 9-29 Development of the first RC:CNIC configuration of Yanagisawa.

[38] B. R. Myers, "Transistor-RC network synthesis," *IRE WESCON Conv. Rec.*, **3** (Part 2), 65–74 (August 1959).

[39] R. E. Thomas, Tech. Note No. 8, Circuit Theory Group, Univ. of Illinois, Urbana, Ill., 1959.

SYNTHESIS OF TRANSFER FUNCTIONS / 387

The similarity of expression 9.59 with expression 7.51 should be noted. In addition, the method of development is analogous to the development of Figure 8-32 in Section 8-3.

Identification of Y_1, Y_2, Y_3, and Y_4 in the form of RC admittances forms a specified voltage transfer ratio $N(s)D(s)$ is straightforward and we write it

(9.60) $$\frac{V_2}{V_1} = \frac{N(s)}{D(s)} = \frac{1}{1 + \dfrac{D(s) - N(s)}{N(s)}}$$

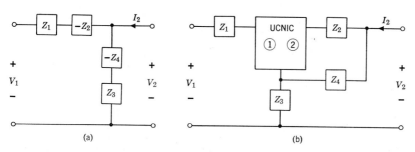

Figure 9-30 Development of the second RC:CNIC configuration of Yanagisawa.

Now we choose a polynomial $Q(s)$ with simple negative real roots such that

(9.61) $$Q(s)^0 + 1 \geq \max\,[N(s)^0,\,D(s)^0]$$

From Theorem 3-5, it is seen that we can always write

(9.62) $$\frac{D(s) - N(s)}{Q(s)} = Y_{RC}^{(3)} - Y_{RC}^{(4)}$$
$$\frac{N(s)}{Q(s)} = Y_{RC}^{(1)} - Y_{RC}^{(2)}$$

The remaining steps should be obvious. From the method of synthesis, it is seen that any real rational function can be realized in the form of Figure 9-29c. Because of the use of CNIC in realizing both zeros and poles, the final realization, in general, has fewer resistances and capacitances. Even though the method does not impose any restriction on the transfer function, from the sensitivity point of view, it is better to realize second-order stages and cascade them with buffer amplifiers.

The development of the second Yanagisawa configuration is shown in Figure 9-30, where the intermediate steps have been omitted for brevity. It is apparent that the voltage transfer ratio of this RC:NIC structure is given as

(9.63) $$\left.\frac{V_2}{V_1}\right|_{I_2=0} = \frac{Z_3 - Z_4}{(Z_1 - Z_2) + (Z_3 - Z_4)} = \frac{1}{1 + \dfrac{Z_1 - Z_2}{Z_3 - Z_4}}$$

Synthesis steps follow a similar pattern to that outlined for the first Yanagisawa configuration (Figure 9-29c), and will be omitted. The active two-port in the second case does not have a common ground with the input and output ports, thus limiting its practical usefulness.

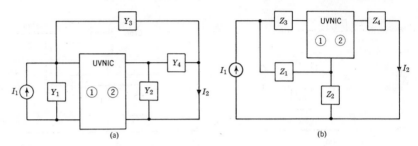

Figure 9-31 Yanagisawa type RC:VNIC configurations as advanced by Thomas.

The dual structures to Yanagisawa's configurations were suggested by Thomas.[40] These networks, shown in Figure 9-31, employ the VNIC as the active device and are suitable for the realization of current transfer ratio. Synthesis methods are left as an exercise. These structures are easily developed from the basic RC:NIC configurations.

Example 9-7 Let us realize the maximally flat delay filter of Example 9-6 with a multiplier constant of value 6.

$$t_v = \frac{N(s)}{D(s)} = \frac{6}{s^2 + 3s + 3}$$

Choose $Q(s) = (s + 1)$. Then

$$\frac{N(s)}{Q(s)} = 6 - \frac{6s}{s+1}, \qquad \frac{D(s) - N(s)}{Q(s)} = s + \frac{5s}{s+1} - 3$$

As a result,

$$Y_1 = \frac{6s}{s+1}, \qquad Y_2 = 6; \qquad Y_3 = 3; \qquad Y_4 = s + \frac{5s}{s+1}$$

Final realization is shown in Figure 9-32.

[40] R. E. Thomas, *loc. cit.*

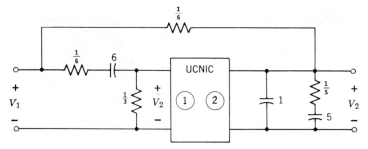

Figure 9-32 Realization of the transfer voltage ratio of Example 9-7, using Yanagisawa's method.

If a load resistance is desired at the output, then the output can be taken across the resistance across the input port of the CNIC.

Synthesis Using Nonideal NIC

A practical NIC, in general, is not an ideal device. If h_{11} and h_{22} are real, Larky's compensation technique can be followed to a certain extent. It is also possible to "precorrect" the companion RC network to compensate for the nonideal behavior of the NIC. To understand the method of precorrection, consider again Example 9-6. Suppose the available NIC is nonideal and is described by

$$h_{11} = \tfrac{1}{2}$$

$$h_{22} = \tfrac{1}{2}s + \tfrac{1}{4}$$

$$h_{12}h_{21} = 1$$

A careful examination of Figure 9-28 reveals that the effect of nonidealness can be precorrected by adjusting the series resistance connected to port 1 of the impedance converter and by adjusting the shunt resistance and capacitance at port 2. The precorrected network is shown in Figure 9-33. From this simple example, it is seen that in some cases it is possible

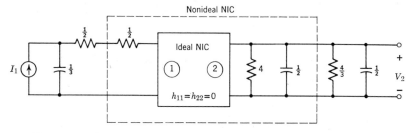

Figure 9-33 Illustration of the use of nonideal NIC in the design of filter.

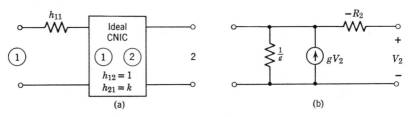

Figure 9-34 A nonideal NIC model and its controlled-source equivalent circuit.

to compensate for a nonideal NIC by precorrecting the element values of some elements of the companion *RC* network. Thus, it is not necessary to use an extra compensating network.

We shall outline a method of precorrection for the cascade *RC*:NIC configuration of Linvill, where the NIC is of the current-inversion type. This implies that the form of the transfer function is given by Equation 9.56. For simplicity, we consider a CNIC with $h_{11} \neq 0$. The model of such a negative-impedance converter and its controlled source equivalent is indicated in Figure 9-34, where

(9.64)
$$g = \frac{(1 + k)}{h_{11}}$$
$$R_2 = \frac{h_{11}}{k}$$

A transistor realization of this active device, suggested by Horowitz,[41] is shown in Figure 9-35. If the transistors are not ideal, i.e., $r_{e1} \neq 0$ and $r_{e2} \neq 0$, then this circuit is almost identical to that of Figure 9-7a. We

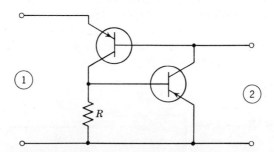

Figure 9-35 A transistorized circuit realization of the nonideal NIC of Figure 9-34.

[41] I. M. Horowitz, "Synthesis of active *RC* transfer functions," MRI Res. Rept. R-507-56, PIB-437, Polytech. Inst. of Brooklyn.

thus obtain

(9.65)
$$k = \frac{R}{r_{e2}}$$

$$h_{11} = r_{e1}$$

In order to guarantee the compensation for h_{11}, it is sufficient to ensure the presence of a positive resistance in series at port 2 of the network $\mathcal{N}_A$. Since the denominator of Equation 9.56 is of the difference type, this can

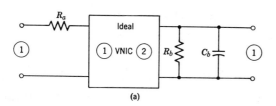

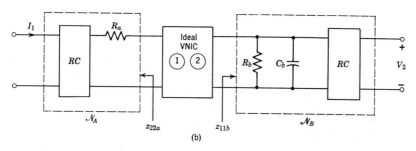

Figure 9-36 (a) A more accurate model of a nonideal VNIC. (b) The precorrection scheme for the Linvill's $RC:NIC$ configuration to enable the use of nonideal VNIC.

always be achieved by adding and subtracting a constant. If this approach is not desirable, a suitable $Q(s)$ can always be chosen to insure $z_{22a}(\infty)$ to be a finite constant.

A general precorrection scheme has been considered by Hurtig,[42] based on a more accurate model of a nonideal NIC as shown in Figure 9-36a. Thus, in order to precorrect the Linvill $RC:VNIC$ configuration, it is desirable to have the final network realization in the form shown in Figure 9-36b. To guarantee the extraction of series and shunt impedance in the realization of $\mathcal{N}_A$ and $\mathcal{N}_B$, it can be shown that the following

---

[42] G. Hurtig, III, "Transfer function realization using non-ideal negative-impedance converter," M.S. Thesis, Cornell University, Ithaca, N.Y., June 1964.

conditions must be satisfied:

(9.66)
(i) $0 < z_{22a}(\infty) < \infty$
(ii) $z_{11b}(\infty) = 0$
(iii) $0 < z_{11b}(0) < \infty$

Conditions in Equation 9.66 can be satisfied by choosing $Q(s)$ so that:

(9.67)
(i) $Q(s)^0 = D(s)^0$
(ii) $\text{sgn}(D_n) = \text{sgn}(Q_n)$
(iii) $\text{sgn}(D_0) = \text{sgn}(Q_1)$

where $D(s) = \sum_{j=0}^{n} D_j s^j$ and $Q(s) = \sum_{i=0}^{n} Q_i s^i$. Observe from Figure 9-36b that the specified transfer impedance $Z_{21}(s)$ must have at least a transmission zero at infinity.

9-4 SENSITIVITY CONSIDERATIONS: THE HOROWITZ DECOMPOSITION

Almost all of the synthesis methods of the last two sections hinge upon the selection of an arbitrary polynomial $Q(s)$ having only distinct negative real roots. Let us now investigate the possibility of selecting a $Q(s)$ that would minimize the sensitivity of the network with respect to the conversion factor of the NIC.

An RC:NIC decomposition is of the form

(9.68)
$$\frac{D(s)}{Q(s)} = \frac{D_a(s)}{Q_a(s)} - k\frac{D_b(s)}{Q_b(s)}$$

which implies that

(9.69) $\quad D(s) = D_a(s)Q_b(s) - kQ_a(s)D_b(s) = A(s) - kB(s)$

where

(9.70) $\qquad A(s) = D_a(s)Q_b(s), \quad B(s) = Q_a(s)D_b(s)$

Let us first repeat the requirements of a proper RC:$-RC$ decomposition of $D(s)$ suitable for RC:NIC type synthesis:[43]

(i) $A(s)$ and $B(s)$ must have only negative real roots with multiplicity not exceeding two;

---

[43] See Section 3-6, p. 100.

SENSITIVITY CONSIDERATIONS: THE HOROWITZ DECOMPOSITION / 393

(ii) If N_A and N_B denote the numbers of zeros of $A(s)$ and $B(s)$, respectively, to the right of any point on the negative real axis, then

$$|N_A - N_B| \leq 1$$

The above two conditions follow from the requirement that $D_a(s)/Q_a(s)$ and $D_b(s)/Q_b(s)$ be RC driving-point impedances.

We have shown in Section 5-3, p. 182, that minimization of the coefficient sensitivities minimizes the root sensitivities. Thus we shall focus our attention on making the coefficient sensitivities small.

Let

(9.71)
$$D(s) = \sum_{i=0}^{n} D_i s^i$$
$$A(s) = \sum_{i=0}^{m} A_i s^i$$
$$B(s) = \sum_{i=0}^{m} B_i s^i$$

Without any loss of generality, let us assume that the nominal value of k is unity. Now

(9.72)
$$\hat{S}_k^{D_i} = -\frac{B_i}{D_i}$$

This implies that $\hat{S}_k^{D_i}$ can be minimized by minimizing B_i and hence by decreasing A_i. Since $B_n, B_{n+1}, \ldots, B_m$ can be made equal to zero without violating the two constraints, from the sensitivity point of view it is better to choose $A(s)$ of degree n [same as $D(s)^0$] and to choose $B(s)$ of degree $n - 1$. For simplicity, assume $D(s)$ is a second-degree polynomial with no negative real roots:

(9.73)
$$D(s) = s^2 + D_1 s + D_0$$

Observe that without violating the constraints we can make $\hat{S}_k^{D_0}$ equal to zero by choosing B_0 equal to zero. Thus there are two possible decompositions of $D(s)$:

(9.74a)
$$D(s) = (s + \alpha_1)(s + \alpha_2) - \beta_1 s$$

(9.74b)
$$= (s + \alpha)^2 - \beta s$$

For the first decomposition, Equation 9.74a:

(9.75a)
$$D_1 = \alpha_1 + \alpha_2 - \beta_1, \qquad D_0 = \alpha_1 \alpha_2$$
$$(\hat{S}_k^{D_1})_I = -\frac{\beta_1}{D_1}$$

and for the second decomposition, Equation 9.74b:

(9.75b)
$$D_1 = 2\alpha - \beta, \qquad D_0 = \alpha^2$$
$$(\hat{S}_k^{D_1})_{II} = -\frac{\beta}{D_1}$$

It is a simple exercise to show that

$$\beta_1 - \beta = \frac{(\alpha_2 - \alpha)^2}{\alpha_2} > 0$$

implying

$$(\hat{S}_k^{D_1})_I > (\hat{S}_k^{D_1})_{II}$$

Therefore, for a second-order polynomial, the second decomposition (9.74b) is optimum. For a higher-order polynomial, the form of the optimum decomposition is of the same form.

Following a root-locus approach, Horowitz[44] has proved that, for a polynomial $D(s)$ with no negative real roots, there exists a unique $RC:NIC$ type decomposition that minimizes the coefficient sensitivity with respect to a variation of the conversion factor of the NIC, and the desired optimum decomposition is given as:

(9.76a) $\qquad D(s) = a^2(s) - sb^2(s) \qquad$ if $\qquad D(s)^0$ is even

(9.76b) $\qquad\qquad = sb^2(s) - a^2(s) \qquad$ if $\qquad D(s)^0$ is odd

where $a(s)/sb(s)$ is an RC driving-point impedance. For future reference we shall refer to the decomposition of $D(s)$ as shown in Equation 9.76 as the *Horowitz Decomposition*. It can be shown[44] that the Horowitz Decomposition also minimizes $S_k^{D(j\omega)}$ and is usually the optimum with respect to the passive structures.

In some of the synthesis techniques using other types of active devices, e.g., Hakim's method of Section 8-3, p. 318, $RC:NIC$ type decomposition is used. As a result, the Horowitz Decomposition also represents the best possible for these synthesis techniques.

A simple algebraic procedure[45] can be followed to derive the Horowitz Decomposition. The main steps of the procedure are outlined next. Assume $D(s)$ does not have any negative real roots. Hence the roots of $D(s^2)$ will have a quadrantal symmetry, i.e., we can express;

(9.77a) $\qquad D(s^2) = F(s)F(-s) \qquad$ if $\qquad D(s^2)^0$ is even

(9.77b) $\qquad\qquad\;\; = -F(s)F(-s) \qquad$ if $\qquad D(s^2)^0$ is odd

---
[44] I. M. Horowitz, "Optimization of negative-impedance conversion methods of active RC synthesis," *IRE Trans. on Circuit Theory*, CT-6, 296–303 (September 1959).
[45] E. A. Guilemin, *Passive network synthesis*, Wiley, New York, p. 559.

where $F(s)$ is the Hurwitz polynomial obtained by grouping together the left-half s plane zeros of $D(s^2)$ and $F(-s)$ is the corresponding anti-Hurwitz polynomial. Now it is well known that Ev $F(s)$/Od $F(s)$ is always an LC driving-point impedance function.[46] Let

(9.78)
$$\text{Ev } F(s) = a(s^2)$$
$$\text{Od } F(s) = sb(s^2)$$

Therefore, by applying $LC:RC$ transformation (Theorem 3-1), we observe that $a(s)/sb(s)$ is an RC driving-point impedance.

Now from expressions 9.77 and 9.78, we have

$$D(s^2) = \pm\{a(s^2) + sb(s^2)\}\{a(s^2) - sb(s^2)\}$$
$$= \pm\{a^2(s^2) - s^2b^2(s^2)\}$$

This implies that

(9.79)
$$D(s) = \pm\{a^2(s) - sb^2(s)\}$$

which is the Horowitz Decomposition.[47]

If negative real roots are present, then the optimum decomposition of $D(s)$ is obtained by expressing

$$D(s) = D'(s)D''(s)$$

where $D'(s)$ has all the negative real roots and $D''(s)$ has no negative real roots, and deriving the Horowitz Decomposition of $D''(s)$.

Example 9-8 Suppose

$$D(s) = s^3 - 4s^2 + 4s - 16$$

The roots of $D(s^2) = s^6 - 4s^4 + 4s^2 - 16$ are shown in Figure 9-37. Therefore, we can write

$$D(s^2) = (s + 2)(s + 1 + j1)(s - 2)(s - 1 + j1)(s - 1 - j1)$$
$$= (4s^2 + 4 + s^3 + 6s)(4s^2 + 4 - s^3 - 6s)$$
$$= -\{(4s^2 + 4)^2 - (s^3 + 6s)^2\}$$

Hence, the Horowitz Decomposition of $D(s)$ is given as

$$D(s) = s(s + 6)^2 - 16(s + 1)^2$$

---

[46] E. S. Kuh and D. O. Pederson, *Principles of circuit synthesis*, McGraw-Hill, New York, p. 97.
[47] This algebraic procedure of obtaining the Horowitz Decomposition has also been proposed by: D. A. Calahan. "Notes on Horowitz optimization procedure," *IRE Trans. on Circuit Theory*, **CT-7**, 352–354 (September 1960) and R. E. Thomas, "Polynomial decomposition in active RC network synthesis," *IRE Trans. on Circuit Theory*, **CT-9**, 270–274 (September 1961).

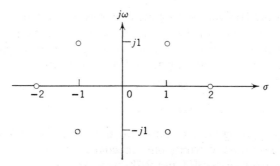

Figure 9-37 The locations of the roots of the polynomial $D(s^2)$ of Example 9-8.

9-5 SUMMARY

Several transistor circuit realizations of the negative-impedance converter (NIC) are developed in Section 9-1, p. 351, from their nullator-norator models given in Tables 9-1 and 9-2. The basic two-transistor circuits of Figures 9-4 through 9-7 represent the current-inversion type NIC, and those of Figures 9-8 through 9-11 represent the voltage-inversion type NIC.

A practical NIC circuit is examined from a sensitivity point of view in Section 9-1, p. 359. It is shown that in order to minimize the sensitivity, a NIC circuit must have very small h_{11} and h_{22}, and the variation of the conversion factor $h_{12}h_{21}$ over the frequency range of interest should be small. The conversion-factor-sensitivities of several typical NIC circuits are also derived for comparison purposes.

An NIC in practice is open-circuit-stable (OCS) at one-port and is short-circuit-stable (SCS) at the other port (Section 9-1, p. 362). Hence when used with *RLC* networks, the impedance seen at the OCS port and the admittance seen at the SCS port must each be very high.

Three driving-point synthesis methods of *RC*:NIC one-port networks are outlined in Section 9-2. Out of the three methods, Sandberg's method is most attractive because of easy to use realization steps.

Two methods of transfer function synthesis are reviewed in Section 9-3. Linvill's method (Section 9-3, p. 383), which employs two *RC* two-ports and an NIC in cascade, basically realizes an open-circuit transfer impedance. However, by suitably developing the *RC* two-ports, other types of transfer functions can easily be realized. The cascaded configuration allows certain amount of independent adjustment of the transmission zeros. There are essentially three minor drawbacks of Linvill's method: (1) since the transmission zeros are realized as the transmission zeros of the

RC two-ports, realization of complex transmission zeros pose a somewhat difficult problem, (2) transfer function is always realized within a multiplicative constant, (3) positive real transmission zeros cannot be realized if a common ground is desired between the input and output ports. In contrast to Linvill's method, Yanagisawa's method (Section 9-3, p. 385), which employs four RC one-ports and an NIC, does not impose any restriction on the realizability of the transfer function other than that it be a real rational function. The synthesis procedure is also extremely simple in this case. A major drawback of both methods is the necessity of using additional buffer amplifiers when cascading more than one stage.

Section 9-4 considers the optimum decomposition for minimizing the pole sensitivities of the transfer function with respect to the conversion factor $h_{12}h_{21}$. It is shown that the best decomposition of the denominator polynomial is given by expression 9.76.

Additional References

Antoniou, A., "New RC-active-network synthesis using negative-impedance converters," *Proc. IEE (London)*, **114**, 894–902 (July 1967).

Antreich, K., and E. Gleissner, "Über die realisierung von impulsefilteren durch aktive RC-netzwerke," *Archiv der Elektrischen Übertragung*, **19**, 309–316 (June 1965).

Barranger, J., "Voltage transfer function matrix realization using current negative-immittance converters," *IEEE Trans. on Circuit Theory*, **CT-13**, 97 (March 1966).

Bitzer, W., "Über negativ-impedanc-konverter," *Bull. SEV*, **10**, 373–379 (1965).

Blecher, F. H., "Application of synthesis techniques to electronic circuit design," *IRE Trans. on Circuit Theory*, **CT-7** (Special Supplement), 79–91 (August 1960).

Butler, F., "Active impedance converters," *Wireless World*, **71**, 600–604 (December 1965).

Chien, R. T., "On the synthesis of active networks with one negative-impedance converter," *Proc. NEC*, **16**, 405–411 (1960).

Cruz, J. B., Jr., "A synthesis procedure based on Linvill's RC active structure," *IRE Trans. on Circuit Theory*, **CT-6**, 133–134 (March 1959).

Drew, A. J., and J. Gorski-Popiel, "Directly-coupled negative-impedance converter," *Proc. IEE (London)*, **111**, 1282 (1964).

Franklin, D. P., "Direct-coupled negative-impedance converter," *Electronics Letters*, **1**, 1 (March 1965).

Gorski-Popiel, J., and A. J. Drew, "RC active ladder networks," *Proc. IEE (London)*, **112**, 2213–2219 (December 1965).

Gorski-Popiel, J., "RC active networks," *Electronics Letters*, **1**, 288–289 (December 1965).

Hirasaki, T., and H. Date, "An approximation method of prescribed amplitude and phase characteristics and realization by active network," *Summaries of Papers, I.C.M.C.I.*, Part 2, Circuit Theory, Tokyo, Japan, pp. 81–82 (September 1964).

Holt, A. G. J., and F. W. Stephenson, "Design tables for active filters having 2nd and 4th order Chebyshev responses in pass and stop bands," *Proc. IEE (London)*, **111**, 1807–1820 (November 1964).

Holt, A. G. J., and F. W. Stephenson, "The effects of errors in the element values and the converter transfer characteristic on the responses of active *RC* filters," *Radio Electronic Engr.*, **26**, 449–457 (1963).

Holt, A. G. J., and F. W. Stephenson, "A specialization of the Yanagisawa synthesis procedure to obtain *RC* active networks with optimum pole sensitivity," *Proc. IEE (London)*, **112**, 2227–2234 (1965).

Holt, A. G. J., and M. L. Canning, "Synthesis of an active 2-port network using n-port synthesis," *Proc. IEE (London)*, **112**, 1689–1694 (September 1965).

Horowitz, I. M., "Negative Impedance Converters," *IRE Trans. on Component Parts*, **CT-9**, 33–38 (March 1962).

Hove, R. G., "Matched filter design with *RC* active delay networks for low data rate signal processing," Proc. 1st Annual Communications Conf., and Globecom VII, Univ. of Colorado, June 1965.

Hudson, F. J., "Synthesis of transfer admittance functions using active components," *IBM J. Res. and Dev.*, **7**, 40–43 (January 1963).

Huelsman, L. P., "Active *RC* synthesis with prescribed sensitivities," *Proc. N.E.C.*, **16**, 412–426 (1960).

Huelsman, L. P., "Use of two negative-impedance converters to synthesize *RC* transfer function," *IRE Trans. on Circuit Theory*, **CT-8**, 357 (September 1961).

Huelsman, L. P., "Optimum sensitivity active *RC* band pass filters," *Electronics Letters*, **1**, 226 (October 1965).

Huelsman, L. P., "The compensation of negative-immittance converters," *Proc. I.E.E.E.*, **54**, 1015–1016 (July 1966).

Jost, W., "Theoretische und experimentelle untersuchung eines negativ impedanz konverters," *Arch. Elekt. Übertragung*, **20**, 281–288 (1966).

Kallmann, H. E., "A Simple DC–AC negative impedance converter," *Proc. IEEE*, **52**, 199–200 (February 1964).

Keen, A. W., "Derivation of some basic negative-immittance converter circuits," *Electronics Letters*, **2**, 113–115 (March 1966).

Keen, A. W., "Transfer function synthesis with active unbalanced equivalents of the lattice," *Radio and Electronic Engr.*, **32**, 101–111 (August 1966).

Kuo, F. F., "Transfer function synthesis with active elements," "*Proc. N.E.C.*, **13**, 1049–1056 (1957).

Lee, R. F., "A comment on Sandberg's graphical method for the synthesis of biquadratic driving-point impedance functions," *IEEE Trans. on Circuit Theory*, **CT-10**, 441–442 (September 1963).

Linvill, J. G., "Synthesis techniques and active *RC* filters," *IRE Nat'l Conv. Rec.*, **5** (Part 2), 90–93 (1957).

Marshak, A. H., "Direct coupled negative-impedance balanced converter," *Electronics Letters*, **1**, 142–143 (July 1965).

Sipress, J. M., "Synthesis of multi-parameter active *RC* networks," NEREM Record, November 1960.

Storey, D. J, and W. J. Cullyer, "Active low-pass linear phase filters for pulse transmission," *Proc. IEE (London)* **112**, 661–668 (April 1965).

Storey, D. J., and W. J. Cullyer, "Network synthesis using negative-impedance converters," *Proc. IEE (London)*, **111**, 891–906 (May 1964).

Thomas, R. E., "The use of the active lattice to optimize transfer function sensitivities," Proc. Symp. Active Networks and Feedback Systems, MRI Symposia Series, Vol. 10, Polytech. Inst. of Brooklyn, pp. 179–188 (1960).

Thomas, R. E., "The active constant-resistance lattice," *Proc. N.E.C.*, **15**, pp. 727–737 (1959).

Thomas, R. E., "Miyata's Method applied to active *RC* network synthesis," *Proc. N.E.C.*, **18**, 40–44 (1962).

Todd, C. D., "A versatile negative impedance converter," *Semi-conductor Products*, **6**, 25–29 (May 1963).

Tsirel'son, D. A., "Theory of active impedance converters," *Telecommunications and Radio Engr.*, Part 1, No. 6, pp. 50–54 (June 1966).

Valand, J., "One technique of *RC* active filter synthesis," *Proc. IEEE*, **54**, 1121–1122 (1966).

Vlach, J., and J. Bendik, "Active filters with low sensitivity to element changes," *Radio and Electronic Engr.*, **33**, 305–316 (May 1967).

Vlach, J., "Improving sensitivity of negative impedance converter active filters by approximation," Proc. Asilomar Conf. on Circuits and Systems, Monterey, Calif. pp. 89–99, November 1967.

Problems

9.1 The circuit of Figure 2-65a is a CNIC[48] and the circuit of Figure 2-65b is a VNIC.[49] Compute the conversion-factor-sensitivity $S_k^{h_{12}h_{21}}$ and compare the two circuits.

9.2 Show that any real rational function can be realized as the driving-point function of either of the two active one-ports of Figure 9-16.

9.3 Using the technique outlined in Section 3-4, realize the following voltage transfer ratios by Kinariwala's method (Section 9-2, p. 366):

(a) $\dfrac{H(s^2 + 1)}{s^2 + s + 1}$

(b) $\dfrac{H}{s^2 + s + 3}$

9.4 Realize the following driving-point functions using Kinariwala's method:

(a) s^2

(b) $\dfrac{s^2 + s + 1}{s^2 + s + 2}$

(c) $\dfrac{s^2 - 3s + 2}{s^2 + s + 1}$

---

[48] A. I. Larky, *loc. cit.*
[49] B. R. Myers, *Electronics Letters*, **1**, 68–70 (May 1965).

9.5 Realize the functions of Problem 9.4 by means of the method given in Section 9-2, p. 370.

9.6 Show that a biquadratic impedance function can be realized in the form of Figure 9-21a (or Figure 9-21c) with Z_1 and Z_2 as simple resistors and, at most, two capacitors in the realization of Z_3 and Z_4.[50]

9.7 Realize using Sipress' method the driving-point functions of Problem 9.4.

9.8 Figures 9-38a and b show two RC:NIC configurations that are suitable for transfer-function synthesis. Develop an appropriate synthesis procedure for each configuration. *Note:* Figure 9-38 is

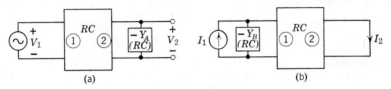

Figure 9-38

essentially a modification of Kuh's structure of Figure 8-22. Similarly, Figure 9-38b is a modified version of Figure 8-29.

9.9 It is desired to realize a notch filter using Linvill's method. One approach to this problem is to realize the $j\omega$-axis transmission zeros by means of a twin-T RC structure which can then be chosen as either of the two RC two-ports in the Linvill configuration. Using such an approach, synthesize

$$Z_{21} = \frac{H(s^2 + 1)}{s^2 + s + 1}$$

9.10 Using Linvill's method, realize

(a) $\dfrac{Hs^2}{s^2 + s + 1}$

(b) $\dfrac{H(s + 1)}{s^2 + s + 3}$

(c) Hs^2

(d) $\dfrac{H}{s^3 + 2s^2 + 2s + 1}$

---

[50] I. W. Sandberg, *loc. cit.*; see also R. F. Lee, *loc. cit.*

Use Horowitz decomposition for the denominator polynomials where possible.

9.11 Obtain the Horowitz decomposition for the following polynomials:

(a) $s^2 + 1.414s + 1$
(b) $s^3 + 2s^2 + 2s + 1$
(c) $(s^2 + 1.8477s + 1)(s^2 + 0.7653s + 1)$
(d) $3s^3 + 10s^2 + 9s + 2$
(e) $s^6 - 1$

9.12 Realize the following transfer functions in the form of an $RC:NIC$ structure:

(a) $\dfrac{s^2 - s + 1}{s^2 + s + 1}$

(b) $\dfrac{(s - 1)}{s^2 + 2s + 3}$

(c) $\dfrac{4s^2}{s^2 + s + 1}$

9.13 The Yanagisawa configuration of Figure 9-29c can be generalized for the synthesis of transfer voltage ratio vector of a multi-input single output $RC:NIC$ network. Such a generalization is shown in Figure 9-39.[51] Show that for this structure the transfer voltage ratio vector is given as

$$\mathbf{t}_v = [t_{31}, t_{32}]$$

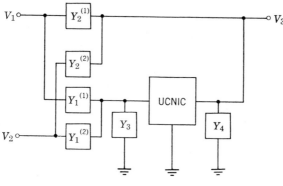

Figure 9-39

[51] J. Barranger, *loc. cit.*

where

$$t_{31} = \left.\frac{V_3}{V_1}\right|_{\substack{V_2=0\\I_3=0}} = \frac{Y_2^{(1)} - Y_1^{(1)}}{Y_2^{(1)} + Y_4 + Y_2^{(2)} - Y_3 - Y_1^{(1)} - Y_1^{(2)}}$$

$$t_{32} = \left.\frac{V_3}{V_2}\right|_{\substack{V_2=0\\I_3=0}} = \frac{Y_2^{(2)} - Y_1^{(2)}}{Y_2^{(1)} + Y_4 + Y_2^{(2)} - Y_3 - Y_1^{(1)} - Y_1^{(2)}}$$

Are there any realizability restrictions? Develop a suitable synthesis procedure and realize the following transfer vector:

$$\mathbf{t}_v = \left[\frac{3}{s^2 + 3s + 3}, \frac{s^2 - 3s + 3}{s^2 + 3s + 3}\right]$$

10 / Gyrator as a Network Element

The ideal gyrator, which is a lossless, nonreciprocal two-port described by the short-circuit admittance parameters

$$y_{11} = y_{22} = 0$$
$$y_{12} = -y_{21} = G$$

was introduced as a network element by Tellegen[1] in 1948. The ideal gyrator is attractive to the network designer for several reasons. An RLC:gyrator network is passive; hence, the stability of the realized network does not pose any problem. In addition, as will be illustrated later in this chapter, the networks employing gyrators can be made less sensitive with respect to parameter variation compared to other active devices.

Even though the ideal gyrator is an unconditionally stable two-port, practical gyrators are not. This is because of the nonidealness of the practical gyrators. However, with careful design, the instability problem can be avoided.

In this chapter, we are primarily interested in the use of the gyrator as a network element. Section 10-1 is concerned with the realization of the

---

[1] B. D. H. Tellegen, "The gyrator: a new network element," *Philips Res. Rept.*, **3**, 81–101 (April 1948).

gyrator itself. In the next two sections, synthesis of driving-point and transfer functions with the aid of gyrators is treated. Section 10-4 is concerned with the sensitivity problem. Inductor simulation using gyrators is discussed in the following section. The chapter concludes with a summary of main results.

10-1 PRACTICAL CONSIDERATIONS

Before the gyrator was introduced by Tellegen, McMillan[2] suggested a combination of electromechanical and magnetomechanical transducers to form antireciprocal elements. In later years, many novel approaches have been proposed to construct gyrators. Nonideal gyrators for low-frequency applications has been constructed using Hall effect devices.[3] In the microwave region, ferrites have been used to produce the gyrator effect.[4] Gyrators employing electromechanical devices have been recently investigated by Silverman, Schoefler, and Curan.[5]

The development of basic gyrators and impedance inverter circuits using transistors is considered in this section. A systematic approach based on fundamental considerations is followed to construct these basic circuits.

Controlled Source Representations of the Ideal Gyrator

The y matrix of an ideal gyrator can be expressed as the sum of two y matrices:

$$(10.1) \quad \begin{bmatrix} 0 & G \\ -G & 0 \end{bmatrix} = \begin{bmatrix} 0 & G \\ 0 & 0 \end{bmatrix} + \begin{bmatrix} 0 & 0 \\ -G & 0 \end{bmatrix}$$

If we denote

$$(10.2) \quad [y_A] = \begin{bmatrix} 0 & G \\ 0 & 0 \end{bmatrix}, \quad [y_B] = \begin{bmatrix} 0 & 0 \\ -G & 0 \end{bmatrix}$$

then realization of the gyrator is achieved by realizing $[y_A]$ and $[y_B]$ separately, and then connecting them in parallel. Observe that $[y_A]$ and $[y_B]$ actually represent voltage-controlled current sources as shown in Figure 10-1a. The gyrator connection is shown in Figure 10-1b (compare

---

[2] E. M. McMillan, "Violation of the reciprocity theorem in linear passive electromechanical systems," *J. Acoust. Soc. Am.*, **18**, 344–347 (1946).

[3] R. F. Wick, "Solution of the field problem of the germanium gyrator," *J. Appl. Phys.*, **25**, 741–756 (June 1954).

[4] C. L. Hogan, "The ferromagnetic Faraday effect at microwave frequencies and its applications: the microwave gyrator," *Bell System Tech. J.*, **31**, 1–31 (January 1952).

[5] J. H. Silverman, J. D. Schoefler, and D. R. Curan, "Passive electromechanical gyrators and isolators," *Proc. NEC*, **17**, 521–529 (1961).

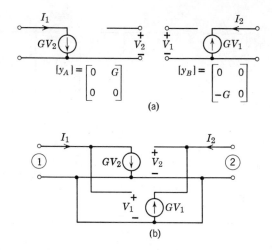

Figure 10-1 Controlled-source model of a grounded gyrator obtained by paralleling a positive gain VCT and a negative gain VCT.

with Figure 2-44b).[6] Hence, to construct a gyrator we need two VCT's described by transfer constants that are equal and opposite to each other.

An interesting modification of the representation of Figure 10-1b is shown in Figure 10-2d.[7] The steps involved in the development of this alternate representation are indicated in Figure 10-2, and should be self-explanatory. This modified representation can also be used to construct gyrators. In this case we need a unity gain noninverting type VVT and a VCT having a negative real transadmittance.

A controlled-source representation of the gyrator can also be obtained by splitting the z-matrix of the ideal gyrator. However, this approach will lead to a balanced model of the gyrator (see Figure 2-44a).

Three-Transistor Gyrator Realizations[8]

Transistor realization of a gyrator are easily obtained from the controlled source representations of Figures 10-1b and 10-2d. The first step is to replace each controlled source by its nullator-norator equivalent.

The VCT's of Figure 2-51d and 2-53b (which satisfy the polarity requirements stated earlier) can be connected in the way indicated in Figure 10-1b to form an antireciprocal two-port. The complete circuit is indicated in

---

[6] L. M. Vallese, "Understanding the gyrator," *Proc. IRE*, **43**, 483 (April 1955).
[7] F. J. Witt, U.S. Patent No. 3,109,147 (October 29, 1963).
[8] S. K. Mitra, "Equivalent circuits of gyrator," *Electronics Letters*, **3**, 333–334 (July 1967).

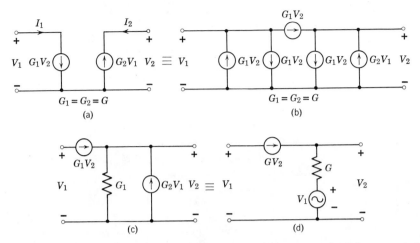

Figure 10-2 The development of alternate controlled-source model of a grounded gyrator using negative gain VCT and a positive gain VVT from the model of Figure 10-1c.

Figure 10-3a. Two possible transistorized conversions of this model are shown in Figures 10-3b and 10-3c, which have been obtained by making use of the nullator-norator model of the ideal transistor indicated earlier in Figure 8-2. These circuits were originally advanced by Sipress and Witt.[9] In a recent paper Shenoi[10] discussed in detail the practical operation of the circuit of Figure 10-3b, which has almost negligible open-circuit input and output impedances. As a result, for most purposes, additional compensation is not necessary.

In a similar manner, the alternate nullator-norator model of a gyrator shown in Figure 10-4a has been obtained from Figure 10-2d by using the VCT of Figure 2-53b and the VVT of Figure 2-52b. Two possible transistorized versions of this model are sketched in Figures 10-4b and 10-4c. The gyrator circuit of Figure 10-4b has been proposed by Witt.[11] The structure of Figure 10-4c is a new circuit.

The duals of Figures 10-3a and 10-4a can also be used to construct gyrators. The main problem with the dual models is that they yield four terminal gyrators (no common ground). A novel modification of

---

[9] J. M. Sipress and F. J. Witt, U.S. Patents No. 3,001,157 (September 19, 1961) and 3,120,645 (February 4, 1964).

[10] B. A. Shenoi, "Practical realization of a gyrator circuit and RC gyrator filters," *IEEE Trans. on Circuit Theory*, **CT-12**, 374–380 (1965).

[11] F. J. Witt, *loc. cit.*

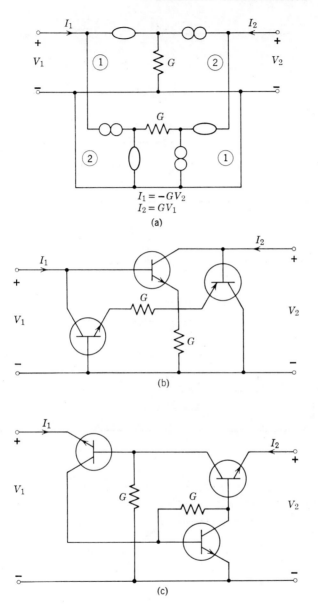

Figure 10-3 Nullator-norator model of gyrator based on Figure 10-1c and its corresponding three-transistor realizations.

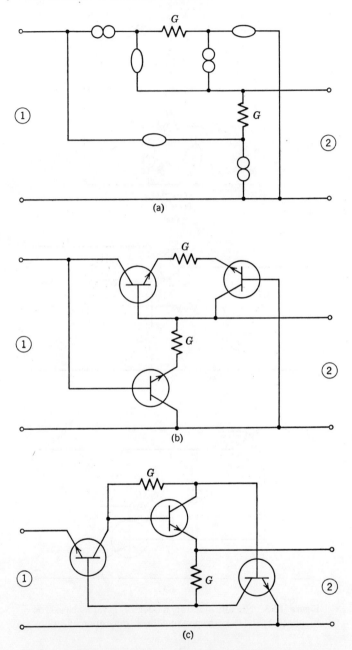

Figure 10-4 Nullator-norator model of gyrator based on Figure 10-2d and its corresponding three-transistor realizations.

one of these duals, which results in a three-terminal version, is suggested in Problem 10.2.

An alternate nullator-norator representation of Figure 10-1b is presented next. To obtain a VCT of opposite polarity to that of the VCT of Figure 2-51d, we connect two such VCT's in cascade along with a shunt resistor as shown in Figure 10-5a. Use of Figure 10-5a and Figure 2-51d in Figure 10-1b results in the gyrator model of Figure 10-5b. Two transistorized versions of this nullator-norator model are indicated in Figures

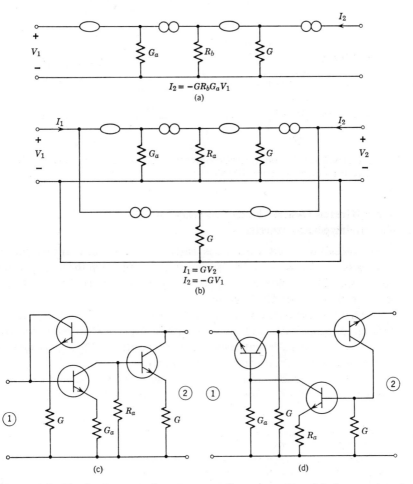

Figure 10-5 The development of an alternate nullator-norator model of gyrator based on Figure 10-1c and its corresponding three-transistor realizations.

10-5c and 10-5d. The gyrator circuit of Figure 10-5c has been suggested elsewhere.[12] On the other hand, the circuit of Figure 10-5d is new.

The nullator-norator models of Figures 10-3a and 10-5b along with their transistorized versions were recently independently advanced by Bendík.[13]

The performance of the above circuits can be improved by using more complicated nullator-norator models of the VCT's and the VVT. This, of course, implies that more transistors will be needed in the actual construction.

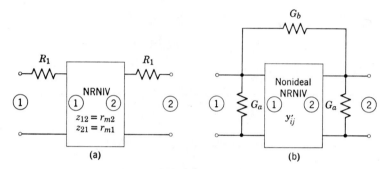

Figure 10-6 (a) A nonideal nonreciprocal negative-impedance inverter. (b) Conversion of the nonideal NRNIV to an active gyrator.

Active Gyrator Design Using Nonreciprocal Negative-Impedance Inverter

The transistor circuits developed earlier are essentially antireciprocal two-ports (i.e., $y_{12} = -y_{21}$) and can be made to approach the ideal gyrator operation. The development of these circuits are based on controlled source models of the ideal gyrator.

An alternate approach, which is applicable only to active gyrator design, has been proposed by Yanagisawa and Kawashima.[14] Their method make use of a nonreciprocal negative-impedance inverter (NRNIV) having real and positive open-circuit transfer impedances.

Consider the circuits of Figure 10-6a, where we have added a resistor in series to each port of a NRNIV. The y-parameters of this two-port

---

[12] W. New and R. Newcomb, "An integrable time-variable gyrator," *Proc. IEEE*, **53**, 2161–2162 (December 1965).

[13] J. Bendík, "Equivalent gyrator networks with nullators and norators," *IEEE Trans. on Circuit Theory*, **CT-14**, 98 (March 1967).

[14] T. Yanagisawa and Y. Kawashima, "Active gyrator," *Electronics Letters*, **3**, 105–107 (March 1967).

are given as:

$$y'_{11} = y'_{22} = \frac{-R_1}{r_{m1}r_{m2} - R_1^2}$$

(10.3)
$$y'_{12} = \frac{r_{m2}}{r_{m1}r_{m2} - R_1^2}$$

$$y'_{21} = \frac{r_{m1}}{r_{m1}r_{m2} - R_1^2}$$

If we choose

(10.4)
$$r_{m1}r_{m2} > R_1^2$$
$$R_1 > r_{m2}$$

then it is seen from Equation 10.3 that

(10.5)
$$y'_{11} = y'_{22} < 0$$
$$y'_{12} > 0$$
$$y'_{21} > 0$$

and also

(10.6)
$$y'_{11} + y'_{12} < 0$$
$$y'_{21} > y'_{12}$$

Now let us connect a resistive Π-network in parallel with the active two-port of Figure 10-6a, as indicated in Figure 10-6b. Our aim is to choose the various components of this composite two-port to make it a gyrator. If G_a and G_b are selected so that

(10.7)
$$y'_{11} = y'_{22} = -(G_a + G_b)$$
$$y'_{12} = G_b - g_1$$
$$y'_{21} = G_b + g_2$$

then the short-circuit admittance parameters of the composite two-port are given as:

(10.8)
$$y_{11} = y_{22} = 0$$
$$y_{12} = -g_1$$
$$y_{21} = g_2$$

which represents an active gyrator.

Knowing G_a and G_b, the parameters of the network of Figure 10-6a can be computed from:

(10.9)
$$r_{m1} = \frac{y'_{21}}{y'_{12}y'_{21} - y'^{2}_{11}}$$

$$r_{m2} = \frac{y'_{12}}{y'_{12}y'_{21} - y'^{2}_{11}}$$

$$R_1 = \frac{y'_{11}}{y'_{12}y'_{21} - y'^{2}_{11}}$$

In order to have positive values for the above parameters, it is seen from Equations 10.9 and 10.5 that $(y'^{2}_{11} - y'_{12}y'_{21})$ must be negative, i.e., the condition

(10.10)
$$G_b(g_2 - g_1) - g_2 g_1 > G_a(G_a + 2G_b)$$

must be satisfied. This indicates that we cannot construct a lossless gyrator ($g_1 = g_2$) using this approach.

Let us illustrate the method by means of an example.

Example 10-1 Consider the realization of

$$[y] = \begin{bmatrix} 0 & -0.2 \\ 4.0 & 0 \end{bmatrix}$$

Thus, $g_1 = 0.2$ mhos and $g_2 = 4.0$ mhos. Let us choose $G_a = 0.1$ mhos, then from inequality 10.10 we note that G_b must be chosen to satisfy

$$G_b > 0.23 \text{ mhos}$$

Choosing $G_b = 0.3$ mhos, we obtain from Equations 10.7 and 10.9,

$$r_{m1} = 15.926 \, \Omega, \qquad r_{m2} = 0.37 \, \Omega, \qquad R_1 = 1.48 \, \Omega$$

Transistor realization of the NRNIV is considered next. The development of these circuits follow the pattern outlined for the other type of active elements. Two controlled-source equivalent circuits of the NRNIV having real and positive open-circuit transfer impedances are shown in Figure 10-7. Replacing the controlled sources of Figure 10-7 by their nullator-norator models, nullator-norator representations of the NRNIV are readily obtained. Two such representations are indicated in Figure 10-8. Using the equivalence relation of a transistor and its nullator-norator model (Figure 8-2), transistor realizations of the NRNIV are easily obtained.[15]

[15] S. K. Mitra, "Non-reciprocal negative impedance inverter," *Electronics Letters*, 3, 388 (August 1967).

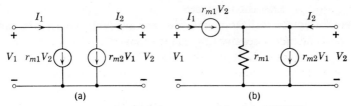

Figure 10-7 Controlled-source models of the NRNIV.

Other Circuits

Practical gyrator circuits employing more than three transistors and based on parallel connection of VCT's of opposite polarity (Figure 10-1) have also been advanced by many authors. Of these the gyrator circuits of Rao and Newcomb,[16] Orchard and Sheahan,[17] Chua and Newcomb,[18] and Holmes, Gruetzmann, and Heinlein[19] appear more promising. Details on the actual design and experimental results will be found in the original papers.

Two novel methods of gyrator circuit design have been proposed by Yanagisawa.[20] His methods allow the adjustment of internal resistors to

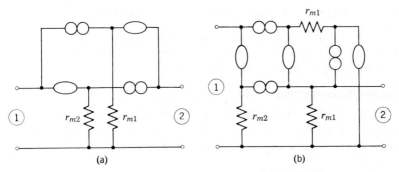

Figure 10-8 Nullator-norator models of the NRNIV.

[16] T. N. Rao and R. W. Newcomb, "A direct coupled gyrator suitable for integrated circuits and time variation," *Electronics Letters*, **2**, 250–251 (July 1966).

[17] D. F. Sheahan and H. J. Orchard, "Integrable gyrator using m.o.s. and bi-polar transistors," *Electronics Letters*, **2**, 390–391 (October 1966).
D. F. Sheahan and H. J. Orchard, "High-quality transistorized gyrator," *Electronics Letters*, **2**, 274 (1966).

[18] H. T. Chua and R. W. Newcomb, "Integrated direct-coupled gyrator," *Electronics Letters*, **3**, 182–184 (May 1967).

[19] W. H. Holmes, S. Gruetzmann, and W. E. Heinlein, "High-performance direct-coupled gyrators," *Electronics Letters*, **3**, 45–46 (February 1967).

[20] T. Yanagisawa, "Realization of a lossless transistor gyrator," *Electronics Letters*, **3**, 167–168 (April 1967). T. Yanagisawa, "Direct-coupled transistor lossless gyrator," *Electronics Letters*, **3**, 311–312 (July 1967).

make y_{11} and y_{22} each equal to zero.

Several alternate realization schemes are available. A simple method would be to cascade a negative impedance converter (NIC) with a negative impedance inverter (NIV) (see Problem 10.3). Nonideal gyrators can also be constructed by means of feedback amplifiers. This is illustrated in Problem 10.4.

References to additional gyrator circuits will be found at the end of the chapter.

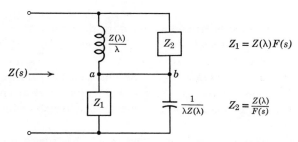

Figure 10-9 The Bott-Duffin cycle in the realization of a positive real function.

10-2 SYNTHESIS OF DRIVING-POINT FUNCTIONS

In this section, we shall outline two extensions of well-known passive RLC driving-point function synthesis techniques. The final realizations are in the form of RLC:gyrator networks.

Cascade Representation of Bott-Duffin Cycle

The Bott-Duffin method[21] enables one to realize positive real driving-point function $Z(s)$ without transformers. (The Bott-Duffin representation of $Z(s)$ is shown in Figure 10-9). The key to the method is *Richard's theorem:*

Theorem 10-1 *If $Z(s)$ is a p.r. function, and λ is a positive real constant, then*

$$(10.11) \qquad F(s) = \frac{\lambda Z(s) - s Z(\lambda)}{\lambda Z(\lambda) - s Z(s)}$$

is also p.r. and is not of higher order than $Z(s)$.

Expression 10.11 can be solved for $Z(s)$ to yield

$$(10.12) \qquad Z(s) = Z(\lambda) \frac{s + \lambda F(s)}{\lambda + s F(s)}$$

---

[21] R. Bott and R. J. Duffin, "Impedance synthesis without use of transformers," *J. Appl. Phys.*, **20**, 816 (August 1949).

To enable further degree reduction in $F(s)$, the following assumptions are made regarding $Z(s)$: $Z(s)$ is minimum resistive and $Z(\infty)$ is constant. Under the assumptions, which are easily met, it is possible to select λ in Equation 10.11 so that $F(s)$ has either a pole or a zero on the $j\omega$-axis, which can be removed subsequently. To this end, we note that $\text{Re}\, Z(j\omega)$ is zero for some value of ω, say ω_0. This implies that $Z(j\omega_0)$ is imaginary. Moreover, $Z(\lambda)$ is bounded and

$$0 < \lambda Z(\lambda) \leq \infty$$

Two possible cases must be examined.

Case A $\qquad\qquad j\omega_0 Z(j\omega_0) > 0$

For this case, a positive λ can be chosen to make

(10.13) $\qquad\qquad \lambda Z(\lambda) - j\omega_0 Z(j\omega_0) = 0$

which in turn implies that $F(s)$ will have a pole at $s = j\omega_0$.

Case B $\qquad\qquad \dfrac{Z(j\omega_0)}{j\omega_0} > 0$

For this case, we can choose a positive λ to make

(10.14) $\qquad\qquad \lambda Z(j\omega_0) - j\omega_0 Z(\lambda) = 0$

implying that $F(s)$ will have a zero at $s = j\omega_0$.

Observe that the representation of Figure 10-9 is a balanced bridge. Hence any impedance can be connected between nodes a and b, without altering the balance.

An elegant modification of the Bott-Duffin method has been suggested by Hazony and Schott.[22] They have shown that if gyrators are used in the realization process, alternate representation of the Bott-Duffin cycle will be one of the two networks of Figure 10-10. We shall illustrate the method by means of two examples.

Example 10-2 Realize

(10.15) $\qquad\qquad Z(s) = \dfrac{s^2 + s + 1}{4s^2 + s + 1}$

We first note that $\text{Re}\, Z(j\omega)$ is zero at $\omega_0 = 1/\sqrt{2}$, i.e., $\text{Re}\, Z(j/\sqrt{2}) = 0$. From Equation 10.15 we have $Z(j/\sqrt{2}) = -j/\sqrt{2}$, indicating Case A

---

[22] D. Hazony and F. W. Schott, "A cascade representation of the Bott-Duffin synthesis," *IRE Trans. on Circuit Theory*, **CT-5**, 144 (June 1958).

416 / GYRATOR AS A NETWORK ELEMENT

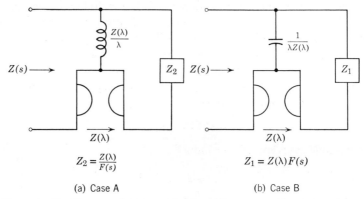

(a) Case A (b) Case B

Figure 10-10 The cascade representations of the Bott-Duffin cycle, using a gyrator.

must be followed. To select λ, we set

$$\lambda Z(\lambda) - j\omega_0 Z(j\omega_0) = \lambda\left(\frac{\lambda^2 + \lambda + 1}{4\lambda^2 + \lambda + 1}\right) - \frac{j}{\sqrt{2}}\left(-\frac{j}{\sqrt{2}}\right) = 0,$$

which simplifies to

$$2\lambda^3 - 2\lambda^2 + \lambda - 1 = (\lambda - 1)(2\lambda^2 + 1) = 0$$

Hence $\lambda = 1$. This implies that

$$Z(\lambda) = Z(1) = \tfrac{1}{2}$$

Therefore

$$F(s) = \frac{Z(s) - sZ(1)}{Z(1) - sZ(s)} = \frac{4s^2 + 3s + 2}{2s^2 + 1}$$

which has a pair of $j\omega$-axis poles at $s = \pm j/\sqrt{2}$. Now

$$Z_2(s) = \frac{Z(\lambda)}{F(s)} = \frac{\tfrac{1}{2}}{\dfrac{4s^2 + 3s + 2}{2s^2 + 1}} = \frac{1}{4 + \dfrac{3s}{s^2 + \tfrac{1}{2}}}$$

The complete realization is shown in Figure 10-11.

Example 10-3 Consider next the realization of

(10.16) $$Z(s) \doteq \frac{2s^2 + s + 1}{s^2 + s + 2}$$

It can be shown that Re $Z(j\omega)$ has a zero at $\omega = \omega_0 = 1$ and $Z(j1) = +j$. This indicates that Case B must be followed. We solve

$$\lambda Z(j\omega_0) - j\omega_0 Z(\lambda) = j\lambda - j\left(\frac{2\lambda^2 + \lambda + 1}{\lambda^2 + \lambda + 2}\right) = 0$$

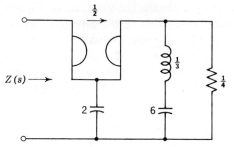

Figure 10-11 The realization of the positive real driving-point function of Example 10-2, using a gyrator.

for λ, which yields $\lambda = 1$. Hence

$$Z(\lambda) = Z(1) = 1$$

As a result,

$$F(s) = \frac{Z(s) - sZ(1)}{Z(\lambda) - sZ(s)} = \frac{s^2 + 1}{2s^2 + 2s + 2}$$

Note that $F(s)$ has a pair of zeros at $s = \pm j1$ as expected. We now compute Z_1,

$$Z_1(s) = Z(\lambda)F(s) = \frac{s^2 + 1}{2s^2 + 2s + 2} = \frac{1}{2 + \dfrac{2s}{s^2 + 1}}$$

The final realization is shown in Figure 10-12a.

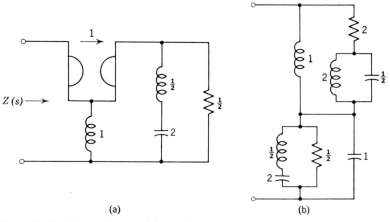

(a) (b)

Figure 10-12 The realization of the positive real driving-point function of Example 10-3 with and without a gyrator.

418 / GYRATOR AS A NETWORK ELEMENT

It is interesting to compare the network of Figure 10-12a with the original Bott-Duffin realization of the driving-point function of (10.16), shown in Figure 10-12b. We observe that use of a gyrator has resulted in the saving of passive components. The network of Figure 10-12a employs one resistance, two inductances and one capacitance in addition to a gyrator. Whereas the network of Figure 10-12b utilizes 2 resistances, 3 inductances, and 3 capacitances.

Nonreciprocal Darlington Synthesis[23]

The purpose of the Darlington method of synthesis is to realize a specified positive real driving-point function as the input immittance of a

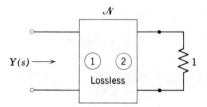

Figure 10-13 The configuration proposed by Darlington for the realization of a positive real driving-point function.

lossless two-port terminated at the output port by a resistance (Figure 10-13). If gyrators are used as network elements, the lossless two-port realization is usually considerably simplified.

The method of synthesis begins with the decomposition and partitioning of the specified driving-point admittance $Y(s) = N(s)/D(s)$ as follows:

$$Y(s) = \frac{m_1 + n_1}{m_2 + n_2}$$

(10.17a)
$$= \frac{m_1}{n_2} - \frac{\dfrac{m_1 m_2 - n_1 n_2}{n_2^2}}{\dfrac{m_2}{n_2} + 1} \quad \text{(Case A)}$$

(10.17b)
$$= \frac{n_1}{m_2} - \frac{\dfrac{n_1 n_2 - m_1 m_2}{m_2^2}}{\dfrac{n_2}{m_2} + 1} \quad \text{(Case B)}$$

[23] D. Hazony, "Two extensions of the Darlington synthesis procedure," *IRE Trans. on Circuit Theory*, **CT-9**, 284–288 (September 1961).

SYNTHESIS OF DRIVING-POINT FUNCTIONS / 419

In Equations 10.17, m_1 and n_1 are the even and odd parts of $N(s)$, respectively. Likewise, m_2 and n_2 are the even and odd parts of $D(s)$.

The input admittance of the one port of Figure 10-13 is

(10.18) $$Y(s) = y_{11} - \frac{y_{12}y_{21}}{y_{22} + 1}$$

where $\{y_{ij}\}$ denote the short-circuit parameters of the lossless two-port $\mathcal{N}$. In order to be able to compare Equations 10.17 with 10.18, further decomposition of Equations 10.17 is necessary. Note that $(m_1m_2 - n_1n_2)$ is the numerator of the even part of $Y(s)$; hence, it is an even polynomial with quadrantal symmetry. Thus we can always write

(10.19) $$(m_1m_2 - n_1n_2) = P(s)P(-s) = (M + N)(M - N)$$

where $P(s) = (M + N)$ is a polynomial with real coefficients and $M = \mathrm{Ev}\ P(s)$ and $N = \mathrm{Od}\ P(s)$. Using Equation 10.19 in Equation 10.17 and comparing it with Equation 10.18, we obtain the following identifications for the two cases:

Case A

(10.20a)
$$y_{11} = \frac{m_1}{n_2}$$
$$y_{22} = \frac{m_2}{n_2}$$
$$y_{12} = \pm \frac{(M + N)}{n_2}$$
$$y_{21} = \pm \frac{(M - N)}{n_2}$$

Case B

(10.20b)
$$y_{11} = \frac{n_1}{m_2}$$
$$y_{22} = \frac{n_2}{m_2}$$
$$y_{12} = \pm \frac{(M + N)}{m_2}$$
$$y_{21} = \pm \frac{(M - N)}{m_2}$$

We have now to show that the y parameters given by expressions 10.20 form a realizable set in both cases. To this end, we state the following theorem:

Theorem 10-2 *The necessary and sufficient conditions satisfied by the short-circuit admittance parameters, y_{mn}, of a lossless nonreciprocal two-port are:*

(i) y_{11} and y_{22} be LC driving-point functions,
(ii) The residues, k_{mn}, at each $j\omega$-axis pole must satisfy the residue condition,

$$k_{11}, k_{22} \text{ real}$$
$$k_{11} \geq 0$$
$$k_{12} = k_{21}{}^*$$
$$k_{11}k_{22} \geq k_{12}k_{21}$$

(iii) $y_{12}(j\omega) = -y_{21}(j\omega)^*$

The necessity follows from Theorem 6-7. Sufficiency will be evident from subsequent discussions.

The problem now is to prove that the y parameters of expression 10.20 satisfy the conditions of Theorem 10-2. We shall prove it for Case A only. The proof for Case B follows parallel lines and will be omitted.

From expression 10.20a we first observe that conditions (i) and (iii) are obviously satisfied. It remains to test condition (ii). At a finite pole at $s = j\omega_i$ (which is a zero of n_2) we have:

(10.21)
$$k_{11i} = \left.\frac{m_1}{\dfrac{d}{ds}(n_2)}\right|_{s=j\omega_i} \qquad k_{22i} = \left.\frac{m_2}{\dfrac{d}{ds}(n_2)}\right|_{s=j\omega_i}$$

$$k_{12i} = \left.\frac{\pm(M+N)}{\dfrac{d}{ds}(n_2)}\right|_{s=j\omega_i} \qquad k_{21i} = \left.\frac{\pm(M-N)}{\dfrac{d}{ds}(n_2)}\right|_{s=j\omega_i}$$

From expression 10.21, it is noted that $k_{12i} = k_{21i}^*$. Since

$$(M^2 - N^2)\big|_{s=j\omega_i} = (m_1 m_2 - n_1 n_2)\big|_{s=j\omega_i} = (m_1 m_2)\big|_{s=j\omega_i}$$

we conclude from expression 10.21 that the residue condition is satisfied with an equal sign at all finite poles. If $Y(s)$ has poles on the $j\omega$-axis (which implies that m_2 and n_2 have common factors), $(M^2 - N^2)\big|_{s=j\omega_i} = 0$ and hence the residue condition is also satisfied at these poles. A possible pole at infinity must be treated separately. Let us write

(10.22)
$$Y(s) = \frac{A_p s^p + A_{p-1} s^{p-1} + \cdots}{B_q s^q + B_{q-1} s^{q-1} + \cdots}$$

Using the notations of Equation 10.22, the residues of y_{mn} at a possible pole at infinity, k_{mn}^∞, can be calculated. For convenience, we have tabulated them in Table 10-1. In deriving Table 10-1, we have used the fact that if the residue condition is not satisfied with an equal sign at a pole at $s = j\omega_i$, then $Y(s)$ must have a pole at $s = j\omega_i$. We conclude from the entries of Table 10-1 that condition (ii) is also satisfied at infinity.

Table 10-1

Case A	k_{11}^∞	k_{22}^∞	$k_{12}^\infty k_{21}^\infty$	Case B
p even, $p = q + 1$	$\dfrac{A_p}{B_q}$	0	0	p odd, $p = q + 1$
p even, $p = q$	$\dfrac{A_p}{B_{q-1}}$	$\dfrac{B_q}{B_{q-1}}$	$\dfrac{A_p B_q}{(B_{q-1})^2}$	p odd, $p = q$
p even, $p = q - 1$	0	0	0	p odd, $p = q - 1$
p odd, $p = q + 1$	$\dfrac{A_{p-1}}{B_{q-1}}$	$\dfrac{B_q}{B_{q-1}}$	$\dfrac{A_{p-1}B_q - A_p B_{q-1}}{(B_{q-1})^2}$	p even, $p = q + 1$
p odd, $p = q$	0	0	0	p even, $p = q$
p odd, $p = q - 1$	0	$\dfrac{B_q}{B_{q-1}}$	0	p even, $p = q - 1$

This proves our assertion on the realizability of the y parameter of expression 10.20a.

Let us illustrate the method by means of an example.

Example 10-4 Realize

(10.23) $$Z(s) = \frac{9s^2 + 26s + 1}{4s^2 + s + 1}$$

Form the numerator of the even part:

(10.24)
$$\begin{aligned} m_1 m_2 - n_1 n_2 &= (9s^2 + 1)(4s^2 + 1) - (26s)(s) \\ &= 36s^4 - 13s^2 + 1 \\ &= (2s + 1)(3s + 1)(1 - 2s)(1 - 3s) \end{aligned}$$

Comparing Equations 10.24 and 10.19, we note that there are two possible identifications for $P(s) = (M + N)$:

(10.25a) $(M + N) = (2s + 1)(3s + 1) = 6s^2 + 5s + 1$

or

(10.25b) $(M + N) = (1 - 2s)(3s + 1) = -6s^2 + s + 1$

Choosing $(M + N)$ as given by Equation 10.25a and using the identification of Case A [Equation 10.17a], we arrive at

$$y_{11} = \frac{9s^2 + 1}{s} = \overbrace{\left[\frac{1}{s} + 9s\right.}^{\mathcal{N}'} + \left.\vphantom{\frac{1}{s}}0\right]$$

$$y_{22} = \frac{4s^2 + 1}{s} = \left[\frac{1}{s} + 4s\ +\ 0\right]$$

$$y_{12} = -\frac{6s^2 + 5s + 1}{s} = \left[-\frac{1}{s} - 6s\ +\ -5\right]$$

$$y_{21} = -\frac{6s^2 - 5s + 1}{s} = \left[-\frac{1}{s} - 6s\ +\ 5\right]$$

Lossless   Gyrator
Reciprocal

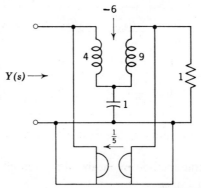

Figure 10-14 The realization of the driving-point function of Example 10-4, using the nonreciprocal Darlington synthesis technique.

The realization of the lossless reciprocal two-port $\mathcal{N}'$ can be accomplished by Cauer's method.[24] It can be shown that in the realization of $\mathcal{N}'$ on the y basis, an ideal transformer is needed. One solution to the problem of elimination of ideal transformers is scaling, which sometimes gives a satisfactory answer. Let us rewrite expression 10.18 as

(10.26) $$Y(s) = y_{11} - \frac{(Ky_{12})(Ky_{21})}{K^2 y_{22} + K^2}$$

[24] See, for example, M. E. Van Valkenburg, *Modern network synthesis*, Wiley, New York, Section 11.4.

SYNTHESIS OF DRIVING-POINT FUNCTIONS / 423

Expression 10.26 indicates how the scaling constant K can be used to advantage. An alternate solution is to realize the two-port $\mathcal{N}'$ on the z basis; the complete realization is shown in Figure 10-14.

An equivalent realization of Equation 10.23 is illustrated next. Let us choose instead the $(M + N)$ polynomial shown in Equation 10.25b and use the identification of Case B. We then have

$$y_{11} = \frac{26s}{4s^2 + 1} = \begin{vmatrix} \frac{26s}{4s^2+1} & + & 0 \end{vmatrix}$$

$$y_{22} = \frac{s}{4s^2 + 1} = \begin{vmatrix} \frac{s}{4s^2+1} & + & 0 \end{vmatrix}$$

$$y_{12} = -\frac{-6s^2 + s + 1}{4s^2 + 1} = \begin{vmatrix} -\frac{s + \frac{5}{2}}{4s^2+1} & + & \frac{3}{2} \end{vmatrix}$$

$$y_{21} = -\frac{6s^2 + s - 1}{4s^2 + 1} = \begin{vmatrix} -\frac{s - \frac{5}{2}}{4s^2+1} & + & -\frac{3}{2} \end{vmatrix}$$

$$\qquad\qquad\qquad\qquad\qquad\qquad \mathcal{N}'' \qquad\quad \text{Gyrator}$$

Note that the two-port $\mathcal{N}'$ is still nonreciprocal. To enable realization of $\mathcal{N}'$, we invert its y parameters to derive the z parameters:

$$z'_{11} = \begin{vmatrix} \frac{4}{25s} & + & 0 \end{vmatrix}$$

$$z'_{22} = \begin{vmatrix} \frac{104}{25s} & + & 0 \end{vmatrix}$$

$$z'_{12} = \begin{vmatrix} \frac{4}{25s} & + & \frac{2}{5} \end{vmatrix}$$

$$z'_{21} = \begin{vmatrix} \frac{4}{25s} & + & -\frac{2}{5} \end{vmatrix}$$

Lossless Gyrator
Reciprocal

The complete realization is shown in Figure 10-15.

Note that the main problem of this method is the decomposition of the numerator of the even part in the form of Equation 10.19. This type of decomposition has been investigated by Delansky.[25] The various possible decompositions, along with the two cases of identification, lead to several equivalent network realizations of a specified driving-point function.

[25] J. F. Delansky, M. S. Thesis, Cornell University, June 1964.

424 / GYRATOR AS A NETWORK ELEMENT

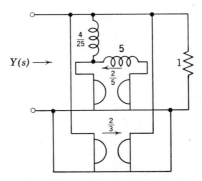

Figure 10-15 An alternate realization of the driving-point function of Example 10-4.

$RLC:t_d^{(1)}$ One-Port[26]

Let us now digress for a while and present an interesting extension of the previous section. Our aim is to outline the sufficiency proof of Theorem 6-14. We plan to use gyrators in the synthesis; hence condition (iii) of Theorem 6-14 is not necessary.

Let $Y(s)$ be the driving-point admittance, which satisfies the necessary conditions (i), (ii), and (iv) of Theorem 6-14. Since $-Y(-s)$ is positive real, we can realize $-Y(-s)$ by a resistance terminated lossless two-port. Using the notations of Equation 10.22, we represent $-Y(-s)$ as

$$-Y(-s) = \frac{A_p s^p + A_{p-1} s^{p-1} + \cdots}{B_q s^q + B_{q-1} s^{q-1} + \cdots}$$

Condition (ii) implies that the degrees of numerator and denominator polynomial are related as follows:

(10.27a) $\qquad p = q + 1$

or

(10.27b) $\qquad p = q - 1$

Furthermore, due to condition (iv), we have

(10.28a) $\qquad \dfrac{A_p}{A_{p-1}} \geq \dfrac{1}{\delta_1}$

---

[26] The treatment of this section is similar to B. K. Kinariwala, "The Esaki diode as a network element," *IRE Trans. on Circuit Theory*, **CT-8**, 389–397 (December 1961).

and

(10.28b)
$$\frac{B_q}{B_{q-1}} \geq \frac{1}{\delta_1}$$

where δ_1 is the dissipation factor of the tunnel diode.

Thus, sufficiency is proved if we can show that a capacitor of value $1/\delta_1$ can be extracted across the terminating resistor as the first step of realization of $-Y(-s)$. Observe that from the specified driving-point function, the y parameters of the lossless two-port are obtained by making use of either expression 10.20a or 10.20b. To guarantee the extraction of shunt capacitor across port 2, it is clear that $k_{22\infty}$, the residue of y_{22} at infinity, must be greater than $1/\delta_1$, and that the modified residue condition

(10.29)
$$k_{11\infty}\left(k_{22\infty} - \frac{1}{\delta_1}\right) - k_{12\infty}k_{21\infty} \geq 0$$

must be satisfied. Using the entries of Table 10-1, in the above equation and making use of expressions 10.27 and 10.28, it is easy to show that condition 10.29 is indeed satisfied. This completes the sufficiency proof.

10-3 TRANSFER FUNCTION SYNTHESIS

Not much work has been done in the area of transfer function realization using a gyrator as a network element. The most important result in this direction is the transfer function realization by the RC:gyrator cascade configuration and will be the subject of discussion of this section. In the latter part of this section, we shall indicate a procedure to realize transfer functions using a reactive gyrator as the active element.

Another approach of transfer function realization will be first to realize by conventional RLC synthesis procedure and then replace the inductors by capacitor terminated gyrators (Section 10-5).

RC:Gyrator Two-Port Realization

The aim of this method, which is a generalization of Horowitz's method,[27] is to realize a given transfer function $T(s) = N(s)/D(s)$ in the form of the network shown in Figure 10-16. For the purpose of illustration, let us consider the realization of a voltage transfer ratio V_2/V_1. For the two-port of Figure 10-16, it can be shown that

(10.30)
$$t_v = \frac{V_2}{V_1} = \frac{-\alpha z_{21b}y_{21a}}{z_{11b} + \alpha^2 y_{22a}}$$

---

[27] I. M. Horowitz, "Synthesis of active RC transfer functions by means of cascaded RC and RL structures," *IRE Nat'l Convention Record*, **5**, (Part 2) 190–195 (August 1957).

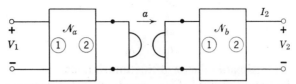

Figure 10-16 The RC:Gyrator cascade configuration for transfer voltage ratio realization.

Realization of t_v begins with the selection of a polynomial $Q(s)$ with distinct negative real zeros such that $Q(s)^0 = D(s)^0$ or $D(s)^0 - 1$.

Let us make an $RC:RL$ type decomposition[28] of $D(s)/Q(s)$ (assuming this is possible). Then we can write

$$(10.31) \qquad \frac{D(s)}{Q(s)} = \frac{D_a(s)}{Q_a(s)} + \frac{D_b(s)}{Q_b(s)}$$

where $D_a(s)/Q_a(s)$ is an RC driving-point impedance and $D_b(s)/Q_b(s)$ is an RL driving-point impedance, i.e., an RC driving-point admittance. Note that the decomposition of $D(s)/Q(s)$ in the form of Equation 10.31 is achieved by a partial fraction expansion (Equation 3.55) and by grouping the terms accordingly (Equations 3.57 and 3.58). A sufficient condition for the feasibility of the above decomposition is that condition 3.56 be satisfied.[29] $N(s)$ can be properly grouped to yield $N(s) = N_a(s) \cdot N_b(s)$. We thus have

$$(10.32) \qquad t_v = \frac{N(s)}{D(s)} = \frac{\dfrac{N_a(s)}{Q_a(s)} \cdot \dfrac{N_b(s)}{Q_b(s)}}{\dfrac{D_a(s)}{Q_a(s)} + \dfrac{D_b(s)}{Q_b(s)}}$$

Comparing Equation 10.32 with Equation 10.30, we identify

$$(10.33) \qquad \begin{aligned} z_{11b} &= \frac{D_a(s)}{Q_a(s)}; & z_{21b} &= \frac{N_a(s)}{Q_a(s)} \\ y_{22a} &= \frac{D_b(s)}{\alpha^2 Q_b(s)}; & -y_{21a} &= \frac{N_b(s)}{Q_b(s)} \end{aligned}$$

We observe from Equation 10.33 that the decomposition of $N(s)$ into $N_a(s)N_b(s)$ is arbitrary except that the two-port parameters given by Equation 10.33 be RC realizable. The following example will help to clarify the method.

---

[28] See Section 3-6, p. 102.
[29] More explicit conditions on $D(s)$ are given by Theorem 10-3 on p. 429.

Example 10-5 Let us realize

$$t_v = \frac{H}{s^2 + 3s + 3}$$

Choose $Q(s) = (s+1)(s+2)$. This yields

$$\frac{D(s)}{Q(s)} = \frac{s^2 + 3s + 3}{(s+1)(s+2)} = 1 + \frac{1}{s+1} - \frac{1}{s+2}$$

Observe that condition 3.56 on the residues is satisfied. We thus identify:

(10.34)
$$z_{11b} = \frac{1}{s+1}; \qquad z_{21b} = \frac{H_b}{s+1}$$

$$y_{22a} = 1 - \frac{1}{s+2} = \left(\frac{s+2}{s+1}\right)^{-1}; \qquad -y_{21a} = \frac{H_a}{s+2}$$

where we have taken $\alpha = 1$ without any loss of generality. The networks $\mathcal{N}_a$ and $\mathcal{N}_b$ are realized following standard RC two-port synthesis procedure, and the complete network is shown in Figure 10-17. The realized value of the gain constants can be easily computed from the two companion RC networks, and we find $H_a = 2$ and $H_b = 1$. Hence the realized value of H is 2.

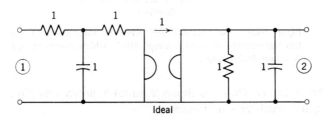

Figure 10-17 RC:Gyrator cascade realization of the transfer voltage ratio of Example 10-5.

It should be noted that the structural configuration is similar to that of Linvill's cascade RC:NIC structure (Figure 9-27). As a result, this method has the same drawbacks as Linvill's procedure. The maximum value of the constant H is limited by the maximum obtainable by passive companion networks. If a common ground is desired between the input and the output, the transfer function must have no positive real transmission zeros.

An interesting observation can be made here on the results of Example 10-5. Note from expression 10.34 that $y_{22a}(0)$ and $y_{22a}(\infty)$ are finite and positive. Hence in the development of the network $\mathcal{N}_a$ we can first remove a part of $y_{22a}(0)$ and then develop the remainder by the continued fraction expansion. This leads to an alternate network shown in Figure 10-18, where the gyrator is shunted by positive resistances. If the gyrator is not ideal ($y_{11} \neq 0$, $y_{22} \neq 0$), the nonideal behavior can be compensated by precorrecting the shunt resistances R_A and R_B.[30] It should be clear from Equation 10.30 that the direction of gyration is immaterial in the transfer function realization.

Conditions for $RC:RL$ Decomposition

The success of the previous synthesis method hinges upon the $RC:RL$ decomposition of the denominator polynomial $D(s)$ of the transfer function. Calahan[31] has derived the condition which determines whether an $RC:RL$ decomposition is possible.

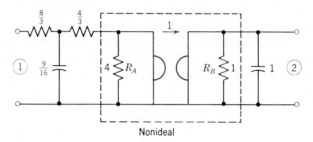

Figure 10-18 An alternate RC:Gyrator realization of the transfer voltage ratio of Example 10-5, which allows the use of a nonideal gyrator.

For the specified $D(s)$, we choose a suitable polynomial $Q(s)$ with distinct negative real zeros and express

$$\frac{D(s)}{Q(s)} = \frac{D_a(s)}{Q_a(s)} + \frac{D_b(s)}{Q_b(s)}$$

where $D_a(s)/Q_a(s)$ is an RC driving-point impedance and $D_b(s)/Q_b(s)$ is an RL driving-point impedance. This implies that $D(s)$ must be expressed as a sum of two polynomials

(10.35) $$D(s) = A(s) + B(s)$$

---

[30] Some interesting results have been obtained on the "precorrection" of nonideal gyrator behavior. See W. J. Howard, "Transfer function synthesis using non-ideal gyrators," Elec. Engg. Res. Lab., Cornell Univ., Ithaca, N.Y., Tech. Rep. 87, July 1965.

[31] D. A. Calahan, "Restrictions on the natural frequencies of an RC-RL network," J. Franklin Inst., 272, 112–133 (August 1961).

where

(10.36)
$$A(s) = Q_a(s)D_b(s)$$
$$B(s) = Q_b(s)D_a(s)$$

Since $D(s)/Q(s)$ is a positive real function, $D(s)$ must necessarily be Hurwitz.

Let η_A and η_B be the number of zeros of $A(s)$ and $B(s)$ to the right of any point $s = \sigma$ on the negative real axis. From the properties of $N_a(s)/Q_a(s)$ and $N_b(s)/Q_b(s)$, it follows that a necessary condition is

(10.37)
$$2 \geq \eta_A - \eta_B \geq 0$$

For a successful $RC:RL$ decomposition the following necessary conditions must also be satisfied in addition to condition 10.37:

(i) Leading coefficients of $A(s)$ and $B(s)$ are positive;
(ii) Roots of $A(s)$ and $B(s)$ are real and negative.

Note that condition 10.37 restricts the degrees of the various polynomials as follows:

(10.38)
$$A(s)^0 = D(s)^0$$
$$B(s)^0 = A(s)^0 \quad \text{or} \quad A(s)^0 - 2$$

Theorem 10-3, which gives the necessary and sufficient conditions for an $RC:RL$ decomposition of a polynomial $D(s)$, has been advanced by Calahan[32]; it is stated next without the proof.

Theorem 10-3 *The $RC:RL$ decomposition of a polynomial $D(s)$*

(10.39)
$$D(s) = \prod_{i=1}^{m} (s + p_i)(s + p_i^*)$$

is possible if and only if

(10.40)
$$\sum_{i=1}^{m} \arg p_i \leq \frac{\pi}{2}$$

In expression 10.40, $\arg p_i$ is the angle subtended by the vector p_i with the negative real axis. Consider the following example.

Example 10-6 The polynomial of interest is given as

(10.41)
$$D(s) = s^4 + 5s^3 + \tfrac{21}{2}s^2 + 11s + 5$$
$$= (s^2 + 2s + 2)(s^2 + 3s + \tfrac{5}{2})$$

[32] D. A. Calahan, *loc. cit.*

The roots of $N(s)$ in the upper left-half plane are at
$$p_1 = -1 + j1, \qquad p_2 = -\tfrac{3}{2} + j\tfrac{1}{2}$$
Therefore
$$\arg p_1 = \tan^{-1} 1 = 45°$$
$$\arg p_2 = \tan^{-1} \tfrac{1}{3} = 18° \; 26'$$
Observe that
$$\arg p_1 + \arg p_2 = 63°26' < 90°$$
and hence the angle condition (10.40) is satisfied. This implies that an $RC:RL$ decomposition of (10.41) is possible.

A suitable $Q(s)$ is given as[33]
$$Q(s) = (s + \tfrac{1}{2})(s + \tfrac{4}{3})(s + 2)$$
A partial fraction expansion of $N(s)/Q(s)$ yields

(10.42) $$\frac{D(s)}{Q(s)} = \frac{\tfrac{5}{4}}{s + \tfrac{1}{2}} + \frac{1}{s + 2} + \frac{7}{6} + s - \frac{\tfrac{5}{9}}{s + \tfrac{4}{3}}$$

Note that condition 3.56 is satisfied. We group the terms in Equation 10.42 as

$$\left(\frac{\tfrac{5}{4}}{s + \tfrac{1}{2}} + \frac{1}{s + 2} \right) + \left(s + \frac{7}{6} - \frac{\tfrac{5}{9}}{s + \tfrac{4}{3}} \right)$$
$$= \left[\frac{\tfrac{9}{4}s + 3}{(s + \tfrac{1}{2})(s + 2)} \right] + \left[\frac{s^2 + \tfrac{5}{2}s + 1}{s + \tfrac{4}{3}} \right]$$

It is seen that $\tfrac{9}{4}(s + \tfrac{4}{3})/(s + \tfrac{1}{2})(s + 2)$ is an RC impedance and that $(s + \tfrac{1}{2})(s + 2)/(s + \tfrac{4}{3})$ is an RL impedance.

It should be noted that it is always possible to find an $RC:RL$ decomposition for a second order Hurwitz polynomial. However, because of the angle condition (10.40), it may not be possible to obtain an $RC:RL$ decomposition of a higher-order polynomial having a high Q pole pair. This is not a serious problem, due to the fact that from a sensitivity point of view, it is usually preferable to realize a higher-order transfer function by cascading second-order stages.

The Reactive Gyrator as an Active Element

It is possible to modify the synthesis method of Section 10-3, p. 425, to enable the use of a restricted class of reactive gyrators.[34] We shall illustrate

---

[33] A systematic procedure for obtaining $Q(s)$ is given in Section 10-4.
[34] S. K. Mitra and W. G. Howard, Jr., "The reactive gyrator—a new concept and its application to active network syhthesis," *IEEE Int'l Conv. Rec.*, **14** (Part 7), 319–326 (1966).

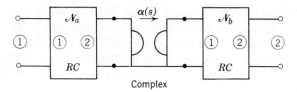

Figure 10-19 The RC:Reactive-Gyrator cascade configuration.

the technique for the case when the reactive gyrator is described by a gyration impedance $\alpha(s) = 1/(s + a)$.

Consider the network of Figure 10-19, for which the overall transfer-voltage ratio is given by

(10.43) $$t_v = \frac{-y_{21a}z_{21b}}{\alpha(s)y_{22a} + \dfrac{1}{\alpha(s)}z_{11b}}$$

Our problem is to decompose a specified voltage transfer ratio $N(s)/D(s)$ and cast it into the form of Equation 10.43 for a given $\alpha(s)$. Following the method outlined in Section 10-3, p. 426, let us select a suitable polynomial $Q(s)$ having distinct negative real zeros and express $N(s)/D(s)$ as

(10.44) $$\frac{N(s)/Q(s)}{D(s)/Q(s)} = \frac{(N_a/Q_a) \cdot (N_b/Q_b)}{(D_a/Q_a) + (D_b/Q_b)} ; \quad Q(s) = Q_a Q_b$$

where $Z_1(s) = D_a/Q_a$ is an RC impedance and $Y_2(s) = D_b/Q_b$ is an RC admittance. $Q(s)$ is selected so that $Z_1^\circ(s) = \alpha(s)/Z_1(s)$, the generalized inverse of $Z_1(s)$ with respect to $\alpha(s)$, is an RC realizable impedance and similarly $Y_2^\circ(s) = 1/\alpha(s)Y_2(s)$ is an RC admittance.[35] A necessary condition for the RC realizability of Z_1° and Y_2° is that zeros of $Q(s)$ be to the left of $s = -a$.

We rewrite expression 10.44 as

(10.45) $$\frac{N(s)/Q(s)}{D(s)/Q(s)} = \frac{\dfrac{N_a}{\alpha(s)Q_a} \cdot \dfrac{\alpha(s)N_b}{Q_b}}{\alpha(s) \cdot \dfrac{D_a}{\alpha(s)Q_a} + \dfrac{1}{\alpha(s)} \cdot \dfrac{\alpha(s)D_b}{Q_b}}$$

Comparing Equation 10.43 with expression 10.45, we identify

(10.46) $$y_{22a} = \frac{D_a}{\alpha(s)Q_a} ; \quad -y_{21a} = \frac{N_a}{\alpha(s)Q_a}$$

$$z_{11b} = \frac{\alpha(s)D_b}{Q_b} ; \quad z_{21b} = \frac{\alpha(s)N_b}{Q_b}$$

[35] See Section 3-3, p. 90.

432 / GYRATOR AS A NETWORK ELEMENT

From the parameters as given above, the RC networks $\mathcal{N}_a$ and $\mathcal{N}_b$ are realized following standard synthesis procedures. The sufficient condition on $D(s)$ for guaranteeing the realizability of the RC two-ports can be easily derived from the results of Theorem 10-3. If $D(s)$ is of the form of Equation 10.39, then the identification of (10.46) is possible for $\alpha(s) = 1/(s + a)$ if

(10.47) $$\sum_{i=1}^{m} \arg(p_i - a) \leq \frac{\pi}{2}$$

Example 10.7 will help to clarify the above method.

Example 10-7 Let us realize

$$t_v = \frac{N(s)}{D(s)} = \frac{H}{s^2 + 3s + 3}$$

using a reactive gyrator described by $\alpha(s) = 1/(s + \frac{1}{2})$. We note that $\arg(p_i - \frac{1}{2}) = \tan^{-1}\sqrt{3/2} = 40°54'$, which is definitely less than 90°. This implies that the angle condition (10.47) is satisfied.

Choose $Q(s) = (s + 1)(s + 2)$. Then

$$\frac{D(s)}{Q(s)} = \frac{s^2 + 3s + 3}{(s+1)(s+2)} = 1 + \frac{1}{s+1} - \frac{1}{s+2} = \frac{1}{s+1} + \frac{s+1}{s+2}$$

We thus have

(10.48) $$\frac{D_a}{Q_a} = \frac{1}{s+1}; \quad \frac{D_b}{Q_b} = \frac{s+1}{s+2}$$

Using Equation 10.48 in 10.46, we obtain

$$y_{22a} = \frac{s + \frac{1}{2}}{s + 1}; \quad -y_{21a} = H_a \frac{(s + \frac{1}{2})}{(s + 1)}$$

$$z_{11b} = \frac{(s + 1)}{(s + \frac{1}{2})(s + 2)}; \quad z_{21b} = \frac{H_b}{(s + \frac{1}{2})(s + 2)}$$

The two-port $\mathcal{N}_b$ is realized by a continued fraction expansion of z_{11b}. Network $\mathcal{N}_a$ is realized easily by choosing H_a equal to unity. Final network realizations are shown in Figure 10-20. The value of H_b is found to be equal to 1/3, which in this case is also the value of the multiplier constant H.

Synthesis steps for other types of $\alpha(s)$ are described elsewhere.[36] Of course, to keep the companion network RC realizable, $\alpha(s)$ must be an RC driving-point impedance.

---

[36] S. K. Mitra and W. G. Howard, *loc. cit.*

SENSITIVITY CONSIDERATIONS: THE CALAHAN DECOMPOSITIONS / 433

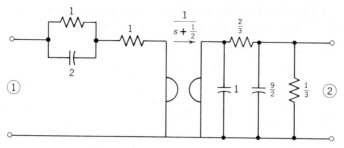

Figure 10-20 Realization of the transfer-voltage ratio of Example 10-7, using a reactive gyrator.

Reactive gyrators in general are potentially unstable devices, so care must be taken in the actual network design to prevent oscillation.

10-4 SENSITIVITY CONSIDERATIONS: THE CALAHAN DECOMPOSITIONS

From our previous discussions on $RC:RL$ decomposition we observe that, even though this type of decomposition is possible for a restricted class of polynomials satisfying the angle condition 10.40, there still exist for a decomposable polynomial many such decompositions. We naturally expect that out of all possible $RC:RL$ decompositions of a specified polynomial $D(s)$, at least one would be "optimum" from the sensitivity point of view. A reasonable sensitivity criteria would be the minimization of $\hat{S}_{\alpha^2}^{p_i}$, where $s = p_i$ is a root of $D(s)$ and α is the gyration constant of the gyrator. Since minimization of root sensitivity is achieved by minimizing the coefficient sensitivities, we base our next discussions on the latter approach.

The Optimum Decomposition

Consider for simplicity the $RC:RL$ decomposition of a second-order Hurwitz polynomial $D(s)$. In general, we can write

(10.49) $\quad D(s) = D_2 s^2 + D_1 s + D_0$
$\qquad = \alpha^2 (A_2 s^2 + A_1 s + A_0) + (B_2 s^2 + B_1 s + B_0)$

From Equation 10.49 it is clear that the coefficient sensitivity $\hat{S}_{\alpha^2}^{D_i}$ is minimized by minimizing A_i, and hence B_i. Let us attempt to obtain $RC:RL$ decomposition of $D(s)$ by setting $A_1 = A_0 = 0$. Since the degree condition (10.38) is satisfied, $RC:RL$ decomposition is possible. $A_1 = A_0 = 0$ implies that

(10.50) $\qquad \hat{S}_{\alpha^2}^{D_1} = \hat{S}_{\alpha^2}^{D_0} = 0$

Two possible decompositions of $D(s)$ under this condition are given in Equations 10.51, where without any loss of generality we have set D_0 equal to unity.

(10.51a) $\quad D(s) = D_2 s^2 + D_1 s + 1 = (bs + 1)^2 + (\alpha^2)as^2$

(10.51b) $\quad\quad\quad\quad\quad\quad\quad\quad\quad = (b_1 s + 1)(b_2 s + 1) + (\alpha^2)a_1 s^2$

In Equations 10.51, α^2 is nominally unity for convenience. The coefficient sensitivities of interest are

(10.52) $\quad\quad (S_{\alpha^2}^{D_2})_I = \dfrac{a}{D_2}, \quad (S_{\alpha^2}^{D_2})_{II} = \dfrac{a_1}{D_2}$

Equating like coefficients of Equations 10.51, we have

$$D_2 = b^2 + a = b_1 b_2 + a_1$$
$$D_1 = 2b = b_1 + b_2$$

which can be solved to yield

$$b_1 - b_2 = \pm 2\sqrt{a_1 - a}$$

Since b_1 and b_2 are real and positive numbers, it follows that a_1 must be greater than a or, in other words, the decomposition of Equation 10.51a is less sensitive than (10.51b).

As a matter of fact, Calahan[37] has shown that the "optimum" $RC:RL$ decomposition of a polynomial $D(s)$ having only complex roots is of the form

(10.53) $\quad\quad\quad\quad D(s) = a^2(s) + b^2(s)$

where $b(s)/a(s)$ is an RC driving-point impedance function. In contrast to the "optimum" $RC:RC$ decomposition (Equation 9.76), decomposition 10.53 is not unique. This can be seen by observing that the degree condition (10.38) implies $b(s)^0$ can be either equal to $a(s)^0$ or one less. Now if $b(s)^0 = a(s)^0 = n$, then there are $(2n + 2)$ unknown quantities and $(2n + 1)$ known numbers, indicating nonuniqueness of the solution. The decomposition is unique only when we restrict $b(s)^0$ equal to $a(s)^0 - 1$. The most important characteristic of all of the "optimum" $RC:RL$ decompositions of a specified $D(s)$ is that the root sensitivities in each case are identical in magnitude.[38] The following example will illustrate our point.

[37] D. A. Calahan, "Sensitivity minimization in active RC synthesis," *IRE Trans. on Circuit Theory*, **CT-9**, 38–42 (March 1962).

[38] For a proof, see D. A. Calahan, "Transfer function synthesis based on cascaded RC and RL networks," Tech. Note No. 15, Circuit Theory Group, Elec. Engg. Res. Lab., Univ. of Illinois, Urbana, Ill., September 1, 1960.

SENSITIVITY CONSIDERATIONS: THE CALAHAN DECOMPOSITIONS / 435

Example 10-8 Three "optimum" $RC:RL$ decompositions of $D(s) = (s^2 + 6s + 10)$ are given below:

(10.54a) $$D(s) = s^2 + 6s + 10 = (s+3)^2 + 1$$
(10.54b) $$= \tfrac{1}{2}(s+2)^2 + \tfrac{1}{2}(s+4)^2$$
(10.54c) $$= \frac{s^2}{10} + 10\left(1 + \frac{3}{10}s\right)^2$$

The root sensitivities for the root of $D(s)$ at $p_1 = -3 + j$ can be calculated using Equation 5.66 and are given as:

Decomposition 10.54a
$$S_{\alpha^2}^{p_1} = \frac{1}{s+3+j}\bigg|_{s=-3+j} = \frac{1}{2j} = 0.5\,\underline{/270°}$$

Decomposition 10.54b
$$S_{\alpha^2}^{p_1} = \frac{\tfrac{1}{2}(s+4)^2}{s+3+j}\bigg|_{s=-3+j} = \tfrac{1}{2} = 0.5\,\underline{/0°}$$

Decomposition 10.54c
$$S_{\alpha^2}^{p_1} = \frac{10(1+\tfrac{3}{10}s)^2}{s+3+j}\bigg|_{s=-3+j} = 0.3 + j0.4 = 0.5\,\underline{/53°8'}$$

Observe that all three decompositions have identical $|S_{\alpha^2}^{p_1}|$.

This useful characteristic of the "optimum" $RC:RL$ decompositions offers the designer a freedom in choosing an "optimum" decomposition that may satisfy other design criteria, e.g., stability margin, number of resistances and capacitances, spread of element values. For example, decomposition (10.54a) has the largest stability margin[39] and will, in general, yield a network with less components. (Why?)

For future reference, we shall designate the $RC:RL$ decompositions of the form of Equation 10.53 as the *Calahan decompositions*. We next intend to compare the Calahan and Horowitz decompositions from the point of view of root sensitivity.

A Comparison of Calahan and Horowitz Decompositions[40]

Consider first a second order polynomial which we can write as

(10.55) $$D(s) = (s - p_1)(s - p_1^*) = s^2 + 2\beta_1 \cos\theta_1 + \beta_1^2$$

---

[39] See Section 5-5, p. 186.
[40] D. A. Calahan, *loc. cit.*

where
(10.56) $$p_1 = -\beta_1 \cos\theta_1 + j\beta_1 \sin\theta_1$$
A Calahan decomposition of (10.55) can easily be shown to be
(10.57) $$D(s) = \alpha^2(s + \beta_1 \cos\theta_1)^2 + (\beta_1 \sin\theta_1)^2$$
where α, the gyration constant, is nominally taken as unity. The root sensitivity of the above decomposition can be shown to be equal to

(10.58) $$S_{\alpha^2}^{p_1} = \frac{(\beta_1 \sin\theta_1)^2}{s + \beta_1 \cos\theta_1 + j\beta_1 \sin\theta_1}\bigg|_{s=-\beta_1 \cos\theta_1 + j\beta_1 \sin\theta_1} = \frac{\beta_1 \sin\theta_1}{2j}$$

On the other hand, the Horowitz decomposition of $D(s)$ is simply given as
(10.59) $$D(s) = (s + \beta_1)^2 - 2k\beta_1(1 - \cos\theta_1)s$$
where k, the NIC conversion factor, is nominally unity for convenience. For the second decomposition we have

(10.60) $$S_k^{p_1} = \frac{(s + \beta_1)^2}{s + \beta_1 \cos\theta_1 + j\beta_1 \sin\theta_1}\bigg|_{s=-\beta_1 \cos\theta_1 + j\beta_1 \sin\theta_1}$$
$$= \frac{2\beta_1^2(1 - \cos\theta_1)(-\cos\theta_1 + j\sin\theta_1)}{2j\beta_1 \sin\theta_1}$$

From Equations 10.58 and 10.60 we obtain

(10.61) $$|S_{\alpha^2}^{p_1}| = \frac{\beta_1 \sin\theta_1}{2}$$
$$|S_k^{p_1}| = \frac{\beta_1(1 - \cos\theta_1)}{\sin\theta_1}$$

Since $0 < |\theta_1| \leq 90°$, it can be seen that $|S_{\alpha^2}^{p_1}| < |S_k^{p_1}|$.

For a fourth-order polynomial of the form
(10.62) $$D(s) = (s^2 + s\beta_1 \cos\theta_1 s + \beta_1^2)(s^2 + 2\beta_2 \cos\theta_2 s + \beta_2^2)$$
it can be shown[41] that the root sensitivities for the "optimum" $RC:RL$ and $RC:-RC$ decompositions are given as:

(10.63) $$|S_{\alpha^2}^{p_1}| = \frac{\beta_1 \sin\theta_1}{2}\left[\frac{\beta_1^2 + \beta_2^2 - 2\beta_1\beta_2 \cos(\theta_1 + \theta_2)}{\beta_1^2 + \beta_2^2 - 2\beta_1\beta_2 \cos(\theta_1 - \theta_2)}\right]^{1/2}$$
$$|S_k^{p_1}| = \frac{2\beta_1(1 - \cos\theta_1)}{\xi \sin\theta_1}\bigg[\frac{\beta_2^2 + \beta_1^2 - 2\beta_1\beta_2 \cos\theta_1}{2}$$
$$+ \beta_1^2(1 - \cos\theta_2) - \beta_1(\beta_1 + \beta_2)\Omega\bigg]$$

[41] D. A. Calahan, *loc. cit.*

where

$$\xi = [\beta_1^2 + \beta_2^2 - 2\beta_1\beta_2 \cos(\theta_1 - \theta_2)]^{1/2}$$
$$\times [\beta_1^2 + \beta_2^2 - 2\beta_1\beta_2 \cos(\theta_1 + \theta_2)]^{1/2}$$
(10.64) $\quad \Omega = \{(1 - \cos\theta_1)(1 - \cos\theta_2)\}^{1/2}$
$$p_1 = -\beta_1 \cos\theta_1 + j\beta_1 \sin\theta_1$$

In deriving expression 10.63, it has been assumed that the nominal values of α and k are unity. We leave it as an exercise to the reader to prove that $|S_k^{p_1}| > |S_{\alpha^2}^{p_1}|$.

Using a root-locus approach, Calahan[41] has shown that "optimum" $RC:RL$ decomposition in general yields a lower-root sensitivity.

Construction of Optimum Decompositions

There exists an elegant algebraic method for obtaining a class of Calahan decompositions. The method is as follows.[42]

Consider $D(s)$ as given by Equation 10.39, which can be rewritten as

(10.65) $\quad D(s) = \prod_{i=1}^{m}(s + \sigma_i + j\omega_i)\prod_{i=1}^{m}(s + \sigma_i - j\omega_i), \quad \sigma_i > 0, \quad \omega_i > 0$

It is clear that one can express

(10.66) $\quad \prod_{i=1}^{m}(s + \sigma_i + j\omega_i) = E(s) + jF(s)$

which leads to

(10.67) $\quad D(s) = \{E(s) + jF(s)\}\{E(s) - jF(s)\} = E^2(s) + F^2(s)$

It can be shown[43] that $F(s)/E(s)$ is an RC impedance and hence Equation 10.67 is a Calahan decomposition. Example 10.9 illustrates the method along with a variation that yields an alternate Calahan decomposition.

Example 10-9 Let us consider the polynomial $D(s)$ given by Equation 10.41, which we rewrite as

$$D(s) = (s + 1 + j)(s + 1 - j)(s + \tfrac{3}{2} + j\tfrac{1}{2})(s + \tfrac{3}{2} - j\tfrac{1}{2})$$

We identify

$$E(s) + jF(s) = (s + 1 + j)(s + \tfrac{3}{2} + j\tfrac{1}{2})$$
$$= \{(s+1)(s+\tfrac{3}{2}) - \tfrac{1}{2}\} + j\{(s+\tfrac{3}{2}) + \tfrac{1}{2}(s+1)\}$$
$$= (s^2 + \tfrac{5}{2}s + 1) + j(\tfrac{3}{2}s + 2)$$

---

[42] D. A. Calahan, *loc. cit.*
[43] See Problem 10.12.

Therefore

(10.68) $D(s) = \{(s^2 + \frac{5}{2}s + 1) + j(\frac{3}{2}s + 2)\}\{(s^2 + \frac{5}{2}s + 1) - j(\frac{3}{2}s + 2)\}$

$= (s^2 + \frac{5}{2}s + 1)^2 + (\frac{3}{2}s + 2)^2$

Observe that

$$\frac{(\frac{3}{2}s + 2)}{(s^2 + \frac{5}{2}s + 1)} = \frac{\frac{3}{2}(s + \frac{4}{3})}{(s + 2)(s + \frac{1}{2})}$$

is an *RC* driving-point impedance function.

An alternate Calahan decomposition is obtained as follows. We first form $D(1/s)$:

$$D\left(\frac{1}{s}\right) = \frac{5}{s^4}(s^2 + s + \frac{1}{2})(s^2 + \frac{6}{5}s + \frac{2}{5})$$

$$= \frac{5}{s^4}(s + \frac{1}{2} + j\frac{1}{2})(s + \frac{1}{2} - j\frac{1}{2})(s + \frac{3}{5} + j\frac{1}{5})(s + \frac{3}{5} - j\frac{1}{5})$$

which is regrouped to yield

$$\frac{5}{s^4}[(s + \frac{1}{2} + j\frac{1}{2})(s + \frac{3}{5} + j\frac{1}{5})][(s + \frac{1}{2} - j\frac{1}{2})(s + \frac{3}{5} - j\frac{1}{5})]$$

$$= \frac{5}{s^4}[(s^2 + \frac{11}{10}s + \frac{1}{5}) + j(\frac{7}{10}s + \frac{2}{5})][(s^2 + \frac{11}{10}s + \frac{1}{5}) - j(\frac{7}{10}s + \frac{2}{5})]$$

$$= \frac{5}{s^4}[(s^2 + \frac{11}{10}s + \frac{1}{5})^2 + (\frac{7}{10}s + \frac{2}{5})^2]$$

This implies that

$$D(s) = 5[(\frac{1}{5}s^2 + \frac{11}{10}s + 1)^2 + s^2(\frac{2}{5}s + \frac{7}{10})^2]$$

Once again, we observe that

$$\frac{s(\frac{2}{5}s + \frac{7}{10})}{(\frac{1}{5}s^2 + \frac{11}{10}s + 1)} = \frac{0.4s(s + 1.75)}{0.2(s + 1.15)(s + 4.35)}$$

is an *RC* admittance function.

10-5 INDUCTANCE SIMULATION USING GYRATOR

Probably one of the most attractive features of the gyrator is that an inductor can simply be simulated by terminating the gyrator by a capacitor. Thus an easy method of inductorless filter design would be to replace the inductors in a standard passive *RLC* filter by capacitor-loaded gyrators.

There are two advantages of this approach: (1) low sensitivity and (2) tunability.

We have shown in Section 5-7, p. 196, that passive RLC resonators have an extremely low sensitivity (Σ_Q equal to 2 and independent of Q). Low sensitivity feature of doubly-loaded LC filters has also been pointed out by Orchard.[44] On the other hand, active RC filters employing a NIC or a noninverting type VVT have a very high sensitivity (Σ_Q of the order of $2Q$). Quantitative comparison of second-order doubly-loaded LC filters and RC:NIC filters have been recently provided by Gorski-Popiel,[45] and Woodard and Newcomb,[46] leading to the conclusion of superiority of LC filters from the point of view of sensitivity.

An additional feature of the LC filters is the possibility of adjusting the inductor values to bring certain resonating frequencies to their design values. This feature is equally available for the simulated inductors obtained by capacitor-loaded gyrators by connecting trimmer capacitances.

A major problem of the gyrator circuits discussed earlier is that they can produce only grounded inductors. However, in many practical LC filters, floating inductors are present. One way to circumvent this problem would be to use two grounded gyrators per floating inductor in the way suggested by Holt and Taylor[47] (see Section 2-5, p. 49). This method requires very careful matching of the gyrator parameters as pointed out by Sheahan,[48] who has suggested using a novel circuit to convert a grounded gyrator into a floating gyrator. A third possibility is to construct gyrators with floating ports.[49]

It can be pointed out here that active filters of relatively large degree have been successfully constructed using simulated inductors by several authors.[50]

---

[44] H. J. Orchard, "Inductorless filters," *Electronics Letters*, **2**, 224 (September 1966).

[45] J. Gorski-Popiel, "Horowitz minimum-sensitivity decomposition," *Electronics Letters*, **2**, 334–335 (September 1966).

[46] J. Woodard and R. W. Newcomb, "Sensitivity improvement of inductorless filters," *Electronics Letters*, **2**, 349–350 (September 1966).

[47] A. G. J. Holt and J. Taylor, "Method of replacing ungrounded inductors by grounded gyrators," *Electronics Letters*, **1**, 105 (1965).

[48] D. F. Sheahan, "Gyrator-flotation circuit," *Electronics Letters*, **3**, 39–40 (January 1967).

[49] W. H. Holmes, S. Gruetzmann, and W. E. Heinlein, "Direct-coupled gyrators with floating ports," *Electronics Letters*, **3**, 46–47 (February 1967).

[50] D. F. Sheahan and H. J. Orchard, "Bandpass-filter realization using gyrators," *Electronics Letters*, **3**, 40–42 (January 1967); W. H. Holmes, S. Gruetzmann, and W. E. Heinlein, "High-quality active C filters using gyrators," *Int'l Solid-State-Circuit Conf. Digest*, Philadelphia, Pa., **10**, 122–123 (1967).

Let us now compute the effective inductance and the Q factor of a simulated inductor formed by terminating a practical gyrator by a capacitor.[51] As a first approximation a practical gyrator, which is a nonideal device, can be represented by a short-circuit admittance matrix:

$$[y] = \begin{bmatrix} G_1 & g_a \\ -g_b & G_2 \end{bmatrix} \tag{10.69}$$

where G_1 and G_2 account for nonzero y_{11} and y_{22}, and are very small compared to the gyration admittances g_a and g_b. If we terminate the port 2 of the above gyrator by a capacitor of value C farads, the input impedance seen at port 1 will be:

$$\begin{aligned} Z_{in}(j\omega) &= \frac{G_2 + j\omega C}{G_1 G_2 + g_a g_b + j\omega G_1 C} \\ &= \frac{(G_1 G_2^2 + g_a g_b G_2 + \omega^2 G_1 C^2) + j\omega C g_a g_b}{(G_1 G_2 + g_a g_b)^2 + \omega^2 G_1^2 C^2} \end{aligned} \tag{10.70}$$

Hence, the capacitor terminated practical gyrator behaves like a lossy inductor having an inductance,

$$L_{eq}(\omega) = \frac{C g_a g_b}{(G_1 G_2 + g_a g_b)^2 + \omega^2 G_1^2 C^2} \text{ henries} \tag{10.71}$$

and a Q given by

$$Q = \frac{\omega C g_a g_b}{G_1 G_2^2 + g_a g_b G_2 + \omega^2 G_1 C^2} \tag{10.72}$$

Note from expression 10.72 that to obtain high Q inductors, the gyrator must be designed to have very small G_1 and G_2 in comparison to the gyration admittances. Usually a practical gyrator is designed to have $g_a = g_b$ and $G_1 = G_2$. In this case

$$Q = \frac{\omega C g_a^2}{G_1(G_1^2 + g_a^2 + \omega^2 C^2)} \simeq \frac{\omega C g_a^2}{G_1(g_a^2 + \omega^2 C^2)} \tag{10.73}$$

for $G_1 \ll g_a$.

The maximum value of the Q is at $\omega = \omega_{max} = g_a/C$ and is equal to $Q_{max} = g_a/2G_1$. Note that these values agree with the result obtained by Orchard.[52]

We have pointed out earlier that in order to simulate high Q inductors, the parasitic admittances G_1 and G_2 of a practical gyrator must be very

---

[51] T. N. Rao, P. Gary, and R. W. Newcomb, "Equivalent inductance and Q of a capacitor-loaded gyrator," *IEEE J. of Solid-State Circuits*, SC-2, 32–33 (March 1967).
[52] H. J. Orchard, *loc. cit.*

small compared to the gyration admittances g_a and g_b. In many design approaches this is achieved by using negative admittances for cancellation purposes. However, such an approach is unsatisfactory for sensitivity reasons. An inductance simulated using such a gyrator will have an effective impedance given as:

$$j\omega L_{eq} + r_1 - r_2 \tag{10.74}$$

where r_1 (or r_2) is adjusted later to make $r_1 \to r_2$. The effective Q can then be made large. On the other hand, the Q-sensitivity with respect to r_2 is

$$S_{r_2}^Q = \frac{r_2}{r_1 - r_2}$$

which becomes extremely large as r_1 approaches r_2.

Thus for practical reasons, it appears that a high quality gyrator can only be designed by paralleling two VCTs whose input and output admittances are extremely low. Unfortunately at high frequencies, the phase shift introduced by the transistors due to delay of charge carriers and internal capacitors create a problem. The effect of phase shift is to introduce a negative resistance in series with the simulated inductor and as a consequence, makes the inductor Q highly sensitive to variation of gyrator parameters.[53]

10-6 SUMMARY

Six three-transistor basic gyrator circuits are developed from their nullator-norator models (Section 10-1, p. 405). A method of active gyrator design which makes use of the nonreciprocal negative-impedance inverter has been presented (Section 10-1, p. 410).

Two methods of synthesis of positive real driving-point functions using RLC:gyrator networks are outlined in Section 10-2. The first method is a modification of the well-known Bott-Duffin representation. The second one is an extension of Darlington's method.

A realization method of transfer voltage ratio by means of RC:gyrator two-port is outlined (Section 10-3, p. 425). The final realization is in the form of a cascade configuration of two RC two-ports and a gyrator. The synthesis method makes use of the RC:RL decomposition of the denominator polynomial which in turn must satisfy the angle condition (10.40). An extension of this technique allows one to use reactive gyrators having RC gyration impedances (Section 10-3, p. 430).

---

[53] M. Bialko, "On Q factor and Q sensitivity of an inductor simulated by a practical gyrator," *Electronics Letters*, 3, 168–169 (April 1967).

Optimum $RC:RL$ decomposition (Equation 10.53), which minimizes the root sensitivity with respect to the gyration impedance, is discussed in Section 10-4. It is shown that the optimum $RC:RL$ decomposition (if it exists) has a lower sensitivity than the optimum $RC:-RC$ decomposition. (Section 10-4, p. 435). An algebraic method of construction of the optimum $RC:RL$ decomposition is outlined (Section 10-4, p. 437).

Various aspects of inductance simulation using gyrator are discussed in Section 10-5.

Additional References

Anderson, B. D., W. New, and R. W. Newcomb, "Proposed adjustable tuned circuits for micro-electronic structures," *Proc. IEEE*, **54**, 411 (March 1966).

Bach, R. E., Jr., and A. W. Carlson, "Practical realization of direct-coupled gyrator circuits," Proc. Asilomar Conf. on Circuits and Systems, Monterey, Calif., November 1967.

Bermano, C. P., and D. R. Curan, "Low frequency gyrator," *Proc. Electronic Components Conf.*, San Francisco, Calif., p. 24 (May 1961).

Bhusan, M., and R. W. Newcomb, "Grounding of capacitors in integrated circuits," *Electronics Letters*, **3**, 148–149 (April 1967).

Bialko, M., "Selective network with nonideal gyrator," *Electronics Letters*, **2**, 471–472 (December 1966).

Bogert, B. P., "Some gyrator and impedance inverter circuits," *Proc. IRE*, **43**, 793–796 (July 1955).

Butler, F., "Gyrators—using direct-coupled transistor circuits," *Wireless World*, **73**, 89–93 (February 1967).

Carlin, H. J., "Principles of gyrator networks," Proc. Symp. on Modern Advances in Microwave Technology, MRI Symp. Series, Polytech. Inst. of Brooklyn, **4**, 175–204 (1955).

Carlin, H. J., "Synthesis of non-reciprocal networks," Proc. Symp. on Modern Network Synthesis, MRI Symp. Series, Polytech. Inst. of Brooklyn, **5**, 11—44 (April 1956).

Ford, R. L., and F. E. J. Girling, "Active filters and oscillators using simulated inductances," *Electronics Letters*, **2**, 52 (1966).

Garg, J. M., and H. J. Carlin, "Network theory of semi-conductor Hall-plate circuits," *IEEE Trans. on Circuit Theory*, **CT-12**, 59–73 (March 1965).

Gensel, J., "Negative widerstände und gyratoren," *Nachrichtentechnik*, **7**, 249–256 (June 1957).

Gensel, J., "Der gyratorverstarker als element zum aufbau spulenfreir siebketten," *Wiss. Z. Hochsch Elektrotech. Ilemenau*, **8**, 49–64 (1966).

Gensel, J., "Eine methode, mach der man negativ immittanz konverter und gyratorschaltungen vergleichen kann," *Electronics Letters*, **2**, 361–362 (October 1966).

Gensel, J., "Ein quantitativer vergleich zwischess negativ immittanz konverter und gyrator schaltungen," *Electronics Letters*, **2**, 362–364 (October 1966).

Ghausi, M. S., and F. D. McCarthy, "A realization of transistor gyrators," *Semiconductor Products and Solid State Technol.*, pp. 13–17 (1964).

Gorski-Popiel, J., "RC-active synthesis using positive-immittance converters," *Electronics Letters*, **3**, 381–382 (August 1967).

Grubbs, W. J., "Hall effect gyrators, isolators, and circulators with high efficiency," *Proc. IRE*, **47**, 528–535 (April 1959).

Hakim, S. S., "RC-gyrator low-pass filter," *Proc. IEE (London)*, **113**, 1504–1506 (September 1966).

Harrison, T. J., "A gyrator realization," *IEEE Trans. on Circuit Theory*, **10**, 303 (June 1963).

Hazony, D., "Zero cancellation synthesis using impedance operators," *IRE Trans. on Circuit Theory*, **CT-8**, 114–120 (June 1961).

Ho, C., and N. Balabanian, "Synthesis of active and passive compatible impedances," *IEEE Trans. on Circuit Theory*, **CT-14**, 118–128 (June 1967).

Huelsman, L. P., "Brune realization using gyrator-capacitor sections," *IEEE Trans. on Circuit Theory*, **CT-12**, 439 (September 1965).

Johnson, A. K., and P. M. Chirlian, "A cryotron gyrator," *Proc. IEEE*, **54**, 806–807 (May 1966).

Keen, A. W., and J. Peters, "Nonreciprocal representation of the floating inductor with grounded-amplifier realizations," *Electronics Letters*, **3**, 369–371 (August 1967).

Murdoch, J. B., "RC-gyrator cascade synthesis," *IEEE Trans. on Circuit Theory*, **CT-11**, 268–271 (June 1964).

Murdoch, J. D., and D. Hazony, "Cascade driving-point impedance synthesis by removal of sections containing arbitrary constants," *IRE Trans. on Circuit Theory*, **CT-9**, 56–61 (March 1962).

Myers, B. R., "Contributions to transistor-RC network synthesis," Circuit Theory Group Technical Note No. 7, Elec. Engg. Res. Lab., U. of Illinois, Urbana, Illinois, February 20, 1959.

Nair, K. K., and M. N. S. Swamy, "On gyrator networks," *J. Inst. Telecomm. Engrs. (India)*, **7**, 111–116 (May 1961).

Nonnenmacher, W., and F. Schreiber, "Der zweidrahtverstäarker als gyrator und als vierpol zur herstellung ungewöhnlicher scheinwiderstände," *Frequenz*, **8**, 201–204 (September 1954).

Prudhon, M., "Gyrators and non-reciprocal systems," *Cables et Transmission*, **11**, 66–73 (January 1957).

Sharpe, G. E., "The pentode gyrator," *IRE Trans. on Circuit Theory*, **CT-4**, 321–323 (December 1957).

Shekel, J., "The gyrator as a three-terminal element," *Proc. IRE*, **41**, 1014–1016 (August 1953).

Silverman, J. H., "The electromechanical circulator," *IRE Trans. on Component Parts*, **CP-9**, 81–85 (June 1962).

Silverman, J. H., "Now you can use non-reciprocal devices at low frequencies," *Electronics*, pp. 56–59 (February 22, 1963).

Soohoo, R. F., *Theory and Applications of Ferrites*, Prentice-Hall, Englewood Cliffs, N.J., 159–161 (1960).

Su, K. L., "A Transistor-circuit realization of the inductance," *Proc. IEEE*, **54**, 2025–2027 (1966).

Su, K. L., "F.E.T.–circuit realization of the inductance," *Electronics Letters*, **2**, 469–470 (December 1966).

Tellegen, B. D. H., "The synthesis of passive, resistance-less four-poles that may violate the reciprocity relation," *Philips Res. Rept.*, **3**, 321–337 (1948).

Tellegen, B. D. H., "The synthesis of two-poles by means of networks containing gyrators," *Philips Res. Rept.*, **4**, 31–37 (February 1949).

Tsirel'son, D. A., "Methods of making gyrators," *Telecommns. and Radio Engg.*, 12–15 (April 1966).

Vasudeva, M. S., "Gyrator realization," *Electronics Letters*, **2**, 201–202 (1966).

Problems

10.1 Form the dual of Figure 10-3a. Construct several transistor realizations of the dual circuit.

10.2 Show that the nullator-norator circuit of Figure 10-21a is the dual

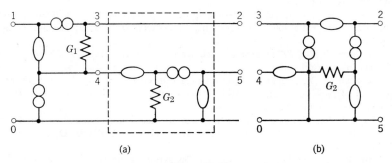

Figure 10-21

of Figure 10-4a. To obtain a 3-terminal gyrator from the circuit of Figure 10-21a the portion inside the dotted box can be replaced by the equivalent circuit of Figure 10-21b.[54] Verify the equivalence and then construct several transistor realizations.

10.3 Show that by cascading an NIC and any NIV, an ideal gyrator can be formed.[55] Realize several gyrators using this approach and at most two NICs. How would you modify the resulting circuits to incorporate nonideal NICs?

10.4 Use of feedback amplifiers to produce nonideal gyrators was proposed by Bogert.[56] The proposed arrangements are shown in Figure 10-22. Compute the open-circuit impedance matrix of both circuits and show that gyrator action is obtained for $A = 2$. How would you compensate these circuits?

[54] Based on J. M. Sipress, U.S. Patent no. 3,098,978, July 23, 1963.
[55] T. J. Harrison, *loc. cit.*
[56] B. P. Bogert, *loc. cit.*

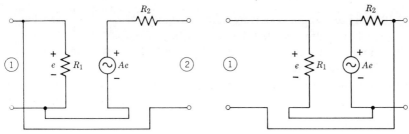

Figure 10-22

10.5 Using the method outlined in Section 10-2, p. 414, realize the following driving-point impedances:

(a) $\dfrac{s^3 + 2s^2 + s + 1}{s^3 + s^2 + 2s + 1}$

(b) $\dfrac{s^2 + s + 2}{2s^2 + s + 1}$

10.6 Realize by the method of nonreciprocal Darlington synthesis (Section 10-2, p. 418) the impedance functions of Problem 10.5.

10.7 Synthesize by means of an RC:gyrator network the following transfer voltage ratios:

(a) $\dfrac{Hs^2}{3s^2 + 3s + 1}$

(b) $\dfrac{H(s + 1)}{s^2 + 2s + 6}$

Compute the value of H and compare it with the maximum H obtainable by a passive RLC unbalanced network in each case.

10.8 Is it possible to obtain an $RC:RL$ decomposition of the following polynomials?

(a) $s^2 + 1.4142s + 1$
(b) $s^4 + 2.6131s^3 + 3.1412s^2 + 2.6131s + 1$
(c) $s^4 + 2s^3 + s^2 + 3$
(d) $(s^2 + 1.969s + 1)(s^2 + 3.76s + 4)(s^2 + 5.196s + 9)$

10.9 Develop a synthesis procedure to realize a transfer voltage ratio in the form of Figure 10-19 using a reactive gyrator described by a gyration impedance $\alpha(s) = (s + b)/(s + a)$ where $b > a$.[57]

---

[57] S. K. Mitra and W. G. Howard, *loc. cit.*

10.10 Realize using a reactive gyrator having $\alpha(s) = \dfrac{(s+1)}{(s+2)}$ the transfer voltage ratio $t_v = H/(s^2 + 7s + 13)$.

10.11 Derive the expressions for $|S_{\alpha^2}^{p_1}|$ and $|S_k^{p_1}|$ given in expression 10.63 and show that $|S_{\alpha^2}^{p_1}| < |S_k^{p_1}|$.

10.12 Following the method outlined in Section 10-4, p. 437, show that the "optimum" $RC:RL$ decomposition of a fourth-order polynomial $D(s)$,

$$D(s) = (s + a_1 + jb_1)(s + a_1 - jb_1)(s + a_2 + jb_2)(s + a_2 - jb_2)$$

is given as

$$D(s) = E^2(s) + F^2(s)$$

where

$$E(s) = (s + a_1)(s + a_2) - b_1 b_2$$

$$F(s) = b_1(s + a_2) + b_2(s + a_1)$$

and $F(s)/E(s)$ is an RC impedance if $D(s)$ satisfies the angle condition. How would you attempt to prove the validity of the decomposition procedure for a higher-order polynomial?[58]

10.13 Show that the RC:gyrator two port of Figure 10-23[59] realizes the

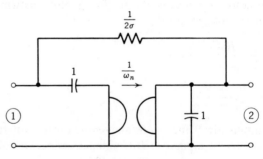

Figure 10-23

bandpass transfer function:

$$\frac{V_2}{V_1} = \frac{Hs}{s^2 + 2\sigma s + \omega_n^{\,2}}$$

Determine the value of H.

10.14 Obtain several Calahan decompositions of each of the polynomials of Problem 10.8, where possible.

---

[58] See D. A. Calahan, *loc. cit.*
[59] W. H. Holmes, *loc. cit.*

11 / *The Operational Amplifier as a Network Element*

The operational amplifier, essentially a very high-gain voltage-controlled voltage source, is an extremely versatile commercially manufactured active device. Even though it started as the most basic unit of an analog computer, it is now being used in many widely different areas, such as control engineering, communication systems, network design, etc. A brief introduction to the operational amplifier was given in Section 2-1, p. 29.

The main aim of this chapter is to outline the use of the operational amplifier as a network component in the synthesis of driving-point and transfer functions. We could have discussed these devices in Chapter 8, where we considered the controlled sources and their applications. However, these high-gain d-c voltage amplifiers have several distinctive features which, coupled with their ready availability in monolithic integrated form as off-the-shelf items, do justify their separate treatment.

This chapter is divided into nine sections. The first two sections are concerned with the physical characteristics of a practical (nonideal) operational amplifier and the advantages of its use in a feedback circuit. Construction of other active elements using operational amplifiers is considered in the next section. In Sections 11-4 and 11-5, several techniques for realizing driving-point and transfer functions are discussed. Network design by means of the coefficient matching approach is considered in the following two sections. Section 11-8 outlines several methods for realizing inductances. Main results of this chapter are summarized in the last section.

11-1 PRACTICAL CONSIDERATIONS

Since the operational amplifiers are easily available in packaged form, we do not intend to study any aspects of their actual physical construction.[1] The purpose of this section is to discuss the characteristics of a practical operational amplifier and compare it with the idealized version. It will be shown that these devices, when used with proper companion networks, can approach the idealized behavior. Our discussion, which will be based on two very simple circuits, will also point out some fundamental problems associated with these devices, along with various schemes to improve the performance of an actual circuit.

The Idealized Device

To facilitate the understanding of a real operational amplifier and its applications, we postulate an idealized device that is best defined in terms of its controlled source equivalent circuit as shown in Figure 11-1a. The principal characteristics of this amplifier are as follows.

The output voltage is always of the same polarity with the input voltage applied at the terminal marked "+" (to be called the *noninverting* input terminal). On the other hand, the output voltage is always of opposite polarity with the input signal applied at the *inverting* input terminal (marked "−"). The output voltage responds only to the difference of the voltages between the two terminals with a gain that is a real constant approaching infinity for all frequencies (implying infinite bandwidth). The input admittance and output impedance of the ideal amplifier are zero. The idealized device is also said to have a zero "offset," which implies that the output voltage is zero when the voltage difference between the input terminals is zero.

The usual representation of the differential input operational amplifier is as shown in Figure 11-1b. Often, for many applications, the amplifier is used "single-ended" with the noninverting input terminal grounded. The circuit symbol of a single-ended operational amplifier is shown in Figure 11-1c.

---

[1] For readers interested in the design of operational amplifiers, the following references may be useful. *RCA Linear Integrated Circuit Fundamentals*, Technical Series IC-40, RCA, Harrison, New Jersey, 1966; R. J. Widlar, "A monolithic high gain dc amplifier," *Proc. NEC*, **20**, 169–174 (1964); and R. J. Widlar, "A unique circuit design for a high performance operational amplifier especially suited to monolithic construction," *Proc. NEC*, **21**, 169–174 (1965).

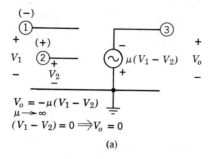

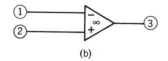

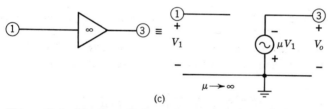

Figure 11-1 (*a*) Controlled-source representation of the ideal differential input operational amplifier. (*b*) Circuit symbol of a differential input operational amplifier. (*c*) The single-ended operational amplifier.

Characteristics of a Practical Operational Amplifier

In practice, the operational amplifier is a nonideal device. It is characterized by a frequency-dependent voltage gain whose magnitude starts from a very high value at dc (usually in the range of 80 to 120 db) and then monotonically decreases for higher frequencies. Likewise, the phase of the voltage gain is a monotonically decreasing function starting from 0° at d-c. The magnitude and phase responses of a typical amplifier are shown in Figure 11-2.

A convenient representation of the magnitude response is the straight-line approximation known as the Bode plot.[2] The Bode plot of the magnitude function of our typical amplifier is shown in Figure 11-2 by the dotted lines. Essentially, the Bode plot is a representation of the gain

[2] See for example, M. E. Van Valkenburg, *Network analysis*, 2nd ed., Prentice-Hall, Englewood Cliffs, N.J., 1964.

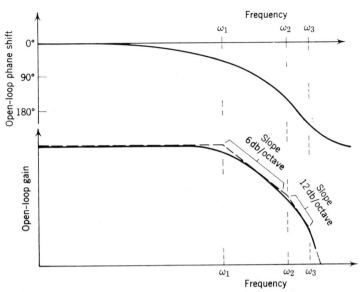

Figure 11-2 The frequency response of a practical operational amplifier.

function by a rational function of real poles and zeros. For the amplifier represented in Figure 11-2, we can thus express the gain as

$$(11.1) \qquad \mu(s) \cong \frac{\mu_0}{(s + \omega_1)(s + \omega_2)(s + \omega_3)}$$

in the band of interest (which for this case is the frequency interval from zero to unity gain crossover frequency). Observe from Figure 11-2, i.e., expression 11.1, that the approximated magnitude of the voltage gain rolls off at the first break frequency ω_1 with a slope of 6 db/octave; then, at the second break frequency ω_2, the slope increases to 12 db/octave, and after the third break frequency, the magnitude rolls off at 18 db/octave. A 90° phase shift occurs between ω_1 and ω_2, 180° phase shift is somewhere between ω_2 and ω_3, and so on. It is clear that the Bode plot alone gives a rough estimate of the phase response.

Most of the practical amplifiers have a fairly large bandwidth. For example, the unity gain crossover frequency of an uncompensated integrated operational amplifier (which can be used as a rough measure of available bandwidth) is typically 10 MHz or more.

The input admittance and output impedance are finite and nonzero quantities. Typical values of the input and output impedances are 100 kΩ and 100 Ω, respectively. The controlled-source representation of a nonideal operational amplifier is shown in Figure 11-3. The feedback impedance in most cases is extremely large and can be neglected.

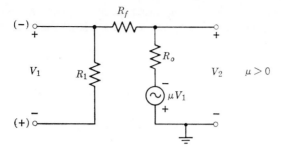

Figure 11-3 The controlled-source model of a nonideal operational amplifier.

In addition to the above characteristics, a practical operational amplifier, like all physical devices, also has maximum limits on the input and output signal, beyond which the input-output relationships become nonlinear. One particular voltage limit, which should be observed in a differential input operational amplifier, is the "common-mode voltage limit." The maximum peak input voltage that can be applied to either input terminal of the amplifier without driving the input transistors to saturation is the "common-mode voltage limit." Usually, the amplifiers are designed to limit the output voltage at the saturation voltage without causing any damage.

Even though the figures cited above may appear a little discouraging at first glance, the practical operational amplifiers, in general, yield satisfactory results for most purposes. The main reason for this is that for linear applications these devices are always used with negative feedback, i.e., feedback paths from the output to the inverting input terminal, which improves the actual performance of these high-gain amplifiers.

At this point, we would like to mention a few more sources of errors in the nonideal operational amplifier. These are: (a) finite input offset voltage and its variation or "drift" due to a change of external environment, e.g., temperature, time, and supply voltage; (b) finite unequal common-mode impedances of the two input terminals; (c) noise; and (d) common-mode rejection error. A discussion on the effect of these errors and of the methods used to minimize their effect is beyond the scope of the book. We refer the reader to several excellent sources published by manufacturers.[3]

---

[3] See for example: *Applications manual for computing amplifiers for modeling, measuring, manipulating, and much else*, George A. Philbrick Researches Inc., Dedham, Massachusetts, Second Edition, June 1966; *Handbook of operational amplifier applications*, Burr-Brown Reserach Corp., Tucson, Arizona, 1963; and R. Stata, "Operational amplifiers," Parts I, II, and IV, Application Notes, Analog Devices, Inc., Cambridge, Mass.

452 / THE OPERATIONAL AMPLIFIER AS A NETWORK ELEMENT

A comparative study of many practical operational amplifiers has been made by Stata,[4] which may be helpful in selecting the right amplifier for a particular use.

The Effect of Feedback

Operational amplifiers are invariably used with negative feedback to achieve linearity and gain stability. The output impedance and input admittances are also reduced due to feedback to some extent.

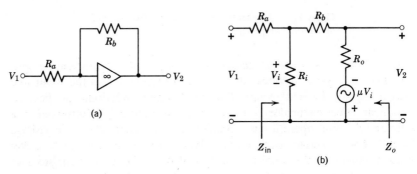

Figure 11-4 An inverting-type VVT realization and its controlled-source representation.

To illustrate these points, let us first consider for simplicity the circuit of Figure 11-4a. Replacing the amplifier by its nonideal model of Figure 11-3, we arrive at the circuit of Figure 11-4b. Analysis yields the following expressions for the closed loop gain (V_2/V_1), the differential input voltage (V_i), and the output impedance (Z_o) of the composite circuit:

$$(11.2) \qquad \frac{V_2}{V_1} = -\frac{G_a}{G_b} \cdot \left[\frac{1}{1 + \dfrac{(G_a + G_b + G_i)(G_o + G_b)}{G_b(\mu G_o - G_b)}} \right]$$

$$(11.3) \qquad V_i = -\left(\frac{G_o + G_b}{\mu G_o - G_b}\right) V_2$$

$$(11.4) \qquad Z_o = \frac{G_a + G_b + G_i}{(G_b + G_o)(G_a + G_i) + (1 + \mu)G_b G_o}$$

where $G_a = 1/R_a$, $G_b = 1/R_b$, $G_i = 1/R_i$, and $G_o = 1/R_o$. In deriving the last three expressions we have assumed R_f to be infinite. For convenience,

---

[4] R. Stata, "Operational amplifiers, Part III, survey of commercially available operational amplifiers," Application Notes, Analog Devices, Inc., Cambridge, Mass.

let us define

(11.5)
$$A = \frac{\mu G_o - G_b}{G_b + G_o}$$

(11.6)
$$\beta = \frac{G_b}{G_a + G_b + G_i}$$

Observe that A is actually the output voltage V_2, when 1 volt is applied to the input of the operational amplifier with V_1 set equal to zero. β, called the feedback factor, is the fraction of unity output voltage being fed back to the amplifier input. Substituting Equations 11.5 and 11.6 in Equations 11.2 through 11.4, we automatically get

(11.7)
$$\frac{V_2}{V_1} = -\frac{G_a}{G_b}\left[\frac{1}{1 + \frac{1}{A\beta}}\right]$$

(11.8)
$$V_i = -\frac{1}{A}V_2$$

(11.9)
$$Z_o = \frac{\frac{1}{G_o + G_b}}{1 + A\beta} \simeq \frac{R_o}{1 + A\beta}$$

The quantity $A\beta$ in Equations 11.7 and 11.9 will be referred to as the "loop gain." For an ideal operational amplifier,

(11.10) $\quad G_o \to \infty, \quad G_i = 0, \quad G_f = 0, \quad \mu \to \infty$

and hence

$$A = \mu \to \infty$$

(11.11)
$$\beta = \frac{G_b}{G_a + G_b}$$

Using this information in Equations 11.7 through 11.9, we have

(11.12)
$$\frac{V_2}{V_1} = -\frac{G_a}{G_b}$$

(11.13) $\quad V_i = 0$

(11.14) $\quad Z_o = 0$

The above expressions indicate that with an ideal operational amplifier, the circuit of Figure 11-4 behaves as an inverting type VVT with an input impedance R_a and zero output impedance.

454 / THE OPERATIONAL AMPLIFIER AS A NETWORK ELEMENT

Note from Equation 11.13 that V_i equal to zero implies that the voltages at both the input terminals are equal. Since the noninverting terminal is at ground, the common mode voltage is zero for this circuit.

To get a feeling for the magnitude of actual departures from the idealized values, consider the following example.

Example 11-1 A typical practical operational amplifier is described by
$$\mu = 10^5, \quad R_i = 100 \text{ k}\Omega, \quad R_o = 100 \text{ }\Omega$$
Let
$$R_a = 10 \text{ k}\Omega, \quad R_b = 100 \text{ k}\Omega.$$

Using these values, we obtain from Equations 11.5 and 11.6

$$A \cong \frac{\mu G_o}{G_b + G_o} \cong \mu\left(1 - \frac{G_b}{G_o}\right) = 0.999 \times 10^5$$

$$\beta = 0.0833$$

Note that $|A\beta| = 83.25 \times 10^2 \gg 1$. (This is usually the case.) Hence, from Equation 11.7 we obtain

$$\frac{V_2}{V_1} \cong -\frac{G_a}{G_b}\left(1 - \frac{1}{A\beta}\right) \cong -10(1 - 1.2 \times 10^{-4}) = -9.9988$$

which is very close to the ideal value of ten. Similarly from Equation 11.8, the magnitude of error voltage is

$$|V_i| \doteq 0.1 \text{ mv}$$

for an output voltage of 10 volts. This causes a current of 1 nA going into the operational amplifier. These values of input error voltage and current indicate the usual assumption that zero-input voltage and zero-input current for an ideal amplifier will cause little error even in the case of a practical amplifier.

The output impedance of the closed-loop amplifier is obtained from Equation 11.9 as

$$Z_o \cong \frac{R_o}{A\beta} = 0.12 \text{ ohm}$$

which is almost equal to the idealized value of zero.

The sensitivity of the closed-loop gain V_2/V_1 due to an incremental variation of the amplifier gain μ can be easily calculated as

(11.15) $$S_\mu^{V_2/V_1} \cong \frac{1}{A\beta}$$

Thus, if $A\beta$ is fairly large, the variation of the closed-loop gain is almost negligible. Note from Equation 11.9 that the output impedance is roughly inversely proportional to $A\beta$. Thus, a *measure* of improvement of the performance of a practical operational amplifier is the amount of available loop gain. Since $A\beta$ is also frequency-dependent, we can conclude that reasonable idealized performance of a practical operational amplifier is restricted to those frequencies where $A\beta$ is much greater than 1.

Next, we consider a commonly used circuit that makes use of the differential inputs of the operational amplifier. The circuit is shown in

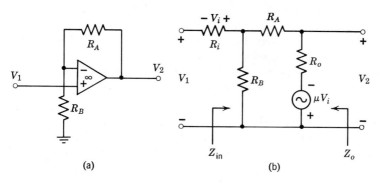

Figure 11-5 A noninverting-type VVT realization and its controlled-source representation.

Figure 11-5a. Replacing the operational amplifier by its nonideal representation of Figure 11-3 (with R_f infinite), we arrive at the circuit of Figure 11-5b, which upon analysis yields the following expressions for closed-loop voltage gain, input impedance, and output impedance:

$$(11.16) \qquad \frac{V_2}{V_2} \cong \frac{R_A + R_B}{R_B}\left[\frac{1}{1 + \dfrac{1}{\mu\beta}}\right]$$

$$(11.17) \qquad Z_{\text{in}} \cong R_i(1 + \mu\beta)$$

$$(11.18) \qquad Z_o \cong \frac{R_o}{1 + \mu\beta}$$

where

$$(11.19) \qquad \beta = \frac{R_i R_B}{R_i(R_A + R_B) + R_A R_B} \cong \frac{R_B}{R_A + R_B}$$

In deriving the above equations, it has been assumed that R_o is very small and $R_i(R_A + R_B) > R_A R_B$.

For an ideal operational amplifier (Equation 11.10), the above equations reduce to:

(11.20)
$$\frac{V_2}{V_1} = \frac{R_A + R_B}{R_B}$$
$$Z_{in} \to \infty$$
$$Z_o \to 0$$

Equation (11.20) indicates that the circuit of Figure 11-5a ideally is an ideal noninverting type VVT. Even with a practical operational amplifier, this circuit is extremely close to an ideal VVT (see Problem 11.1). In reality, the common-mode input impedance limits the input impedance Z_{in}, which is also affected by the frequency dependence of the open-loop gain and stray capacitances at the input.

It should be noted that the input voltage V_1 appears at both the input terminals. Hence, with this circuit the common-mode voltage limit must be observed.

One major problem always associated with feedback is the stability of the closed-loop system. This problem arises due to the frequency dependence of the open-loop gain μ of the practical operational amplifier, and is considered in the next section.

11-2 STABILITY CONSIDERATIONS

We shall first derive the Bode criterion of stability for a single-loop feedback system and then apply the criterion to a typical operational amplifier circuit. Several typical corrective measures to insure stability will also be discussed.

The Bode Criterion[5]

A single-loop negative feedback amplifier can be represented by the signal-flow graph of Figure 11-6. The overall closed-loop gain is

(11.21) $$A_{CL} = \frac{V_2}{V_1} = -\frac{A_{OL}}{1 + \beta A_{OL}}$$

where A_{OL} is the open-loop gain of the amplifier and β is the feedback factor. In general, $\beta A_{OL} \gg 1$; therefore, A_{CL} is equal to $-1/\beta$. βA_{OL} is called the "loop gain."

Now for a practical circuit, A_{OL} and β are functions of frequency, which implies that the closed-loop gain will also be a frequency-dependent

[5] Our development here is based on P. J. Granata and H. B. Aasnaes, "A modified Bode criterion for feedback-system stability," EEE, **14**, 97–99 (October 1966).

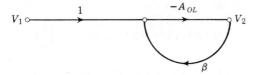

Figure 11-6 Signal flow-graph representation of a single loop feedback amplifier.

function. Thus, if at some frequency ω_0, $(1 + \beta A_{OL})$ becomes equal to zero, A_{CL} will have a $j\omega$-axis pole at $\omega = \omega_0$ indicating instability of the closed-loop system. Rephrasing the previous sentence, we observe that if

(11.22a) $\qquad\qquad |\beta A_{OL}| = 1$

(11.22b) $\qquad\qquad \arg(\beta A_{OL}) = 180°$

for some $\omega = \omega_0$, the feedback circuit will oscillate. Equation 11.22 states that at $\omega = \omega_0$, the loop gain in db is zero. We can now state the Bode criterion for strict stability as a theorem:

Theorem 11-1 *A single-loop feedback circuit is strictly stable if at the unity crossover frequency ω_0 of the loop gain, the phase angle of the loop gain is less than 180°.*

It has been tacitly assumed in Theorem 11-1 that the magnitude of the loop gain is less than one beyond ω_0, and A_{OL} and β are also strictly stable. It should be noted that the feedback amplifier is usually designed with a phase margin of 45° or more to allow for the effect of parasitics and variations in circuit parameters. This means that $\arg(\beta A_{OL})$ at $\omega = \omega_0$ should be less than 135°.

We have observed before that if the Bode plot rolls off at a slope of 12 db/octave between two frequencies ω_x and ω_y, the phase shift of 180° occurs somewhere between ω_x and ω_y. Using this information, we can rephrase the Bode criterion of Theorem 11-1 as:

Corollary *A single-loop feedback circuit is strictly stable, if at the 0-db crossover frequency ω_0 of the loop gain, the loop gain has a slope of less than 12 db/octave.*

In order to have a good stability margin, it is the usual practice to require a 6 db/octave roll off of the loop gain βA_{OL} at ω_0.[6]

---

[6] Note that in almost all of the available integrated operational amplifiers the gain rolls off faster than 12 db/octave thus requiring compensation for stable operation.

Since

(11.23) $$\log(\beta A_{OL}) = \log(A_{OL}) - \log\left(\frac{1}{\beta}\right)$$

the Bode criterion of stability can be alternately written as:

Theorem 11-2 *For strict stability of a single-loop feedback circuit, the difference of slopes of the Bode plots of open loop gain A_{OL} and the reciprocal of the feedback factor β should be less than 12 db/octave at the frequency where the two plots intersect (ω_0).*

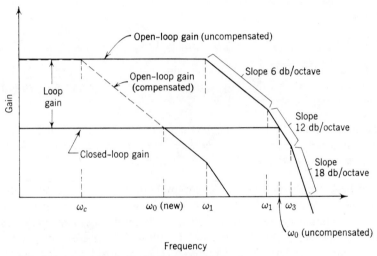

Note 1. Rate of closure at ω_0 (uncompensated)
12 db/octave $\Longrightarrow$ marginally stable

Note 2. Rate of closure at ω_0 (new)
6 db/octave $\Longrightarrow$ stable

Figure 11-7 An illustration of the effect of the first method of internal compensation of a practical operational amplifier.

The slope difference is also known as "rate of closure." For a good stability margin, the rate of closure should be approximately 6 db/octave.

If the rate of closure is 12 db/octave or more, there are basically two ways by which the rate of closure can be decreased. One approach would be to modify the frequency response of the open-loop gain; it will be called *internal compensation*. The second approach is the *external compensation* which is achieved by modifying the feedback network, i.e., β.

We consider both of these approaches next.

Compensation Techniques

The main features of both types of compensation are best illustrated by considering the simple inverting amplifier circuit of Figure 11-4. From Expression 11.7 we note that the loop gain is given by $A\beta$. The quantity A is approximately equal to μ, the loop gain of the operational amplifier, if μ is very large and G_o negligibly small. Furthermore, if $G_a \gg (G_b + G_i)$, $1/\beta$ is approximately equal in magnitude to the closed-loop gain. For most practical purposes, these assumptions are valid, and the closed-loop stability can be established from the Bode plots of μ and the closed-loop gain G_a/G_b. If these assumptions are not valid, then the actual frequency response of A and β as given by Equations 11.5 and 11.6 should be used.

We base our discussion on the assumption that the above conditions are satisfied. Let the uncompensated open-loop frequency response of the operational amplifier be the one shown in Figure 11-2. We have redrawn in Figure 11-7 the Bode plot of the frequency response of the open-loop gain and also the Bode plot of a typical closed-loop gain response. Note that the critical point is the frequency ω_0, where the open-loop gain is equal to the closed-loop gain, i.e., the loop gain is 0-db. Beyond ω_0, the closed-loop gain is limited to the open-loop gain. The rate of closure of the two plots is seen to be 12 db/octave, which indicates that the inverting amplifier is marginally stable. To make it strictly stable, we have to either modify the feedback factor, i.e., the closed-loop gain, or compensate the open-loop gain function $\mu(s)$.[7]

Consider the internal compensation approach first. The practical operational amplifier is in general provided with at least one terminal where additional network elements can be connected for internal compensation purposes. Figure 11-8a indicates a possible representation of the operational amplifier with the compensation terminal shown as node "x."

If we connect a capacitor C_1 to ground at terminal x (Figure 11-8b), the open-loop gain of the compensated amplifier becomes

$$\mu_{\text{compensated}} = \frac{\omega_C \mu(s)}{s + \omega_C}$$

where $\omega_C = 1/R_1 C_1$. C_1 is chosen to make ω_C less than ω_1, the first break frequency of the open-loop gain. The closed-loop amplifier will be stable, if the 0-db loop gain crossover frequency ($\omega_{0\text{new}}$) lies between ω_C

---

[7] An excellent discussion on compensation techniques will be found in J. N. Giles, "Integrated operational amplifier frequency compensation," EDN, X, 24–34 (October 1965). See also RCA Handbook, *loc. cit.*

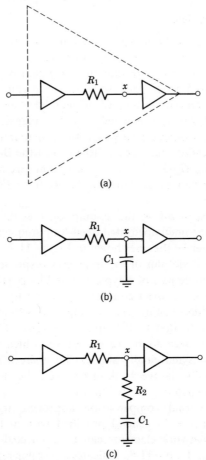

Figure 11-8 Possible circuit arrangements for internal compensation.

and ω_1. This is illustrated in Figure 11-7, where the dotted lines indicate the response after compensation.

Several comments are here in order. Note from Figure 11-7 that due to compensation, less loop gain is available at higher frequencies. This means that the performance of the closed-loop amplifier will not be satisfactory at higher frequencies where enough loop gain is not available. Compensation thus reduces the effective closed-loop bandwidth. On the other hand, smaller closed-loop bandwidth means that broadband noise is also reduced. In addition, there is less possibility of oscillation due to capacitive loading or parasitics.

The closed-loop bandwidth can be increased maintaining stable operation by connecting a series RC tuned circuit to the compensating terminal

STABILITY CONSIDERATIONS / 461

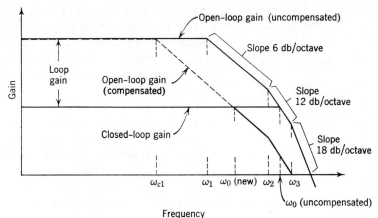

Note 1. Rate of closure at ω_0 (uncompensated)
12 db/octave $\Longrightarrow$ marginally stable

Note 2. Rate of closure at ω_0 (new) 6 db/octave $\Longrightarrow$ stable

Figure 11-9 Illustration of the effect of the second method of internal compensation of a practical operational amplifier.

as indicated in Figure 11-8c. The compensated loop gain can then be expressed as

$$\mu_{\text{compensated}} = \frac{R_2}{R_1 + R_2} \cdot \mu(s) \cdot \frac{s + \omega_{C2}}{s + \omega_{C1}}$$

where $\omega_{C2} = 1/R_2 C_1$ and $\omega_{C1} = 1/(R_1 + R_2)C_1$. If ω_{C2} is chosen to be equal to ω_1, the open-loop frequency response will have a slope of 6 db/octave between ω_{C1} and ω_2. The new 0-db loop gain crossover frequency $\omega_{0_{\text{new}}}$ can be made to lie between ω_{C1} and ω_2 by choosing proper ω_{C1}. (See Figure 11-9.) This will lead to stable operation when feedback is applied.

Next, we consider the external compensation approach to obtain stable closed-loop operation. For illustrative purposes, consider the inverting VVT of Figure 11-4, whose frequency response is shown in Figure 11-7, along with the open-loop frequency response of the uncompensated operational amplifier. We observe that at ω_0 (the 0-db loop gain crossover frequency), the rate of closure is 12 db/octave, indicating marginal stability. A novel corrective procedure is to connect a small capacitor C_b in parallel with R_c (Figure 11-10). The new closed loop gain is

$$\frac{V_2}{V_1} = -\frac{G_a}{sC_b + G_b} = -\frac{G_a}{G_b} \cdot \frac{G_b/C_b}{(s + \omega_b)}$$

where $\omega_b = 1/R_b C_b$. If C_b is chosen to introduce a break in the closed-loop frequency response at ω_b, which is slightly smaller than ω_0, the rate

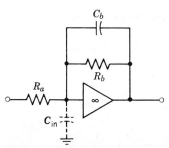

Figure 11-10 Externally compensated inverting-type VVT circuit.

of closure is then reduced to 6 db/octave, leading to stable closed-loop operation. An approximate value of C_b is equal to $R_a C_{in}/R_b$, where C_{in} is equal to the total stray capacitances at the input of the operational amplifier.[8]

Another cause for oscillation is the presence of excessive capacitive load at the output. In this case, it is preferable to isolate the capacitance by placing a small resistor R_y in the feedback loop as indicated in Figure 11-11.[8] The value of R_y is usually of the order of the open-loop output impedance of the amplifier.

For the noninverting VVT of Figure 11-5, the situation is similar. The methods of internal compensation discussed earlier can be used effectively. The external compensation technique may sometimes prove helpful. For example, if for the noninverting VVT of Figure 11-5 the rate of closure is 12 db/octave, stability can be achieved by modifying the feedback circuit as indicated in Figure 11-12.[9]

There are many other factors (such as layout, selection of companion network elements, power supply) that affect the performance of the closed-loop circuit. A discussion on each of these topics is beyond the scope of the book. We refer the reader to the application manuals available from manufacturers, a few of which have been cited earlier.

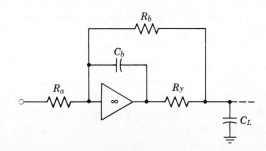

Figure 11-11 Externally compensated inverting-type VVT driving a capacitive load.

---

[8] Philbrick Manual, *loc. cit.*

[9] *Advantages of a 12 db/octave operational amplifier and their applications*, Application Note, Analog Device, Inc., Cambridge, Massachusetts.

11-3 REALIZATION OF IDEAL ACTIVE DEVICES

An interesting aspect of the operational amplifier is that most of the "ideal" active devices discussed in the last four chapters can be constructed with these high-gain amplifiers. As a result, the synthesis techniques considered in the previous chapters are easily adaptable to the use of operational amplifiers as the active elements. The basis of the development of some of the circuits is the nullator-norator equivalent circuit of the operational amplifier, which we develop next.

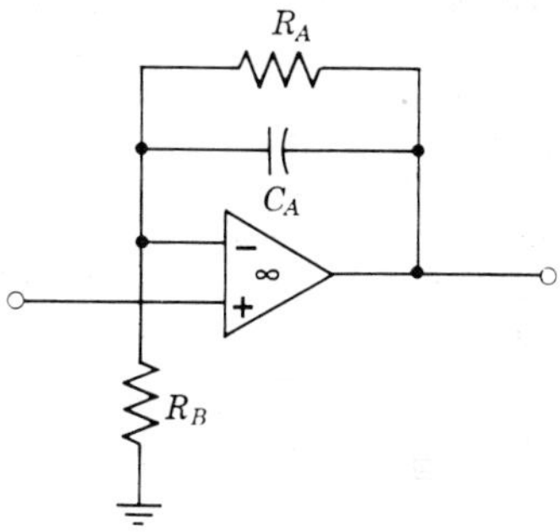

Figure 11-12 Externally compensated noninverting-type VVT.

Equivalent Representation of the Operational Amplifier[10]

Consider the two-port of Figure 11-13. Because of the nullator at the input, we have

$$V_1 = I_1 = 0$$

On the other hand, at the output end we have

$$V_2, I_2 \text{ arbitrary}$$

due to the presence of the norator at the output. Thus the input-output relationship of this two-port can be written as

$$(11.24) \qquad \begin{bmatrix} V_1 \\ I_1 \end{bmatrix} = \begin{bmatrix} 0 & 0 \\ 0 & 0 \end{bmatrix} \begin{bmatrix} V_2 \\ -I_2 \end{bmatrix}$$

or, in other words, it is described by a null transmission matrix.

We shall now show that the transmission matrix of the ideal operational amplifier is also a null matrix. The transmission matrix of the nonideal operational amplifier of Figure 11-3 with the noninverting terminal grounded can be shown to be:

$$(11.25) \qquad F = \begin{bmatrix} \dfrac{R_o + R_f}{R_o - \mu R_f} & \dfrac{R_o R_f}{R_o - \mu R_f} \\ \dfrac{(R_i + R_f)(R_o + R_f)}{R_i R_f (R_o - \mu R_f)} - \dfrac{1}{R_f} & \dfrac{R_o(R_i + R_f)}{R_i(R_o - \mu R_f)} \end{bmatrix}$$

[10] E. Butler and S. K Mitra, "An equivalent circuit for the operational amplifier," Proc. 3rd Allerton Conference on Circuit and System Theory, Univ. of Illinois, Urbana, Ill., October 1965.

464 / THE OPERATIONAL AMPLIFIER AS A NETWORK ELEMENT

In the limiting case of an ideal operational amplifier, we have

$$R_i \to \infty, \quad R_f \to \infty, \quad R_o \to 0, \quad \mu \to \infty$$

and, consequently, the transmission matrix of the ideal operational amplifier becomes a null matrix. Since the first and second rows of expression 11.25 approach their limiting values independently, the ideal operational amplifier is a "normal" network.[11] Thus, the network of Figure 11-13 represents an operational amplifier.[12]

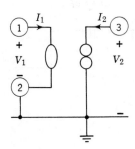

Figure 11-13 The nullator-norator model of an ideal operational amplifier.

The usefulness of the nullator-norator circuit in forming duals and in obtaining transistorized realizations has been shown in earlier chapters. In a similar manner, the nullator-norator equivalent representation of Figure 11-13 is attractive for constructing duals of operational amplifier circuits and realizing other nullator-norator circuits with the aid of operational amplifiers. Since, in most of the practical operational amplifiers, one of the output terminals is always grounded, conversion of nullator-norator circuits into operational amplifier circuits can be achieved if one of the terminals of each of the norators is grounded. Of course, if "floating" power supplies are used, this restriction can be lifted.

Realization of Controlled Sources

The circuit of Figure 2-51c, which represents a CVT, is easily converted into an operational amplifier circuit as indicated in Figure 11-14a. It should be noted that by connecting a resistance R_a in series with the input of this circuit, the inverting VVT of Figure 11-4a is obtained.

Direct conversion of the circuit of Figure 2-51d is not possible without a floating supply. A VCT having a nongrounded output is easily obtained from Figure 2-51d by interchanging the load and the norator. The final realization is shown in Figure 11-14b.

The noninverting VVT of Figure 11-5 is a direct conversion of the nullator-norator representation of Figure 2-51b. Setting R_B equal to zero and R_A infinite, the unity gain voltage-follower of Figure 11-15a is

---

[11] H. J. Carlin, "Singular network elements," *IEEE Trans. on Circuit Theory*, **CT-11**, 67–72 (March 1964).

[12] Independently proposed by G. Martinelli, "Comments on 'nullor model of the transistor,'" *Proc. IEEE*, **53**, 1965 (July 1965).

REALIZATION OF IDEAL ACTIVE DEVICES / 465

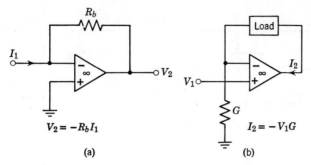

Figure 11-14 (a) A negative-gain CVT circuit. (b) A negative-gain VCT circuit.

easily obtained from it. A similar development of Figure 2-51c leads to the CCT circuit of Figure 11-15b with ungrounded output.

A circuit for a differential input type VCT will be found in Problem 11.2.

Realization of a Negative-Impedance Converter

The nullator-norator representations of NIC are given in Tables 9-1 and 9-2. From them, we observe that only the Type I-CNIC can be directly converted into an operational amplifier circuit. Such a conversion leads to the CNIC circuit of Figure 11-16a.

It was noted in Chapter 9 that the negative-impedance converter is a potentially unstable two-port. Thus, we would like to know the stability characteristics of each port of the CNIC of Figure 11-16a. To this end, we replace the operational amplifier by its controlled source equivalent circuit as shown in Figure 11-16b, where for simplicity we have neglected the input, output, and feedback impedances. The input admittance seen at port 1 when port 2 is terminated by an impedance Z_L is given as

$$(11.26) \qquad Y_{11} = \left(\frac{R_1 R_2 + R_2 Z_L (1 - \mu)}{Z_L + R_1 (1 + \mu)} \right)^{-1}$$

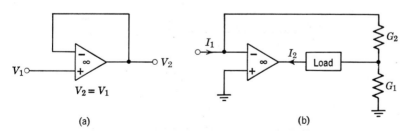

Figure 11-15 (a) A voltage-follower circuit. (b) A CCT circuit.

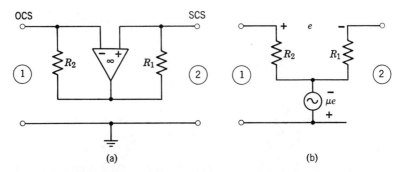

Figure 11-16 A grounded CNIC circuit and its controlled-source model.

The instability is caused by the frequency dependence of the open-loop gain μ. Observe that as the magnitude of μ approaches

$$\left|1 + \frac{R_1}{Z_L}\right|$$

the input admittance approaches infinite value. This indicates that port 1 is open-circuit stable (OCS). In a similar manner, we examine the input impedance seen at port 2 with port 1 terminated in a load Z_s, which is given as

(11.27) $$Z_{22} = \frac{R_1 R_2 + Z_s R_1 (1 + \mu)}{Z_s - R_2(\mu - 1)}$$

expression 11.27 indicates that for some value of μ, Z_{22} will tend to infinite value. As a result, port 2 is short-circuit stable (SCS).

Next, we consider the realization of a VNIC. One possible realization of the Type I-VNIC is shown in Figure 11-17. Note that the final realization is essentially a four terminal device, since the common terminal of the input and output ports is not connected to the system ground. If a floating supply is used for the operational amplifier, then the VNIC can be

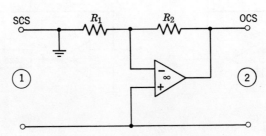

Figure 11-17 A four-terminal VNIC circuit.

employed as a three-terminal device. Assuming the gain of the operational amplifier to be μ, we compute the input impedance seen at ports 1 and 2, when ports 2 and 1 are terminated, respectively, with R_L and R_s:

(11.28)
$$Z_{11} = \frac{R_1 + R_2 + \mu R_1}{R_1 + R_2 - \mu R_1} R_L$$

$$Z_{22} = \frac{R_1 + R_2 - \mu R_1}{R_1 + R_2 + \mu R_1} R_s$$

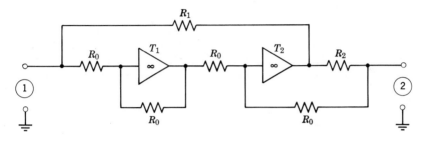

Figure 11-18 A grounded CNIC circuit using two single-ended operational amplifiers.

From the expressions above, we conclude that port 1 of the VNIC of Figure 11-17 is short-circuit stable (SCS) and port 2 is open-circuit stable (OCS).

For analog computer simulation, it is preferable to have a circuit using single-ended operational amplifiers. One such CNIC realization is shown in Figure 11-18 and was advanced by Antoniou.[13] For this circuit, analysis yields the following g parameters:

(11.29)
$$g_{11} \cong \frac{(2R_0 + R_1)(R_0 + R_1)}{R_1 R_0^2 \mu_2} \cong 0$$

$$g_{12} \cong 1 + \frac{R_1(2R_0 + R_1)}{R_1 R \mu_2} \cong 1$$

$$g_{21} \cong 1 + \frac{2(R_1 + R_0)}{R_0 \mu_2} \cong 1$$

$$g_{22} \cong \frac{2R_1}{\mu_2} \cong 0$$

[13] A. Antoniou, "Negative-impedance converters using operational amplifiers," *Electronics Letters*, **1**, 88–89 (June 1965).

where μ_2 is the voltage gain of the second amplifier T_2. Computation of the input impedances with the ports terminated reveals that port 1 is short-circuit stable and port 2 is open-circuit stable.

Realization of a four-terminal VNIC using single-ended operational amplifiers is suggested as a problem (see Problem 11.3).

Realization of a Gyrator

One way to realize an "ideal" gyrator is to cascade a negative-impedance converter (NIC) with a negative-impedance inverter (NIV).[14] Two such

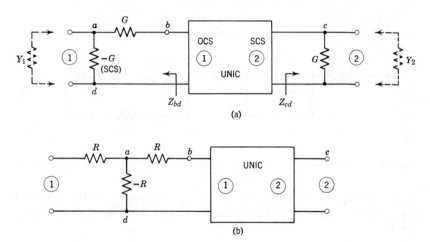

Figure 11-19 Gyrator realizations by cascading a CNIC with an NIV.

realizations are indicated in Figure 11-19. Since negative-impedance converters (and hence negative resistances) are associated with open-circuit and short-circuit stability, it is necessary to determine the proper way to connect the active devices in order to build the gyrator as indicated in Figure 11-19.

Consider the network of Figure 11-19a first. The admittance Y_{ad} seen by the negative conductance $-G$, when port 1 is terminated by an admittance Y_1 and port 2 is terminated by an admittance Y_2, is given as

$$Y_{ad} = Y_1 + G + \frac{G^2}{Y_2}$$

The values Y_{ad} are bounded between G and ∞, i.e.,

$$0 \leq Z_{ad} \leq R$$

---
[14] Problem 10.3.

This implies that the negative resistance connected between nodes (a, d) must be short-circuit stable. In a similar manner, we observe the variation of the impedances Z_{bd} and Z_{cd}, seen by the negative-impedance converter

$$Z_{bd} = R + \frac{1}{Y_1 - G}$$

$$Z_{cd} = \frac{1}{G + Y_2}$$

From the above, we conclude that port 2 of the NIC should be short-circuit stable. Since port 1 of the NIC is open-circuit stable, the impedance Z_{bd} should be large. On the other hand, Z_{bd} varies from 0 to ∞. Hence, the realized gyrator might oscillate if Y_1 is very small or very large. A complete realization of Figure 11-19a is shown in Figure 11-20.

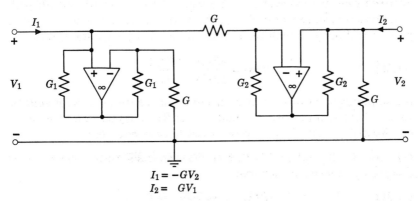

$$I_1 = -GV_2$$
$$I_2 = GV_1$$

Figure 11-20 A gyrator circuit using two operational amplifiers.

Practical realization of the second model of gyrator as indicated in Figure 11-19b is difficult for stability reasons. The pertinent impedances Z_{ad}, Z_{cd}, and Z_{bd} can each vary from 0 to ∞. As a result, stability cannot be achieved for a large range of terminating impedances Z_1 and Z_2.

We would like to point out here an interesting aspect of the circuit of Figure 11-20. If we eliminate the negative resistance connected across the input terminals, the remaining circuit can be used as a current source whose output current I_2 is directly proportional to the voltage applied at the input port.[15]

A circuit for realizing a three-terminal gyrator using four single-ended operational amplifiers will be found in Problem 11.2.

---

[15] D. H. Sheingold, "Constant current source for analog computer use," *IEEE Trans. on Electronic Computers*, **EC-12**, 324 (June 1963).

11-4 SYNTHESIS OF DRIVING-POINT FUNCTIONS

Use of operational amplifiers as an active element in the realization of driving-point functions is considered in this section. For example, operational amplifier can be used to reduce the number of passive R, L, C, elements in the Bott-Duffin synthesis. Driving-point function synthesis methods outlined in previous chapters can also be modified for operational amplifier use.

Active Realization of the Bott-Duffin Cycle

A brief introduction to the Bott-Duffin method of synthesis of p.r. driving-point functions was given in Section 10-2, p. 414, where a modification of the Bott-Duffin structure utilizing gyrators was presented. An alternate elegant modification was suggested by Lampard and Stuart.[16]

We observe from Section 10-2, p. 414, that if $Z(s)$ is a minimum-resistive positive real function with the restriction that $Z(\infty)$ is constant, then

(11.30) $$F(s) = \frac{\lambda Z(s) - sZ(\lambda)}{\lambda Z(\lambda) - sZ(s)}$$

is also positive real with either a zero or a pole on the $j\omega$-axis for suitable values of the real positive constant λ. Let Re $Z(j\omega)$ be zero at $\omega = \omega_0$, i.e., Re $Z(j\omega_0) = 0$. There are two cases to be considered:

(1) *Case A.* If $j\omega_0 Z(j\omega_0) > 0$, then $F(s)$ can be made to have a pole at $s = j\omega_0$ by choosing λ such that

(11.31) $$\lambda Z(\lambda) - j\omega_0 Z(j\omega_0) = 0$$

(2) *Case B.* If $Z(j\omega_0)/j\omega_0 > 0$, then $F(s)$ can be made to have a zero at $s = j\omega_0$ by selecting λ to make

(11.32) $$\lambda Z(j\omega_0) - j\omega_0 Z(\lambda) = 0$$

The original Bott-Duffin structure is shown again in Figure 11-21. Since the representation of Figure 11-21 is a balanced bridge, any impedance can be connected between nodes a and b without altering the balance.

It is possible to use an active element to obtain the balance and thus reduce the number of passive elements. The active element can be an operational amplifier.

---

[16] A. G. Stuart and D. G. Lampard, "Bridge networks incorporating active elements and application to network synthesis," *IEEE Trans. on Circuit Theory*, **CT-10**, 357–362 (September 1963).

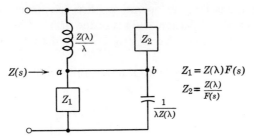

Figure 11-21 The circuit arrangement of a Bott-Duffin cycle in the realization of a positive real driving-point function.

The proposed realization scheme for a balanced bridge is shown in Figure 11-22a. Routine analysis yields

$$(11.33) \quad Z_{bd} = \frac{Z_B Z_C - \dfrac{R_o(Z_A + Z_B)}{\mu} - \dfrac{R_o Z_B(Z_A + Z_C)}{\mu R_i}}{Z_A - \dfrac{Z_A + 2Z_B}{\mu} - \dfrac{Z_B(Z_A + Z_C)}{\mu R_i}}$$

where μ is the gain of the amplifier, R_i, its input resistance and R_o, its output resistance. Thus

$$(11.34) \quad Z_{bd} \to \frac{Z_B Z_C}{Z_A} \quad \text{as} \quad \mu \to -\infty$$

Therefore, we note that if the amplifier gain is large enough, the impedance Z_{bd} realized by the amplifier is a good approximation to the desired value

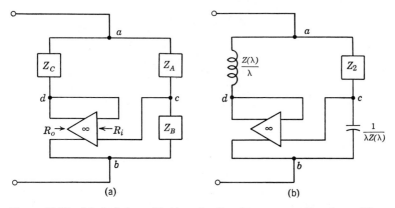

Figure 11-22 (a) A balanced-bridge circuit using an operational amplifier. (b) A modified Bott-Duffin circuit using an operational amplifier.

$Z_B Z_C / Z_A$ for balanced-bridge operation. It is seen from Equation 11.33 that the error term in Equation 11.34 is of the order of $1/\mu$. For large μ, moreover, a second-order effect is produced by the input impedance R_i (R_i is usually large).

The active network representation of the Bott-Duffin cycle is shown in Figure 11-22b. The input impedance of this one-port is given as:

(11.35) $$Z(s) = \frac{\dfrac{Z(\lambda)}{\lambda} s \left(Z_2(s) + \dfrac{\lambda Z(\lambda)}{s} \right)}{Z_2(s) + \dfrac{Z(\lambda)}{\lambda} s}$$

$$= Z(\lambda) \frac{s + \lambda F(s)}{\lambda + s F(s)}$$

which is the desired input impedance (see Equation 10.12).

Example 11-2 Consider the realization of the driving-point impedance of Example 10-3:

(11.36) $$Z(s) = \frac{2s^2 + s + 1}{s^2 + s + 2}$$

The passive RLC and the RLC:gyrator realizations are shown in Figure 10-12.

Observe that $\operatorname{Re} Z(j\omega)$ has a zero at $\omega = \omega_0 = 1$ and $Z(j1) = +j$. This indicates that Case B must be followed. We solve

$$\lambda Z(j\omega_0) - j\omega_0 Z(\lambda) = j\left[\lambda - \frac{2\lambda^2 + \lambda + 1}{\lambda^2 + \lambda + 2} \right] = 0$$

for λ, which yields $\lambda = 1$. Hence

$$Z(\lambda) = Z(1) = 1$$

As a result,

$$F(s) = \frac{Z(s) - sZ(1)}{Z(\lambda) - sZ(s)} = \frac{s^2 + 1}{2s^2 + 2s + 2}$$

Note that $F(s)$ has a pair of $j\omega$ axis zeros at $s = \pm j1$ as expected. Therefore

$$Z_2(s) = \frac{Z(\lambda)}{F(s)} = 2 + \frac{2s}{s^2 + 1}$$

The final realization is shown in Figure 11-23.

Note that the total number of passive elements needed in the standard Bott-Duffin realization [Figure 10-12b] is eight, whereas the total number needed in the active realization is five.

Recently Parsons[17] has shown that in balancing the bridge it is possible to use a unity gain VVT instead of an operational amplifier. Consider the scheme of Figure 11-24, for which the impedance seen between nodes b and d is

(11.37)
$$Z_{bd} = \frac{Z_A R_o + \dfrac{Z_B R_i}{Z_B + R_i}(Z_C + R_o)}{Z_A}$$

If the VVT is an ideal unity gain voltage amplifier, then

(11.38)
$$Z_{bd} \to \frac{Z_B Z_C}{Z_A} \quad \text{as} \quad R_o \to 0,\ R_i \to \infty$$

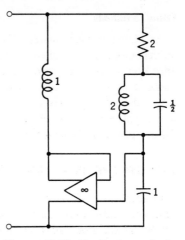

Figure 11-23 Realization of the positive real driving-point impedance of Example 11-2.

as desired.

Note that in the active bridge of Figure 11-24, everything can be referred to point b as ground; hence, "floating" power supply is not required. The main disadvantage is that the deviation of R_i and R_o from idealized values has first-order effect on Z_{bd}.

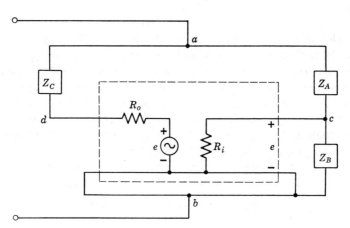

Figure 11-24 An alternate active Bott-Duffin circuit arrangement.

---

[17] T. W. Parsons, "Comment on 'bridge networks incorporating active elements and application to network synthesis,'" *IEEE Trans. on Circuit Theory*, **CT-11**, 509 (December 1964).

Other Methods

As stated earlier, many of the synthesis techniques discussed earlier (which employ controlled sources, negative resistances, negative-impedance converters, or gyrators) can be considered as techniques for synthesis using operational amplifiers. To this end, we replace the respective active devices by their operational amplifier realizations. For driving-point synthesis, we would like to point out that the methods of Sections 9-2 are particularly suitable.

11-5 SYNTHESIS OF TRANSFER FUNCTIONS

In addition to the design of transmission networks, methods of transfer function realization using resistances, capacitances, and operational amplifiers are also useful in simulation of complex systems in an analog computer. Some typical synthesis techniques will be discussed in this section.

Synthesis Using a Single-Ended Operational Amplifier

One of the earliest and most frequently used active RC configuration using a single operational amplifier is the one shown in Figure 11-25a. The voltage transfer ratio of this circuit is given as

(11.39) $$\frac{V_2}{V_1} = -\frac{y_{21A}}{y_{12B}}$$

where the subscripts A and B refer to the networks $\mathcal{N}_A$ and $\mathcal{N}_B$, respectively. Note that the RC networks being reciprocal, $y_{21i} = y_{12i}$.

We note from expression 11.39 that if we restrict the RC two-port to have no transformers, then the short-circuit transfer admittances, y_{21A} and y_{12B}, can have no positive real transmission zeros. As a result, we can only realize voltage transfer ratios having no positive real zeros and poles in this configuration. This is not a major restriction, even though it excludes all-pass type functions of odd order.

To realize a specified voltage transfer ratio $t_v = -N(s)/D(s)$, we choose a polynomial $Q(s)$ having only distinct negative real roots so that

$$Q(s)^0 + 1 \geq \max\,[N(s)^0, D(s)^0]$$

Next, we identify

(11.40)
$$-y_{21A} = \frac{N(s)}{Q(s)}$$

$$-y_{12B} = \frac{D(s)}{Q(s)}$$

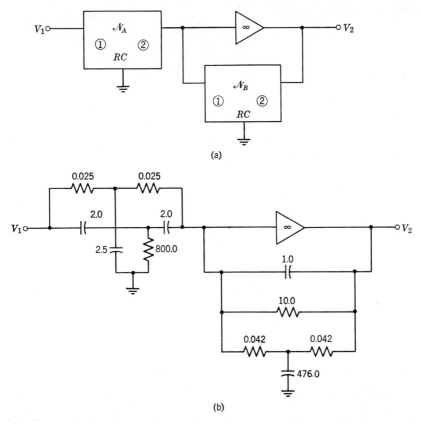

Figure 11-25 Transfer voltage ratio realization using a single-ended operational amplifier.

Realization of the *RC* two-ports is achieved following conventional synthesis procedures. One drawback of this structure is that if t_v has complex poles and zeros, then the *RC* two-port must be realized following parallel-ladder, Dasher, or Fialkow-Gerst techniques, which are in general not simple to use. This problem can be circumvented for transfer functions of low order. For example, for second-order transfer admittances there exist catalogs of *RC* two-ports,[18] which can be used to advantage in most cases. This is illustrated in the next example.

Example 11-3 Let us realize

$$t_v = -\frac{s^2 + 2.0}{s^2 + 0.1s + 1.2}$$

[18] A table of such networks is given in Appendix D.

We thus have to realize

$$-y_{21A} = \frac{s^2 + 2.0}{s + p_1}$$

$$-y_{12B} = \frac{s^2 + 0.2s + 1.2}{s + p_1}$$

An examination of the entries of Table D-1 reveals that the network N_7 will realize y_{21A} and the network N_{20} will realize y_{12B} provided $p_1 < 0.2$. Let us choose $p_1 = 0.1$. Then the element values of RC two-port $\mathcal{N}_A$ are obtained from Table D-1 as:

$$R_{1A} = 0.025$$
$$R_{2A} = 2.5$$
$$C_{1A} = 800$$
$$C_{2A} = 2$$

There are two possible choices for N_{20}. Choosing the top network, we obtain the element values of the RC two-port $\mathcal{N}_B$ from Table D-1 as:

$$R_{1B} = 0.042$$
$$R_{2B} = 10$$
$$C_{1B} = 476$$
$$C_{2B} = 1$$

The final realization is shown in Figure 11-25b.

Mathews-Seifert's Approach

In the structure of Figure 11-25a, the poles and zeros of the transfer function are realized by the transmission zeros of three-terminal RC networks $\mathcal{N}_A$ and $\mathcal{N}_B$. This is complicated when the transmission zeros are complex. This problem can be avoided if simple active zero-producing networks are used instead. The active configuration of Figure 8.21a, which employs an inverting type VVT, is suitable for this purpose. We observe from expression (8.20) that the forward short-circuit transfer admittance of this two-port is given as

$$y_{21} = Y_3 - \mu Y_4$$

where Y_3 and Y_4 represent two RC one-port admittances. Replacing $\mathcal{N}_A$ and $\mathcal{N}_B$ by this type of zero-producing sections, we arrive at the configuration of Figure 11-26, originally suggested by Mathews and Seifert.[19] It follows that the voltage transfer ratio of this structure is given

---

[19] M. V. Mathews and W. W. Seifert, "Transfer-function synthesis with computer amplifiers and passive networks," Proc. Western Joint Computer Conference, pp. 7–12 (March 1955).

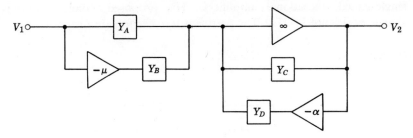

Figure 11-26 Mathew-Seifert's configuration for transfer-voltage ratio realization.

as

(11.41) $$\frac{V_2}{V_1} = -\frac{Y_A - \mu Y_B}{Y_C - \alpha Y_D}$$

The form of the transfer function indicates that any real rational function can be realized by this structure.

Realization of a specified transfer function $t_v = -N(s)/D(s)$ is achieved very easily. We first choose a polynomial $Q(s)$ having all simple negative real roots and of degree

$$Q(s)^0 + 1 \geq \max\,[N(s)^0,\,D(s)^0]$$

By Theorem 3-5, we can express

(11.42) $$\frac{N(s)}{Q(s)} = Y_{RC}^{(1)} - Y_{RC}^{(2)}$$

$$\frac{D(s)}{Q(s)} = Y_{RC}^{(3)} - Y_{RC}^{(4)}$$

where $Y_{RC}^{(j)}$ is a passive RC driving-point admittance. Thus we identify

(11.43) $\quad Y_A = Y_{RC}^{(1)}, \quad \mu Y_B = Y_{RC}^{(2)}, \quad Y_C = Y_{RC}^{(3)}, \quad \alpha Y_D = Y_{RC}^{(4)}$

Note that because of the structure assumed, poles and zeros can be controlled separately by changing the gains of the inverting amplifiers.

Lovering's Method

Although the previous method imposes no restriction on the transfer function and is simple to use, it is preferable in some cases to minimize the number of active elements at the expense of more passive elements if necessary. Lovering[20] recently advanced a method which uses two

---

[20] W. F. Lovering, "Analog computer simulation of transfer function," *Proc. IEEE*, **53**, 306 (March 1965).

single-ended operational amplifiers. His proposed configuration is shown in Figure 11-27a. The pertinent transfer function is given as

$$\text{(11.44)} \qquad \frac{V_2}{V_1} = -\frac{Y_A - Y_B}{Y_C - Y_D}$$

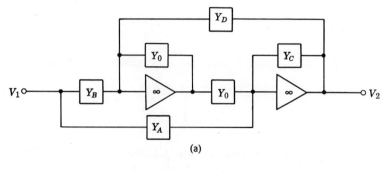

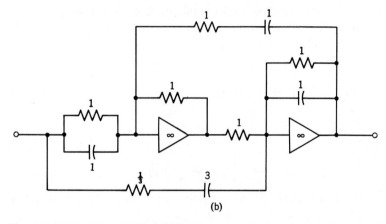

Figure 11-27 (a) Lovering's configuration for transfer-voltage ratio realization. (b) Realization of the transfer-voltage ratio of Example 11-4, using Lovering's method.

Note that the form of expression 11.44 is identical to Equation 11.41. Hence the synthesis can be carried out in the same fashion. Since Y_0 does not appear in Equation 11.44, any suitable preselected nonzero value can be used. We illustrate the method by means of an example.

Example 11-4 Realize the all-pass function

$$t_v = \frac{N(s)}{D(s)} = \frac{s^2 - s + 1}{s^2 + s + 1}$$

Choose $Q(s) = (s + 1)$. Then

(11.45)
$$\frac{N(s)}{Q(s)} = \frac{s^2 - s + 1}{s + 1} = (s + 1) - \left(\frac{3s}{s + 1}\right)$$

$$\frac{D(s)}{Q(s)} = \frac{s^2 + s + 1}{s + 1} = (s + 1) - \left(\frac{s}{s + 1}\right)$$

Comparing Equation 11.45 with Equation 11.44, we identify

$$Y_A = \frac{3s}{s + 1}, \quad Y_B = s + 1, \quad Y_D = s + 1, \quad Y_C = \frac{s}{s + 1}$$

Final realization is indicated in Figure 11-27b, where we have set Y_0 equal to 1 mho. It should be noted that alternate identification can be obtained by interchanging Y_B with Y_A, and Y_C with Y_D, respectively.

A General Method Using a Differential-Input Amplifier

The previous three methods used the single-ended operational amplifiers as the active elements. The method outlined in Section 11-5, p. 474, which used one single-ended operational amplifier, is restricted to transfer functions having no positive real zeros and poles. These restrictions were eliminated by making use of additional single-ended amplifiers (Section 11-5, p. 476, and 11-5, p. 477). Note that these three methods are also suitable to simulate transfer functions on the analog computer.

The integrated operational amplifiers available in the market are usually of the differential-input type. Thus it is profitable to develop a synthesis technique which makes use of the differential-input amplifiers. Such a method is described next.[21]

Consider the active RC configuration of Figure 11-28. Routine analysis yields the following expression for the transfer voltage ratio:

(11.46) $$t_v = \frac{V_2}{V_1} = \frac{Y_A(Y_B + Y_D + Y_F) - Y_B(Y_A + Y_C + Y_E)}{Y_F(Y_A + Y_C + Y_E) - Y_E(Y_B + T_D + Y_F)}$$

In order to realize a specified transfer voltage ratio $t_v = N(s)/D(s)$ in the form of Figure 11-28, it is convenient to choose the RC components such that

(11.47) $$Y_A + Y_C + Y_E = Y_B + Y_D + Y_F$$

---

[21] S. K. Mitra, "Active RC filters employing a single operational amplifier as the active element," Proceedings of Hawaii International Conference on System Sciences, Honolulu, Hawaii, January 1968.

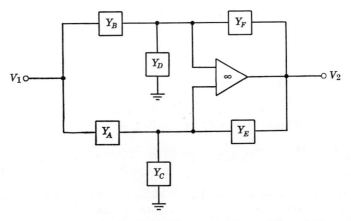

Figure 11-28 An active RC configuration using a differential input operational amplifier.

then Equation 11.46 reduces to

$$(11.48) \qquad t_v = \frac{Y_A - Y_B}{Y_F - Y_E}$$

Observe that the form of the transfer function is identical to those of Mathew-Seifert's structure (Equation 11.41) and of Lovering's configuration (Equation 11.44). Hence identical synthesis procedure follows.

We first choose a polynomial $Q(s)$ having distinct negative real roots such that

$$Q(s)^0 + 1 \geq \max\,[N(s)^0,\, D(s)^0]$$

and then express $N(s)/D(s)$ as

$$(11.49) \qquad t_v = \frac{N(s)/Q(s)}{D(s)/Q(s)} = \frac{Y_{RC}^{(1)} - Y_{RC}^{(2)}}{Y_{RC}^{(3)} - Y_{RC}^{(4)}}$$

where $Y_{RC}^{(j)}$ is a passive RC driving-point admittance (Theorem 3-5). Comparing Equations 11.48 and 11.49 we easily obtain

$$(11.50) \qquad \begin{aligned} Y_A &= Y_{RC}^{(1)} \\ Y_B &= Y_{RC}^{(2)} \\ Y_F &= Y_{RC}^{(3)} \\ Y_E &= Y_{RC}^{(4)} \end{aligned}$$

Identification of the remaining RC one-ports is obtained by making use of Equation 11.47, which we can rewrite as

$$(11.51) \quad Y_C - Y_D = (Y_F - Y_E) - (Y_A - Y_B) = \frac{D(s) - N(s)}{Q(s)}$$

Again using Theorem 3-5, $[D(s) - N(s)]/Q(s)$ can be expressed as the difference of two RC driving-point admittances;

$$(11.52) \quad Y_C - Y_D = \frac{D(s) - N(s)}{Q(s)} = Y_{RC}^{(5)} - Y_{RC}^{(6)}$$

where $Y_{RC}^{(5)}$ and $Y_{RC}^{(6)}$ are passive RC driving-point admittances. Hence

$$(11.53) \quad \begin{aligned} Y_C &= Y_{RC}^{(5)} \\ Y_D &= Y_{RC}^{(6)} \end{aligned}$$

We now consider an example to illustrate the method.

Example 11-5 Let us realize

$$t_v = \frac{N(s)}{D(s)} = \frac{s^2 + 2.0}{s^2 + 0.1s + 1.2}$$

Choosing $Q(s) = (s + 1)$ we easily obtain

$$\frac{N(s)}{Q(s)} = \frac{s^2 + 2}{s + 1} = (s + 2) - \left(\frac{3s}{s+1}\right)$$

$$\frac{D(s)}{Q(s)} = \frac{s^2 + 0.1s + 1.2}{s + 1} = (s + 1.2) - \left(\frac{2.1s}{s+1}\right)$$

$$\frac{D(s) - N(s)}{Q(s)} = \frac{0.1s - 0.8}{s + 1} = \left(\frac{0.9s}{s+1}\right) - 0.8$$

Thus

$$Y_A = s + 2 \qquad Y_B = \frac{3s}{s+1}$$

$$Y_F = s + 1.2 \qquad Y_E = \frac{2.1s}{s+1}$$

$$Y_C = \frac{0.9s}{s+1} \qquad Y_D = 0.8$$

The final realization is sketched in Figure 11-29.

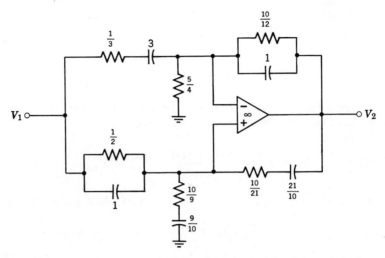

Figure 11-29 Realization of the transfer voltage ratio of Example 11-5.

Several comments are here in order. Since choice of $Q(s)$ is not unique, it can be chosen either to minimize the root sensitivity (Horowitz Decomposition) or the numbers of passive components or the spread of the element values. The input impedance of the filter can be made high by impedance scaling. Output impedance of the filter is always low. As a result, the active filter can be easily cascaded without additional buffer amplifiers. Finally, we observe that the method does not impose any restriction on the realizability of the transfer function, other than that it be a real rational function.

Realization of All-Pass Function

Each of the previous methods lead to inductorless filters. However, in some cases, use of active elements along with RLC elements may be advantageous for many reasons. We now outline a method which can be followed to realize all-pass transfer functions by means of an RLC: operational-amplifier configuration.

A strictly stable all-pass transfer voltage ratio can be expressed as

$$(11.54) \qquad t_v = \pm \frac{N(-s)}{N(s)} = \pm \left(\frac{m-n}{m+n}\right)$$

where $N(s) = m + n$ is a Hurwitz polynomial and

$$m = \text{Ev } N(s)$$
$$n = \text{Od } N(s)$$

SYNTHESIS OF TRANSFER FUNCTIONS / 483

Realization of Equation 11.54 can be achieved by means of the network of Figure 11-30a,[22] which has the following transfer voltage ratio:

(11.55) $$\frac{V_2}{V_1} = \frac{Z_2 - Z_1}{Z_2 + Z_1}$$

If we make $Z_1 = 1$ and Z_2 an LC driving-point impedance,[23] the voltage transfer ratio becomes an all-pass function. From the specified function

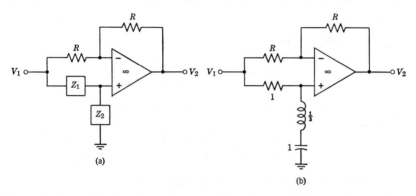

Figure 11-30 Realization of an all-pass transfer voltage ratio by an active RLC configuration containing a single differential input operational amplifier.

of Equation 11.54, Z_2 is identified as follows: t_v is expressed as

(11.56a) $$t_v = \frac{m-n}{m+n} = \frac{\dfrac{m}{n} - 1}{\dfrac{m}{n} + 1} \quad \text{(Case A)}$$

(11.56b) $$t_v = \frac{n-m}{n+m} = \frac{\dfrac{n}{m} - 1}{\dfrac{n}{m} + 1} \quad \text{(Case B)}$$

---

[22] Observe that the network of Figure 11-30a is essentially a special case of that of Figure 11-28.
[23] This identification was advanced by J. Toffler, in Appln. Bulletin APP-114, Fairchild Semiconductor, California, May 1965. Compare with the circuit of Problem 11.8.

Comparing Equations 11.56 and 11.55, we identify

(11.57) $$Z_2 = \frac{m}{n} \quad \text{(Case A)}$$

$$= \frac{n}{m} \quad \text{(Case B)}$$

Note that if Z_1 and Z_2 identifications are interchanged, the allpass character of the circuit remains invariant. An important property of this circuit is that it can be cascaded at the output without an additional isolation stage.

Example 11-6 Let us synthesize

$$t_v = \frac{s^2 - 3s + 3}{s^2 + 3s + 3}$$

Hence $m = (s^2 + 3)$ and $n = 3s$. We use the identification for Case A. Therefore

$$Z_2 = \frac{m}{n} = \frac{s^2 + 3}{3s} = \frac{s}{3} + \frac{1}{s}$$

The complete realization is sketched in Figure 11-30b.

Other Methods

It should be noted that many of the synthesis methods outlined in previous chapters are easily adaptable to operational amplifier use. In particular, we would like to mention the methods discussed in Section 8-3, pp. 311–327, Section 9-3, pp. 382–389, Section 10-3, pp. 425–428.

In addition we refer the reader to the synthesis technique of Pande and Shukla.[24] The method can realize voltage transfer ratios with no positive real critical frequencies, within a multiplicative constant.

11-6 NETWORK DESIGN BY COEFFICIENT MATCHING TECHNIQUES

A brief introduction to active network design by coefficient matching approach was given in Section 8-4. This approach is very efficient for simple transfer functions, in particular second-order functions. One inherent advantage of this type of network design is that the network topology is known a priori to the network designer. In addition, the

---

[24] H. C. Pande and R. S. Shukla, "Synthesis of transfer functions using an operational amplifier," *Proc. IEE* (*London*), **112**, 2208–2212 (December 1965).

design equations in general do not lead to unique solutions for the element values. This gives the designer some freedom in fast alternation of the final network, if necessary. Of course, regular active network synthesis is also not unique. But the network designer using the synthesis approach has no knowledge of the final topology in most cases. Alteration of final design is not simple in many cases.

In this section, we plan to review several design techniques that employ a single operational amplifier and are based on the coefficient matching approach. In each case, the techniques are developed for a biquadratic transfer function of the form

$$t_v = -\frac{a_2 s^2 + a_1 s + a_0}{b_2 s^2 + b_1 s + b_0} \tag{11.58}$$

where the coefficients a_i and b_j are real and nonnegative.

Bohn's Method

The configuration proposed by Bohn[25] is indicated in Figure 11-31, for which the transfer voltage ratio is given as

$$\frac{V_2}{V_1} = -\frac{Y_3 Y_1 + Y_6(Y_1 + Y_2 + Y_3 + Y_4)}{Y_3 Y_4 + Y_5(Y_1 + Y_2 + Y_3 + Y_4)} \tag{11.59}$$

Let us assume the following identifications:

$$Y_1 = G_1, \quad Y_2 = C_2 s, \quad Y_3 = G_3$$
$$Y_4 = C_4 s + G_4, \quad Y_5 = Y_6 = C_0 s \tag{11.60}$$

Substituting Equation 11.60 in Equation 11.59, we obtain

$$\frac{V_2}{V_1} = \frac{C_0(C_2 + C_4)s^2 + C_0(G_1 + G_3 + G_4)s + G_1 G_3}{C_0(C_2 + C_4)s^2 + [G_3 C_4 + C_0(G_1 + G_3 + G_4)]s + G_3 G_4} \tag{11.61}$$

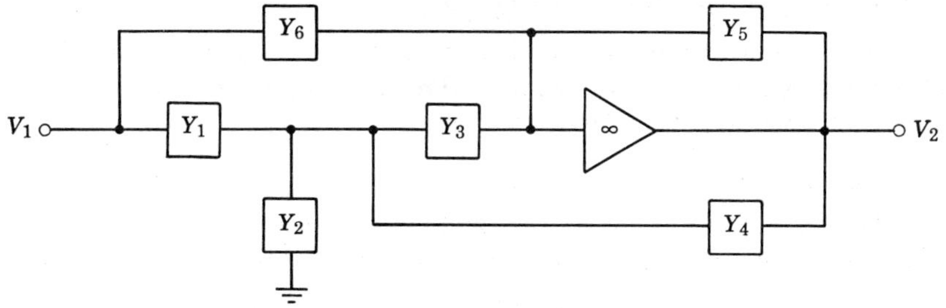

Figure 11-31 A multiloop configuration proposed by Bohn.

---

[25] E. V. Bohn, *Transform Analysis of Linear Systems*, Addison-Wesley, Reading, Mass., p. 71 (1963).

Comparing Equations 11.61 and 11.58, we observe that the specified transfer function must also satisfy the following conditions:

$$a_2 = b_2$$
$$b_1 > a_1$$

The element values are obtained by equating like coefficients of Equations 11.61 and 11.58:

(11.62)
$$a_2 = C_0(C_2 + C_4)$$
$$a_1 = C_0(G_1 + G_3 + G_4)$$
$$b_1 - a_1 = G_3 C_4$$
$$a_0 = G_1 G_3$$
$$b_0 = G_3 G_4$$

Equations 11.62 can be solved for the element values. Since there are six unknowns and five equations, the solution is not unique. Consider the following example for illustrative purposes.

Example 11-7 Realize

(11.63)
$$t_v = -\frac{s^2 + s + 1}{s^2 + 2s + 2}$$

From Equations 11.63 and 11.62, we obtain

$$C_0(C_2 + C_4) = 1$$
$$C_0(G_1 + G_3 + G_4) = 1$$
$$G_3 C_4 = 1$$
$$G_3 G_4 = 2$$

Letting $C_0 = \frac{1}{4}$, the above equations yield

$$G_3 = 1, \quad G_1 = 1, \quad G_4 = 2, \quad C_4 = 1, \quad C_2 = 3$$

Final realization is shown in Figure 11-32.

By choosing various *RC* configurations for the one-port networks of Figure 11-31, many types of transfer functions can be realized. Holt and Sewell[26] have cataloged eleven networks with Y_2 equal to zero.

Brennan and Bridgman's Method

The configuration proposed by Brennan and Bridgman[27] was analyzed in Section 4-9. It has been redrawn in Figure 11-33 for convenience. For

---

[26] A. G. J. Holt and J. I. Sewell, "Active *RC* filters employing a single operational amplifier to obtain biquadratic responses," *Proc. IEE (London)*, **112**, 2227–2234 (December 1965).

[27] A. Bridgman and R. Brennan, "Simulation of transfer function using only one operational amplifier," *Proc. WESCON Convention Record*, **1** (Part 4), 273–278 (1957).

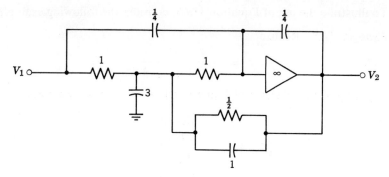

Figure 11-32 Realization of the transfer voltage ratio of Example 11-7, using Bohn's method.

this circuit we have

$$\text{(11.64)} \qquad \frac{V_2}{V_1} = -\frac{Y_1 Y_3}{Y_5(Y_1 + Y_2 + Y_3 + Y_4) + Y_3 Y_4}$$

It is evident from Equation 11.64 that complex transmission zeros cannot be realized by this method.

To realize a low-pass transfer voltage ratio we set

$$\text{(11.65)} \qquad \begin{array}{ccc} Y_1 = 1/R_1, & Y_3 = 1/R_3, & Y_4 = 1/R_4 \\ Y_2 = C_2 s, & Y_5 = C_5 s \end{array}$$

Substitution of Equation 11.65 in Equation 11.64 yields

$$\text{(11.66)} \qquad \frac{V_2}{V_1} = -\frac{1}{(R_1 R_3 C_2 C_5)s^2 + \left(R_3 + R_1 + \dfrac{R_3 R_1}{R_4}\right) C_5 s + \dfrac{R_1}{R_4}}$$

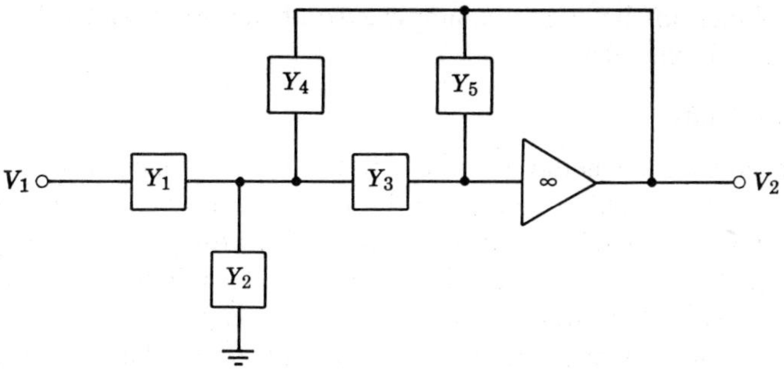

Figure 11-33 A multiloop configuration proposed by Bridgman and Brennan.

To illustrate the use of Equation 11.66, consider the following example.

Example 11-8 Realize

$$t_v = -\frac{1}{s^2 + 3s + 3} \tag{11.67}$$

Equating like coefficients of Equations 11.66 and 11.67, we obtain

$$R_1 R_3 C_2 C_5 = 1$$

$$\left(R_3 + R_1 + \frac{R_1 R_3}{R_4}\right) C_5 = 3$$

$$\frac{R_1}{R_4} = 3$$

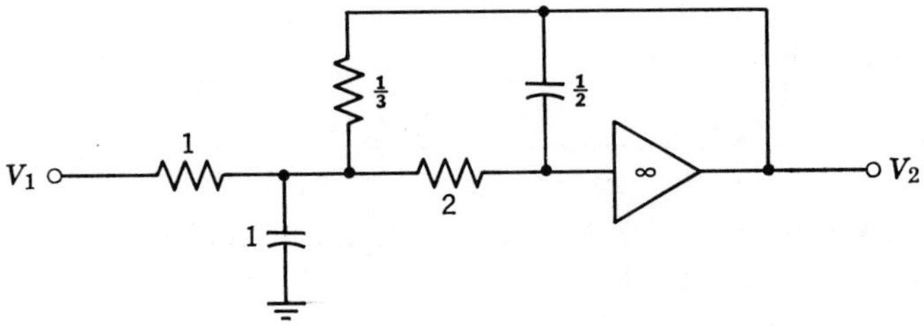

Figure 11-34 Realization of the transfer-voltage ratio of Example 11-8, using Bridgman-Brennan's approach.

One possible solution of the above set of equations is:

$$R_1 = 1, \qquad R_3 = \tfrac{1}{2}, \qquad R_4 = \tfrac{1}{3}, \qquad C_2 = 2, \qquad C_5 = 1$$

which yields the complete realization indicated in Figure 11-34.

Other choices for the one-port admittances have been considered by Brennan and Bridgman, leading to the realization of second-order transfer functions having (i) a simple zero at the origin and (ii) a simple negative real axis zero.

Other Methods

It is clear that the methods outlined in Section 8-4 are equally applicable for operational amplifier use. In addition, there are several other methods for the realization of second-order and third-order transfer functions, which are based on the coefficient matching technique. References to these methods will be found at the end of this chapter.

In the following section, we outline a recent method of design of second-order filters which appears to have very low sensitivity with respect to passive and active components.

11-7 A MINIMUM SENSITIVE REALIZATION OF TRANSFER FUNCTION

The synthesis techniques presented earlier in this chapter and in Chapters 8 and 9 have an inherent drawback—they are not suitable for realization of transfer functions having high Q pole-pair. The main reason for this is that complex poles are generated by taking the difference of two RC network functions and, as a result, they depend on the cancellation of terms. For very high Q pole-pair, these terms are of the same order, which leads to the high Q-sensitivity.

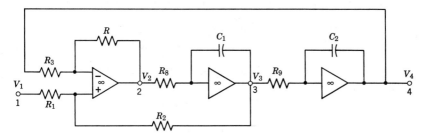

Figure 11-35 A minimum-sensitive configuration suitable for low-pass, band-pass and high-pass second-order transfer-voltage ratio realization.

The above problem theoretically does not arise in the RC:gyrator type synthesis; however, practical gyrators being nonideal devices do lead to many practical problems including the sensitivity.

A recent method proposed by Kerwin, Huelsman, and Newcomb[28] appears to have excellent properties with respect to the sensitivity. Their proposed configuration is shown in Figure 11-35. If output is taken from terminal 4, then the transfer voltage ratio is given as

$$(11.68) \quad \frac{V_4}{V_1} = \frac{R_2(R + R_3)}{(R_1 + R_2)R_3 \left(s^2 C_1 C_2 R_8 R_9 + s \frac{C_2 R_9 R_1 (R + R_3)}{R_3(R_1 + R_2)} + \frac{R}{R_3} \right)}$$

Note that V_4/V_1 as given above is a low-pass transfer function. For convenience in design, we can let

$$(11.69) \quad R_1 = R_3 = R_8 C_1 = R_9 C_2 = 1$$

---

[28] W. J. Kerwin, L. P. Huelsman, and R. W. Newcomb, "State variable synthesis for insensitive integrated circuit transfer functions," *IEEE J. Solid-State Circuits*, **SC-2**, 114–116 (September 1967).

which simplifies expression 11.68 to

$$\text{(11.70)} \qquad \frac{V_4}{V_1} = \frac{R_2(1+R)}{(1+R_2)\left(s^2 + s\dfrac{R+1}{R_2+1} + R\right)}$$

Observe that for high Q pole-pair, the resonant frequency is approximately given as

$$\omega_0 \cong \sqrt{R}$$

Examination of expression 11.68 also reveals that no difference terms appear in the denominator; hence we expect the sensitivity to be low. To show this, let us compute the Q of the pole-pair, which is given as

$$\text{(11.71)} \qquad Q = \sqrt{\frac{RR_3 C_1 R_8}{C_2 R_9} \cdot \frac{R_1 + R_2}{R_1(R+R_3)}}$$

The various Q-sensitivities can be calculated from Equation 11.71 and they are given as:

$$S_{C_1}^Q = \tfrac{1}{2}; \qquad S_{C_2}^Q = -\tfrac{1}{2}$$
$$S_{R_8}^Q = \tfrac{1}{2}; \qquad S_{R_9}^Q = -\tfrac{1}{2}$$

$$\text{(11.72)} \qquad S_{R_2}^Q = \frac{R_2}{R_1 + R_2} < 1; \qquad S_{R_1}^Q = -S_{R_2}^Q$$

$$S_{R_3}^Q = \frac{1}{2}\frac{R-1}{R+1} < \frac{1}{2}; \qquad S_R^Q = -S_{R_3}^Q$$

Note that all of the Q-sensitivities are less than one and are independent of Q. Moreover, Σ_Q is comparable to that of LC filters. In addition, if the passive components are built with thin film technology, the percentage variations of all resistors due to temperature are almost equal and similarly the percentage variations of all capacitors due to temperature are almost equal. As a result, the percentage change in system Q due to temperature will be negligibly small provided effect of other variations are neglected.

If a high-pass filter is desired, the output can be taken from terminal 2. For a band-pass response, output must be taken from terminal 3. In each case, outputs coincide with the output terminals of an operational amplifier and hence provide a very low-impedance output.

For realizing a general transfer function, the outputs from terminals 2, 3, and 4 can be added by using a summing amplifier. Since the potential at terminal 3 is of opposite sign to that of terminals 2 and 4, it may be necessary to use an inverting-amplifier to change the sign. Such an arrangement is shown in Figure 11-36. The transfer voltage ratio V_5/V_1 is given as

(11.73)
$$\frac{V_5}{V_1} = \frac{R_2(R+R_3)R_5(R_6+R_7)}{(R_1+R_2)R_3(R_4+R_5)R_7} \left[\frac{R_8C_1R_9C_2s^2 + \frac{(R_4+R_5)R_6R_9C_2}{R_5(R_6+R_7)}s + \frac{R_4}{R_5}}{R_8C_1R_9C_2s^2 + \frac{R_1(R+R_3)R_9C_2}{(R_1+R_2)R_3}s + \frac{R}{R_3}} \right]$$

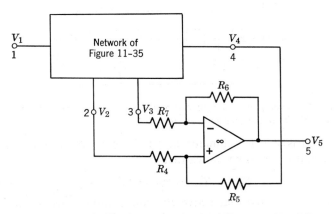

Figure 11-36 Modification of the circuit of Figure 11-35 to allow the realization of a general second-order transfer voltage ratio.

Using the component values given by expression 11.69 we have

(11.74)
$$\frac{V_5}{V_1} = \frac{R_2(1+R)(1+R_7)}{(1+R_2)(1+R_4)R_7} \left[\frac{s^2 + \frac{1+R_4}{1+R_7}s + R_4}{s^2 + \frac{1+R}{1+R_2}s + R} \right]$$

It has been shown[29] that at the resonant frequency $s = j\omega_0 \simeq j\sqrt{R}$,

$$S_R^{|(V_5/V_1)(j\omega_0)|} = S_{R_3}^{|(V_5/V_1)(j\omega_0)|} = 0$$

$$S_{R_2}^{|(V_5/V_1)(j\omega_0)|} = S_{R_3}^{|(V_5/V_1)(j\omega_0)|} = 1$$

$$S_{R_8}^{|D(j\omega_0)|} = S_{C_1}^{|D(j\omega_0)|} = 0$$

$$S_{R_9}^{|D(j\omega_0)|} = S_{C_2}^{|D(j\omega_0)|} = 1$$

---

[29] W. J. Kerwin et al., *loc. cit.*

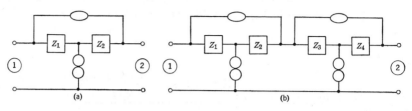

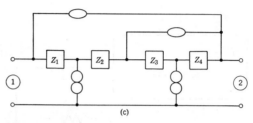

Figure 11-37 Development of nullator-norator models of the generalized active transformer suitable for inductance simulation.

where $D(s)$ is the denominator of all the pertinent transfer functions. Moreover, it can also be shown that

$$S_{R_8}^{\omega_0} = S_{R_9}^{\omega_0} = S_{C_1}^{\omega_0} = S_{C_2}^{\omega_0} = -\tfrac{1}{2}$$

It should be noted that active RC configurations, using three operational amplifiers similar to that of Figure 11-35, have also been proposed by Sutcliffe[30] and Geffe[31] for transfer function realization.

11-8 SIMULATION OF INDUCTANCE

One of the simplest ways to build an inductorless filter is to design a conventional LC filter and replace each inductor by simulated inductances. Due to the low sensitivity feature of the LC ladder filters, this approach appears to be very promising.

Several methods have been proposed to simulate inductances by using operational amplifiers. We shall outline below two such methods. Remaining methods will be found in the list of references at the end of this chapter.

Realization of Grounded Inductance

Consider the type I-CGIC nullator-norator model of Table 9-1, which has been redrawn in Figure 11-37a for convenience. A cascade of two

---

[30] H. Sutcliffe, "Tunable filter for low frequencies using operational amplifiers," *Electronic Engg.*, **36**, 399–403 (June 1964).

[31] P. Geffe, "Make a filter out of an oscillator," *Electronic Design* 10, pp. 56–58 (May 10, 1967).

such CGIC's (Figure 11-37b) will result in a generalized active transformer described by the following transmission matrix:

$$\begin{bmatrix} 1 & 0 \\ 0 & \dfrac{Z_2 Z_4}{Z_1 Z_3} \end{bmatrix}$$

If we now terminate the port 2 by a resistance R_L, the input impedance seen at port 1 is

(11.75) $$Z_{11} = \frac{Z_1 Z_3 R_L}{Z_2 Z_4}$$

It is seen from Equation 11.75 that if either Z_2 is a capacitance and Z_4 is a resistance or vice versa, the input impedance is identical to that of an inductance provided that Z_1 and Z_3 are resistances. In a similar manner, an inductance can be simulated by terminating port 1 by a resistance and making either Z_1 or Z_3 a capacitance and the remaining Z_i resistances.

Using the properties of the nullator, the circuit of Figure 11-37b can be easily modified to that shown in Figure 11-37c. The behavior of the modified two-port is still unchanged, and hence it can also be used for inductance simulation.

Conversion of the nullator-norator models of Figure 11-37b and c to operational amplifier circuits can now be effected by making use of the nullator-norator model of the operational amplifier as given in Figure 11-13. This results in the circuits of Figure 11-38 which are suitable for simulation of grounded inductance. The circuits of Figure 11-38a, b, and c are new and the circuit of Figure 11-38d is due to Riordan.[32]

Realization of Ungrounded Inductance

In many applications, it is often required to use "floating" inductors i.e., inductors having both terminals above ground. One simple way to realize an ungrounded inductance would be to connect the circuits of Figure 11-38 in a back-to-back fashion as indicated by Riodran.[32] However, the effective inductances of the two circuits must be exactly equal; otherwise, any unbalance would appear as a parasitic inductance to ground. An alternate solution would be to use any of the circuit of Figure 11-38 with a "floating" power supply.

Another novel solution to the problem of simulating floating inductance has been advanced recently.[33] The proposed circuit is shown in Figure

[32] R. H. S. Riordan, "Simulated inductors using differential amplifiers," *Electronics Letters*, **3**, 50–51 (February 1967).
[33] G. J. Deboo, "Application of a gyrator-type circuit to realize ungrounded inductors," *IEEE Trans. on Circuit Theory*, **CT-14**, 101–102 (March 1967).

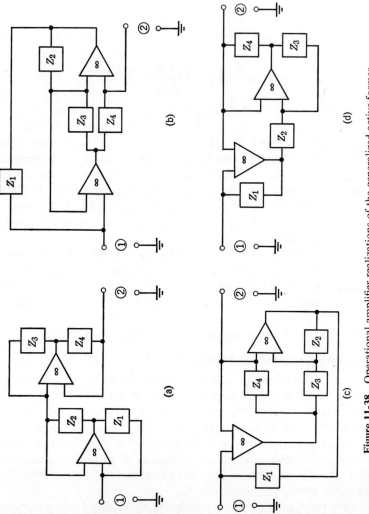

Figure 11-38 Operational amplifier realizations of the generalized active former.

11-39. The y-matrix of this two-port is given as

(11.76) $$[y] = \begin{bmatrix} 1/sCR^2 & -1/sCR^2 \\ -1/sCR^2 & 1/sCR^2 \end{bmatrix}$$

which is identical to that of a floating inductance of value CR^2 henries.

The circuit of Figure 11-39 can easily be converted into a grounded gyrator (see Problem 11.17).

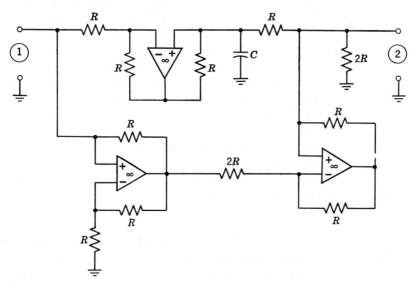

Figure 11-39 Realization of a floating inductance, using three operational amplifiers.

11-9 SUMMARY

A practical operational amplifier is a nonideal device having finite input and output impedance and a very high gain, which is frequency dependent. However, in most linear applications an operational amplifier is used with negative feedback, resulting in satisfactory performance. The physical characteristics of a practical operational amplifier are discussed in Section 11-1. The stability of a circuit using an operational amplifier is discussed in Section 11-2. It is shown that, since the gain roll-off of an integrated operational amplifier is usually more than 12 db/octave, for a stable closed-loop operation the amplifier must be compensated. Two most commonly used compensation techniques are described.

Various circuits to construct controlled sources, negative-impedance converter and gyrator are developed in Section 11-3.

An operational amplifier can be used to balance a bridge circuit and, as a result, when used in the Bott-Duffin cycle it results in a considerable saving of passive components in the realization of a positive real driving-point function (Section 11-4, p. 470).

Several transfer function realization schemes are outlined in Section 11-5. The method described in Section 11-5, p. 474, which uses a single-ended operational amplifier, is probably the most commonly used method because of its simplicity. This method also allows the use of tables of RC two-ports cataloged in Appendix D, thus enabling a fast realization of second-order transfer functions. A completely general method of synthesis, which makes use of a single differential-input amplifier and six RC one-ports, is advanced in Section 11-5, p. 479. This method is extremely simple to use and does not impose any restriction on the realizability of a real rational transfer voltage ratio. The configuration used allows the possibility of cascading without additional buffer amplifiers.

Two design methods based on the coefficient matching technique are reviewed in Sections 11-6 and 11-7. The active RC configuration used in Section 11-7 employs three operational amplifiers and has the lowest Q-sensitivity (comparable to LC ladder filters) in comparison to the active RC configurations of other synthesis methods. For high Q realization, the technique outlined here appears to be most suitable.

Simulation of a grounded and a floating inductance is considered in Section 11-8.

It should be noted that the driving-point and transfer function synthesis techniques outlined in the previous chapters (Chapters 7 through 10) are easily adaptable for operational amplifier use. This can be simply achieved by replacing the pertinent active element by its operational amplifier realization.

Additional References

Aggarwal, G. K., "On fourth-order simulation by one amplifier," *J. Electronics and Control*, **15**, 449 (1963).

Aggarwal, G. K., "A single operational amplifier simulates general third order linear systems," *Proc. Indian Acad. Sci.*, **58**, 257 (1963).

Antoniou, A., "Gyrators using operational amplifiers," *Electronics Letters*, **3**, 350–352 (August 1967).

Blair, K., "Getting more value out of an integrated operational amplifier data sheet," Integrated Circuits Application Note AN-273, Motorola Semiconductor Products Inc., Phoenix, Arizona.

Butler, E., "The operational amplifier as a network element," M.S. Thesis, School of Elec. Engg., Cornell University, Ithaca, N.Y., June 1965.

Cunningham, W. J., "Time-delay networks for an analog computer," *IRE Trans. on Electronic Computers*, **EC-3**, 16 (1954).

Electro-Technology Staff Report, "Packaged operational amplifiers," *Electro-Technology*, **79**, 73–77 (January 1967).

Field, R. K., "The tiny, exploding world of linear microcircuits," *Electronic Design*, **15**, 49–66, 68–72 (July 19, 1967).

Fowler, W. J., "A look at linear integrated circuits," *Electronic Industries*, **24**, 64–69, 194 (September 1965).

Giles, J. N., "Linear integrated circuits applications handbook," Fairchild Semiconductor, Mountain View, Calif., 1967.

"Handbook of operational amplifier active RC networks," Burr-Brown Research Corp., Tucson, Arizona, 1966.

Holt, A. G. J., and J. I. Sewell, "Table for the voltage transfer function of single amplifier double ladder parallel feedback system," *IEEE Trans. on Circuit Theory*, **CT-13**, 326–329 (September 1966).

"Integrated operational amplifiers," Technical Information and Applications, Microelectronic Div., Radiation Inc., Melbourne, Fla., 1st Ed., December 1966.

Karplus, W. J., "Synthesis of non-p.r. driving-point impedance functions using analog computer units," *IRE Trans. on Circuit Theory*, **CT-4**, 877–879 (September 1957).

Keen, A. W., and J. L. Peters, "Inductance simulation with a single differential-input operational amplifier," *Electronics Letters*, **3**, 136–137 (April 1967).

Keen, A. W., and J. L. Peters, "Nonreciprocal representation of the floating inductor with grounded amplifier realizations," *Electronics Letters*, **3**, 369–371 (August 1967).

McVey, P. J., "Synthesis of transfer functions by RC Networks with two or three computing amplifiers," *International J. on Control*, **2**, 125–133 (August 1965).

Mitra, S. K., "Transfer function synthesis using a single operational amplifier," *Electronics Letters*, **3**, 333–334 (July 1967).

Morrill, C. D., "A sub-audio time delay circuit," *IRE Trans. on Electronic Computers*, **EC-3**, 45 (1954).

Morrison, C. F., Jr., "A primer in the art of using operational amplifiers in general utility instrumentation," Washington State University, Pullman, Washington.

Morse, A. S., and L. P. Huelsman, "A gyrator realization using operational amplifiers," *IEEE Trans. on Circuit Theory*, **CT-11**, 277–278 (June 1964).

Morse, A. S., "The use of operational amplifiers in active network theory," *Proc. NEC*, **20**, 748 (1964).

Morse, A. S., "The use of operational amplifiers in active network theory," M.S. Thesis, University of Arizona, Tucson, Arizona, January 1964.

Paul, R. J. A., "Realization of fourth-order rational transfer functions with adjustable coefficients," *Proc. IEE (London)*, **111**, 877–882 (May 1964).

Paul, R. J. A., "Simulation of rational transfer functions with adjustable coefficients," *Proc. IEE (London)*, **110**, 671–679 (1963).

Prescott, A. J., "Loss-compensated active gyrator using differential input operational amplifiers," *Electronics Letters*, **2**, 283–284 (1966).

Sommerville, M. J., and G. H. Tomlinson, "Filter synthesis using active RC networks," *J. Electronics and Control*, **12**, 401–420 (1962).

Tomlinson, G. H., "Synthesis of delay networks for an analog computer," *Proc. IEE (London)*, **112**, 1806–1814 (September 1965).

VanDenBerg, R. J., "The care and feeding of the μA709 integrated linear amplifier," Honeywell Aero Division, Minneapolis, Minn., October 1966.

498 / THE OPERATIONAL AMPLIFIER AS A NETWORK ELEMENT

Wadhwa, L. K., "Simulation by a single operational amplifier of third order transfer functions having a pole at the origin," *Radio Electronic Engr.*, **27**, 373 (1964).

Wadhwa, L. K., "Simulation of third order systems by a single operational amplifier," *IEEE Trans. on Electronic Computers*, **EC-13**, 128 (1964).

Wadhwa, L. K., "Simulation of third order systems with double lead using one operational amplifier," *Proc. IRE*, **50**, 1538–1539 (June 1962).

Wissemann, L., and J. Robertson, "Designing with monolithic operational amplifiers," *Electro-Technology*, **79**, 50–52 (February 1967).

Problems

11.1 Using the model of Figure 11-5*b*, compute the *g* matrix of the noninverting VVT of Figure 11-5*a*. Show that for a typical practical operational amplifier, the noninverting VVT approaches very closely an ideal VVT.

11.2 Show that the circuit of Figure 11-40 represents a differential input VCT. Construct a gyrator using this type of VCT.[34]

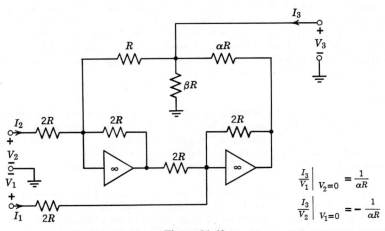

Figure 11-40

11.3 Using the inverting VVT of Figure 11-4*a* and Gintzon's approach outlined in Section 7-1, p. 256, construct a short-circuit stable and an open-circuit stable negative resistance.[35]

11.4 Verify the stability constraints of the input and output ports of the NIC circuits of Figure 11-8.

---

[34] A. S. Morse and L. P. Huelsman, *loc. cit.*
[35] W. J. Karplus, *loc. cit.*

11.5 Using the active realization method of the Bott-Duffin cycle (Section 11-4, p. 470), synthesize the following positive real driving-point admittances:

(a) $\dfrac{s^3 + 3s^2 + 3s + 1}{s^3 + 3s^2 + 2s + 2}$

(b) $\dfrac{s^4 + 2s^3 + 3s^2 + s + 1}{s^4 + s^3 + 3s^2 + 2s + 1}$

(c) $\dfrac{s^2 + s + 1}{s^2 + s + 4}$

(d) $\dfrac{s^3 + 2s^2 + s + 1}{s^3 + s^2 + 2s + 1}$

11.6 Realize the following transfer voltage ratios in the form of Figure 11-25a. Synthesize the two-port $\mathcal{N}_B$ using the parallel-ladder realization technique for complex transmission zeros.

(a) $\dfrac{-H}{s^2 + s + 1}$

(b) $\dfrac{-Hs}{s^2 + s + 1}$

(c) $\dfrac{-Hs^2}{s^2 + s + 1}$

(d) $\dfrac{-H(s^2 + 2)}{s^2 + s + 1}$

(e) $\dfrac{-H(s^2 + 2s + 3)}{s^2 + s + 2}$

11.7 Realize the transfer functions of Problem 11.6 in the form of Figure 11-25a by making use of tables of the RC two-port of Table D-1 (Appendix D).

11.8 Realize the transfer functions of Problems 11.6 following Mathew-Seifert's approach (Section 11-5, p. 476). How would you now realize these functions with a positive gain?

11.9 Repeat Problem 11.8 by following Lovering's approach (Section 11-5, p. 477).

11.10 Synthesize the following all-pass functions in the form of Figure 1130a.

(a) $\dfrac{s^2 - s + 1}{s^2 + s + 1}$

(b) $-\dfrac{s^2 - s + 1}{s^2 + s + 1}$

(c) $\dfrac{s^3 - 6s^2 + 15s - 15}{s^3 + 6s^2 + 15s - 15}$

(d) $\dfrac{-s^3 + 6s^2 - 15s + 15}{s^3 + 6s^2 + 15s + 15}$

11.11 Synthesize the transfer voltage ratio of Problem 11.6 and Problem 11.10 by following the method of Section 11-5, p. 479, for various values of the multiplier constant.

11.12 If it is desired not to use inductors in the realization of the transfer functions of Problems 11.10, which of the other methods of this chapter will you use?

11.13 Develop a design procedure to realize

$$t_v = -\dfrac{s^2 + 2s + 2}{s^2 + s + 1}$$

by setting $Y_1 = C_1 s + G_1$, $Y_4 = G_4$ and keeping the remaining identifications of Equation 11.60 for the structure of Figure 11-31.

11.14 Realize the transfer functions of Problem 11.6 by means of the structures of Figures 11-35 and 11-36.

11.15 Constant resistance networks are useful in the design of higher-order transfer functions by cascading the networks realizing simpler transfer functions. An active equivalent to the constant resistance bridged T network is shown in Figure 11-41.[36] Show that as $\mu \to \infty$

$$Z_{BE} \to R_0^2/Z_a$$

$$Z_{AB} \to R_0$$

$$\dfrac{V_2}{V_1} \to \dfrac{R_0}{R_0 + Z_a}$$

11.16 Realize the following voltage transfer functions by means of the network of Figure 11-41. Use more than one constant resistance

[36] A. G. Stuart and D. G. Lampard, *loc. cit.*

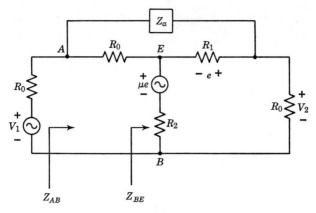

Figure 11-41

section if necessary. If possible, maximize the gain constant H in each case.

(a) $\dfrac{H(s^2 + 1)}{(s + 1)^2}$

(b) $\dfrac{H(s + 1)}{s^2 + 2s + 2}$

(c) $\dfrac{H(s^2 + 2)(s + 1)}{(s^2 + s + 2)(s^2 + 2s + 2)}$

11.17 Develop grounded gyrator realization from the circuit of Figure 11-39.

12 / Polynomial Decomposition Theorems and Their Applications

In Section 1-3, it has been mentioned that a major step in the synthesis of active networks is the decomposition and partitioning of the numerator and denominator polynomials of the specified network function. The method of decomposition, as followed in most of the conventional active RC network synthesis procedures, starts with the selection of a polynomial $Q(s)$ with distinct negative real roots only. Then the polynomial (denominator and/or numerator) of the given function to be realized is divided by $Q(s)$. The next step involves a partial fraction expansion of the resulting functions. The terms with positive residues are grouped together into a subfunction, and the terms with negative residues are similarly grouped. These subfunctions are usually in the form of RC immittances.

In this chapter we shall present three new methods of decomposition and partitioning. These methods are simple and straightforward and yield subfunctions that are always in the form of RC immittances. These decompositions are always in suitable forms for direct identification of the companion network parameters for many active synthesis procedures. A few of these applications will be considered.

12-1 A UNIQUE DECOMPOSITION[1]

It will be shown in this section that any real rational function can always be decomposed and partitioned into *unique* subfunctions. The basic idea behind the method of decomposition is based on the symmetry of zeros and poles of the reflection coefficient derived from the equivalent active LC immittance.

Let $Y(s) = N(s)/D(s)$ be an arbitrary real rational admittance function with poles and zeros anywhere in the s-plane.

The equivalent LC admittance obtained by an $RC:LC$ transformation (Equation 3.8) is given as

$$\hat{Y}(s) = \frac{N(s^2)}{sD(s^2)} \tag{12.1}$$

We assume here that if $N(s)$ has any zero at the origin, it is not cancelled with the factor s in the denominator of $\hat{Y}(s)$. The reflection coefficient corresponding to the admittance function $\hat{Y}(s)$ is

$$\hat{\rho}(s) = \frac{1 - \hat{Y}(s)}{1 + \hat{Y}(s)} \tag{12.2}$$

which upon substitution of (12.1) becomes

$$\hat{\rho}(s) = \frac{sD(s^2) - N(s^2)}{sD(s^2) + N(s^2)} = -\frac{P(-s)}{P(s)} \tag{12.3}$$

where we have set

$$P(s) = N(s^2) + sD(s^2)$$

Observe from Equation 12.3 that to every zero of $\hat{\rho}(s)$ in the right-half s plane there corresponds a pole of $\hat{\rho}(s)$ in the left-half s plane situated symmetrically with respect to the origin and vice versa. A typical pole-zero distribution is shown in Figure 12-1. Note that $\hat{\rho}(s)$ cannot have any zeros and poles on the $j\omega$-axis if $N(s)$ is assumed to have no common factors with $D(s)$, with the exception of a possible pole and a zero at the origin. Let $(m_1 - n_1)$ denote the *unique* anti-Hurwitz polynomial obtained by grouping together the right-half plane zeros of $\hat{\rho}(s)$ and let $(m_2 + n_2)$ denote the *unique* Hurwitz polynomial formed by grouping together the left-half plane zeros of $\hat{\rho}(s)$. The zero at the origin, if present, can be grouped with either of the two polynomials. We can thus write

$$\hat{\rho}(s) = -\frac{(m_1 - n_1)}{(m_1 + n_1)} \cdot \frac{(m_2 + n_2)}{(m_2 - n_2)} \tag{12.4}$$

---
[1] S. K. Mitra, "A unique synthesis method of transformerless active RC networks," *J. Franklin Inst.*, **274**, 115–129 (August 1962).

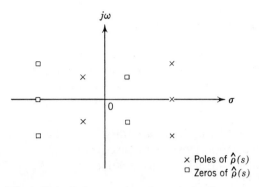

Figure 12-1 Pole-zero plot of a typical reflection coefficient $\hat{\rho}(s)$.

From Equations 12.4 and 12.1, we obtain

(12.5) $$\hat{Y}(s) = \frac{1 - \hat{\rho}(s)}{1 + \hat{\rho}(s)} = \frac{m_1 m_2 - n_1 n_2}{m_2 n_1 - m_1 n_2}$$

Since m_1 and m_2 are even polynomials and n_1 and n_2 are odd polynomials, we can express

(12.6) $$\begin{aligned} m_i &= a_i(s^2) \\ n_i &= s b_i(s^2) \end{aligned}$$

Note that since $(m_i + n_i)$ is a Hurwitz polynomial, $m_i/n_i = a_i(s^2)/sb_i(s^2)$ is a passive LC driving-point impedance. The function $a_i(s)/sb_i(s)$, obtained from m_i/n_i by $LC:RC$ transformation (Equation 3.9), is therefore a passive RC driving-point impedance. Substituting Equation 12.6 in Equation 12.5, we have

(12.7) $$\hat{Y}(s) = \frac{a_1(s^2)a_2(s^2) - sb_1(s^2)sb_2(s^2)}{a_2(s^2)sb_1(s^2) - a_1(s^2)sb_2(s^2)}$$

From Equation 12.7, a similar decomposition for $Y(s)$ can be derived by using the inverse transformation

(12.8) $$Y(s) = \sqrt{s}\hat{Y}(\sqrt{s})$$

This yields the following theorem:

Theorem 12-1 *A real rational admittance function $Y(s)$ can always be decomposed and partitioned as*

(12.9) $$Y(s) = \frac{a_1(s)a_2(s) - sb_1(s)b_2(s)}{a_2(s)b_1(s) - a_1(s)b_2(s)}$$

where $a_1(s)/b_1(s)$ and $a_2(s)/b_2(s)$ are unique *subfunctions satisfying the properties of passive RC driving-point admittances.*

The decomposition of Equation 12.9 will be referred to as the *Unique Decomposition*.[2] Application of the unique decomposition will be considered in Section 12-4.

Example 12-1 Let

(12.10) $$Y(s) = \frac{1}{8s}$$

This implies that $N(s) = 1$ and $D(s) = 8s$. Hence

$$sD(s^2) - N(s^2) = 8s^3 - 1 = (2s - 1)(4s^2 + 2s + 1)$$

Therefore

$$m_1 - n_1 = 1 - 2s$$
$$m_2 + n_2 = 4s^2 + 2s + 1$$

from which one obtains

(12.11) $$a_1(s) = 1, \qquad b_1(s) = 2$$
$$a_2(s) = 4s + 1, \qquad b_2(s) = 2$$

Thus the unique decomposition of $Y(s) = 1/8s$ is given as

(12.12) $$Y(s) = \frac{1}{8s} = \frac{(1)(4s + 1) - s(2)(2)}{(4s + 1)(2) - (1)(2)}$$

Observe that

$$\frac{b_1(s)}{a_1(s)} = \frac{2}{1} = 2$$

and

$$\frac{b_2(s)}{a_2(s)} = \frac{2}{4s + 1}$$

are passive *RC* driving-point impedances as expected.

12-2 THE ODD-PART PARTITIONING[3]

The second type of decomposition and partitioning to be considered in this section is based on the simple observation that a class of real rational

---

[2] Somewhat similar results have been independently obtained by W. Saraga, in *Recent Developments in Network Theory*, S. R. Deards, Ed., Pergamon Press, New York, pp. 199–210 (1963).

[3] R. E. Thomas, "Polynomial decomposition in active network synthesis," *IRE Trans. on Circuit Theory*, **CT-8,** 270–274 (September 1961); S. K. Mitra, "A new approach to active *RC* network synthesis," *J. Franklin Inst.*, **274,** 185–197 (September 1962).

functions, upon $RC:LC$ transformation, can be related to the odd part of a positive real function.

Let

(12.13) $$G(s) = \frac{m_1 + n_1}{m_2 + n_2}$$

be a positive real function, where $m_1(m_2)$ is the even part of the numerator (denominator) of $G(s)$ and $n_1(n_2)$ is the odd part of the numerator (denominator) of $G(s)$. From the properties of positive real function, it can be shown[4] that m_1/n_1, m_2/n_2, m_1/n_2, and m_2/n_1 are passive LC driving-point functions.

The odd part of $G(s)$ is given as

(12.14) $$\text{Od } G(s) \triangleq \tfrac{1}{2}[G(s) - G(-s)] = \frac{m_2 n_1 - m_1 n_2}{m_2{}^2 - n_2{}^2}$$

We now state below without a proof the properties of Od $G(s)$, which can be derived from the known properties of a positive real function.

Theorem 12-2[5] *The necessary and sufficient conditions satisfied by a real rational function $P(s)/Q(s)$ to be the odd part of a positive real function are:*

(12.15)
(i) $P(s)^0 \leq Q(s)^0 + 1$;

(ii) $P(s)$ is an odd polynomial and $Q(s)$ is an even polynomial;

(iii) $j\omega$-axis poles of $P(s)/Q(s)$ are simple with real and positive residues.

A real rational function $F(s)$ satisfying conditions (12.15) of Theorem 12-2 can be related to the odd part of a positive real function $G(s)$. In constructing $G(s)$, any of the known methods can be employed.[6] Note that since the even part is not specified, any arbitrary positive real constant can be added to $G(s)$ without changing its odd part. Thus, determination of $G(s)$ is not unique. Once $G(s)$ has been determined, $F(s)$ [which is Od $G(s)$] can be decomposed and partitioned in the form of Equation 12.14.

The decomposition of Equation 12.14 can be related to the active RC case by means of an $LC:RC$ transformation. To this end, we introduce the following notations:

(12.16) $$m_i = a_i(s^2), \qquad n_i = sb_i(s^2)$$

[4] E. A. Guillemin, *Passive network synthesis*, Wiley, New York, Ch. 1 (1956).
[5] D. F. Tuttle, *Network synthesis*, Wiley, New York, **1**, 1958.
[6] H. Ruston and J. Bordogna, *Electric networks—functions, filters, analysis*, McGraw-Hill, New York, 1966.

Observe that since $a_1(s^2)/sb_1(s^2)$ is a passive LC admittance, $a_1(s)/b_1(s)$ is a passive RC admittance function. Similarly, we note that $a_1(s)/b_2(s)$, $a_2(s)/b_2(s)$, and $a_2(s)/b_1(s)$ are passive RC driving-point admittances.

Using expressions 12.16 in Equation 12.14 we note that a real rational function $F(s)$, satisfying the properties of the odd part of a positive real function, can be expressed as

$$(12.17) \qquad F(s) = s\left[\frac{a_2(s^2)b_1(s^2) - a_1(s^2)b_2(s^2)}{a_2^2(s) - s^2 b_2^2(s)}\right]$$

There are two possible cases to be examined. If $F(s)$ is considered as an impedance function, we obtain the following theorem by applying $LC:RC$ transformation to expression 12.17 and conditions 12.15:

Theorem 12-3 *If a real rational function $Z(s) = N(s)/D(s)$ satisfies the following conditions:*

(12.18)
(i) $N(s)^0 \leq D(s)^0$;

(ii) *poles and zeros of $Z(s)$ anywhere in the s plane, with the exception that the poles on the negative real axis are simple with real and positive residues;*

(iii) $Z(\infty) \geq 0$;

it can always be decomposed and partitioned as

$$(12.19) \qquad Z(s) = \frac{a_2(s)b_1(s) - a_1(s)b_2(s)}{a_2^2(s) - sb_2^2(s)}$$

where

$$\frac{a_i(s)}{sb_j(s)} \qquad (i = 1, 2; j = 1, 2)$$

are all passive RC driving-point impedance functions.

The decomposition and partitioning of an impedance function in the form of Equation 12.19 will be referred to as *odd part partitioning*. In a dual manner, by considering $F(s)$ as an admittance function and using $LC:RC$ transformation one obtains the *odd-part partitioning* of an admittance function as given by the following theorem.

Theorem 12-4[7] *If a real rational function $Y(s) = A(s)/B(s)$ satisfies the following conditions:*

(12.20)
(i) $A(s)^0 \leq B(s)^0 + 1$;

(ii) *poles and zeros of $Y(s)$ anywhere in the s plane, with the restriction that negative real axis poles of $Y(s)/s$ are simple with real and positive residues;*

(iii) $Y(0) \geq 0$;

[7] S. K. Mitra, *loc. cit.*

it can always be decomposed and partitioned as

(12.21) $$Y(s) = s\left[\frac{a_2(s)b_1(s) - a_1(s)b_2(s)}{a_2^2(s) - sb_2^2(s)}\right]$$

where

$$\frac{sb_i(s)}{a_j(s)} \quad (i, j = 1, 2)$$

are all passive RC driving-point admittance functions.

Example 12-2 Obtain an odd-part partitioning of the impedance function

(12.22) $$Z(s) = \frac{1 - s}{s^2 + 3s + 4}$$

Note that $Z(s)$ satisfied conditions 12.18 of Theorem 12-3.

We first form the equivalent LC impedance:

(12.23) $$\hat{Z}(s) = sZ(s^2) = \frac{s(1 - s^2)}{s^4 + 3s^2 + 4}$$

Our problem is now to find a positive real function

$$G(s) = (m_1 + n_1)/(m_2 + n_2)$$

so that

(12.24) $$\hat{Z}(s) = \text{Od } G(s) = \frac{m_2 n_1 - m_1 n_2}{m_2^2 - n_2^2}$$

We shall follow Gewartz's approach to determine $G(s)$ from the specified Od $G(s)$. From the poles of $\hat{Z}(s)$, $(m_2 + n_2)$ is first determined by grouping together the left-half s-plane poles. Thus, from Equation 12.23, we obtain

$$m_2 + n_2 = s^2 + s + 2$$

Let us arbitrarily choose

$$G(s) = \frac{m_1 + n_1}{m_2 + n_2} = \frac{a_2 s^2 + a_1 s + a_0}{s^2 + s + 2}$$

which implies that

$$m_2 n_1 - m_1 n_2 = (a_1 - a_2)s^3 + (2a_1 - a_0)s$$

Comparing the above expression with the numerator of Od $G(s)$, as given in Equation 12.23, we have the following relations:

$$a_2 - a_1 = 1$$
$$2a_1 - a_0 = 1$$

THE EVEN-PART PARTITIONING / 509

Note that there are many solutions of the above equations, many of which will not lead to positive real $G(s)$. If we set $a_2 = 0$, we get

$$G(s) = \frac{-s - 3}{s^2 + s + 2}$$

which definitely is not positive real. The minimum value of the Re $G(j\omega)$ is -2, and hence the minimum resistive p.r. function, whose odd part is Equation 12.23, is given by $G(s) + 2$, i.e.,

(12.25) $$\frac{-s - 3}{s^2 + s + 2} + 2 = \frac{2s^2 + s + 1}{s^2 + s + 2}$$

From the above, we observe that $m_1 = (2s^2 + 1)$, $n_1 = s$, $m_2 = s^2 + 2$, and $n_2 = s$. Thus:

$$a_1(s) = 2s + 1, \quad b_1(s) = 1$$
$$a_2(s) = s + 1, \quad b_2(s) = 1$$

Using these values in Equation 12.19, we obtain the odd part partitioning of the impedance of Equation 12.22 as

(12.26) $$Z(s) = \frac{1 - s}{s^2 + 3s + 4} = \frac{(s + 2)(1) - (1)(2s + 1)}{(s + 2)^2 - s(1)^2}$$

By adding any positive real constant to Equation 12.25, we can form many other p.r. functions whose odd parts are identical. For example, consider the following:

$$\frac{2s^2 + s + 1}{s^2 + s + 2} + 1 = \frac{3s^2 + 2s + 3}{s^2 + s + 2}$$

from which we identify $m_1 = (3s^2 + 3)$, $n_1 = 2s$. Thus

$$a_1(s) = 3s + 3, \quad b_1(s) = 2$$

Hence an alternate odd-part partitioning of Equation 12.22 is

(12.27) $$\frac{(s + 2)(2) - (1)(3s + 3)}{(s + 2)^2 - s(1)^2}$$

12-3 THE EVEN-PART PARTITIONING

An immediate consequence of the previous section will be to consider the even part of positive real function $G(s)$ in its decomposed and partitioned form, and to put it in an appropriate form for the active RC case by means of a suitable transformation.

First we consider the properties of the even part of a positive real function, which is presented without a proof in the form of Theorem 12-5.

Theorem 12-5[8] *The necessary and sufficient conditions for a real rational function $H(s) = P(s)/Q(s)$ to be the even part of a positive real function are:*

(12.28)
(i) $P(s)^0 \leq Q(s)^0$.

(ii) *Poles of $H(s)$ lie in quadrantal symmetry excluding the $j\omega$-axis.*

(iii) *Zeros of $H(s)$ lie in a quadrantal symmetry and $j\omega$-axis zeros are of even multiplicity.*

(iv) *If $\frac{1}{2}[P(s)^0 + Q(s)^0]$ is even, the leading coefficients of $P(s)$ and $Q(s)$ are of same sign; otherwise, they are of opposite sign.*

Given an even-part $H(s)$ of a p.r. function $G(s)$ satisfying conditions 12.28, the corresponding minimum reactive p.r. function $G(s)$ can be determined uniquely,[9] and we can then write

$$(12.29) \qquad H(s) = \text{Ev } G(s) = \frac{m_1 m_2 - n_1 n_2}{m_2^2 - n_2^2}$$

where

$$(12.30) \qquad G(s) = \frac{m_1 + n_1}{m_2 + n_2}$$

Since m_1 and m_2 are even polynomials and n_1 and n_2 are odd polynomials, we can express Equation 12.29 as

$$(12.31) \qquad H(s) = \frac{a_1(s^2)a_2(s^2) - s^2 b_1(s^2) b_2(s^2)}{a_2^2(s^2) - s^2 b_2^2(s^2)}$$

by using the identifications of expression (12.16). Applying a transformation

$$(12.32) \qquad s^2 \to s$$

to Equations 12.31 and 12.28, we arrive at the following theorem:

---

[8] D. F. Tuttle, *loc. cit.*
[9] For methods of constructing a p.r. function from a specified even part, see any text on network synthesis, e.g., H. Ruston and J. Bordogna, *loc. cit.*

Theorem 12-6[10] *If a real rational function $E(s) = N(s)/D(s)$ of total degree*[11] *r satisfies the following conditions:*

(12.33)
(i) $N(s)^0 \leq D(s)^0$;

(ii) $E(s)$ has no poles on the negative real axis and zeros of $E(s)$ on the negative real axis are of even multiplicity;

(iii) the leading coefficients of $N(s)$ and $D(s)$ are of same sign;

it can always be decomposed and partitioned as

(12.34a) $$E(s) = \frac{a_1(s)a_2(s) - sb_1(s)b_2(s)}{a_2^2(s) - sb_2^2(s)} \quad \text{if } r \text{ is even}$$

(12.34b) $$= -\frac{a_1(s)a_2(s) - sb_1(s)b_2(s)}{a_2^2(s) - sb_2^2(s)} \quad \text{if } r \text{ is odd}$$

where

$$\frac{a_i(s)}{sb_j(s)} \quad (i = 1, 2; j = 1, 2)$$

are passive RC driving-point impedances.

It should be pointed out here that conditions 12.33 are sufficient to decompose $E(s)$ in the form of Equation 12.34 but not necessary. We shall designate decomposition 12.34 appropriately as the *even-part partitioning*.

Example 12-3 Let

(12.35) $$E(s) = \frac{1}{s^3 - 1} = \frac{1}{(s-1)(s^2 + s + 1)}$$

Note that $E(s)$ satisfies conditions 12.33. Since the leading coefficients are of the same sign and the sum of numerator degree plus the denominator degree is odd, it is clear that $-E(s^2)$ satisfies the properties of the even part of a positive real function.

It can be shown that the minimum reactive p.r. function $G(s)$ corresponding to the even-part $-E(s^2)$ is given as

$$G(s) = \frac{\tfrac{2}{3}s^2 + \tfrac{4}{3}s + 1}{s^3 + 2s^2 + 2s + 1}$$

from which we identify

$$m_1 = \tfrac{2}{3}s^2 + 1, \quad n_1 = \tfrac{4}{3}s$$
$$m_2 = 2s^2 + 1, \quad n_2 = s^3 + 2s$$

---

[10] R. E. Thomas, *loc. cit.*
[11] Total degree $= N(s)^0 + D(s)^0$.

This implies that
$$a_1(s) = \tfrac{2}{3}s + 1, \qquad b_1(s) = \tfrac{4}{3}$$
$$a_2(s) = 2s + 1, \qquad b_2(s) = s + 2$$

Using the above quantities in expression 12.34b, we obtain the desired even-part partitioning of $E(s)$ of Equation 12.35 as

$$(12.36) \qquad E(s) = \frac{1}{s^3 - 1} = -\frac{(\tfrac{2}{3}s + 1)(2s + 1) - s(\tfrac{4}{3})(s + 2)}{(2s + 1)^2 - s(s + 2)^2}$$

12-4 SYNTHESIS OF DRIVING-POINT FUNCTIONS

The unique decomposition and the odd-part partitioning are in proper form for driving-point function realization. The identification of the

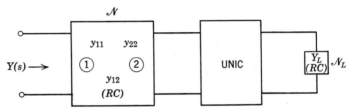

Figure 12-2 The RC:NIC cascade configuration.

parameters of the companion RC networks is straightforward and in general readily yields the network realizations. The active element in the synthesis in each case is the ideal impedance converter. We first consider the application of the unique decomposition.

Application of Unique Decomposition[12]

The synthesis procedures will be based on the cascade RC:NIC configuration of Kinariwala (Figure 9-17) and the one-port RC:GIC structures of Figures 9-21 and 9-22. A major aim of this section is to offer an alternate proof of Theorem 9-1 by showing that any real rational function can be realized in the form of Figure 9-17 without transformers.

Cascade RC:NIC Configuration. The input admittance of the cascade RC:NIC configuration (redrawn in Figure 12-2 for convenience) is given as

$$(12.37) \qquad Y(s) = y_{11} - \frac{y_{12}^2}{y_{22} - Y_L}$$

---

[12] Based on S. K. Mitra, "A unique synthesis method of transformerless active RC networks," *J. Franklin Inst.*, **274**, 115–129 (August 1962).

where y_{ij} represent the short-circuit parameters of the RC two-port $\mathcal{N}$ and Y_L is the admittance of the RC one-port $\mathcal{N}_L$. A specified driving-point admittance $Y(s)$ can be uniquely decomposed (Theorem 12-1) as

(12.9) $$Y(s) = \frac{a_1(s)a_2(s) - sb_1(s)b_2(s)}{a_2(s)b_1(s) - a_1(s)b_2(s)}$$

which can be reexpressed as

(12.38) $$Y(s) = \frac{a_1(s)}{b_1(s)} - \frac{\dfrac{a_1^2(s) - sb_1^2(s)}{b_1^2(s)}}{\dfrac{a_1(s)}{b_1(s)} - \dfrac{a_2(s)}{b_2(s)}}$$

Comparison of expressions 12.37 and 12.38 yields

$$y_{11} = y_{22} = \frac{a_1(s)}{b_1(s)}$$

(12.39) $$y_{12} = \pm \frac{\sqrt{a_1^2(s) - sb_1^2(s)}}{b_1(s)}$$

$$Y_L = \frac{a_2(s)}{b_2(s)}$$

Note that because of the properties of unique decomposition, y_{11}, y_{22}, and Y_L as given in expression 12.39 are passive RC driving-point admittances. For the realizability of the RC two-port $\mathcal{N}$, the residue condition must be satisfied. The residues of y_{11}, y_{22}, and y_{12} in a pole at $s = p_i$ [which is a zero of $b_1(s)$] are given as

(12.40) $$k_{11} = k_{22} = \frac{a_1(s)}{\dfrac{d}{ds}[b_1(s)]}\bigg|_{s=p_i}$$

$$k_{12} = \pm \frac{\sqrt{a_1^2(s) - sb_1^2(s)}}{\dfrac{d}{ds}[b_1(s)]}\bigg|_{s=p_i} = \pm \frac{a_1(s)}{\dfrac{d}{ds}[b_1(s)]}\bigg|_{s=p_i}$$

Observe that the residue condition is satisfied with an equal sign at each pole. Next, we must show that y_{12} can be made rational if required. In general, the numerator of y_{12}^2 will not be a perfect square. However, if it is not a perfect square, we can make it one by augmentation. To show this, we go back to the derivation of the unique decomposition as given in Section 12-1. Let us augment the numerator and the denominator of $\hat{Y}(s)$ of Equation 12.5 with a polynomial $(m_0^2 - n_0^2)$ where $(m_0 + n_0)$

is a Hurwitz polynomial. We thus have

$$\hat{Y}(s) = \frac{m_1 m_2 - n_1 n_2}{m_2 n_1 - m_1 n_2} \cdot \frac{m_0^2 - n_0^2}{m_0^2 - n_0^2} \tag{12.41}$$

$$= \frac{m_1' m_2' - n_1' n_2'}{m_2' n_1' - m_1' n_2'}$$

where

$$\begin{aligned} m_i' &= m_i m_0 + n_i n_0 \\ n_i' &= n_i m_0 + m_i n_0 \end{aligned} \qquad (i = 1, 2) \tag{12.42}$$

It is evident that $(m_i' + n_i') = (m_i + n_i)(m_0 + n_0)$ is also a Hurwitz polynomial. Hence m_i'/n_i' is a passive LC admittance function. The augmented unique decomposition is then given by

$$Y(s) = \frac{a_1'(s)a_2'(s) - sb_1'(s)b_2'(s)}{a_2'(s)b_1'(s) - a_1'(s)b_2'(s)} \tag{12.43}$$

where now

$$\begin{aligned} a_i'(s) &= a_i(s)a_0(s) + sb_i(s)b_0(s) \\ b_i'(s) &= b_i(s)a_0(s) + a_i(s)b_0(s) \end{aligned} \tag{12.44}$$

As a result, the y parameters and Y_L as given by expression 12.39 are modified as:

$$\begin{aligned} y_{11} &= y_{22} = \frac{a_1(s)a_0(s) + sb_1(s)b_0(s)}{b_1(s)a_0(s) + a_1(s)b_0(s)} \\ y_{12} &= \pm \frac{\sqrt{[a_1^2(s) - sb_1^2(s)][a_0^2(s) - sb_0^2(s)]}}{b_1(s)a_0(s) + a_1(s)b_0(s)} \\ Y_L &= \frac{a_2(s)a_0(s) + sb_2(s)b_0(s)}{b_2(s)a_0(s) + b_0(s)a_2(s)} \end{aligned} \tag{12.45}$$

The factor $[a_0^2(s) - sb_0^2(s)]$ in the expression for y_{12} in expression 12.45 can be chosen equal to that factor of $[a_1^2(s) - sb_1^2(s)]$, which is not a perfect square, and thus make y_{12} a rational function.

A lattice realization (Figure 12-3) of $\mathcal{N}$ is always possible, since the residues of the y parameters are all equal in magnitude. This proves Theorem 9-1.

Example 12-4 We consider the realization of an inductance of 8 henries, i.e., $Y(s) = 1/8s$. The unique decomposition of this function is given in Equation 12.12. We first proceed to check whether $a_1^2(s) - sb_1^2(s)$ is a perfect square. From expression 12.11, we have

$$a_1^2(s) - sb_1^2(s) = 1 - 4s$$

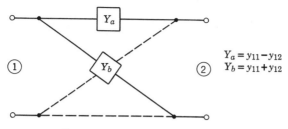

Figure 12-3 The symmetrical lattice.

which is not a perfect square. We thus choose

$$a_0^2(s) - sb_0^2(s) = 1 - 4s$$

i.e.,

(12.46)
$$a_0(s) = a_1(s) = 1$$
$$b_0(s) = b_1(s) = 2$$

Using expressions 12.46 and 12.11 in expressions 12.45, we obtain the desired expressions for the companion network parameters:

(12.47)
$$y_{11} = y_{22} = \left(\frac{4s+1}{4}\right)$$
$$y_{12} = \pm\left(\frac{4s-1}{4}\right)$$
$$Y_L = \frac{8s+1}{4(2s+1)}$$

Using the plus sign for y_{12}, we obtain the following identifications for the series and cross branches of the lattice:

$$Y_a = y_{11} - y_{12} = \frac{4s+1}{4} - \frac{4s-1}{4} = \tfrac{1}{2}$$

$$Y_b = y_{11} + y_{12} = \frac{4s+1}{4} + \frac{4s-1}{4} = 2s$$

The final complete realization is indicated in Figure 12-4.

One drawback of the final configuration is the lattice realization of the RC two-port N. As a result, input terminals may not have a common ground with the active element. However, in some cases, instead of realizing N in a lattice structure, an unbalanced realization is possible by following Ozaki's method.[13]

[13] N. Balabanian, *Network Synthesis*, Prentice-Hall, Englewood Cliffs, N.J., 1958.

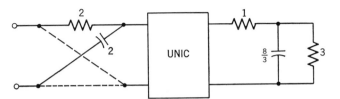

Figure 12-4 Realization of the driving-point impedance of Example 12-1 using the unique decomposition.

RC:*GIC Configuration*. The *RC*:Generalized-Impedance-Converter (*RC*:*GIC*) structures of Section 9-2, p. 370, are also suitable for synthesizing an arbitrary real rational function with the aid of unique decomposition.[14] The companion *RC* networks, which are one-port *RC* networks, are always guaranteed realizable without any augmentation.

Consider the first *RC*:*GIC* structure, which has been redrawn in Figure 12-5 for convenience. The input admittance of this active network is given as

(12.48) $$Y(s) = \frac{Y_1 Y_3 - Y_2 Y_4}{Y_3 - Y_4}$$

A specified driving-point admittance $Y(s)$ can be uniquely decomposed in the form

(12.9) $$Y(s) = \frac{a_1(s)a_2(s) - sb_1(s)b_2(s)}{a_2(s)b_1(s) - a_1(s)b_2(s)}$$

which can be reorganized as

(12.49) $$Y(s) = \frac{\dfrac{a_1(s)}{b_1(s)} \cdot \dfrac{a_2(s)}{b_2(s)} - s}{\dfrac{a_2(s)}{b_2(s)} - \dfrac{a_1(s)}{b_1(s)}}$$

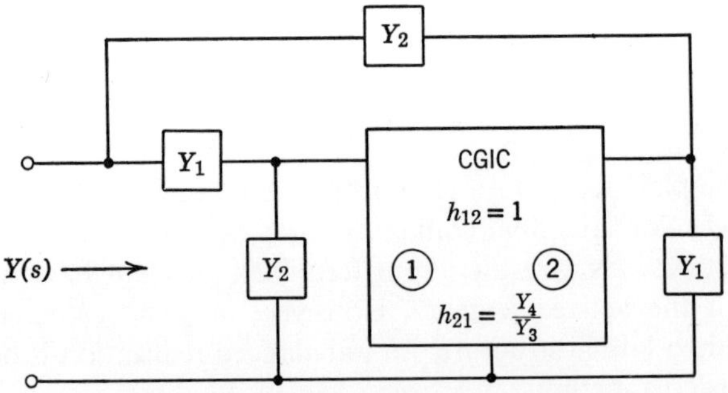

Figure 12-5 The *RC*:*GIC* configuration.

[14] See also W. Saraga, *loc. cit.*

Comparing Equations 12.48 and 12.49, we identify

(12.50)
$$Y_1 = \frac{a_1(s)}{b_1(s)}$$
$$Y_2 = \frac{sb_1(s)}{a_1(s)}$$
$$Y_3 = \frac{a_2(s)}{b_2(s)}$$
$$Y_4 = \frac{a_1(s)}{b_1(s)}$$

Because of unique decomposition, Y_1 through Y_4 as given in Equation 12.50 are *RC* realizable driving-point admittances. We illustrate the approach by means of an example.

Example 12-5 Realize

(12.51)
$$Y(s) = \frac{N(s)}{D(s)} = s^2$$

We first form

$$sD(s^2) - N(s^2) = s - s^4 = s(1 - s)(s^2 + s + 1)$$

Two possible identifications are possible depending on whether the factor s is considered a part of $(m_1 - n_1)$ or $(m_2 + n_2)$, i.e.,

Case A.

(12.52a) $m_1 - n_1 = 1 - s,$ $m_2 + n_2 = s(s^2 + s + 1)$

Case B.

(12.52b) $m_1 - n_1 = s^2 - s,$ $m_2 + n_2 = s^2 + s + 1$

Consider Case A first. From expression 12.52a we obtain

$$m_1 = a_1(s^2) = 1$$
$$n_1 = sb_1(s^2) = s$$
$$m_2 = a_2(s^2) = s^2$$
$$n_2 = sb_2(s^2) = s^3 + s$$

This implies that

(12.53)
$$a_1(s) = 1$$
$$b_1(s) = 1$$
$$a_2(s) = s$$
$$b_2(s) = s + 1$$

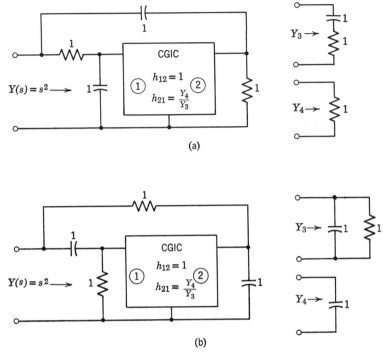

Figure 12-6 Two RC:GIC realizations of the driving-point impedance of Example 12-2.

Using these identifications in expression 12.50, we obtain

(12.54)
$$Y_1 = 1$$
$$Y_2 = s$$
$$Y_3 = \frac{s}{s+1}$$
$$Y_4 = 1$$

and the corresponding realization of $Y(s)$ is shown in Figure 12-6a.

Next, we consider Case B of expression 12.52b. We identify

$$m_1 = a_1(s^2) = s^2$$
$$n_1 = sb_1(s^2) = s$$
$$m_2 = a_2(s^2) = s^2 + 1$$
$$n_2 = sb_2(s^2) = s$$

Hence

(12.55)
$$a_1(s) = s$$
$$b_1(s) = 1$$
$$a_2(s) = s + 1$$
$$b_2(s) = 1$$

The expressions for the companion RC networks are easily obtained from expression 12.50 and are given as:

(12.56)
$$Y_1 = s$$
$$Y_2 = 1$$
$$Y_3 = s + 1$$
$$Y_4 = s$$

The alternate realization of $Y(s) = s^2$ is indicated in Figure 12-6b. Comparing the two realizations, we observe that with the Case A identifications, the final realization of the admittance function of Equation 12.51 has one less capacitor and is preferable from a practical point of view.

The second RC:GIC structure of Figure 9-22 is also suitable for synthesis, using unique decomposition technique. To show this, we note that the input admittance of these structures is given by Equation 9.38, repeated below for convenience

(12.57)
$$Y(s) = \frac{Y_7 - Y_8}{\dfrac{Y_7}{Y_6} - \dfrac{Y_8}{Y_5}}$$

To facilitate the realization, we rewrite the unique decomposition of a specified $Y(s)$ in the following form:

(12.58)
$$Y(s) = \frac{\dfrac{a_2(s)}{b_2(s)} - \dfrac{sb_1(s)}{a_1(s)}}{\dfrac{a_2(s)}{b_2(s)} \cdot \dfrac{b_1(s)}{a_1(s)} - 1}$$

Comparing Equations 12.57 and 12.58, we identify:

(12.59)
$$Y_7 = \frac{a_2(s)}{b_2(s)}$$
$$Y_8 = \frac{sb_1(s)}{a_1(s)}$$
$$Y_6 = \frac{a_1(s)}{b_1(s)}$$
$$Y_5 = Y_8 = \frac{sb_1(s)}{a_1(s)}$$

As before, unique decomposition guarantees the *RC* realizability of Y_5, Y_6, Y_7, and Y_8.

Application of Odd-Part Partitioning

The odd-part partitioning (Equations 12.19 and 12.21) is also directly applicable to the cascade *RC*:NIC configuration and the *RC*:GIC structure.

Cascade RC:NIC Configuration.[15] We shall first show that a real rational function that is equal to the odd part of a p.r. function can be realized as the input impedance (admittance) of a cascade *LC*:NIC network. Then, applying the *LC*:*RC* transformation, desired results for the active *RC* case will be derived.

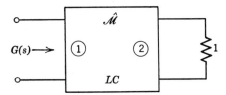

Figure 12-7 The Darlington configuration for realization of a positive real impedance $G(s)$.

Consider the Darlington realization of a p.r. impedance $G(s)$:[16]

$$(12.13) \qquad G(s) = \frac{m_1 + n_1}{m_2 + n_2}$$

as shown in Figure 12-7. In terms of the z parameters of the lossless two-port $\hat{\mathcal{M}}$ we can write

$$(12.60) \qquad G(s) = z_{11}(s) - \frac{z_{12}^2(s)}{z_{22}(s) + 1}$$

The parameters of the lossless two-port are given as follows:

Case A.

$$(12.61) \qquad \begin{aligned} z_{11} &= \frac{m_1}{n_2} \\ z_{22} &= \frac{m_2}{n_2} \\ z_{12} &= \frac{\pm\sqrt{m_1 m_2 - n_1 n_2}}{n_2} \end{aligned}$$

[15] Based on S. K. Mitra, "A new approach to active *RC* network synthesis," *J. Franklin Inst.*, **274**, 185–197 (September 1962). See also R. E. Thomas, *loc. cit.*

[16] See, for example, M. E. Van Valkenburg, *Modern network synthesis*, Wiley, New York, 1962.

Case B.

(12.62)
$$z_{11} = \frac{n_1}{m_2}$$
$$z_{22} = \frac{n_2}{m_2}$$
$$z_{12} = \frac{\pm\sqrt{n_1 n_2 - m_1 m_2}}{m_2}$$

The Case A identification is followed if the polynomial $\sqrt{m_1 m_2 - n_1 n_2}$ is an even polynomial; Case B identification is used if it is odd.

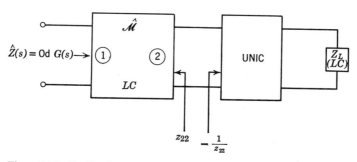

Figure 12-8 Realization of odd part of a positive real impedance by means of a cascade LC:NIC configuration.

The odd part of $G(s)$ can be computed from Equation 12.60 and is given as

(12.63)
$$\text{Od } G(s) = \tfrac{1}{2}[G(s) - G(-s)]$$
$$= z_{11}(s) - \frac{z_{12}^2(s) z_{22}(s)}{z_{22}^2(s) - 1}$$

where we made use of the fact that $z_{ij}(s)$ are odd functions.

Let us replace the 1-ohm resistor of Figure 12-7 with a negative LC impedance, $-\hat{Z}_L(s)$, where

(12.64)
$$-\hat{Z}_L(s) = -\frac{1}{z_{22}(s)}$$

The input impedance of this new network (Figure 12-8) is then given by

(12.65)
$$\hat{Z}(s) = z_{11}(s) - \frac{z_{12}^2(s)}{z_{22}(s) - \dfrac{1}{z_{22}(s)}}$$
$$= z_{11}(s) - \frac{z_{12}^2(s) z_{22}(s)}{z_{22}^2(s) - 1}$$

which is identical to Od $G(s)$ as given by Equation 12.63. In terms of the polynomials of $G(s)$, we thus have

(12.66) $$\hat{Z}(s) = \text{Od } G(s) = \frac{m_2 n_1 - m_1 n_2}{m_2{}^2 - n_2{}^2}$$

We can summarize our results as follows. Given a function $\hat{Z}(s)$, which can be identified as the odd part of a p.r. impedance $G(s)$, the realization of $\hat{Z}(s)$ is arrived at by first realizing $G(s)$ by the Darlington procedure and then replacing the 1-ohm resistor by a negative LC impedance equal to $-1/z_{22}(s)$.

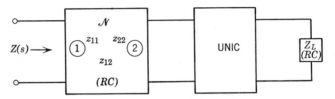

Figure 12-9 Cascade RC:NIC configuration.

The cascade RC:NIC realization of an odd-part partitionable impedance $Z(s)$, i.e., $Z(s)$ satisfies condition 12.18 of Theorem 12-3, is now evident. The desired realization Figure 12-9 is obtained by replacing each inductor in the realization of $\hat{Z}(s) = \text{Od } G(s)$ by a resistor of equal value. The parameters characterizing the companion RC network can be derived from expressions 12.61 and 12.62 by an LC:RC transformation and using the following notations:

(12.16) $\qquad m_i = a_i(s^2), \qquad n_i = sb_i(s^2) \qquad i = 1, 2$

We thus have

Case A.

(12.67)
$$z_{11} = \frac{a_1(s)}{sb_2(s)}$$
$$z_{22} = \frac{a_2(s)}{sb_2(s)}$$
$$z_{12} = \frac{\pm\sqrt{a_1(s)a_2(s) - sb_1(s)b_2(s)}}{sb_2(s)}$$
$$Z_L = \frac{b_2(s)}{a_2(s)}$$

Case B.

$$z_{11} = \frac{b_1(s)}{a_2(s)}$$

$$z_{22} = \frac{b_2(s)}{a_2(s)}$$

(12.68)

$$z_{12} = \frac{\pm\sqrt{b_1(s)b_2(s) - \frac{1}{s}a_1(s)a_2(s)}}{a_2(s)}$$

$$Z_L = \frac{a_2(s)}{sb_2(s)}$$

The mutually exclusive cases, Case A and Case B, are determined by the evenness and oddness of the polynomial $\sqrt{a_1(s^2)a_2(s^2) - s^2b_1(s^2)b_2(s^2)}$. The usual difficulty of making the numerator of z_{12}^2 a perfect square applies. In general, augmentation of the p.r. function $G(s)$ can always be carried out to alleviate this problem. However, an alternate solution may be followed in some instances. Since the odd part is only specified, any arbitrary positive and real constant can be added to $G(s)$ without modifying its odd part. For a suitable constant, often the numerator of the new even part can be made a perfect square.

A similar development can be carried out on an admittance basis. Thus, a real rational function $Y(s)$ satisfying conditions 12.20 of Theorem 12-4 can be realized as the input admittance of the cascaded RC:NIC structure shown in Figure 12-2. The pertinent parameters are given as follows:

Case A.

$$y_{11} = \frac{a_1(s)}{b_2(s)}$$

$$y_{22} = \frac{a_2(s)}{b_2(s)}$$

(12.69)

$$y_{12} = \frac{\pm\sqrt{a_1(s)a_2(s) - sb_1(s)b_2(s)}}{b_2(s)}$$

$$Y_L = \frac{sb_2(s)}{a_2(s)}$$

Case B.

(12.70)
$$y_{11} = \frac{sb_1(s)}{a_2(s)}$$
$$y_{22} = \frac{sb_2(s)}{a_2(s)}$$
$$y_{12} = \frac{\pm\sqrt{s\{sb_1(s)b_2(s) - a_1(s)a_2(s)\}}}{a_2(s)}$$
$$Y_L = \frac{a_2(s)}{b_2(s)}$$

In general, the two-port network realized using the parameters of expression 12.67, 12.68, 12.69, or 12.70 will contain an ideal transformer. However, in some cases, the ideal transformer can be eliminated by following the technique outlined in Section 9-2, p. 368. We illustrate the method by means of an example.

Example 12-6 Obtain a cascade RC:NIC realization of the impedance function

(12.22)
$$Z(s) = \frac{1-s}{s^2 + 3s + 4}$$

Two odd-part partitionings of $Z(s)$ have been calculated in Example 12-2. Consider the decomposition of $Z(s)$ in the way indicated in Equation 12.26. Thus
$$a_1(s) = 2s + 1, \quad b_1(s) = 1$$
$$a_2(s) = s + 2, \quad b_2(s) = 1$$

First we must determine whether a rational z_{12} can be obtained. Note that
$$A(s^2) = a_1(s^2)a_2(s^2) - s^2 b_1(s^2)b_2(s^2)$$
$$= (2s^2 + 1)(s^2 + 2) - s^2 = 2(s^2 + 1)^2$$

is a perfect square; hence no augmentation is necessary. In addition, $\sqrt{A(s^2)}$ is an even polynomial, implying Case A identification must be used. Thus, from expression 12.67 we have

$$z_{11} = \frac{2s+1}{s}$$
$$z_{22} = \frac{s+2}{s}$$
$$z_{12} = \frac{\sqrt{2}(s+1)}{s}$$
$$Z_L = \frac{1}{s+2}$$

To eliminate the transformer,[17] z_{11} and z_{12} are first developed into an *RC* two-port network as shown in Figure 12-10a. The resultant two-port $\mathcal{N}'$ has the following open-circuit parameters:

$$z_{11} = \frac{2s+1}{s}$$

$$z'_{22} = \frac{s+1}{s}$$

$$z'_{12} = \frac{s+1}{s} = \alpha z_{12}$$

where $\alpha = 1/\sqrt{2}$. This indicates that by connecting in series at the output port of $\mathcal{N}'$ an impedance

$$\alpha^2 z_{22} - z'_{22} = \frac{s+2}{2s} - \frac{s+1}{s} = -\frac{1}{2}$$

we shall have a two-port having the parameters z_{11}, αz_{12}, and $\alpha^2 z_{22}$. Terminating this new two-port by a negative load impedance, $-\alpha^2 Z_L$, we shall have the realization of Equation 12.22. These steps are sketched in Figure 12-10b. Finally, by moving the negative resistance to the opposite side of the NIC, we have the desired realization as shown in Figure 12-10c.

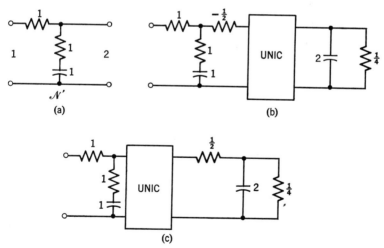

Figure 12-10 Realization of the driving-point impedance of Example 12-6, using odd-part partitioning.

---

[17] See pp. 368 and 370.

An interesting property of the odd-part partitioning is the form of the decomposition of the denominator polynomial. When applied to cascade RC:NIC configuration, it is clear that the denominator has the form of Horowitz decomposition (Equation 9.76). Hence, if the network is realized directly as indicated by expression 12.67, 12.68, 12.69, or 12.70, the final network has a minimum pole sensitivity with respect to the conversion factor of the NIC.

RC:GIC Structures. We now consider the application of odd-part partitioning to the RC:GIC one-port structures.[18] An odd-part partitionable admittance can be realized in the form of Figure 12-5 as follows. The decomposition of $Y(s)$, as given by Equation 12.21, is recast as

(12.71) $$Y(s) = \frac{\dfrac{sb_1(s)}{a_2(s)} \cdot \dfrac{a_2(s)}{b_2(s)} - \dfrac{a_1(s)}{b_2(s)} \cdot \dfrac{sb_2(s)}{a_2(s)}}{\dfrac{a_2(s)}{b_2(s)} - \dfrac{sb_2(s)}{a_2(s)}}$$

Comparing the above with the input admittance (expression 12.48) we identify

(12.72)
$$Y_3 = \frac{a_2(s)}{b_2(s)}$$
$$Y_4 = \frac{sb_2(s)}{a_2(s)}$$
$$Y_1 = \frac{sb_1(s)}{a_2(s)}$$
$$Y_2 = \frac{a_1(s)}{b_2(s)}$$

Note that the one-ports are all guaranteed RC realizable.

A similar comparison of the odd-part partitioning of an impedance with the input impedance of the second configuration (Figure 9-21) can be carried out. This is left as an exercise.

Cascade-Feedback RC:NIC Configuration. Rohrer[19] has shown that an odd part partitionable admittance can be easily realized in the form of the cascade-feedback RC:NIC configuration (Figure 9-23) of Sipress.

---

[18] S. K. Mitra, "Notes on Sandberg's method of active RC one-port synthesis," *IRE Trans. on Circuit Theory*, **CT-9**, 422–423 (December 1962).

[19] R. A. Rohrer, "Minimum sensitivity RC-NIC driving-point synthesis," *IEEE Trans. on Circuit Theory*, **CT-10**, 442–443 (September 1963).

We refer the interested reader to the original paper for the method of identification.

12-5 SYNTHESIS OF TRANSFER FUNCTIONS

Several interesting applications of the three types of decomposition will be considered in this section.

Application of the Unique Decomposition

The unique decomposition can be easily used to realize transfer impedance by means of a modified Linvill type $RC:NIC$ configuration.[20]

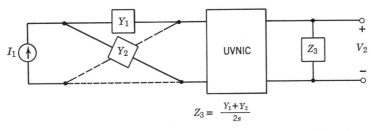

Figure 12-11 An alternate form of Linvill's cascade $RC:NIC$ configuration for transfer-impedance realization.

The proposed modification is shown in Figure 12-11. The transfer impedance of this two-port is

$$\text{(12.73)} \qquad \frac{V_2}{I_1} = \frac{1}{2}\left(\frac{Y_1 - Y_2}{s - Y_1 Y_2}\right)$$

where Y_1 and Y_2 are required to be passive RC admittances.

Let $Z_{21}(s)$ be a specified real rational transfer impedance function. Following the methods outlined in Section 12-1, p. 503, we can uniquely decompose $2Z_{21}(s)$ as

$$\text{(12.74)} \qquad 2Z_{21}(s) = \frac{a_2(s)b_1(s) - a_1(s)b_2(s)}{a_1(s)a_2(s) - sb_1(s)b_2(s)}$$

which can be reexpressed as

$$\text{(12.75)} \qquad Z_{21}(s) = \frac{1}{2}\left[\frac{\dfrac{a_1(s)}{b_1(s)} - \dfrac{a_2(s)}{b_2(s)}}{s - \dfrac{a_1(s)}{b_1(s)} \cdot \dfrac{a_2(s)}{b_2(s)}}\right]$$

[20] First pointed out by W. Saraga, *loc. cit.*

Comparing expressions 12.73 and 12.75, we obtain

$$Y_1 = \frac{a_1(s)}{b_1(s)}$$

(12.76)

$$Y_2 = \frac{a_2(s)}{b_2(s)}$$

Note that

(12.77) $$Z_3 = \frac{1}{2}\left(\frac{Y_1}{s} + \frac{Y_2}{s}\right) = \frac{1}{2}\left(\frac{a_1(s)}{sb_1(s)} + \frac{a_2(s)}{sb_2(s)}\right)$$

is also a passive RC realizable impedance function.

Application of Odd-Part Partitioning

A drawback of the structure of Figure 12-11 is that input and output parts do not have a common terminal. This situation does not arise in the modified Yanagisawa type RC:NIC configuration, which we consider next.[21]

For the modified Yanagisawa configuration of Figure 12-12, the voltage-transfer ratio is given as

(12.78) $$t_v = \frac{V_2}{V_1}\bigg|_{I_2=0} = \frac{-y_{12a} + y_{12b}}{y_{22a} - y_{22b}}$$

A specified voltage transfer ratio $t_v = N(s)/D(s)$ satisfying conditions 12.18 of Theorem 12-3 can be decomposed and partitioned in the form of Equation 12.19:

$$t_v = \frac{a_2(s)b_1(s) - a_1(s)b_2(s)}{a_2^2(s) - sb_2^2(s)}$$

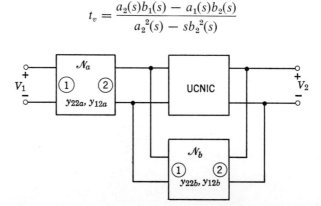

Figure 12-12 The generalized Yanagisawa RC:NIC configuration.

---

[21] G. Branner, "On active networks using current inversion type negative-impedance converters," *IEEE Trans. on Circuit Theory*, **CT-10**, 290 (June 1963).

which can be rearranged as

(12.79)
$$t_v = \frac{\dfrac{b_1(s)}{b_2(s)} - \dfrac{a_1(s)}{a_2(s)}}{\dfrac{a_2(s)}{b_2(s)} - \dfrac{sb_2(s)}{a_2(s)}}$$

Comparing Equations 12.78 and 12.79, we can identify

$$y_{22a} = \frac{a_2(s)}{b_2(s)}, \qquad -y_{12a} = \frac{b_1(s)}{b_2(s)}$$

$$y_{22b} = \frac{sb_2(s)}{a_2(s)}, \qquad -y_{12b} = \frac{a_1(s)}{a_2(s)}$$

Since the zeros of y_{12a} and y_{12b} are on the negative real axis, the networks $\mathcal{N}_a$ and $\mathcal{N}_b$ can each be developed as ladder networks. Realization of $t_v = N(s)/D(s)$ is completed by connecting these two networks as indicated in Figure 12-11 after proper admittance level adjustments have been made.

Application of Even-Part Partitioning

In a similar manner, we can show that a certain class of functions can be realized in the form of the modified Yanagisawa configuration by applying the even-part partitioning.[22]

Let $t_v = N(s)/D(s)$ be a real rational function of even total degree and satisfying conditions 12.33. Then we can express

$$t_v = \frac{a_1(s)a_2(s) - sb_1(s)b_2(s)}{a_2^2(s) - sb_2^2(s)}$$

which can be rearranged as

(12.80)
$$t_v = \frac{\dfrac{a_1(s)}{b_2(s)} - \dfrac{sb_1(s)}{a_2(s)}}{\dfrac{a_2(s)}{b_2(s)} - \dfrac{sb_2(s)}{a_2(s)}}$$

Comparison of Equations 12.78 and 12.80 leads to

$$y_{22a} = \frac{a_2(s)}{b_2(s)}, \qquad -y_{12a} = \frac{a_1(s)}{b_2(s)}$$

$$y_{22b} = \frac{sb_2(s)}{a_2(s)}, \qquad -y_{12b} = \frac{sb_1(s)}{a_2(s)}$$

Ladder realization of $\mathcal{N}_a$ and $\mathcal{N}_b$ is again seen to be possible.

---

[22] G. Branner, *loc. cit.*

Additional References

Branner, G. R., "On methods of polynomial decomposition in active network synthesis," *IEEE Trans. on Circuit Theory*, **CT-10,** 525–526 (December 1963).

Branner, G. R., "Active network synthesis—polynomial decomposition approach," Tech. Documentary Rept. ASD-TDR-62-1071, Aeronautical Systems Div., Air Force Systems Command, WPAFB, Ohio.

Ramachandran, V. and M. N. S. Swamy, "Generalized gyrator and driving point function synthesis," Proc. Asilomar Conf. on Circuits and Systems, Monterey, Calif., November 1967.

Thomas, R. E., "Miyata's method applied to active network synthesis," *Proc. NEC*, **18,** 40–44 (1962).

Appendix

A / *Useful Network Descriptions*

Table A-1 TWO-PORT NETWORK PARAMETERS

Description	Matrix Relation	Matrix Designation
Open-circuit impedance or z-parameters	$\begin{bmatrix} V_1 \\ V_2 \end{bmatrix} = \begin{bmatrix} z_{11} & z_{12} \\ z_{21} & z_{22} \end{bmatrix} \begin{bmatrix} I_1 \\ I_2 \end{bmatrix}$	$\mathbf{Z}, z_{ij}$
Short-circuit admittance or y-parameters	$\begin{bmatrix} I_1 \\ I_2 \end{bmatrix} = \begin{bmatrix} y_{11} & y_{12} \\ y_{21} & y_{22} \end{bmatrix} \begin{bmatrix} V_1 \\ V_2 \end{bmatrix}$	$\mathbf{Y}, y_{ij}$
h-parameters	$\begin{bmatrix} V_1 \\ I_2 \end{bmatrix} = \begin{bmatrix} h_{11} & h_{12} \\ h_{21} & h_{22} \end{bmatrix} \begin{bmatrix} I_1 \\ V_2 \end{bmatrix}$	$\mathbf{H}, h_{ij}$
g-parameters	$\begin{bmatrix} I_1 \\ V_2 \end{bmatrix} = \begin{bmatrix} g_{11} & g_{12} \\ g_{21} & g_{22} \end{bmatrix} \begin{bmatrix} V_1 \\ I_2 \end{bmatrix}$	$\mathbf{G}, g_{ij}$
Transmission parameters	$\begin{bmatrix} V_1 \\ I_1 \end{bmatrix} = \begin{bmatrix} A & B \\ C & D \end{bmatrix} \begin{bmatrix} V_2 \\ -I_2 \end{bmatrix}$	$\mathbf{F}$

Table A-2 TWO-PORT PARAMETER RELATIONS

	Y		H		G		F	
Given $\mathbf{Z} = \begin{bmatrix} z_{11} & z_{12} \\ z_{21} & z_{22} \end{bmatrix}$	$\dfrac{z_{22}}{\Delta z}$	$-\dfrac{z_{12}}{\Delta z}$	$\dfrac{\Delta z}{z_{22}}$	$\dfrac{z_{12}}{z_{22}}$	$\dfrac{1}{z_{11}}$	$-\dfrac{z_{12}}{z_{11}}$	$\dfrac{z_{11}}{z_{21}}$	$\dfrac{\Delta z}{z_{21}}$
	$-\dfrac{z_{21}}{\Delta z}$	$\dfrac{z_{11}}{\Delta z}$	$-\dfrac{z_{21}}{z_{22}}$	$\dfrac{1}{z_{22}}$	$\dfrac{z_{21}}{z_{11}}$	$\dfrac{\Delta z}{z_{11}}$	$\dfrac{1}{z_{21}}$	$\dfrac{z_{22}}{z_{21}}$

	Z		H		G		F	
Given $\mathbf{Y} = \begin{bmatrix} y_{11} & y_{12} \\ y_{21} & y_{22} \end{bmatrix}$	$\dfrac{y_{22}}{\Delta y}$	$-\dfrac{y_{12}}{\Delta y}$	$\dfrac{1}{y_{11}}$	$-\dfrac{y_{12}}{y_{11}}$	$\dfrac{\Delta y}{y_{22}}$	$\dfrac{y_{12}}{y_{22}}$	$-\dfrac{y_{22}}{y_{21}}$	$-\dfrac{1}{y_{21}}$
	$-\dfrac{y_{21}}{\Delta y}$	$\dfrac{y_{11}}{\Delta y}$	$\dfrac{y_{21}}{y_{11}}$	$\dfrac{\Delta y}{y_{11}}$	$-\dfrac{y_{21}}{y_{22}}$	$\dfrac{1}{y_{22}}$	$-\dfrac{\Delta y}{y_{21}}$	$-\dfrac{y_{11}}{y_{21}}$

	Z		Y		G		F	
Given $\mathbf{H} = \begin{bmatrix} h_{11} & h_{12} \\ h_{21} & h_{22} \end{bmatrix}$	$\dfrac{\Delta h}{h_{22}}$	$\dfrac{h_{12}}{h_{22}}$	$\dfrac{1}{h_{11}}$	$-\dfrac{h_{12}}{h_{11}}$	$\dfrac{h_{22}}{\Delta h}$	$-\dfrac{h_{12}}{\Delta h}$	$-\dfrac{\Delta h}{h_{21}}$	$-\dfrac{h_{11}}{h_{21}}$
	$-\dfrac{h_{21}}{h_{22}}$	$\dfrac{1}{h_{22}}$	$\dfrac{h_{21}}{h_{11}}$	$\dfrac{\Delta h}{h_{11}}$	$-\dfrac{h_{21}}{\Delta h}$	$\dfrac{h_{11}}{\Delta h}$	$-\dfrac{h_{22}}{h_{21}}$	$-\dfrac{1}{h_{21}}$

	Z		Y		H		F	
Given $\mathbf{G} = \begin{bmatrix} g_{11} & g_{12} \\ g_{21} & g_{22} \end{bmatrix}$	$\dfrac{1}{g_{11}}$	$-\dfrac{g_{12}}{g_{11}}$	$\dfrac{\Delta g}{g_{22}}$	$-\dfrac{g_{12}}{g_{22}}$	$\dfrac{g_{22}}{\Delta g}$	$-\dfrac{g_{12}}{\Delta g}$	$\dfrac{1}{g_{21}}$	$\dfrac{g_{22}}{g_{21}}$
	$\dfrac{g_{21}}{g_{11}}$	$\dfrac{\Delta g}{g_{11}}$	$-\dfrac{g_{21}}{g_{22}}$	$\dfrac{1}{g_{22}}$	$-\dfrac{g_{21}}{\Delta g}$	$\dfrac{g_{11}}{\Delta g}$	$\dfrac{g_{11}}{g_{21}}$	$\dfrac{\Delta g}{g_{21}}$

	Z		Y		H		G	
Given $\mathbf{F} = \begin{bmatrix} A & B \\ C & D \end{bmatrix}$	$\dfrac{A}{C}$	$\dfrac{\Delta F}{C}$	$\dfrac{D}{B}$	$-\dfrac{\Delta F}{B}$	$\dfrac{B}{D}$	$\dfrac{\Delta F}{D}$	$\dfrac{C}{A}$	$-\dfrac{\Delta F}{A}$
	$\dfrac{1}{C}$	$\dfrac{D}{C}$	$-\dfrac{1}{B}$	$\dfrac{A}{B}$	$-\dfrac{1}{D}$	$\dfrac{C}{D}$	$\dfrac{1}{A}$	$\dfrac{B}{A}$

$\Delta z = z_{11}z_{12} - z_{12}z_{21}; \quad \Delta y = y_{11}y_{22} - y_{12}y_{21}$
$\Delta h = h_{11}h_{22} - h_{12}h_{21}; \quad \Delta g = g_{11}g_{22} - g_{12}g_{21}$
$\Delta F = AD - BC$

Table A-3 SOME USEFUL NETWORK SPECIFICATIONS

Relevant Diagram	Network Function	Relation
	$\dfrac{V_2}{I_1} = Z_{21}$	$\dfrac{z_{21}R}{z_{22} + R}$
	$\dfrac{I_2}{V_1} = Y_{21}$	$\dfrac{y_{21}/R}{y_{22} + \dfrac{1}{R}}$
	$\dfrac{V_2}{V_1} = t_v$	$\dfrac{z_{21}}{z_{11}} = -\dfrac{y_{21}}{y_{22}}$
	$\dfrac{I_2}{I_1} = t_I$	$-\dfrac{z_{21}}{z_{22}} = \dfrac{y_{21}}{y_{11}}$
	$\dfrac{V_2}{V_0} = t_v$	$\dfrac{z_{21}}{z_{11} + R}$
	$\dfrac{I_2}{I_0} = T_I$	$\dfrac{y_{21}}{y_{11} + \dfrac{1}{R}}$
	$\dfrac{V_2}{V_0} = T_V$	$\dfrac{z_{21}R_1}{(z_{11} + R_1)(z_{22} + R_2) - z_{12}z_{21}}$
	$\dfrac{V_1}{I_1} = Z_{\text{in}}$	$z_{11} - \dfrac{z_{12}z_{21}}{z_{22} + Z_L} = z_{11}\dfrac{\dfrac{1}{y_{11}} + Z_L}{z_{22} + Z_L}$

Adapted from E. S. Kuh and D. O. Pederson, *Principles of circuit synthesis*, McGraw-Hill, New York, 1959.

Table A-4 *FORMULAS FOR INTERCONNECTED TWO-PORTS*

Type of Connection	Relevant Diagram	Parameter Relation
Series		$\begin{bmatrix} z_{11} & z_{12} \\ z_{21} & z_{22} \end{bmatrix} = \begin{bmatrix} z_{11a} & z_{12a} \\ z_{21a} & z_{22a} \end{bmatrix} + \begin{bmatrix} z_{11b} & z_{12b} \\ z_{21b} & z_{22b} \end{bmatrix}$
Parallel		$\begin{bmatrix} y_{11} & y_{12} \\ y_{21} & y_{22} \end{bmatrix} = \begin{bmatrix} y_{11a} & y_{12a} \\ y_{21a} & y_{22a} \end{bmatrix} + \begin{bmatrix} y_{11b} & y_{12b} \\ y_{21b} & y_{22b} \end{bmatrix}$
Series-Parallel		$\begin{bmatrix} h_{11} & h_{12} \\ h_{21} & h_{22} \end{bmatrix} = \begin{bmatrix} h_{11a} & h_{12a} \\ h_{21a} & h_{22a} \end{bmatrix} + \begin{bmatrix} h_{11b} & h_{12b} \\ h_{21b} & h_{22b} \end{bmatrix}$
Parallel-Series		$\begin{bmatrix} g_{11} & g_{12} \\ g_{21} & g_{22} \end{bmatrix} = \begin{bmatrix} g_{11a} & g_{12a} \\ g_{21a} & g_{22a} \end{bmatrix} + \begin{bmatrix} g_{11b} & g_{12b} \\ g_{21b} & g_{22b} \end{bmatrix}$
Cascade		$\begin{bmatrix} A & B \\ C & D \end{bmatrix} = \begin{bmatrix} A_a & B_a \\ C_a & D_a \end{bmatrix} \begin{bmatrix} A_b & B_b \\ C_b & D_b \end{bmatrix}$

Note. Under some circumstances the two-ports can be interconnected without ideal transformer so that the formulas in the right-hand column hold.

Table A-5 *SUMMARY OF NORMALIZATION RELATIONS*[a]

Type of Normalization	Relation between the Elements	Relation between the Network Functions	Comments
Impedance	$r_n R_0 = R_n$ $r_n L_0 = L_n$ $C_0/r_n = C_n$	$G_n(s) = r_n G_0(s)$	Does not hold for dimensionless network functions.
Frequency	$R_0 = R_N$ $L_0/\Omega = L_N$ $C_0/\Omega = C_N$	$G_N(s) = G_0\left(\dfrac{s}{\Omega}\right)$	Holds for all network functions.

[a] The subscript 0 refers to the original network; the subscript n or N refers to the normalized network.

Appendix

B / Properties of Passive RLC One-Port Networks

In this appendix, we have summarized the properties of the driving-point impedance, $Z(s)$, and the driving-point admittance, $Y(s)$, of various classes of RLC networks. Methods of synthesis of driving-point function are briefly mentioned in each case. For a proof of these properties and for details on the realization procedures, we refer the reader to the textbooks mentioned at the end of this appendix.

For convenience, the driving-point function of an RLC network will be represented in the following form:[1]

(B.1) $$\frac{N(s)}{D(s)} = \frac{N_m s^m + N_{m-1} s^{m-1} + \cdots + N_1 s + N_0}{D_n s^n + D_{n-1} s^{n-1} + \cdots + D_1 s + D_0}$$

[1] In Expression B.1 it is assumed that $N(s)$ and $D(s)$ do not have any common factors.

B-1 LC ONE-PORT

Inspection Tests for Necessary Conditions.

(a) The coefficients of the polynomials, $N(s)$ and $D(s)$, $(N_m, N_{m-1}, \ldots, N_1, N_0, D_n, D_{n-1}, \ldots, D_1, D_0)$ are real and nonnegative.
(b) $N(s)^0 - D(s)^0 = m - n = \pm 1$.
(c) Either $N_0 = 0$ and $N_1 \neq 0$ or $D_0 = 0$ and $D_1 \neq 0$.
(d) Poles and zeros are all on $j\omega$-axis and are simple.
(e) Either $N(s)$ is an even polynomial and $D(s)$ is an odd polynomial or vice versa.

Necessary and Sufficient Conditions. Either:

(f) All poles of $Z(s)$ [$Y(s)$] are simple and on $j\omega$-axis. Residue of $Z(s)$ [$Y(s)$] at each pole is real and positive.

or

(g) All poles and zeros of $Z(s)$ [$Y(s)$] are simple and on $j\omega$-axis. The poles and zeros alternate with each other.

Additional Properties.

(h) $N(s) + D(s)$ is a Hurwitz polynomial.

(B.2) (i) $\dfrac{dX(\omega)}{d\omega} > 0$ for all values of ω

where $Z(j\omega) = jX(\omega)$

Synthesis Procedures. A specified $Z(s)$ [$Y(s)$] can be realized by Foster's method which makes use of a partial fraction expansion of $Z(s)$ or $Y(s)$ Alternately, Cauer's method of using the continued fraction expansion of $Z(s)$ around infinity or origin can be followed to realize a specified driving-point function.

Example B-1. Consider

(B.3) $$Z(s) = \frac{s(s^2 + 2)}{(s^2 + 1)(s^2 + 3)} = \frac{s^3 + 2s}{s^4 + 4s^2 + 3}$$

Note that $Z(s)$ as given above satisfies all the five inspection tests for necessary conditions. Moreover, condition (g) is satisfied insuring the LC realizability of $Z(s)$. The partial fraction expansion of $Z(s)$ will be of the form:

(B.4) $$Z(s) = \frac{k_1 s}{s^2 + 1} + \frac{k_3 s}{s^2 + 3}$$

where

(B.5) $$k_1 = \frac{(s^2+1)}{s}Z(s)\bigg|_{s^2=-1} = \frac{s^2+2}{s^2+3}\bigg|_{s^2=-1} = \frac{1}{2}$$

(B.6) $$k_3 = \frac{(s^2+3)}{s}Z(s)\bigg|_{s^2=-3} = \frac{s^2+2}{s^2+1}\bigg|_{s^2=-3} = \frac{1}{2}$$

which implies

(B.7) $$Z(s) = \frac{\frac{1}{2}s}{s^2+1} + \frac{\frac{1}{2}s}{s^2+3}$$

From (B.7) we observe that condition (f) is satisfied. A realization of $Z(s)$ as expressed by (B.7) is shown in Figure B-1.

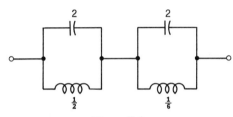

Figure B-1

B-2 RC ONE-PORT

Inspection Tests on $Z(s)$ for Necessary Conditions.

(a) The coefficients of the numerator and denominator polynomials, $N(s)$ and $D(s)$, are real and positive.
(b) $N(s)^0 - D(s)^0 = -1$ or 0.
(c) $N_0 \neq 0$. If $D_0 = 0$, then $D_1 \neq 0$.
(d) Poles and zeros are all on negative-real axis and are simple.
(e) No missing intermediate terms in the numerator and the denominator.

Necessary and Sufficient Conditions for $Z(s)$. Either:

(f) All poles and zeros are on negative real axis and are simple. The poles and zeros alternate with each other. Critical frequency nearest to or at origin is a pole. Critical frequency nearest to or at infinity is a zero.
or
(g) All poles are simple and are on negative real axis. Residue of $Z(s)$ at each pole is real and positive. $Z(\infty)$ is a finite nonnegative constant.

538 / PROPERTIES OF PASSIVE RLC ONE-PORT NETWORKS

Additional Properties of $Z(s)$.

(h) $$\frac{dZ(\sigma)}{d\sigma} < 0$$

(i) $$Z(0) > Z(\infty)$$

(j) $s Z(s^2) = \hat{Z}(s)$ is an LC driving-point impedance.

Conditions for $Y(s)$. Note that $Y(s)/s$ has identical properties to the driving-point impedance of an RC network. Hence $Y(s)/s$ can be considered as an impedance function and the above properties can be tested for its realizability. Alternately, the following conditions hold.

Inspection Tests on $Y(s)$ for Necessary Conditions.

(a) The coefficients of the numerator and denominator polynomials, $N(s)$ and $D(s)$, are real and positive.
(b) $N(s)^0 - D(s)^0 = 1$ or 0.
(c) $D_0 \neq 0$. If $N_0 = 0$, then $N_1 \neq 0$.
(d) Poles and zeros are all on negative-real axis and are simple.
(e) No missing intermediate terms in the numerator and the denominator.

Necessary and Sufficient Conditions for $Y(s)$. Either:

(f) All poles and zeros of $Y(s)$ are simple and on negative-real axis. Poles and zeros alternate with each other. Lowest critical frequency is a zero and the highest critical frequency is a pole.

or

(g) All poles are simple and on negative-real axis. Residue of $Y(s)$ at each finite pole is real and negative. Pole at infinity has a real and positive residue. $Y(0)$ is a finite nonnegative constant.

Additional Properties of $Y(s)$.

(h) $$\frac{dY(\sigma)}{d\sigma} > 0$$

(i) $$Y(0) < Y(\infty)$$

(j) $\hat{Y}(s) = \dfrac{1}{s} Y(s^2)$ is an LC driving-point admittance function.

Synthesis Procedures. A specified RC driving-point function can be realized by Foster's method which makes use of a partial fraction expansion of $Z(s)$ or $Y(s)/s$. Alternately, Cauer's method of using the continued fraction expansion of $Z(s)$ can be followed to realize a specified $Z(s)$.

Example B-2. Consider

(B.8) $$Z(s) = \frac{(s+1)(s+3)}{s(s+2)} = \frac{s^2 + 4s + 3}{(s^2 + 2s)}$$

Observe that $Z(s)$ as given above satisfies all the five inspection tests. Condition (f) for RC impedance function is also seen to be satisfied. Therefore, $Z(s)$ is an RC realizable driving-point impedance.

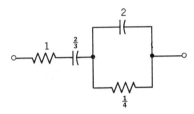

Figure B-2

The partial fraction expansion of $Z(s)$ is given as

(B.9) $$Z(s) = h_\infty + \frac{k_0}{s} + \frac{k_2}{s+2}$$

where

$$h_\infty = Z(\infty) = 1$$

$$k_0 = sZ(s)\bigg|_{s=0} = \frac{(s+1)(s+3)}{(s+2)}\bigg|_{s=0} = \frac{3}{2}$$

$$k_2 = (s+2)Z(s)\bigg|_{s=-2} = \frac{(s+1)(s+3)}{s}\bigg|_{s=-2} = \frac{1}{2}$$

This implies that

(B.10) $$Z(s) = 1 + \frac{\frac{3}{2}}{s} + \frac{\frac{1}{2}}{s+2}$$

(B.10) indicates that condition (g) for RC impedance functions is satisfied. A realization of $Z(s)$ based on the expansion (B.10) is shown in Figure B-2.

B-3 RL ONE-PORT

Conditions for $Z(s)$. Since the dual of an RC network, is an RL network, the driving-point impedance $Z(s)$ of RL network has identical properties to the driving-point admittance of an *RC* network. As a result, the properties of an *RC* driving-point admittance listed in Section B-2 can be used to test the realizability of an *RL* driving-point impedance function. For example, we have the following,

Necessary and Sufficient Conditions for $Z(s)$. Either:

(a) All poles and zeros of $Z(s)$ are simple and on negative-real axis. Poles and zeros alternate with each other. Critical frequency nearest to or at the origin is a zero and the critical frequency nearest to or at infinity is a pole.

or

(b) All poles are simple and on negative real axis. Residue of $Z(s)$ at each finite pole is real and negative. Pole at infinity has a real and positive residue. $Z(0)$ is a finite nonnegative constant.

Conditions for $Y(s)$. In a like manner, because of duality, the *RL* driving-point admittance has identical properties to an *RC* driving-point impedance. Thus, the properties listed in the previous section can be used to test *RL* admittance function. For convenience, we list below only the necessary and sufficient properties.

Necessary and Sufficient Conditions for $Y(s)$. Either:

(a) All poles and zeros are on negative real axis and are simple. The poles and zeros alternate with each other. Critical frequency nearest to or at the origin is a pole. Critical frequency nearest to or at infinity is a zero.

or

(b) All poles are simple and on negative-real axis. Residue of $Y(s)$ at each pole is real and positive. $Y(\infty)$ is a finite nonnegative constant.

B-4 *RLC* ONE-PORT[2]

Inspection Tests for Necessary Conditions.

(a) All coefficients of the numerator and denominator polynomials, N_i and D_j, are real and nonnegative.
(b) $N(s)^0 - D(s)^0 = m - n = 1$ or 0 or -1.
(c) If $N_0 = 0$, $N_1 \neq 0$.
(d) If $D_0 = 0$, $D_1 \neq 0$.
(e) $j\omega$-axis zeros of $N(s)$ and $D(s)$, if any, are simple.
(f) Unless all even and/or all odd terms are missing, $N(s)$ and $D(s)$ do not have any missing terms.

[2] Adapted from M. E. Van Valkenburg, *Introduction to modern network synthesis*, Wiley, New York (1960).

Necessary and Sufficient Conditions.

(g) $N(s) + D(s)$ is a Hurwitz polynomial.
(h) The numerator of the real part of the driving-point function does not have any $j\omega$-axis zeros of odd multiplicity.

References

Balabanian, N., *Network synthesis*, Prentice-Hall, Englewood Cliffs, N.J., Chapters 2 and 3 (1958).

Chen, W. H., *Linear network design and synthesis*, McGraw-Hill, New York, Chapters 3, 4, 5, 6, 7, 8, and 9 (1964).

Guillemin, E. A., *Synthesis of passive networks*, Wiley, New York, Chapters 1, 2, 3, 4, 9, and 10 (1957).

Hazony, D., *Elements of network synthesis*, Reinhold, New York, Chapters 2, 3, 4, 5, and 7 (1963).

Kuh, E. S. and D. O. Perderson, *Principles of circuit synthesis*, McGraw-Hill, New York, Chapters 6 and 10 (1959).

Van Valkenburg, M. E., *Introduction to modern network synthesis*, Wiley, New York, Chapters 3, 4, 5, 6, 7, and 14 (1960).

Yengst, W. C., *Procedures of modern network synthesis*, Macmillan, New York, Chapters 3, 4, 5, and 6 (1964).

Appendix

C / Properties of Passive RLC Two-Port Networks

For a complete description of the external behavior, a passive RLC two-port can be described by any of the two-port parameters tabulated in Table A-1 of Appendix A. The most frequently specified parameter sets are the open-circuit impedance parameters (z_{ij}) and the short-circuit admittance parameter (y_{ij}). In most cases, only a partial description of the external behavior is specified. The commonly used specifications are summarized in Table A-3 of Appendix A.

In this appendix, we have summarized the properties of the impedance and admittance parameters, and the transfer functions of various classes of RLC two-ports. No mention is made on the methods of synthesis of the two-port. For a proof of the properties and for a description of the two-port realization procedures, the reader is referred to the textbooks mentioned at the end of this appendix.

GENERAL TWO-PORT / 543

For convenience, we consider here only the following transfer functions:

$$z_{21}, y_{21}, Z_{21}, Y_{21}, t_v(T_V), t_I(T_I)$$

These transfer functions are defined in Table A-3 of Appendix A.

C-1 GENERAL TWO-PORT

For a general passive RLC two-port, the z-parameters satisfy the conditions of Theorem 6-6 which is repeated below.

Theorem 6-6 *The open-circuit parameters, z_{mn}, of a linear passive RLC two-port satisfy the following conditions:*

(i) *poles of $z_{mn}(s)$ are in the left half s-plane, and $j\omega$-axis poles (if any) are simple;*
(ii) *the residues k_{mn} of z_{mn} at a $j\omega$-axis pole obey the residue condition,*

(6.49)
$$k_{11}, k_{22}, k_{12} \text{ real}$$
$$k_{11}k_{22} - k_{12}^2 \geq 0$$
$$k_{11} \geq 0$$

(iii) *If* $\operatorname{Re} z_{mn}(j\omega) = r_{mn}$, *then*

(6.50)
$$r_{11}r_{22} - r_{12}^2 \geq 0$$
$$r_{11} \geq 0$$

Theorem 6-6 implies that z_{11} and z_{22} are positive real functions.

The y-parameters also satisfy identical conditions. Note that Z_{21}, Y_{21}, y_{21} have exactly the same properties as z_{21}. Several conclusions can be made regarding the properties of these functions from the results of Theorem 6-6. These are summarized next.

Necessary Properties of Transfer Immittance Functions.

(a) Poles on the $j\omega$-axis are simple and the corresponding residues can have positive or negative real values.
(b) No restriction on the location of the zeros. Thus they can be multiple and can be situated anywhere in the s-plane.

Next we observe that the transfer voltage (current) ratio can be expressed as the ratio of z-parameters of y-parameters (see Table A-3). For example,

$$t_v = -\frac{y_{21}}{y_{22}}$$

Consequently, the following necessary properties of the transfer ratios are obtained using the results of Theorem 6-6.

Necessary Properties of Transfer Ratio.

(a) Can not have any poles in the right half s-plane.
(b) No poles at the origin or at infinity.
(c) Poles on the $j\omega$-axis are simple with imaginary residue.
(d) No restriction on the location of the zeros.

C-2 TWO-PORT WITHOUT MUTUAL COUPLING: FIALKOW-GERST CONDITIONS

For an *RLC* grounded two-port without any mutual inductances and ideal transformers, additional restrictions on the z- and y-parameters can be derived. These conditions, given below as *Theorem C-1*, are more commonly known as *Fialkow-Gerst Conditions*.[1]

Theorem C-1 *Let*

$$y_{11} = \frac{A_n s^n + A_{n-1} s^{n-1} + \cdots + A_1 s + A_0}{D(s)}$$

$$y_{22} = \frac{B_n s^n + B_{n-1} s^{n-1} + \cdots + B_1 s + B_0}{D(s)}$$

$$-y_{21} = \frac{C_n s^n + C_{n-1} s^{n-1} + \cdots + C_1 s + C_0}{D(s)}$$

represent the y-parameters of an RLC unbalanced two-port having no mutual coupling, then the following conditions must be satisfied:

$$C_j \geq 0$$
$$B_j \geq C_j$$
$$A_j \geq C_j$$

for $j = 0, 1, \ldots, n$.

It should be noted that the z-parameters of an *RLC* two-port without mutual reactance also satisfy conditions identical to those given in Theorem C-1.

Since transfer voltage and current ratios can be expressed as ratios of z- or y-parameters, we have the following necessary properties for these types of transfer functions.

Necessary Conditions on Transfer Ratio. Let:

$$T(s) = \frac{N_n s^n + N_{n-1} s^{n-1} + \cdots + N_1 s + N_0}{D_n s^n + D_{n-1} s^{n-1} + \cdots + D_1 s + D_0}$$

---

[1] A. D. Fialkow and I. Gerst, "The transfer function of networks without mutual reactance," *Quarterly Appl. Math.*, **12**, 117–131 (1954).

represent a transfer voltage (or current) ratio of an RLC unbalanced two-port with no mutual coupling, then

(a) $T(s)$ can have no positive real transmission zeros.
(b) $0 \leq T(\sigma) \leq 1$ for $\sigma > 0$. Moreover, $T(\sigma)$ is equal to unity only at the origin or at infinity or both.

(c)
$$N_j \geq 0$$
$$N_j \leq D_j$$

for $j = 0, 1, \ldots, n$ [provided that no common factors of the numerator and denominator of $T(s)$ have been cancelled].

C-3 LC TWO-PORT

Properties of the open-circuit impedance parameters and the short-circuit admittance parameters of a lossless reciprocal two-port are next considered. Conditions satisfied by the y-parameters of a lossless non-reciprocal two-port are given in Theorem 10-2, from which one can derive similar conditions for an LC two-port. We have

Theorem C-2 *The necessary and sufficient conditions satisfied by the y-parameters of a lossless reciprocal two-port are:*

(i) *Poles of y_{11}, y_{22}, and y_{22} are simple and on the $j\omega$-axis.*
(ii) *There is either a pole or a zero at the origin and at infinity:*
(iii) *The residues k_{mn} of y_{mn} at a pole obey the residence condition:*

(6.49)
$$k_{11}, k_{22}, k_{12} \text{ are real}$$
$$k_{11}k_{22} - k_{12}^2 \geq 0$$
$$k_{11} \geq 0$$

A consequence of the above theorem is that both y_{11} and y_{22} are LC driving-point functions. Also y_{12} is an odd rational function, having its zeros in a quadrantal symmetry.

As before, the z-parameters have identical properties.

C-4 RC TWO-PORT

Properties of RC two-port can be obtained from the properties of LC two-port by making use of the $LC:RC$ transformation discussed in Section 3-1. We state below the conditions in terms of the z-parameters:

Theorem C-3 *The necessary and sufficient conditions satisfied by the z-parameters of an RC two-port are:*

(i) *Poles of z_{11}, z_{12}, and z_{22} are simple and on the negative real axis.*

(ii) *The critical frequency of z_{11} and z_{22} nearest to or at the origin is a pole.*

(iii) *The critical frequency of z_{11} and z_{22} nearest to or at infinity is a zero.*

(iv) *The residues k_{mn} of z_{mn} at a pole obey the residue condition,*

$$k_{11} \geq 0$$
$$k_{11}k_{22} - k_{12}^2 \geq 0$$

(v) *At infinity*

$$z_{11}(\infty)z_{22}(\infty) - z_{12}^2(\infty) \geq 0$$

It follows that z_{11} and z_{22} must be *RC* driving-point functions. Zeros of z_{12} again can be anywhere in the s-plane.

References

Balabanian, N., *Network synthesis*, Prentice-Hall, Englewood Cliffs, N.J., Chapters 4, 5, 6, 7, and 8 (1958).

Chen, W. K., *Linear network design and synthesis*, McGraw-Hill, New York, Chapters 10, 12, 14, 15, 16, 17, and 18 (1964).

Guillemin, E. A., *Synthesis of passive networks*, Wiley, New York, Chapters 2, 7, 11, 12, and 13 (1957).

Hazony, D., *Elements of network synthesis*, Reinhold, New York, Chapters 6, 7, 8, 9, 10, 11, and 12 (1963).

Kuh, E. S. and D. O. Pederson, *Principles of circuit synthesis*, McGraw-Hill, New York, Chapters 7, 9, 11, and 12 (1959).

Van Valkenburg, M. E., *Introduction to modern network synthesis*, Wiley, New York, Chapters 10, 11, 12, 14, and 15 (1962).

Weinberg, L., *Network analysis and synthesis*, McGraw-Hill, New York, Chapters 7, 8, and 12 (1962).

Yengst, W. Y., *Procedures of modern network synthesis*, Macmillan, New York, Chapters 7, 8, and 9 (1964).

Appendix

D / RC Two-Ports Realizing Second-Order Transfer Functions

Table D-1 REALIZATIONS OF SHORT-CIRCUIT TRANSFER ADMITTANCES

$$-y_{21} = \left.\frac{I_2}{V_1}\right|_{V_2=0}$$

Type	$-y_{21}$	Network Realizations	Element Values
N_1	$(s + a_0)$	(R and C in parallel between ports 1 and 2)	$R = 1/a_0$ $C = 1$

Table D-1 (*Continued*)

Type	$-y_{21}$	Network Realizations	Element Values
N_2	$\dfrac{1}{s+p_1}$		$R = p_1/2$ $C = 4/p_1^2$
N_3	$\dfrac{s+a_0}{s+p_1}$		$R_1 = 1$ $R_2 = (p_1 - a_0)/a_0$ $C = 1/(p_1 - a_0)$
	$p_1 > a_0$		$R_1 = p_1/a_0$ $R_2 = p_1/(p_1 - a_0)$ $C = (p_1 - a_0)$
N_4	$\dfrac{s+a_0}{s+p_1}$		$R_1 = p_1/2(a_0 - p_1)$ $R_2 = 1$ $C = 2/p_1 R_1$
	$a_0 > p_1$		$R_1 = p_1/2a_0$ $R_2 = \dfrac{p_1^2}{4a_0(a_0 - p_1)}$ $C = 1/R_2$
N_5	$\dfrac{s^2 + a_1 s}{(s+p_1)(s+p_2)}$		$R_1 = 1$ $R_2 = p_1 p_2/(a_1 - p_2)(p_1 - a_1)$ $C_1 = a_1/p_1 p_2$ $C_2 = 1/a_1 R_2$
	$p_1 > a_1 > p_2$		$R_1 = 1$ $C_1 = a_1/p_1 p_2$ $R_2 = 1/C_2 a_1$ $C_2 = a_1/(p_2 - a_1)(a_1 - p_1)$
N_6	$\dfrac{s}{(s+p_1)(s+p_2)}$		$C_2 = 1/p_1 p_2$ $R_1 = (p_1 + p_2) - 2R_2$ $R_2 = \sqrt{p_1 p_2}$ $C_1 = C_2 R_2/R_1$
N_7	$\dfrac{s^2 + a_0}{s+p_1}$		$R_1 = p_1/2a_0$ $R_2 = 1/4p_1$ $C_1 = 4a_0/p_1^2$ $C_2 = 2$

Table D-1 (*Continued*)

Type	$-y_{21}$	Network Realization	Element Values
N_8	$\dfrac{s^2}{(s+p_1)(s+p_2)}$ $p_1 \neq p_2$		$R_1 = C_2/C_1$ $R_2 = 1$ $C_1 = 1/(\sqrt{p_2} - \sqrt{p_1})^2$ $C_2 = 1/\sqrt{p_1 p_2}$
N_9	$\dfrac{s}{s+p_1}$		$R = 1$ $C = 1/p_1$
N_{10}	$\dfrac{s^2}{s+p_1}$		$R = 1/4p_1$ $C = 1$
N_{11}	$\dfrac{s^2 + a_1 s}{s + p_1}$ $a_1 > p_1$		$C_1 = a_1/p_1$ $C_2 = a_1/(a_1 - p_1)$ $R = 1/C_2 a_1$
N_{12}	$\dfrac{s^2 + a_1 s + a_0}{s + p_1}$ $a_1 p_1 = a_0$		$R_1 = 1/4p_1$ $R_2 = 1/a_1$ $C = 2$
N_{13}	$\dfrac{s^2 + a_1 s + a_0}{s + p_1}$ $a_1 = p_1$		$R = p_1/2a_0$ $C_1 = 2/Rp_1$ $C_2 = 1$
N_{14}	$\dfrac{1}{(s+p_1)(s+p_2)}$		$R_1 = p_2(p_1 - p_2)/2$ $R_2 = p_2^2$ $C = 1/R_1 p_2$

Table D-1 *(Continued)*

Type	$-y_{21}$	Network Realization	Element Values
N_{15}	$\dfrac{s^2 + a_1 s + a_0}{s + p_1}$		$C_1 = 1$ $R_1 = 1/(a_1 - p_1)$ $C_2 = \dfrac{(a_1 - p_1)^2}{a_1 p_1 - p_1^2 - a_0}$ $R_2 = (a_1 - p_1)/C_2 a_0$
	$a_1^2 > 4a_0$ $(a_1 - p_1)$ $\quad > a_0/p_1$		$C_1 = 1$ $R_1 = a_0/p_1$ $R_2 = p_1/(a_1 p_1 - p_1^2 - a_0)$ $C_2 = 1/p_1 R_2$
N_{16}	$\dfrac{s^2 + a_1 s}{s + p_1}$ $p_1 > a_1$		$C = 2$ $R_1 = 1/2(p_1 - a_1)$ $R_2 = 1/2a_1$
N_{17}	$\dfrac{s^2 + a_1 s + a_0}{s + p_1}$ $a_1^2 \geq 4a_0$ $2p_1 - a_1$ $\quad \geq \sqrt{a_1^2 - 4a_0}$		$R_1 = 1/a_1$ $R_2 = a_0/a_1(a_1 p_1 - 2a_0)$ $C_1 = 2a_1/(a_1 + \sqrt{a_1^2 - 4a_0})$ $C_2 = 2a_1/(a_1 - \sqrt{a_1^2 - 4a_0})$
N_{18}	$\dfrac{s^2 + a_1 s + a_0}{s + p_1}$ $a_1^2 \geq 4a_0$ $a_1 - 2p_1$ $\quad \geq \sqrt{a_1^2 - 4a_0}$		$R_1 = 2p_1/a_1(a_1 + \sqrt{a_1^2 - 4a_0})$ $R_2 = 2p_1/a_1(a_1 - \sqrt{a_1^2 - 4a_0})$ $C_1 = a_1/p_1$ $C_2 = C_1(a_1 - 2p_1)/p_1$
N_{19}	$\dfrac{s^2 + a_1 s + a_0}{s + p_1}$ $a_1^2 < 4a_0$ $p_1 > a_0/a_1$		$R_1 = p_1/2(2p_1^2 - a_1 p_1 + a_0)$ $R_2 = p_1/2(a_0 p_1 - a_0)$ $R_3 = p_1/a_0$ $C = 2$
N_{20}	$\dfrac{s^2 + a_1 s + a_0}{s + p_1}$ $a_1 > p_1$		$R_1 = p_1/2[a_0 - p_1(a_1 - p_1)]$ $R_2 = 1/(a_1 - p_1)$ $C_1 = 2/R_1 p_1$ $C_2 = 1$
	$4a_0 > a_1^2$		$R_1 = p_1/2a_0$ $R_2 = \dfrac{p_1^2(a_1 - p_1)}{4a_0[a_0 - p_1(a_1 - p_1)]}$ $C_1 = (a_1 - p_1)/R_2 a_0$ $C_2 = 1$

Adapted from F. R. Bradley and R. McCoy, "Driftless D-C Amplifier," *Electronics*, pp. 144–148 (April 1952).

Table D-2 REALIZATION OF VOLTAGE TRANSFER RATIO

$$t_v = \left.\frac{V_2}{V_1}\right|_{I_2=0}$$

Type	t_v	Restrictions	Network Realization	Element Values
N_1	$\dfrac{a_1 s + a_0}{s + b_0}$	$a_1 \leq 1$ $a_0 \leq b_0$		$C_1 = a_1$ $C_2 = (1 - a_1)$ $R_1 = 1/a_0$ $R_2 = 1/(b_0 - a_0)$
N_2	$\dfrac{a_0}{s^2 + b_1 s + b_0}$	$a_0 \leq b_0$ $b_1^2 > 4b_0$		$R_2 = 2/\sqrt{b_1^2 - 4b_0}$ $R_1 = 1/a_0 R_2$ $R_3 = 1/(b_0 - a_0)R_2$ $C_1 = 2(1 + b_0 R_2^2)/b_1 R_2$ $C_2 = 1/C_1$
N_3	$\dfrac{a_2 s^2}{s^2 + b_1 s + b_0}$	$a_2 \leq 1$ $b_1^2 > 4b_0$		$C_2 = \sqrt{(b_1^2 - 4b_0)/4b_0}$ $C_1 = a_2/C_2$ $C_3 = (1 - a_2)/C_2$ $R_1 = b_1 C_2/2b_0(C_2^2 + 1)$ $R_2 = 1/R_1 b_0$

Table D-2 (*Continued*)

Type	t_v	Restrictions	Network Realization	Element Values
N_4	$\dfrac{a_2 s^2 + a_1 s + a_0}{s^2 + b_1 s + b_0}$	$(b_1 a_1 a_0 - b_0 a_1^2 - a_0^2) > 0$ $b_1 > a_1 > 0$ $b_1^2 > 4b_0$		$C_2 = \sqrt{b_1 a_1 a_0 - b_0 a_1^2 - a_0^2/a_0}$ $C_1 = a_2/C_2$ $C_3 = (1 - a_2)/C_2$ $R_1 = (C_2^2 + 1)/C_2 a_1$ $R_2 = 1/R_1 a_0$ $R_3 = R_1 a_0/(b_0 - a_0)$
N_5	$\dfrac{a_2 s^2 + a_1 s + a_0}{s^2 + b_1 s + b_0}$	$(b_1 - a_1)(b_0 a_1 - b_1 a_0)$ $- (b_0 - a_0)^2 > 0$ $b_0 > a_0$ $b_1 > a_1 > 0$ $b_1^2 > 4b_0$		$C_2 = \dfrac{\sqrt{(b_1 - a_1)(b_0 a_1 - b_1 a_0) - (b_0 - a_0)^2}}{(b_0 - a_0)}$ $C_1 = (1 - a_2)/C_2$ $C_3 = a_2/C_2$ $R_1 = (C_2^2 + 1)/C_2(b_1 - a_1)$ $R_2 = 1/R_1(b_0 - a_0)$ $R_3 = R_1(b_0 - a_0)/a_0$
N_6	$\dfrac{a_2 s^2 + b_1 s + b_0}{s^2 + b_1 s + b_0}$	$a_2 < 1$ $b_1^2 > 4b_0$		$C_2 = \sqrt{b_1^2 - 4b_0}/4b_0$ $C_1 = (1 - a_2)/C_2$ $C_3 = a_2/C_2$ $R_1 = b_1 C_2/2b_0(C_2^2 + 1)$ $R_2 = 1/R_1 b_0$

N_7	$\dfrac{s^2 + b_1 s + a_0}{s^2 + b_1 s + b_0}$	$b_0 > a_0$ $b_1^2 > 4b_0$	$R_2 = 2/\sqrt{b_1^2 - 4b_0}$ $R_1 = 1/(b_0 - a_0)R_2$ $R_3 = 1/a_0 R_2$ $C_1 = 2(1 + b_0 R_2^2)/b_1 R_2$ $C_2 = 1/C_1$
N_8	$\dfrac{a_2 s^2 + b_0}{s^2 + b_1 s + b_0}$	$a_2 \leq 1$ $b_1^2 > 4b_0$	$R_1 = 4b_1/(b_1^2 + 4b_0)$ $R_2 = b_1(b_1^2 - 4b_0)/2b_0(b_1^2 + 4b_0)$ $R_3 = 2a_2/C_1 C_2 b_1$ $C_1 = (3b_1^2 - 4b_0)/2b_1^2$ $C_2 = a_2(3b_1^2 - 4b_0)/(b_1^2 - 4b_0)/a_2$ $C_3 = (1 - a_2)C_2/a_2$ $C_4 = 1/R_1 R_2 b_0$
N_9	$\dfrac{s^2 + a_0}{s^2 + b_1 s + b_0}$	$b_0 \geq a_0$ $b_1^2 > 4b_0$	$R_1 = 4b_1/(b_1^2 + 4b_0)$ $R_2 = b_1(b_1^2 - 4b_0)/2a_0(b_1^2 + 4b_0)$ $R_3 = 2/C_1 C_2 b_1$ $R_4 = R_2 a_0/(b_0 - a_0)$ $C_1 = (3b_1^2 - 4b_0)/2b_1^2$ $C_2 = (3b_1^2 - 4b_0)/(b_1^2 - 4b_0)$ $C_4 = 1/R_1 R_2 a_0$

The first seven entries of this table have been adapted from S. L. Hakimi and J. B. Cruz, Jr., "On minimal realization of RC two-ports," *Proc. NEC*, **16**, 258–267 (1960); remaining two entries are based on S. K. Mitra, "A note on the design of RC notch networks with maximum gain," *Proc. IEEE (Proceedings Letters)*, **54**, 1487 (October 1966). Additional network realizations will be found in the references at the end of this table.

Additional References

Dutta Roy, S. C., "On the design of parallel-T resistance capacitance networks for maximum selectivity," *J. Inst. Telecomm. Engrs. (India)*, **8,** 218 (September 1962).

Hakimi, S. L., and J. B. Cruz, Jr., *loc. cit.*

Kodali, V. ꝑ., "Note on synthesis of *RC* transfer functions," *IEEE Trans. on Circuit Theory*, **CT-12,** 623–624 (December 1965).

Moschytz, G. S., "Active *RC* filter building blocks using frequency emphasizing networks," *IEEE J. Solid State Circuits*, **SC-2,** 59–62 (June 1967).

Moschytz, G. S., "Two-step precision tuning of twin-T notch filter," *Proc. IEEE*, **54,** 811–812 (May 1966).

Shenoi, B. A., "A new technique for twin-T *RC* network synthesis," *IEEE Trans. on Circuit Theory*, **CT-11,** 435–436 (September 1964).

/ Index

Aasnaes, H. B., 456
ABCD parameters, 531
    controlled-source model based on, 64
Active gyrator, 47
Active network, analysis of, 118–154
    elementary definition, 1, 15
    formal definition, 212
    notations, 226–227
    sensitivity problems in, 9
    stability problems in, 9
Active network elements, 23–75
    impedance converter type, 55
    impedance inverter type, 55
    locations in *AD–BC* plane, 59
    relations between ideal, 55–61
    two-port parameters, 56–71
Active network synthesis, basic steps, 16–18
Active one-port, definition, 212
Active *RC* filters, advantages, 2, 7–8
    comparison with passive *RLC* filter, 19–20
    general properties, 19–20
    problems associated with, 9
    voltage-variable, 10
Active *RC* one-port, basic theorem on synthesis, 374
Active transformer, 42
    generalized, 43
    inductor simulation using, 493
Activity, 212

AD–BC plane, 57
A–D plane, 55
All-pass transfer function, *RC*:operational-amplifier realization, 478
    RC:voltage-amplifier realization, 326
    RLC:operational-amplifier realization, 483
    RLC:voltage-amplifier realization, 344
Amplifier (*see* Controlled sources, Current amplifier, Isolation amplifier, Operational amplifier, Voltage amplifier)
Analog Devices, Inc., 451–452, 462
Analysis of active networks, containing ideal active elements, 143–154
    containing operational amplifier, 148–154
    indefinite admittance matrix approach, 118–139
Anderson, B. D., 49
Antoniou, A., 467
Approximation of transfer function, 3
Armstrong, D.B., 341
Aurell, C. G., 43

Bach, R. E., 336
Balabanian, N., 515
Band-pass filter realization, 89, 332
Bandwidth (3 db), 12
    relation with rise time, 13
Bangert, J., 264

Barranger, J., 397
B–C plane, 57
Bello, P., 42, 220, 266
Bendik, J., 201, 410
Bessel filter, 12, 164, 312, 319, 384, 388, 427, 432, 488
Bessel transfer function, 5
Bialko, M., 441
Bieri, R., 258
Bilinear form of network function, 155–157
  signal flow-graph representation from, 166
Blackman's formula, 206
Blostein, M. L., 189
Bobrow, L. S., 324
Bode, H. W., 156, 219
Bode criterion, 456–458
Bode plot, 449–450
Bodner, H. A., 341
Boesch, F. T., 268
Bogert, B. P., 442
Bohn, E. V., 485
Bohn's method, 485
Bolinder, E. F., 243
Bordogna, J., 506
Bott, R., 414
Bott-Duffin's method, 414–415
  modification using gyrator, 416
  modification using operational amplifier, 470–473
Bown, G. C., 198
Bradley, F. R., 550
Branner, G., 528
Braun, J., 74, 351
Brennan, R., 486
Brennan and Bridgman's method, 486
Bridgman, A., 486
Brown, R. G., 107
Brownlie, J. D., 364
Brune, O., 213
Burr-Brown Research Corp., 451
Butler, E., 463
Butterworth filter, 88–89, 330, 338
Butterworth transfer function, 4

Calahan, D. A., 103, 189, 395, 428, 434
Calahan decomposition, 434
  comparison with Horowitz decomposition, 435–437
  method for obtaining, 437–438

Calfee, R. W., 341
Campbell, G. A., 143
Capacitive gyrator, 240
Capacitive inverse network, 90
Carlin, H. J., 53, 211, 287, 464
Cascade connection of two-ports, 534
Causal network, 211
CCT, definition, 27
  driving-point function synthesis using, 307
  nullator-norator models of, 292–293, 349–350
  relation with CVT and VCT, 28
  transfer function synthesis using, 310–311
  transistorized realizations, 295, 297–300
  two-port parameters, 56
Cederbaum, I., 287
CGIC, 38
  (*see also* Generalized-impedance converter)
Chebyshev transfer function, 4
Chua, H. T., 413
Cimagalli, V., 253
Circulator, 79
  realization using gyrators, 80
CNIC, 38
  (*see also* Negative-impedance converter)
Coefficient matching technique, 18, 328–339, 484–492
Coefficient sensitivity, calculation of, 180
  definition, 172
  relation with root sensitivity, 181
Co-factor, first, 122
  second, 136
Common-base transistor, equivalent representation, 66
Common-emitter transistor, equivalent representation, 66
Common-mode voltage limit, 451, 454, 456
Companion network, 16
Compensation of operational amplifier (*see* Operational amplifier)
Complex transformer, definition, 42
  use of, 220
Constant resistance network using operational amplifier, 500
Controlled source models, of negative resistance, 61
  of two-ports, 62–65

INDEX / **557**

Controlled sources, definition, 26–27
  gain of, 27
  location in $AD-BC$ plane, 59
  nonideal, 31–32
  nullator-norator models of, 72–73, 291–293
  operational amplifier realizations, 464–465
  relations between, 27–28
  transistorized realizations, 293–297
  types of, 26–27
  (*see also* CCT, CVT, VCT, VVT)
Converter, effect of, 44–45
  ideal current, 43–45
  ideal power, 43–45
  ideal voltage, 43–45
  representation of two-ports using ideal, 67–70
  symbolic representations, 44
  (*see also* ICC, Impedance converter, IPC, IVC)
Cooper, R. E., 325
Cruz, J. B., Jr., 88, 553
Curan, D. R., 404
Current amplifier, 27
  (*see also* CCT)
Current-controlled current source (*see* CCT)
Current-controlled voltage source (*see* CVT)
Current-to-current transducer (*see* CCT)
Current-to-voltage transducer (*see* CVT)
CVT, definition, 26
  gyrator:negative-resistance model, 67
  nullator-norator model of, 292–293
  operational amplifier realizations of, 465
  relation with CCT and VVT, 28
  symbolic representation, 26
  transistorized realizations of, 297, 301
  two-port parameters, 56

Darlington compound pair, 305, 361
Darlington synthesis, nonreciprocal, 418–424
Davies, A. C., 71
Deboo, G. J., 493
DeClaris, N., 90, 309
Delansky, J. F., 423
DePian, L., 64, 246
Deviation study, 162
  (*see also* Worst case deviation)

Driving-point function synthesis, basic theorem on active, 308, 374
  using controlled sources, 305–309
  using generalized-impedance converter, 370–374, 516–520, 526
  using gyrators, 414–424
  using negative-impedance converter, 364–381, 512–514, 520–526
  using negative resistances, 266–271
  using operational amplifiers, 470–474
  using tunnel diodes, 266–267
Duffin, R. J., 414

Equivalent circuits, of multipoles, 139–143
  of two-ports, 62–66, 69, 71
Esaki, L., 255
Even part, of a positive real function, 510
  of sensitivity function, 170
Even part partitioning, 509–512
  transfer function synthesis using, 529

Feedback factor, 456
Fialkow, A. D., 544
Fialkow-Gerst conditions, 275, 280, 544
F matrix, 531
Frequency normalization, 13, 534
  factor, 13, 534

Gaash, A., 203
Gafni, H., 287
Gain function, 170
Gain margin of stability, 107
Gain sensitivity, calculation of, 171, 173
  definition, 170
  from sensitivity function, 170–171
  minimization of, 186–187, 207
Ganguly, U. S., 342
Gary, P., 440
Geffe, P., 492
Generalized dual networks, 92–93
Generalized equivalent networks, 114–115
Generalized-impedance converter, current-inversion type, 38
  definition, 37
  driving-point function synthesis using, 370–374, 516–520, 526
  two-port parameters, 56
  voltage-inversion type, 37
Generalized inverse networks, concept of, 90

General inverse networks (*Continued*)
  construction of, 92
  types of, 90
Gerst, I., 544
Gewertz, C. M., 243
Ghausi, M. S., 276
Gibbons, J. F., 305, 361
GIC (*see* Generalized-impedance converter)
Gile, W., 294
Giles, J. N., 459
Gintzon, E. L., 256
Giordano, A. B., 211
Goldstein, A. J., 188
Gorski–Popiel, J., 202, 439
g–parameters of a two-port, 531
  controlled-source model based on, 64
Granata, P. J., 456
Gruetzmann, S., 413, 439
Guillemin, E. A., 394, 506
Guillemin's parallel ladder technique, 315–318
Gyration impedance, 48, 51
Gyrator, active, 47
  design using NRNIV, 410–413
  controlled source model of, 66, 404–406
  driving-point function synthesis using, 414–424
  gyration impedance of a, 48, 51
  ideal, 47
  impedance inversion property of, 48
  inductance simulation using, 48–49, 438–441
  location in B–C plane, 58
  nonideal, 49–51
    compensation of, 49–51
  nullator-norator models of, 405–410
  reactive, 51
    transfer function synthesis using, 430–433
  realization using feedback amplifiers, 444–445
  realization using operational amplifiers, 468–469
  symbolic representation, 47
  transfer function synthesis using, 425–428
  transistorized realizations, 407–409
  two-port parameters, 57
Gyrator:negative-resistance model, of
  norator, 68
  of nullator, 68
  of two-ports, 51, 65–67

Hakim, S. S., 318, 355
Hakim's method, 318–322
Harbout, C. O., 325
Harris, H. E., 287
Harrison, T. J., 51, 443
Hazony, D., 252, 328, 415, 418
Heinlein, W. E., 413, 439
Herbst, N. M., 37, 90, 171, 347, 374
Herold, E. W., 33
Herskowitz, G. J., 276
High-pass filter realization, 89, 330
  from low-pass realization, 88
Hogan, C. L., 404
Holmes, W. H., 47, 413, 439
Holt, A. G. J., 49, 439, 486
Horowitz, I. M., 100, 172, 274, 390, 394, 425
Horowitz decomposition, 394
  comparison with Calahan decomposition, 435–437
  method for obtaining, 394–395
Hoskins, R. F., 363
Howard, W. G., Jr., 90, 91, 428, 430
h–parameters of a two-port, 531
  controlled-source model based on, 63
Huelsman, L. P., 55, 64, 191, 332, 489, 497
Hurtig, G., 391
Hurwitz polynomial, 110
  test for, 110

ICC, 43–45
  model of a transistor using, 82
  model of a two-port using, 71
  symbolic representation, 44
  two-port parameters, 56
Ideal current converter (*see* ICC)
Ideal linear active element, definition, 15
Ideal power converter (*see* IPC)
Ideal transformer, 41–43
  controlled source model of, 65
  location in AD–BC plane, 59
  location in A–D plane, 58
  model using negative resistance, 68
  relations with active elements, 59–60
  symbolic representation, 41
  two-port parameters, 57
Ideal voltage converter, (*see* IVC)
IG (*see* Gyrator)
Impedance converter, 35–45
  controlled source representation of, 64–65

INDEX / 559

Impedance converter (*Continued*)
  conversion factor of, 37
  definition, 35
  necessary and sufficient conditions, 36
  (*see* also Negative-impedance converter, Generalized-impedance converter)
Impedance inverter, 45–53
  definition, 45
  inversion factor of, 46
  necessary and sufficient conditions for, 46
  (*see* also Positive-impedance inverter, Negative-impedance inverter, NRNIV)
Impedance normalization, 13, 534
  factor, 13, 534
Incremental parameter variation, effect of, 9
  pole (zero) displacement estimation, 163–165
Indefinite admittance matrix, definition, 120
  from short-circuit admittance matrix, 124
  network functions from, 135–139
  of a contracted multipole, 127–128
  of a multipole with internal reference terminal, 123–124
  of a transistor, 125
  of ideal active elements, 143–148
  of multipoles connected in parallel, 130–131
  properties, 123
Indiresan, P. V., 258
Inductance, loss compensation of, 264–265
  series compensation of, 265
  shunt compensation of, 265
  simulation of floating, 49
  (*see* also Inductor, Simulated inductance)
Inductive inverse network, 90
Inductor, model, 5
  problems associated with, 2, 5–6
  Q of, 5, 6
  (*see* also Simulated inductance)
Instantaneous power, 210
Integrated microcircuit, advantages, 2
  limitations, 2
Interconnected two-ports, formulas for, 534
Invariant stability factor, definition, 247
  stability of a two-port in terms of, 247

IPC, 43–45
  application of, 69–70
  equivalent circuits using, 69
  symbolic representation, 44
  two-port parameters, 57
Isolation amplifier, 29
Isolator, 48
IT, 42
  (*see* also Ideal transformer)
IVC, 43–45
  equivalent circuit of a transistor using, 82
  equivalent circuit of a two-port using, 71
  symbolic representation, 44
  two-port parameters, 56

Joseph, R. D., 328
$j\omega$–axis transmission zero realization, 315, 320, 333, 475, 481

Karplus, W. J., 497
Kawakami, M., 6, 43, 287
Kawashima, Y., 47, 410
Keen, A. W., 385
Kerwin, W. J., 295, 332, 489
Key, E. L., 328
Khazanov, G. L., 328, 357
Kinariwala, B. K., 95, 235, 268, 308, 366, 424
Kinariwala's method, 366–370
  using odd-part partitioning, 520–526
  using unique decomposition, 512–515
Kuh, E. S., 13, 85, 187, 203, 311, 395, 533
Kuh's method, 311–318
Kuo, F. F., 178, 188

Lampard, D. G., 470
Large parameter sensitivity, 182–183
Larky, A. I., 40, 354
Lattice network, 514–516, 527
Laurent expansion, 164
LC one-port, properties of, 536–537
LC two-port, properties of y–parameters, 545
$LC:-R$ network, definition, 227
  realizability conditions, 235
  synthesis of one-port, 267–268
$LC:RC$ transformation, 84–86, 395, 504, 506–507
LC:tunnel-diode one-port, properties, 237
  synthesis of, 267–268

Leakage transmission, 166
Lee, R. F., 398
Lee, S. C., 200, 340
Lessor, A. E., 2
Lewis, P. H., 299
Linear, lumped, and finite network (see LLF network)
Linear passive reciprocal two-port, properties of z–parameters, 226
Linear passive nonreciprocal two-port,
  properties of y–parameters, 226
  properties of z–parameters, 222–223
Linvill, J. G., 305, 357, 361, 383
Llewellyn, F. B., 246
LLF network, definition, 24, 213
  properties, 213
  types of, 226–227
Loop gain, 166, 184–185, 453, 456
Losee, D. L., 279
Loss compensation of inductors, 264–265
Lovering, W. F., 477
Lovering's method, 477–479
Low-pass filter realization, 12, 88, 164, 312, 319, 336, 388, 427, 432, 488
Low-pass to band-pass transformation, 89
Low-pass to high-pass transformation, 88

Maisel, L. I., 2
Martinelli, G., 181, 263, 286, 464
Mason, S. J., 166
Mathews, M. V., 476
Mathew-Seifert's approach, 476
Maupin, J. T., 343
Maximally flat delay transfer function, 11
Maximally flat magnitude transfer function, 4
McCoy, R., 550
McMillan, E. M., 223, 404
Merill, J. L., Jr., 38
Mitra, S. K., 52, 70, 90, 228, 279, 309, 374, 405, 412, 430, 463, 479, 503, 505, 526, 553
Moreda, H., 6
Morse, A. S., 153, 497
Moschytz, G. S., 10, 186, 203, 340
Mulligan, J. H., Jr., 9
Multiparameter pole (zero) sensitivity, 191–195
  minimization of, 202–203
Multiparameter sensitivity function, 188–190

Multipole(s), classification of, 134–135
  contraction of terminals of, 127–128
  definition, 118
  equivalent representation of, 139–143
  network functions of, 135–139
  nongenerative, 134
  parallel connection of, 130–131
  self-generative, 134
  suppression of terminals of, 129–130
Myers, B. R., 74, 355, 357, 362, 370, 386

Nagata, M., 260
Nathan, A., 150
Negative capacitance, 34
Negative-impedance converter, controlled source model of, 64–65
  current inversion type, 38
  definition, 38
  driving-point function synthesis using, 364–381, 512–514, 520–526
  gyrator:negative-resistance model of, 67–68
  location in AD–BC plane, 59
  location in A–D plane, 58
  negative resistance using, 263
  nonideal, definition, 38–39
    compensation of, 40–41
    synthesis using, 389–392
  nullator-norator model of, 352–353
  operational amplifier realization of, 465–468
  sensitivity considerations, 359–362
  stability considerations, 362–364
  transfer function synthesis using, 381–392, 527–529
  transistorized realizations of, 354–358
  two-port parameters, 56
  voltage inversion type, 38
Negative-impedance inverter, definition, 52
  negative resistance using, 264
  nonideal, 53
  nonreciprocal, 52
    controlled source model of, 413
    gyrator design using, 410–413
    nullator-norator model of, 413
  reactive, 53
  reciprocal, 52
    representation, 52
    two-port parameters, 57
Negative inductance, 34

INDEX / **561**

Negative resistance, 32–34, 254–289
  current-controlled, 255
  driving-point function synthesis using, 266–271
  ideal, controlled source representation, 61
    definition, 32
    model, 33
    $v - i$ characteristic, 33
  nonideal, 33–34, 254–287
  N-type, definition, 33
    model, 34
    realization of, 256, 258–260, 262–263, 288–289
    $v - i$ characteristic, 34
  open-circuit stable, 255
  series type, 255
  short-circuit stable, 255
  shunt type, 255
  S-type, definition, 33
    model, 34
    realization of, 256–258, 260–262, 288–289
    $v - i$ characteristic, 34
  transfer function synthesis using, 272–286
  voltage controlled, 255
Network functions, bilinear form of, 155–157
  of LLF networks, 213
  of multipoles, 135–139
  types of, 533
Network normalization (*see* Normalization)
Network specifications, 533
New, W., 49, 410
Newcomb, R. W., 49, 410, 413, 439–440, 489
Newell, W. E., 2, 194, 196
NIC, 38
  (*see also* Negative-impedance converter)
Nillson, J. W., 107
NIV, 52
  (*see also* Negative-impedance inverter)
Norator, definition, 54
  gyrator:negative resistance model of, 68
  symbolic representation, 54
Normalization, frequency, 13, 534
  impedance, 13, 534
NRNIV, 52
  (*see also* Negative-impedance inverter)
Nullator, definition, 54

Nullator (*Continued*)
  gyrator:negative-resistance model of, 68
  symbolic representation, 54
Nullator-norator model, of controlled sources, 70–75, 291–293
  of gyrator, 405–410
  of negative-impedance converter, 352–353
  of negative-impedance inverter, 413
  of open-circuit, 74, 294
  of operational amplifier, 463–464
  of short-circuit, 74
  transistorized realizations from, 74–75
Nullor, definition, 54
  symbolic representation, 54
  transmission matrix of, 54
Null-return difference, 167
  sensitivity function in terms of, 169

Odd part, of a positive real function, 506
  of sensitivity function, 170
Odd-part partitioning, 505–509
  driving-point function synthesis using, 520–526
  of an admittance function, 507
  of an impedance function, 507
  transfer function synthesis using, 528–529
Open-circuit impedance parameters, 531
  properties of passive two-port in terms of 222–223, 226
Open-loop gain, 456
Operational amplifier, characteristics of a practical, 449–451
  compensation of an, 458–462
  controlled source model of, 30
  controlled source realization using, 452, 455, 464–465
  definition of an ideal, 30, 448–449
  differential-input, 30, 449
  driving-point synthesis using, 470–474
  effect of negative feedback on, 30–31, 452–456
  gyrator realization using, 468–469, 498
  negative-impedance converter using, 465–468
  negative resistance using, 465–468, 498
  nonideal, 449–451
  nullator-norator model of, 463–464
  simulated inductance using, 492–495

562 / INDEX

Operational amplifier (*Continued*)
  single-ended, 30, 449
  stability of, 456–458
  symbolic representation, 30, 449
  transfer function synthesis using, 474–492
  transmission matrix of, 463–464
Orchard, H. J., 9, 413, 439
Order, of a network function, 307

Pande, H. C., 484
Papoulis, A., 163
Parsons, T. W., 473
Passive nonreciprocal two-port, properties of transfer function, 543–544
  properties of z–parameters, 543
Passive one-port, definition, 210
  realizability conditions, 216
  synthesis of, 414–424
Passive RLC network, comparison with active RC network, 19–20
  fundamental restrictions, 6–8
  general properties, 19–20
  limitations of, 3–7
  (*see also* RC network, RL network, and RLC network)
Passive two-port, definition, 212
Passivity, 210
Paul, R. J., 325
Pederson, D. O., 13, 85, 203, 395, 533
Pepper, R., 203
Peterson, L. C., 61
Phase function, 170
Phase sensitivity, 170
  from sensitivity function, 170–171
Philbrick Researches Inc., 451
Phillips, C. L., 276
PIC, 41
  (*see also* Positive-impedance converter)
PIV, 46
  (*see also* Positive-impedance inverter)
Pole sensitivity, calculation of,
  definition, 172
  multiparameter, 191–195
  relation with return difference, 176
  (*see also* Root sensitivity)
Polynomial sensitivity, 171
  minimization of, 184–185
  relation with sensitivity function, 185
Positive-impedance converter, 41–43
  definition, 41

Positive-impedance converter (*Continued*)
  types of, 41–42
  (*see also* Active transformer)
Positive-impedance inverter, 46–51
  definition, 46
  inductance simulation using, 48
  types of, 47
  (*see also* Gyrator)
Positive real function, abbreviation, 216
  definition, 216
  minimum reactive, 252
  real rational, test for, 216–218
Potential instability, 110
p.r. function (*see* Positive real function)
Pseudo-Hurwitz polynomial, 110

Q, of an inductor, 5
  of a resonant circuit, 195
Q–sensitivity, 195
  of RC:NIC type filter, 198
  relation with pole sensitivity, 196
Quality factor, 195

Raisbeck, G., 210
Ramanan, K. V., 262
Rand, A., 6
Rao, T. N., 413, 440
RCA, 448
RC network, definition, 3
  realizability conditions, of one-port, 537–539
    of two-port, 546
  realizing short-circuit transfer admittances, 547–550
  realizing transfer voltage ratios, 551–553
RC:–C one-port, realizability conditions, 239
RC:capacitive gyrator one-port, necessary properties, 240
RC:controlled-source network, driving-point function synthesis, 305–309
  realizability conditions, 239
  transfer function synthesis, 309–340
RC:CR transformation, 86–89
  generalization, 114
  use of, 318, 331, 335
RC:gyrator network, realizability conditions, 237–238, 420
  sensitivity considerations, 433–438
  transfer function synthesis of, 425–428, 446

RC:NIC network, definition, 226
　driving-point function synthesis of, 364–381
　Q–sensitivity, 198
　realizability conditions, 238
　sensitivity considerations, 392–395
　transfer function synthesis of, 381–389, 527–529
　using nonideal NIC, 389–392
　transfer voltage ratio vector realization of, 401
RC:operational amplifier network, design using coefficient matching technique, 484–492
　minimum sensitive realization using, 489–492
　transfer function synthesis using, 474–492
RC:–R network, definition, 226
　driving-point function synthesis of, 266
　realizability conditions, 230–231
　transfer function synthesis of, 272–284
RC:–RC decomposition, 98–102
　method of obtaining, 98–99
　optimum (*see* Horowitz decomposition)
　physical interpretation, 99
　properties, 100
　sensitivity considerations (*see* Horowitz decomposition)
RC:RL decomposition, 102–104
　method of obtaining, 102–103
　optimum (*see* Calahan decomposition)
　physical interpretation, 103
　properties, 103
　sufficiency conditions for, 429
RC:tunnel-diode network, driving-point function synthesis, 266–267
　realizability conditions, 233–234
　transfer functions synthesis, 272–274, 276–284
Reactive gyrator, 51
　gyration impedance of, 51
　symbolic representation, 51
　transfer function synthesis using, 430–433
Real transformer, 41
　(*see also* Ideal transformer)
Reflection coefficient, 503
Residue condition, for a nonreciprocal two-port, 223
　for a reciprocal two-port, 225

Resistive inverse network, 90
Return difference, definition, 167
　physical interpretation of, 167
　pole sensitivity in terms of, 176
　sensitivity function in terms of, 169
Reverse predistortion method, 277
Reza, F. M., 341
Richard's theorem, 414
Riordan, R. H. S., 493
Rise time, 11
　relation with 3-db bandwidth, 13
RLC network, realizability conditions of one-port, 540–541
　realizability conditions of two-port, 543–545
　with no mutual coupling, 544–545
RLC:–R network, definition, 227
　driving-point function synthesis, 268–271
　realizability conditions, 237
　transfer function synthesis, 284, 286
RLC:tunnel-diode network, driving-point function synthesis, 424–425
RL network realizability conditions, 539–540
Rohrer, R. A., 526
Rollett, J. M., 246
Root locus, 107
　relation with root sensitivity, 185–186
Root sensitivity, calculation of, 175–177
　definition, 172
　minimization of, 202–203
　relations between single-parameter, 194
　relation with coefficient sensitivity, 181
　relation with root locus, 185–186
　relation with sensitivity function, 177
　sum of, 177–178
Ruston, H., 506

Sallen, R. P., 328
Sandberg, I. W., 38, 96–97, 115, 187, 237–239, 268, 305, 342, 354, 374–375
Sandberg's method, 370–374
　using odd-part partitioning, 526
　using unique decomposition, 516–520
Saraga, W., 505
Schoeffler, J. D., 189, 404
Schott, F. W., 415
Schwartz, A. F., 364
Scott, H. H., 7

Seifert, W. W., 476
Sensitivity, effect of, 9
  (see also Coefficient sensitivity, Large parameter sensitivity, Pole sensitivity, Polynomial sensitivity, Root sensitivity, Sensitivity function, Sensitivity minimization, Zero sensitivity)
Sensitivity function, calculation of, 173–175
  definition, 168
  multiparameter, 187–191
  relation with pole-zero locations, 200–202
  relation with pole (zero) sensitivity, 177
  relation with return difference, 169
  relation between single-parameter, 189
Sensitivity minimization, cascade approach, 198–200
  computer-based approach, 189
  general remarks on, 183–187, 206
  multiparameter, 200–203
  of RC:gyrator network, 433–438
  of RC:NIC network, 392–396
Series-shunt feedback pair, 295
Sewell, J. I., 486
Sharpe, G. E., 138, 159
Sheahan, D. F., 413, 439
Sheingold, D. H., 469
Shenoi, B. A., 237, 406
Shibayama, H., 287
Short-circuit admittance parameters, 531
  properties of a passive two-port in terms of, 226, 420
Shukla, R. S., 484
Shunt-series feedback pair, 297
Signal flow graph, 166–167
Silverman, J. H., 404
Simulated inductance, Q of, 440
  Q sensitivity of, 441
  using active transformer, 493
  using controlled sources, 25
  using gyrator, 48, 49
  using NIC, 35
  using operational amplifier, 492–495
Sipress, J. M., 198, 375, 406, 444
Sipress' method, 375–381
Spain, B., 138
Stability, concepts, 104–112
Stability conditions, Gewertz's, 246
  Llewellyn's, 246

Stability conditions (*Continued*)
  of a one-port, 240–242
  of a nonreciprocal two-port, 246
  of a reciprocal two-port, 246
  using stability invariant factor, 247
Stability invariant factor, 247
Stable network, absolutely, 111–112
  marginally, 105–106
  open-circuit, 108
  potentially un-, 111–112
  short-circuit, 109
  strictly, 105–106, 116
Stata, R., 451–452
Stewart, J. L., 143
Stuart, A. G., 470
Su, K. L., 53, 255, 260, 264, 276, 309
Sutcliffe, H., 492

Talbot, A., 217
Taylor, J., 49, 439
Tellegen, B. D. H., 47, 223, 250, 403
Tellegen's theorem, 250
Thein, R. R., 2
Thomas, R. E., 326, 395, 505
Toffler, J., 483
Total degree, 511
Total energy input, of an one-port, 210
  of a two-port, 212
Transactor, 26
  (see also Controlled sources, CCT, CVT, VVT, VCT)
Transducer, 26
  (see also Controlled sources, CCT, CVT, VCT, VVT)
Transfer function synthesis, using controlled sources, 309–339
  using gyrator, 425–428
  using negative-impedance converter, 381–392, 527–529
  using negative-resistance, 272–286
  using operational-amplifier, 474–482
  using tunnel-diode, 272–274, 276–284
Transformer (see Active transformer, Complex transformer, Ideal transformer)
Transistor, controlled-source model of, 23–24, 66
  Darlington compound pair, 305, 361
  nullator-norator model of, 293
  unijunction, 33

INDEX / **565**

Transmission matrix parameters, 531
Truxal, J. G., 167, 172, 205, 375
Tsierelson, D. A., 61
Tsuchiya, H., 6
Tunnel diode, dissipation factor of, 232
   driving-point function synthesis using, 266–267
   equivalent circuit, idealized, 23–24
     small-signal, 23, 227, 255
   transfer function synthesis using, 272–284
Tuttle, D. F., 506
Two-port parameters, 531
   controlled source models based on, 61–65
   relations between, 532

UCNIC, 38
   (*see* also Negative-impedance converter)
Unidirectional converter, 26
Unique decomposition, 503–505
   driving-point function synthesis using, 512–519
   transfer function synthesis using, 527–528
Ur, H., 177
UVNIC, 38
   (*see* also Negative-impedance converter)

Vallese, L. M., 405
Van Valkenburg, M. E., 77, 92, 216, 218, 272, 422, 449, 520, 540
Varshney, R. C., 262
VCT, controlled-source model of, 26
   definition, 26
   gyrator:negative-resistance model of, 67
   nullator-norator model of, 292–293, 409
   operational amplifier realization of, 465, 498
   relation with CCT and VVT, 28
   transistorized realization of, 297, 300
VGIC, 37
   (*see* also Generalized-impedance converter)
Virtual ground, 25
Vlach, J., 201
VNIC, 38
   (*see* also Negative-impedance converter)
Voltage amplifier, 15, 27
   (*see* also VVT)
Voltage-controlled current source (*see* VCT)

Voltage-controlled voltage source (*see* VVT)
Voltage follower, 465
Voltage-to-current transducer (*see* VCT)
Voltage-to-voltage transducer (*see* VVT)
VVT, controlled-source model of, 26
   definition, 27
   driving-point function synthesis using, 305–308
   nullator-norator model of, 292–293, 350–351
   operational amplifier realization of, 452, 455
   relation with VCT and CVT, 28
   symbolic representation, 28
   transfer function synthesis using, 309–339
   transistorized realization of, 293–295

Waldmann, L., 258
Warner, R. M., 2
Weinberg, L., 123, 277
Weiss, C. D., 159
Wick, R. F., 404
Widlar, R. J., 448
Witt, F. J., 405, 406
Wohlers, M. R., 268
Woodard, J., 439
Worst case deviation, 162, 190
Wyndrum, R. W., Jr., 340

XIG, 51
XNIV, 53

Yanagisawa, T., 47, 287, 356, 410, 413
Yanagisawa's method, 385–389
Youla, D., 287
y-parameters, 531
   controlled-source model based on, 63
   ideal-converter model based on, 69, 71

Zadeh, L., 118
Zelyakh, E. V., 43
Zero-db crossover frequency, 457
Zero sensitivity, calculation of, 175–177
   definition, 172
   from null-return difference, 176
   relations with sensitivity function, 177
   (*see* also Root sensitivity)
z-parameters, 531
   controlled source model based on, 62
   ideal-converter model based on, 69, 71

/ About the Author

Sanjit Kumar Mitra is Associate Professor of Electrical Engineering at the University of California, Davis. He was born in India and received his B.Sc. and M.Sc. (Tech.) degrees from the Utkal University and Calcutta University, respectively. He earned his M.S. and Ph.D. degrees from the University of California, Berkeley.

Dr. Mitra's professional experience includes the positions of: Assistant Engineer, Electronic Computer Laboratory, Indian Statistical Institute, Calcutta, India; Assistant Professor of Electrical Engineering, Cornell University; and member of technical staff, Bell Telephone Laboratories. He has published a number of papers in the field of active and passive networks and has three patents pending in active circuit design. He is a member of IEEE, Sigma Xi, and Eta Kappa Nu and is listed in the American Men of Science.